Pond Conservation in Europe

Developments in Hydrobiology 210

Series editor

K. Martens

Pond Conservation in Europe

Editors

Beat Oertli[1], Régis Céréghino[2],
Jeremy Biggs[3], Steven Declerck[4], Andrew Hull[5] &
Maria Rosa Miracle[6]

[1] *University of Applied Sciences of Western Switzerland, Geneva, Switzerland*
[2] *University of Toulouse, Toulouse, France*
[3] *Pond Conservation, Oxford, United Kingdom*
[4] *Catholic University of Leuven, Leuven, Belgium*
[5] *Liverpool John Moores University, Liverpool, United Kingdom*
[6] *University of Valencia, Valencia, Spain*

Previously published in *Hydrobiologia,* Volume 597, 2008 and 634, 2009

Springer

Editors
Beat Oertli
University of Applied Sciences of Western Switzerland, Geneva,
Switzerland

Régis Céréghino
University of Toulouse, Toulouse,
France

Jeremy Biggs
Pond Conservation, Oxford,
United Kingdom

Steven Declerck
Catholic University of Leuven, Leuven,
Belgium

Andrew Hull
Liverpool John Moores University,
Liverpool, United Kingdom

Maria Rosa Miracle
University of Valencia, Valencia,
Spain

ISBN 978-90-481-9087-4

Springer Dordrecht Heidelberg London New York

Library of Congress Control Number: 2010923484

Cover illustration: Lowland pond in Western Switzerland (Les Grangettes Nature Reserve). Photograph: Nicola Indermuehle.

Printed on acid-free paper.

Springer is part of Springer Science+Business Media (www.springer.com)

Contents

Hydrobiologia (2008) 597:1–6
DOI 10.1007/s10750-007-9225-8

ECOLOGY OF EUROPEAN PONDS

The ecology of European ponds: defining the characteristics of a neglected freshwater habitat

R. Céréghino · J. Biggs · B. Oertli · S. Declerck

Abstract There is growing awareness in Europe of the importance of ponds, and increasing understanding of the contribution they make to aquatic biodiversity and catchment functions. Collectively, they support considerably more species, and specifically more scarce species, than other freshwater waterbody types. Ponds create links (or stepping stones) between existing aquatic habitats, but also provide ecosystem services such as nutrient interception, hydrological regulation, etc. In addition, ponds are powerful model systems for studies in ecology, evolutionary biology and conservation biology, and can be used as sentinel systems in the monitoring of global change. Ponds have begun to receive greater protection, particularly in the Mediterranean regions of Europe, as a result of the identification of Mediterranean temporary ponds as a priority in the EU Habitats Directive. Despite this, they remain excluded from the provisions of the Water Framework Directive, even though this is intended to ensure the good status of *all* waters. There is now a need to strengthen, develop and coordinate existing initiatives, and to build a common framework in order to establish a sound scientific and practical basis for pond conservation in Europe. The articles presented in this issue are intended to explore scientific problems to be solved in order to increase the understanding and the protection of ponds, to highlight those aspects of pond ecology that are relevant to freshwater science, and to bring out research areas which are likely to prove fruitful for further investigation.

Guest editors: R. Céréghino, J. Biggs, B. Oertli and S. Declerck
The ecology of European ponds: defining the characteristics of a neglected freshwater habitat

R. Céréghino (✉)
EcoLab, Laboratoire d'Ecologie Fonctionnelle, UMR 5245, Université Paul Sabatier, 118 route de Narbonne, 31062 Toulouse cedex 9, France
e-mail: cereghin@cict.fr

J. Biggs
Pond Conservation: The Water Habitats Trust, Oxford Brookes University, Gipsy Lane, Headington, Oxford OX3 0BP, UK

B. Oertli
Department of Nature Management, University of Applied Sciences of Western Switzerland – EIL, 1254 Jussy-Geneva, Switzerland

S. Declerck
Laboratory of Aquatic Ecology, Katholieke Universiteit Leuven, Ch. Deberiotstraat 32, 3000 Leuven, Belgium

Keywords Biodiversity · Conservation · Ecosystem services · European Pond Conservation Network · Small water bodies · Temporary pools · Water policy · Wetlands

Introduction

Ponds are small (1 m^2 to about 5 ha), man-made or natural shallow waterbodies which permanently or temporarily hold water (De Meester et al., 2005). They are numerous, typically outnumbering larger lakes by a ratio of about 100 to 1 (Oertli et al., 2005), and occur in virtually all terrestrial environments, from polar deserts to tropical rainforests. Despite this they have, until recently, been mostly ignored by freshwater biologists or regarded simply as smaller versions of larger lakes. In contrast, practitioners spend considerable amount of effort on the management and creation of ponds, largely without a rigorous scientific framework for their actions (Williams et al., 1999; Pyke, 2005). However, recent research, driven both by the need to improve pond conservation strategies and by increasing interest in fundamental aspects of pond ecology (Biggs et al., 2005; McAbendroth et al., 2005), has started to shed interesting new light on pond ecosystem structure and function. As a result, there is growing evidence that ponds are functionally different from larger lakes (Oertli et al., 2002; Sondergaard et al., 2005) and that, despite their small size, they are collectively exceptionally rich in biodiversity terms (Williams et al., 2004). Thus, ponds often constitute biodiversity "hot spots" within a region or a landscape, challenging conventional applications of species-area models ('big is best') in practical nature conservation (see also Scheffer et al., 2006). Ponds also show greater biotic and environmental amplitudes than rivers and lakes (Davies, 2005). Thus they pose interesting questions about the relationships between waterbody size, the heterogeneity of catchments, the role of small water bodies as refugia, and the existence of networks of aquatic. Ponds also provide an ideal model for investigating metapopulation and metacommunity processes in aquatic systems and the importance of between-waterbody movements, compared to better known within-waterbody movements (Jeffries, 2005). They fit nicely into the basic scheme of metapopulation and metacommunity theory: for obligatory aquatic organisms, ponds are suitable patches in an unsuitable habitat matrix. This in turn plays a significant role in understanding population persistence and recovery from disturbance. Finally, in addition to their inherent biological importance, the small size of ponds and the ease with which they may be manipulated experimentally, makes them ideal models for controlled studies of many basic ecosystem processes (Blaustein & Schwartz, 2001; De Meester et al., 2005), from community assembly rules (Warren & Spencer, 1996) to diversity–productivity relationships (Chase & Ryberg, 2004). In this special issue, therefore, we aim to bring together a set of articles which provide an overview of the developing science describing the ecology of ponds with the objective of (i) exploring the major scientific problems which will need to be solved in order to increase understanding and protection of these vulnerable and neglected habitats, (ii) exploring those aspects of pond ecology that are of relevance to freshwater science generally and (iii) highlighting research areas which are likely to prove fruitful for further investigation.

The second European Pond Workshop

In October 2004, the first European Pond Workshop devoted to the "Conservation and Monitoring of Pond Biodiversity" (Geneva, Switzerland) launched a European network of people and institutions involved in fundamental scientific issues and practical applications needed to protect ponds, the European Pond Conservation Network (EPCN, http://campus.hesge.ch/epcn/, see also Oertli et al., 2005). The second EPCN Workshop was held in Toulouse (France, 23–25 February 2006), under the topic "Conservation of Pond Biodiversity in a Changing European Landscape". The aim of this second workshop was to yield a multi-disciplinary framework on how to maintain ponds and the biodiversity they host, in a landscape subjected to a wide array of potential stressors such as intensification or abandonment of agriculture, socio-economical pressures, climate change. The workshop was divided into plenary sessions and working group meetings. The 55 communications (oral and poster) were related to three sub-topics: (i) Understanding pond ecology (biodiversity, spatial and temporal patterns and ponds as research tools for hypothesis testing), (ii) Added value of ponds (biological indicators, ecosystem services) and (iii) Management of ponds (practical tools for management and monitoring, pond conservation). Three working group meetings were devoted to "The Pond Manifesto" (a publication aiming at presenting the background and the motivations for the EPCN), EPCN management and activities, and joint research programs. The meeting brought together 60 participants from Austria, Belgium, Denmark, France,

Germany, Hungary, Ireland, Italy, Poland, Spain, Switzerland and the UK. This special issue presents a selection of 12 contributions.

Special issue content

Recent studies, mainly in Europe (Williams et al., 2004; Angélibert et al., 2007), have indicated that ponds harbour a significant portion of aquatic biodiversity at the landscape scale. Several contributions in this issue have confirmed and reinforced this idea. For instance, in their comparative study on zooplankton diversity in different freshwater water body types (lakes, rivers, ditches, ponds and wheel tracks), De Bie et al. (2007) found that ponds may disproportionately contribute to total zooplankton species richness at the landscape scale. Ponds also often contain rare, endemic and/or Red Data List species (Oertli et al., 2007) and may as such form an irreplaceable type of habitat for a variety of freshwater biota (Céréghino et al., 2007; Williams et al., 2007).

Owing to their important contribution to aquatic biodiversity, ponds should be considered as an important target system in strategic plans that aim at conserving or developing aquatic biodiversity at the landscape scale. Such plans can only be effective if based on a solid knowledge of the factors that affect pond community structure and diversity. In this issue, several studies document clear associations between the communities of organism groups (macrophytes, zooplankton, macroinvertebrates and waterbirds) and a variety of ecologically relevant gradients, such as hydroperiod (Boix et al., 2007; Della Bella et al., 2007), surface area (Céréghino et al., 2007), salinity (Boix et al., 2007) and among-pond connectivity (Boix et al., 2007; Gascón et al., 2007; Oertli et al., 2007). If these associations are causal, it is clear that the conservation of such environmental gradients at the landscape scale is essential for the conservation of among-pond variability (beta diversity) and total landscape biodiversity (gamma diversity). There is a clear differentiation among communities of macroinvertebrates (and to a lesser extent of macrophytes) between temporary and permanent ponds. Although temporary ponds tend to have lower species richness than permanent ponds, temporary ponds are at least as important as a habitat for uncommon and rare species. In Mediterranean regions, temporary wet habitats are important for conservation. Nevertheless, both temporary and permanent ponds are important for the conservation of regional biodiversity: in Central Italy, both type of ponds present high dissimilarity in the taxonomic composition of aquatic plants (Della Bella et al., 2007), the former containing more annual fast-growing species while in the latter, species with long life-cycles are abundant. Some aquatic species are exclusively found in each pond type.

Essential for the conservation of pond biodiversity is a good knowledge of its threats. Land use practices in the surroundings of ponds may, to an important extent, affect pond characteristics through a diversity of processes that play at the scale of the pond catchment (e.g. nutrient loading, increased erosion, pesticide contamination; Declerck et al., (2006)). Davies et al. (2007) contrasted catchment characteristics among different water body types within a landscape and noted that the small scale of pond catchments combined with their relatively high contribution to landscape scale biodiversity offers a lot of opportunities for cost-efficient conservation strategies. An important reason for this is that deintensification of agriculture at the scale of pond catchments is far more feasible and effective than it is on the catchment scale of larger aquatic systems, such as rivers or lakes. It means that efforts on the pond scale can be, relatively, easily implemented and have the potentiality to yield visible biodiversity benefits on a relatively short term, even in areas where large scale deintensification is not an option.

Due to their small scale, ponds can also be easily created. Pond creation has a lot of potential for nature development plans: new ponds are rapidly colonised by a variety of organisms and well designed and located, pond complexes could be used to significantly enhance freshwater biodiversity within catchments (Williams et al., 2007). Furthermore, pond density in the landscape can be an important factor determining the persistence of metapopulations of rare species. In such development plans, one may also take advantage of the opportunities offered by ponds that are not necessarily created as a part of nature conservation programmes such as ponds that aim at supporting agricultural activities (Céréghino et al., 2007; Williams et al., 2007).

Owing to their small sizes and simple community structure, small aquatic ecosystems may also function as early warning systems for long-term effects on larger aquatic systems. For instance, global warming may lead to higher local and regional richness in high altitude ponds through an increase in the number of colonisation events resulting from the upward shift of geographical ranges of species, while cold stenothermal species may be subject to extinction (Oertli et al., 2007). On the other hand, a survey on crucian carp body condition in ornamental ponds in the UK revealed no correlation with climatic variables (Copp et al., 2007). More direct threats to ponds include habitat destruction (in-filling ponds; deepening of ephemeral pools so that they become permanent) or other forms of strong human impact (e.g. urban runoff, acidification, diffuse agricultural pollution, introduction of exotic species, excessive trampling by livestock). Efficient bioindication metrics based on macroinvertebrate taxa richness and functional feeding groups as well as pollution tolerance are sensitive to nutrient enrichment (Solimini et al., 2007). The species richness of insect and crustacean taxa also respond well to eutrophication (Menetrey et al., 2007, Solimini et al., 2007) or salinity (Boix et al., 2007), while the presence of some indicator species can be associated to the trophic state of the ponds in a given area (Menetrey et al., 2007). However, sampling biodiversity in ponds is still a critical issue, because ponds are rich in microhabitats, often structured by macrophytes. Therefore, standardised methods are required. Becerra et al. (2007) present an effective sampling regime to maximise total taxon richness while minimising sampling effort.

Perspectives

To understand how the biological diversity sustained by ponds is maintained and how ponds function, future research should involve complementary approaches, and focus on the relevant ranges of temporal and spatial scales. Several directions can be identified for relevant research on ponds, ranging from the fields of biological monitoring to evolutionary ecology (reviewed in De Meester et al., 2005). Here, we specifically emphasise those perspectives which call for intense collaborative research at a European level.

Whereas much research has been undertaken at the EU level towards developing robust methodologies and tools for the implementation of the Water Framework Directive (examples include the STAR, AQEM and ECOFRAME projects), small water bodies such as shallow lakes and ponds have not been well represented, despite their ecological role at the landscape—regional scales. Active research into the ecology and conservation of ponds is being undertaken in many European states, addressing different areas of relevance. One of the chief problems is the lack of integration between these research areas, and a general poor level of understanding of patterns of variation in habitats and biota across Europe. Some national environment agencies from countries such as France, the United Kingdom and Switzerland, have recently developed elements of a national strategy for pond conservation, but such efforts remain in the minority across the rest of Europe. Obtaining a typology of small water bodies at a European scale, should be a first step towards optimising the design of new surveys and standardising sampling schemes and monitoring applications. Exploration of fundamental ecological patterns and definition of a typology of European small waterbodies should cover the range of habitats found along broad geographical (North-South, East-West), altitudinal and environmental gradients. The analyses should try to involve the full range of taxonomic groups (i.e. different trophic levels, keystone taxa, umbrella and flagship species). These analyses, in addition to giving a vital understanding of large-scale patterns, are also expected to reveal gaps in existing knowledge. An important issue is the capacity for ponds and pond communities to respond to disturbance and to global change (early-warning systems). This implies that near-pristine systems should be identified as references in the investigated areas, and that long-term monitoring is necessary to assess temporal responses of ponds to local practices and/or global changes. Assessments of responses to various types and/or intensity and frequency of disturbance should preferably be hypothesis-based, in order to reduce (and thus better target) the number of variables that will be assessed. Experimental work should include the main driving forces of community dynamics in ponds, and should thus include both regional (dispersal, external forcing) and local factors (abiotic conditions, biotic interactions). Suggested practical applications should not only be based on the patterns derived from fundamental research, they must be tested and evaluated in the field.

Although there are clear gaps still to fill in our knowledge on pond ecology, this special issue of Hydrobiologia demonstrates that notable improvements have been made these last ten years. For effective conservation of pond biodiversity, this knowledge has now to be communicated to managers, in order to be put into practice. This will be one of the priority tasks for the European Pond Conservation Network, with an important stepping stone at the third European Ponds Workshop, to be held in Valencia (Spain) in May 2008.

Acknowledgements We wish to thank the sponsors of the second European Pond Workshop (CNRS, University Paul Sabatier, Laboratoire d'Ecologie des Hydrosystèmes, Région Midi-Pyrénées, Conseil Général de la Haute-Garonne, French Water Agency).

References

Angélibert, S., N. Indermuehle, D. Luchier, B. Oertli & J. Perfetta, J. 2007. Where hides the aquatic biodiversity in the Canton of Geneva (Switzerland)? Archives des Sciences (in press).

Becerra Jurado, G., M. Masterson, R. Harrington & M. Kelly-Quinn, 2007. Evaluation of sampling methods for macroinvertebrate biodiversity estimation in heavily vegetated ponds. Hydrobiologia doi:10.1007/s10750-007-9217-8.

Biggs, J., P. Williams, P. Whitfield, P. Nicolet & A. Weatherby, 2005. 15 years of pond assessment in Britain: results and lessons learned from the work of Pond Conservation. Aquatic Conservation: Marine and Freshwater Ecosystems 15: 693–714.

Blaustein, L. & S. S. Schwartz, 2001. Why study ecology in temporary pools? Israel Journal of Zoology 47: 303–312.

Boix, D., S. Gascón, J. Sala, A. Badosa, S. Brucet, R. López-Flores, M. Martinoy, J. Gifre & X. D. Quintana, 2007. Patterns of composition and species richness of crustaceans and aquatic insects along environmental gradients in mediterraean water bodies. Hydrobiologia doi: 10.1007/s10750-007-9221-z.

Céréghino, R., A. Ruggiero, P. Marty & S. Angélibert, 2007. Biodiversity and distribution patterns of freshwater invertebrates in farm ponds of a southwestern French agricultural landscape. Hydrobiologia doi:10.1007/s10750-007-9219-6.

Chase, J. M. & W. A. Ryberg, 2004. Connectivity, scale-dependence, and the productivity–diversity relationship. Ecology Letters 7: 676–683.

Copp, G. H., S. Warrington & K. J. Wesley, 2007. Management of an ornamental pond as a conservation site for a threatened native fish species, crucian carp *Carassius carassius*. Hydrobiologia doi:10.1007/s10750-007-9220-0.

Davies, B. R., J. Biggs, P. J. Williams, J. T. Lee & S. Thompson, 2007. A Comparison of the catchment sizes of rivers, streams, ponds, ditches and lakes: implications for protecting aquatic biodiversity in an agricultural landscape. Hydrobiologia doi:10.1007/s10750-007-9227-6.

Davies, B. R., 2005. Developing a Strategic Approach to the Protection of Aquatic Biodiversity. PhD thesis, Oxford Brookes University.

De Bie, T., S. Declerck, K. Martens, L. De Meester & L. Brendonck, 2007. A comparative analysis of cladoceran communities from different water body types: patterns in community composition and diversity. Hydrobiologia doi: 10.1007/s10750-007-9222-y.

De Meester, L., S. Declerck, R. Stoks, G. Louette, F. Van de Meutter, T. De Bie, E. Michels & L. Brendonck, 2005. Ponds and pools as model systems in conservation biology, ecology and evolutionary biology. Aquatic Conservation: Marine and Freshwater Ecosystems 15: 715–726.

Declerck, S., T. De Bie, D. Ercken, H. Hampel, S. Schrijvers, J. Van Wichelen, V. Gillard, R. Mandiki, B. Losson, D. Bauwens, S. Keijers, W. Vyverman, B. Goddeeris, L. De Meester, L. Brendonck & K. Martens, 2006. Ecological characteristics of small ponds: associations with land-use practices at different spatial scales. Biological Conservation 131: 523–532.

Della Bella, V., M. Bazzanti, M. G. Dowgiallo & M. Iberite, 2007. Macrophyte diversity and physico-chemical characteristics of Tyrrhenian coast ponds in central Italy: implications for conservation. Hydrobiologia doi:10.1007/s10750-007-9216-9.

Gascón, S., D. Boix, J. Sala & D. Quintana, 2007. Relation between macroinvertebrate life strategies and habitat traits in Mediterranean salt marsh ponds (Empordà wetlands, NE Iberian Peninsula). Hydrobiologia doi:10.1007/s10750-007-9215-x.

Jeffries, M., 2005. Local-scale turnover of pond insects: intra-pond habitat quality and inter-pond geometry are both important. Hydrobiologia 543: 207–220.

McAbendroth, L., A. Foggo, S. D. Rundle & D. T. Bilton, 2005. Unravelling nestedness and spatial pattern in pond assemblages. Journal of Animal Ecology 74: 41–49.

Menetrey, N., B. Oertli, M. Sartori, A. Wagner & J. B. Lachavanne, 2007. Eutrophication: are mayflies (Ephemeroptera) good bioindicators for ponds? Hydrobiologia doi:10.1007/s10750-007-9223-x.

Oertli B., D. Auderset Joye, E. Castella, R. Juge, D. Cambin & J. B. Lachavanne, 2002. Does size matter? The relationship between pond area and biodiversity. Biological Conservation 104: 59–70.

Oertli, B., J. Biggs, R. Céréghino, P. Grillas, P. Joly & J. B. Lachavanne, 2005. Conservation and monitoring of pond biodiversity: introduction. Aquatic Conservation: Marine and Freshwater Ecosystems 15: 535–540.

Oertli, B., N. Indermuehle, S. Angélibert, H. Hinden & A. Stoll, 2007. Macroinvertebrate assemblages in 25 high alpine ponds of the Swiss National Park (Cirque of Macun) and relation to environmental variables. Hydrobiologia doi:10.1007/s10750-007-9218-7.

Pyke, C. R., 2005. Assessing suitability for conservation action: prioritizing interpond linkages for the California tiger salamander. Conservation Biology 19: 492–503.

Scheffer, M., G. J. van Geest, K. Zimmer, E. Jeppesen, M. Sondergaard, M. G. Butler, M. A. Hanson, S. Declerck & L. De Meester, 2006. Small habitat size and isolation can promote species richness: second-order effects on biodiversity in shallow lakes and ponds. Oikos 112: 227–231.

Solimini, A. G., M. Bazzanti, A. Ruggiero & G. Carchini, 2007. Developing a multimetric index of ecological integrity based on macroinvertebrates of mountain ponds in central Italy. Hydrobiologia doi:10.1007/s10750-007-9226-7.

Sondergaard, M., E. Jeppesen & J. P. Jensen, 2005. Pond or lake: does it make any difference? Archiv fur Hydrobiologie 162: 143–165.

Warren, P. H. & M. Spencer, 1996. Community and food-web responses to the manipulation of energy input and disturbance in small ponds. Oikos 75: 407–418.

Williams, P., M. Whitfield & J. Biggs, 2007. How can we make new ponds biodiverse?—a case study monitored over 8 years. Hydrobiologia doi:10.1007/s10750-007-9224-9.

Williams, P., J. Biggs, M. Whitfield, A. Thorne, S. Bryant, G. Fox & P. Nicolet, 1999. The Pond Book: A Guide to the Management and Creation of Ponds. Ponds Conservation Trust, Oxford.

Williams, P., M. Whitfield, J. Biggs, S. Bray, G. Fox, P. Nicolet & D. Sear, 2004. Comparative biodiversity of rivers, streams, ditches and ponds in an agricultural landscape in Southern England. Biological Conservation 115: 329–341.

Hydrobiologia (2008) 597:7–17
DOI 10.1007/s10750-007-9227-6

ECOLOGY OF EUROPEAN PONDS

A comparison of the catchment sizes of rivers, streams, ponds, ditches and lakes: implications for protecting aquatic biodiversity in an agricultural landscape

B. R. Davies · J. Biggs · P. J. Williams · J. T. Lee · S. Thompson

Abstract In this study we compared the biodiversity of five waterbody types (ditches, lakes, ponds, rivers and streams) within an agricultural study area in lowland England to assess their relative contribution to the plant and macroinvertebrate species richness and rarity of the region. We used a Geographical Information System (GIS) to compare the catchment areas and landuse composition for each of these waterbody types to assess the feasibility of deintensifying land to levels identified in the literature as acceptable for aquatic biota. Ponds supported the highest number of species and had the highest index of species rarity across the study area. Catchment areas associated with the different waterbody types differed significantly, with rivers having the largest average catchment sizes and ponds the smallest. The important contribution made to regional aquatic biodiversity by small waterbodies and in particular ponds, combined with their characteristically small catchment areas, means that they are amongst the most valuable, and potentially amongst the easiest, of waterbody types to protect. Given the limited area of land that may be available for the protection of aquatic biodiversity in agricultural landscapes, the deintensification of such small catchments (which can be termed microcatchments) could be an important addition to the measures used to protect aquatic biodiversity, enabling 'pockets' of high aquatic biodiversity to occur within working agricultural landscapes.

Keywords Watershed · Microcatchment · Aquatic biodiversity · Agri-environment schemes · Diffuse pollution

Guest editors: R. Céréghino, J. Biggs, B. Oertli & S. Declerck
The ecology of European ponds: defining the characteristics of a neglected freshwater habitat

B. R. Davies (✉) · J. Biggs · P. J. Williams
Pond Conservation c/o School of Life Sciences, Oxford Brookes University, Gipsy Lane, Headington, Oxford OX3 0BP, UK
e-mail: brdavies@brookes.ac.uk

B. R. Davies · J. T. Lee · S. Thompson
Spatial Ecology and Landuse Unit, School of Life Sciences, Oxford Brookes University, Gipsy Lane, Headington, Oxford OX3 0BP, UK

Introduction

Pollution from agriculture is recognised as having a significant negative impact on water quality and aquatic biota (Allan, 2004; Foley et al., 2005; Declerck et al., 2006; Donald & Evans, 2006). These pollutants include nutrients and other chemicals used to maximise production on arable land; animal waste and animal health biproducts, e.g. antibiotics and sheep dip, from pastoral land; as well as sediment resulting from eroded soils. They affect aquatic ecosystems both by altering the physicochemical characteristics and quality of a waterbody (e.g. eutrophication, changes to sediment composition) and by direct toxicity impacts on the organisms within it. Many of these pollutants are diffuse in nature, and the broad areas from which they emanate and multiplicity of pathways by which they reach waterbodies, make such pollutants difficult to control and mitigate.

The quantity and concentration of diffuse pollutants reaching waterbodies can potentially be reduced by two mechanisms: (i) source control through reduction in chemical loads, better targeting in the timing of chemical applications and the use of appropriate farming techniques to reduce runoff, e.g. minimum tillage; and (ii) measures to prevent pollutants from reaching waterbodies through deintensifying areas of land, e.g. buffer zones. Although the source control mechanisms go someway towards reducing pollution (e.g. Yates et al., 2006) diffuse pollutants will not be completely eliminated by such means. Thus, methods of land management and pollutant interception are likely to remain important in the effective long-term protection of aquatic biota under current agricultural systems. Typically such methods involve leaving areas adjacent to waterbodies out of agricultural production, through whole field conversion or more commonly, the creation of buffer strips. Although widely used and much tested, this approach has shown mixed results in terms of pollution reduction, e.g. Schmitt et al. (1999), Dosskey (2002), Borin et al. (2004, 2005), and recent evidence (e.g. Wang et al., 1997; Quinn, 2000; Fitzpatrick et al., 2001; Donohue et al., 2006) implies that where such methods have proved relatively ineffective, this has often been because the deintensified area has not included a sufficient proportion of the catchment area of a waterbody.

The catchment of a waterbody is the area over which water, and hence diffuse pollutants, will travel (both by overland and subsurface movement) to enter the waterbody. Catchments have long been recognised as the key to understanding the ecology of freshwaters (Hynes, 1975; Allan et al., 1997; Allan, 2004) and although underlying geology and morphology are the fundamental determinants of water characteristics (Host et al., 1997; Johnson et al., 2004; Wiley et al., 1997; McRae et al., 2004), in areas where landcover is heavily modified, the landuse composition of the catchment will dominate (Hynes, 1975; Lund & Reynolds, 1982; Moss et al., 1996; Allan et al., 1997; Muir, 1999; Cresser et al., 2000; Johnson & Goedkoop, 2002; Tong & Chen, 2002). Thus, the proportion of intensive landuse in a catchment as well as the catchment's size will influence the cost and potential success of a waterbody's protection from diffuse pollution.

Catchment areas are generally perceived as large and are usually described only in the context of rivers or large lakes. For example, in the UK, small river and stream catchments are generally termed 'sub-catchments'. However, in reality all waterbodies, large or small, have a catchment area. The association of catchment areas with larger rivers and lakes is likely to have resulted from the historic use of these waterbodies for navigation, drainage, food supply, water supply, recreation and removal of wastes. This considerable socio-economic value has resulted in a vested interest in the protection of these waterbodies and consequently, both scientific research and environmental protection has tended to be focused at larger waterbodies. The more limited economic value of small waterbodies has meant that until recently, their biodiversity potential has tended to be overlooked with a general presumption that they are inferior versions of their larger equivalents. However, recent evidence has shown that small waterbodies may in fact make a disproportionately large contribution to aquatic biodiversity across landscapes in terms of both their species richness and their species rarity (Biggs et al., 2003, 2007; Williams et al., 2004; De Bie et al., 2008; Davies et al., in press), implying that they are likely to warrant a higher priority in terms of conservation concern. The ease and success of their protection from diffuse agricultural pollution will depend to some extent, as for larger waterbodies, on the proportion of their catchment areas that can be incorporated into protection strategies.

This study investigates the aquatic biodiversity (species richness and rarity) of a suite of waterbody types across an area of UK lowland agricultural landscape in the context of their catchment sizes. The results are used to explore the potential ease and success of the protection of different waterbody types from diffuse agricultural pollution.

Material and methods

The study area and its aquatic biodiversity

A 13 × 11 km study area of lowland agricultural landscape in Britain on the borders of Oxfordshire,

Wiltshire and Gloucestershire was selected. The study area contained three Department for Environment, Food and Rural Affairs (Defra) agricultural landscape classes (Table 1) (Biggs et al., 2003) and was considered typical of lowland agricultural landscapes (Brown et al., 2006). Arable cultivation dominated the landcover (75%), with 9% under woodland, 7% improved grassland, 2% urban and the remaining 7% made up of water, semi-natural grassland and bare rock (Fig. 1). Agricultural land was predominantly arable, comprised mainly of cereals, permanent grass, oil-seed rape, potatoes peas and sugarbeet. There were 205 ha of surface water in the study area comprising 3 rivers, 97 streams, 236 ponds, 8 lakes and 340 ditches. The rivers included lengths of the Thames (c. 16.7 km), Cole (c. 16.8 km) and Coln (c. 4.3 km).

Data on the macrophyte and macroinvertebrate species present in the five dominant waterbody types in the study area (ditches, lakes, ponds, rivers and streams) were collected in two phases. In 2000, a stratified random sample of 20 sites was surveyed for each of four waterbody types (ditches, ponds, rivers and streams), i.e. 80 sites in total (reported in Williams et al., 2004). During 2002–2003, comparable data were collected from a further 20 sites in a fifth waterbody type, lakes. Due to the size division between a lake and a pond (Table 2; Biggs et al., 2003; Williams et al., 2004), data were also collected for a pond to replace one from the existing dataset which would have been categorised as a small lake under Table 2.

Within each waterbody, the sample area and survey methods followed those used by Williams et al. (2004). Each sample area was 75 m^2 to enable direct comparison of data from different waterbody types with inherently different sizes. For linear waterbodies, this area comprised a rectangular section of the waterbody, while for circular waterbodies, the area comprised a triangular wedge with the base following the margin and the apex at the centre of the waterbody. At each site, all wetland macrophytes (marginal, emergent, submerged and floating-leaved plants) were recorded by walking and wading the margin, using a grapnel thrown from the bank and sampling from a boat in the deeper lake sites. A three-minute hand net sample was taken for macroinvertebrates using a standard 1 mm mesh net, with the three minute sample time being divided equally between the mesohabitats in the 75 m^2 area. Macroinvertebrate samples were exhaustively live-sorted in the lab and all individuals (except Diptera larvae and Oligochaeta which were omitted from the analysis) were identified to species level, except very abundant taxa (>100 individuals) which were sub-sampled.

The macrophyte and macroinvertebrate data provided information on aquatic species richness and rarity for each survey site in the ditches, lakes, ponds, rivers and streams (100 sites in total). The species richness of a site was the total number of species found at that site, whilst species rarity was calculated using the Species Rarity Index (SRI). This rarity index follows a process developed by Foster et al. (1990) whereby each species is given a numerical value according to its rarity or threat within Britain, the total for each site is then summed and finally divided by the number of species found at the site, resulting in an index which is not biased towards

Table 1 Defra agricultural landscape classes occurring within the study area

Landscape	Area-km^2 (% of area)	Description	Associated agriculture
LC1—River floodplains and low terraces	16.27 (11.5)	Level to very gently sloping river floodplains and low terraces	Permanent grass, some cereals and oil-seed rape, probably more intensive on terraces
LC6—Pre-quaternary clay landscapes	95.46 (67.1)	Level to gently sloping vales. Slowly permeable, clays (often calcareous) and heavy loams. High base status (Eutrophic)	Permanent grass, cereals (>10–15%), leys, oil-seed rape maize and beans
LC7—Chalk and limestone plateaux and coombe valleys	30.44 (21.4)	Rolling 'Wolds' and plateaux with 'dry' valleys; shallow to moderately deep loams over chalk and limestone	Cereals (and oil-seed rape, beans), sugar beet, potatoes, peas

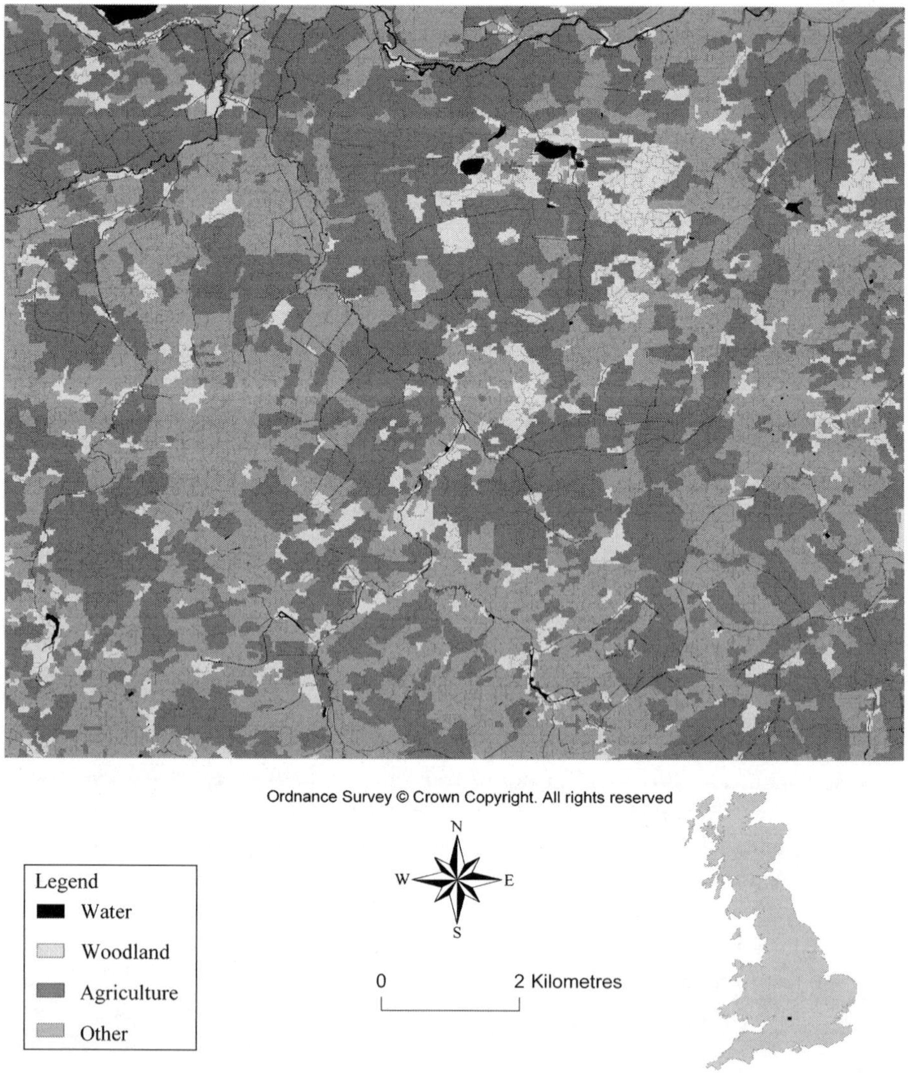

Fig. 1 The study area and its landuse composition

Table 2 Definitions of waterbodies used in this study

Waterbody type	Definition
Lakes	Bodies of water, both natural and man-made, greater than 2 ha in area (Johnes et al., 1994). Includes reservoirs, gravel pits, meres and broads.
Ponds	A body of water, both natural and man-made, between 25 m^2 and 2 ha in area, which may be permanent or seasonal (Collinson et al., 1995).
Rivers	Relatively large lotic waterbodies, created by natural processes. Marked as a double blue line on 1:25,000 OS maps and defined by the OS as greater than 8.25 m in width.
Streams	Relatively small lotic waterbodies, created by natural processes. Marked as a single blue line on 1:25,000 OS maps and defined by the OS as being less than 8.25 m in width. Streams differ from ditches by usually: (i) having a sinuous planform; (ii) not following field boundaries; and (iii) showing a relationship with natural landscape contours, usually by running down valleys.
Ditches	Man-made channels created primarily for agricultural purposes and which usually: (i) have a linear planform; (ii) follow linear field boundaries, often turning at right angles; and (iii) show little relationship with natural landscape contours.

species-rich sites (see Williams et al., 2004 for further details).

Catchment delineation and landcover

Ordnance Survey (OS) Landform Profile and MasterMap data were used to create the underlying Digital Elevation Model (DEM) upon which catchment delineation was based, for an area extending 8 km outside the study area, using the Geographical Information System (GIS) software, ArcGIS 8.2. MasterMap data include topographic information on every landscape feature, each with its own unique identifying code. Landform Profile data comprise height data at 10 m intervals in x and y and recorded to the nearest 0.1 m in z, with a planimetric accuracy of ± 1 m and a vertical accuracy of ± 1.8 m.

All waterbody polygons were extracted from the MasterMap data. Misclassified features were removed and polygons split by overlying features, such as bridges, were joined. Separate data layers were created for ditches, lakes, ponds, rivers and streams according to the definitions in Table 2. A new river or stream was defined at confluence sites and a new ditch was defined at both confluence sites and where it turned by approximately 90°. Results were visually compared with 1:25,000 OS maps, aerial photographs and site visits to ensure that the network of catchments and waterbodies generated from digital data were consistent with the real landscape.

Each waterbody polygon was assigned a constant minimum height value from the OS Landform Profile data and the waterbodies layer converted to a 5 m grid. The 10 m Landform Profile raster data were also converted to a 5 m grid using bilinear resampling, so that it could be combined with the waterbodies whilst retaining their continuity. The waterbodies were 'burnt' into the DEM at their height value minus 10 m to ensure that modelled runoff would flow into the waterbodies and that they would be retained during the catchment delineation process.

The catchments of each waterbody type were delineated separately using the DEM with the depressed waterbodies and the ArcGIS extension ArcHydro Tools (Version 1.1 Beta 2; ESRI, 2001). ArcHydro Tools only modelled surface water flow. Although the study area was predominantly underlain by impermeable clay and so overland flow would have been a dominant process, some throughflow and transport via field drains would have occurred but this was not modelled and is a limitation of the method. ArcHydro Tools also had the underlying assumption that all water will flow to the edge of the DEM. This ignores standing waterbodies which provide natural sinks that retain water within a landscape and so all ponds and lakes were 'seeded' with 'no-data' points at their deepest locations or centre point, to ensure that water flowed towards these points.

River catchments that extended beyond the limit of the data held were estimated using published statistics on catchment size from gauging stations present in the study area (Environment Agency, 2006). Additional manipulation of one river catchment boundary was required by hand due to the very flat nature of the northwest corner of the study area, which meant that ArcHydro Tools was not able to accurately define the catchment boundary in this instance. The landuse composition of the catchments was ascertained by intersecting each catchment with Land Cover Map 2000 data (remotely sensed landuse data in 25 m^2 cells; copyright NERC). For the river catchments that extended beyond the limit of data available, the proportions of different landuse types were taken as those modelled in the GIS for the area over which data was held.

Results

Of the total area of water (205 ha) within the study area (13 × 11 km), lakes comprised the greatest proportion of the water area (38%), followed by rivers (24%), ditches (20%), streams (14%) and ponds (4%) (Fig. 2). There was a broadly inverse relationship between surface water area and number of waterbodies, with rivers and lakes being fewest in number but covering the largest surface area and ponds being one of the most numerous waterbody types but having the smallest total surface area.

Across the study area, the pond sites supported the greatest number of both macrophyte and macroinvertebrate species, followed by rivers, lakes, streams and lastly ditches, which contained no aquatic (submerged or floating) plants (Fig. 3a). Overall, the pond sites supported 238 species, river sites

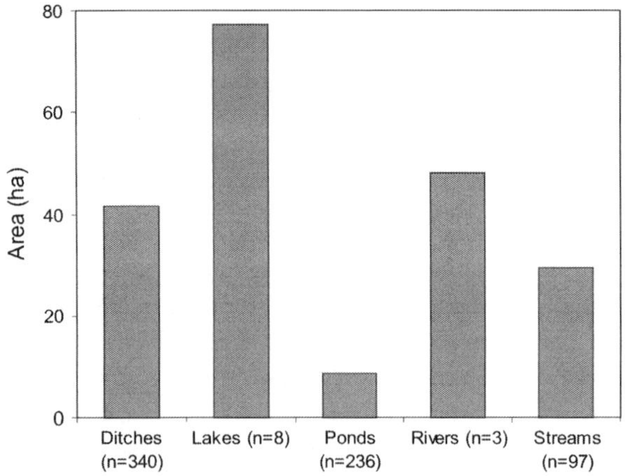

Fig. 2 Surface water area of the different waterbody types within the study area

supported 201 species, lakes 186 species, streams 163 species and finally, ditch sites supported 120 species.

The pond sites had the highest SRI for macrophytes, followed by lakes, rivers, streams and lastly ditches which supported no rare plant species (Fig. 3 b). The ponds also had the highest macroinvertebrate SRI, followed by ditches, lakes, streams and lastly rivers (Fig. 3b).

Catchment sizes were significantly different between waterbody types (Kruskall–Wallis, $P < 0.001$). Rivers had the greatest average catchment areas (43,850 ha), followed by lakes (141 ha), streams (86 ha), ditches (29 ha) and lastly, ponds (18 ha) (Fig. 4). The total catchment areas followed a different pattern: overall, rivers had the greatest total catchment area (131,550 ha), followed by ditches (9,904 ha), streams (8,354 ha), ponds (4,237 ha) and lastly, lakes (1,124 ha) (Fig. 4). This difference arose because ditches had catchments that were small in size but numerous giving a large total catchment area, whilst lakes had large catchments but were few in number. Although river catchments were clearly the largest, there were too few in the sample for this to be confirmed with post hoc Mann–Whitney *U* tests. However, significant differences ($P < 0.001$) in catchment size were seen between ponds and streams, ponds and lakes, ditches and streams and ditches and lakes, whilst no significant difference was observed between the size of pond and ditch catchments and between stream and lake catchments. This showed that ponds and ditches had similarly small

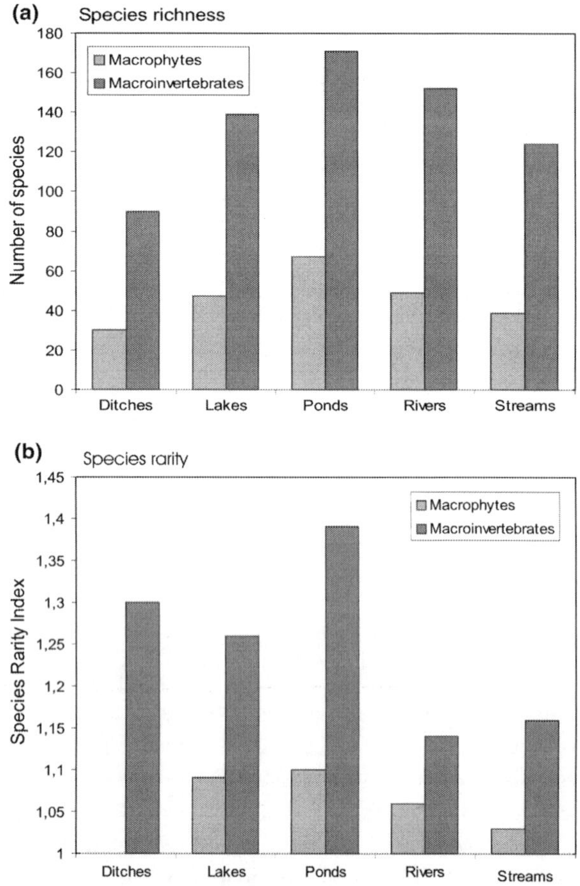

Fig. 3 Macrophyte and macroinvertebrate (**a**) species richness and (**b**) species rarity across the study area

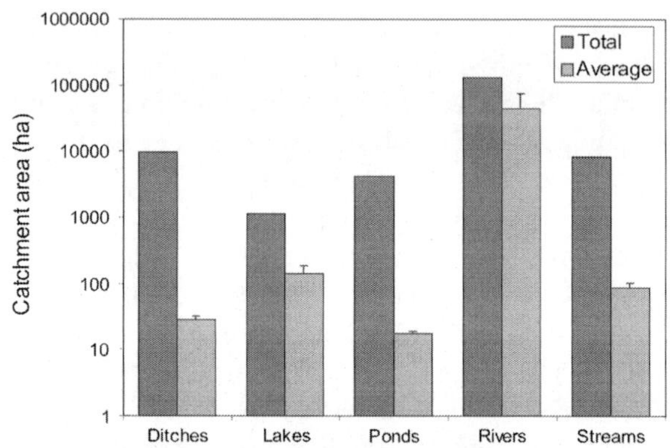

Fig. 4 Average and total catchment areas of the different waterbody types within the study area

catchments, whilst streams and lakes had similarly larger catchments and rivers had the greatest catchment sizes.

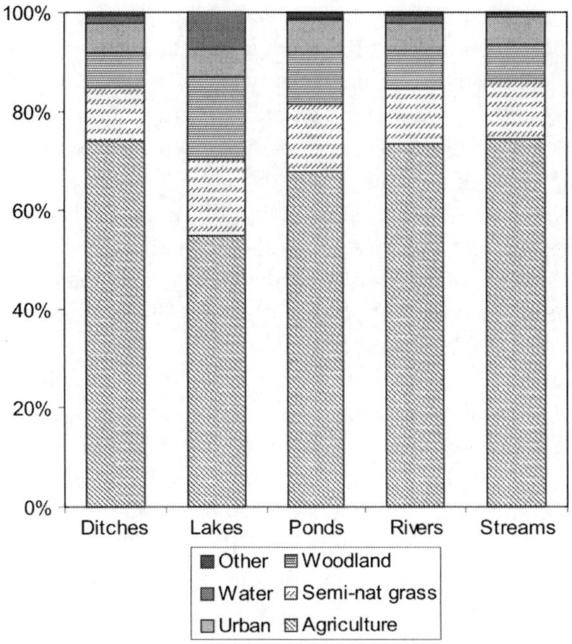

Fig. 5 Proportion of landuses within the total catchment area of each waterbody type

The catchments of all waterbody types were dominated by agricultural landuse, with river, stream, pond and ditch catchments all containing similar proportions of agricultural land (69–74%) together with similar proportions of woodland (7–10%), semi-natural grassland (10–13%) and water (1–2%) (Fig. 5). Lake catchments differed, typically containing less agricultural land (55%) and larger proportions of semi-natural grassland (c. 4% more), woodland (c. 9% more) and water (c. 6% more) than those of the other waterbody types. The proportion of urban land within catchments was similar for all waterbody types (c. 5%).

Discussion

Catchment sizes and the area needed for protection of aquatic biota

Hynes (1975), in his classic work on the ecology of running waters, described rivers as, "*a manifestation of the landscapes that they drain*". Since this pioneering work, many studies have sought to characterise river, stream and sometimes lake catchments, analysing relationships between factors such as catchment area, landuse composition and waterbody ecology. Practically all of these studies have reinforced Hynes' original comments. The extent of agriculture in the developed world is such that it is a dominant component of many waterbodies' catchment areas, with its associated diffuse pollutants having well accepted and largely detrimental impacts on aquatic biodiversity.

Recent research has investigated the importance of such widespread agriculture within catchments, typically finding streams to remain in good condition until agriculture exceeds 30–50% of the catchment (Allan, 2004). For example, working in the US, Wang et al. (1997) found that habitat quality and scores of biotic integrity declined when agricultural landuse exceeded 50% of the catchment. Fitzpatrick et al. (2001) working in the same region, found declines in fish biotic indices above 30% agricultural land, whilst Quinn (2000), working in New Zealand, found 30% agricultural land to be a critical value for macroinvertebrates. More stringent levels were identified by Donohue et al. (2006) who investigated the relationship between the ecological quality of aquatic networks and landcover, identifying thresholds at which 'good' ecological status could be attained in Ireland. They predicted that with more than 1.3% arable land in a catchment or 37.7% pastoral land, ecological status would fall below 'good' levels. Within the current study area, the average landuse composition of the catchments of all the waterbody types exceeded these thresholds. The waterbody type with the least intensive catchment landuse composition was lakes, which were, on average, associated with 55% agricultural land. This implied that within the study area, lakes were already fairly well buffered, largely due to their frequent location on large private estates. This may not, of course, be the situation in other areas.

Agriculture currently covers approximately half of the earth's habitable surface (Clay, 2004) and in many countries covers more than 70% of the land surface. This implies that vast areas would need to be deintensified to reach the maximum thresholds of 30–50% identified as important in the literature. Given the anticipated doubling of food demand forecast for the next 50 years (Donald & Evans, 2006), this may be very difficult to achieve and impractical in landscapes where agricultural production is needed. However, not all waterbodies have large catchment areas and it may be possible to deintensify those with 'microcatchments' to reach the critical thresholds of 30–50%, or even to completely deintensify them. The present study found that, as might be expected, larger waterbodies, (i.e. rivers and lakes), had larger catchments than smaller ponds, streams and ditches. Using the study area as an example, rivers would require a comparatively large amount of land to be deintensified to reach the thresholds of 30 and 50%, compared with the smallest waterbody type, ponds (Table 3). To attain levels of no more than 50% agricultural land in an average catchment, rivers would require 10,086 ha to be deintensified, compared with just 4 ha for ponds. To reach 30% agricultural land, rivers would require 18,856 ha to be deintensified, compared with just 7 ha for ponds.

Therefore, if ponds and other waterbodies with small catchments can be demonstrated to support high levels of biodiversity they may play an important part in a strategy for the protection of aquatic biodiversity because they could be afforded very high levels of protection, e.g. complete deintensification, for a comparatively low land area.

Macrophyte and macroinvertebrate species richness and rarity across the study area

Ponds made a high contribution to both the aquatic macrophyte and macroinvertebrate biodiversity of the study area in terms of both species richness and species rarity. Results for the other waterbody types were more mixed, with rivers supporting a relatively large number of species but low species rarity, whilst ditches supported few species but had a high macroinvertebrate SRI. Studies comparing the aquatic biodiversity of different types of waterbody are few and, as far as the authors are aware, none have compared the range of waterbody types undertaken for this study. However, those studies that have compared aquatic biodiversity for a more limited range of habitats have often found smaller waterbodies, particularly ponds and ditches, to make an important contribution (e.g. Painter, 1999; Armitage et al., 2003; Biggs et al., 2007; Davies et al., in press). These small waterbody types have often been overlooked in biodiversity protection and rarely enjoy the statutory protection afforded to larger waterbodies. The results of this study, supported by other work on comparative biodiversity including smaller waterbody types (Davies, unpublished data; Davies et al., in press), suggest that this may be both a considerable oversight and a missed opportunity. In particular, the valuable contribution of smaller waterbodies to regional aquatic biodiversity means that they could have an important role in the strategic protection of aquatic biota.

Table 3 Areas requiring deintensifying within the study area to reach identified thresholds of 50% and 30% agricultural land

Waterbody type	Average area (ha) requiring deintensification to reach:	
	50% Agricultural land	30% Agricultural land
Rivers	10085.5	18855.5
Streams	20.6	37.8
Lakes	7.1	35.3
Ditches	6.7	12.8
Ponds	3.6	7.2

The potential role for small waterbodies in protection strategies

As identified above, waterbodies with larger catchments require a much greater area to be deintensified to reach the levels that are suggested as needed for the sufficient protection of aquatic biota. In the study area, the largest catchments (associated with rivers) were on average 300 times larger than those of lakes (which had the second largest catchments) and almost 2,500 times larger than those of ponds, which had the smallest mean catchment sizes. Equally, in the area required to deintensify a single average-sized river catchment to 50% agricultural landuse, it would be possible to fully deintensify more than 560 average-sized pond catchments. Thus, the relatively small catchment sizes of smaller waterbodies, and in particular ponds, combined with their important contribution to aquatic biodiversity, means that their inclusion amongst the measures used for biodiversity protection from diffuse agricultural pollution is likely to enhance the effectiveness and economic efficiency of protection across whole landscapes.

The common perception, mentioned above, that catchment areas are associated with larger waterbodies and in particular rivers (and hence cover large areas), may be one of the reasons that whole catchment or large-scale deintensification are not generally proposed as protection measures. Instead, buffer strips (involving much smaller areas) are often employed. The disadvantage of such methods for larger waterbodies is that, depending on the proportion of the catchment that they occupy, they are unlikely to provide sufficient protection for the aquatic biota of the waterbody. For example, under the English Agri-Environment Scheme, Environmental Stewardship, the maximum buffer width offered at the edge of a field to protect a river is six metres. Published data on the effectiveness of buffer zones at varying widths suggest that a six metre buffer is highly unlikely to result in reduced nutrient pollution loads to rivers. Due to the large edge length of a river, the agricultural land-take that would be involved in even a limited buffer area would be considerable, implying that large areas of agricultural land are likely to be taken out of production to protect a river, with very little chemical or ecological gain.

The mechanisms most likely to be used to protect aquatic biodiversity in agricultural landscapes in the UK, are agri-environment schemes (AESs) whose structure would facilitate small catchment deintensification. These schemes remunerate farmers for environmentally sensitive farming, including reducing chemical and nutrient inputs as well as land management to reduce the impacts of diffuse pollution. The scheme is voluntary and is applied to by individual farmers and consequently, the measures offered are at a farm-scale. Such a scale and the potentially ad hoc uptake would favour microcatchment deintensification over that of larger catchments which would require the efforts of many farmers to be coordinated, increasing administrative costs and the complexity of operation, as well as increasing the chances of failure because the lack of cooperation of even a single landowner could jeopardise the success of a waterbody's protection. In contrast, the microcatchment of a smaller waterbody may be wholly encompassed by one land manager. Additionally, microcatchment deintensification has the potential to provide farmers with locally visible, biodiversity benefits. Such local returns are very important for improving satisfaction and a sense of ownership of the AES measures, as opposed to employing measures for broader scale improvements where results are harder to observe at a site level.

Supplementary to the deintensification of small catchments is the possibility of creating small waterbodies. Ponds and ditches are relatively easy to create and can be located so as to have small, completely unintensive catchments with good water quality, providing the best possible chance of supporting high levels of aquatic biodiversity, whilst involving minimum amounts of land (Williams et al., 2008). This is particularly important given evidence of the time lag between landuse change and improved ecological quality (Harding et al., 1998).

Protection of the aquatic biodiversity of smaller waterbodies through catchment deintensification and the creation of small waterbodies can be undertaken relatively quickly. This means that such initiatives could be employed immediately, providing benefits for aquatic biodiversity across a region much more quickly than could be achieved for larger waterbodies through catchment deintensification, whose complex implementation would take some time to execute.

Clearly, the protection of small waterbodies through catchment deintensification and the creation of ponds cannot deliver the complete protection of aquatic biota within agricultural landscapes (Davies, unpublished data). However, their inclusion amongst measures for aquatic biodiversity protection should provide 'pockets' of high biodiversity in working agricultural landscapes, making aquatic biodiversity protection more effective and more economically efficient.

Conclusion

This study appears the first in the literature to compare both the biodiversity and catchment size of such a range of waterbody types. Small waterbodies, and in particular ponds, were found to support a relatively high proportion of the aquatic biodiversity in the study location, which confirmed the results of the few other studies that have made biodiversity comparisons between a more limited range of waterbody types. Smaller waterbodies were also found to have smaller catchment areas than larger waterbodies, which was not a surprising result, but the implications that arise from it are important.

Calculation of the areas of agricultural land within the catchments of the different waterbody types that would need to be deintensified to provide adequate protection, indicated that, for larger waterbodies, such a method of deintensification would be inappropriate and uneconomic due to the scales that are likely to be involved. In contrast, complete catchment protection would be quite feasible for smaller waterbodies. Given that the smaller waterbodies supported higher levels of aquatic biodiversity, deintensification of their relatively small catchments should afford effective protection from many pollutants, enabling pockets of high biodiversity to exist in a working agricultural landscape. Combined with alternative methods to protect waterbodies with larger catchments, and the creation of strategically located new ponds a landscape matrix should result which incorporates minimally impacted aquatic habitats whilst still being economically and socially productive.

Acknowledgements The authors would like to thank Glen Hart and the Ordnance Survey for provision of MasterMap and Landform Profile data and Geoff Smith and CEH Monks Wood for provision of Land Cover Map 2000 data. We are also grateful to Steven Declerck and two anonymous referees for very useful comments on an earlier draft of this text.

References

Allan, J. D., 2004. Landscapes and riverscapes: the influence of land use on stream ecosystems. Annual Review of Ecology, Evolution and Systematics 35: 257–284.

Allan, J. D., D. L Erickson & J. Fay, 1997. The influence of catchment land use on stream integrity across multiple spatial scales. Freshwater Biology 37: 149–161.

Armitage, P. D., K. Szoszkiewicz, J. H. Blackburn & I. Nesbitt, 2003. Ditch communities: a major contributor to floodplain biodiversity. Aquatic Conservation: Marine and Freshwater Ecosystems 13: 165–185.

Biggs, J., P. Williams, M. Whitfield, P. Nicolet, C. Brown, J. Hollis, S. Maund, D. Arnold & T. Pepper, 2003. Aquatic Ecosystems in the UK Agricultural Landscape. Report on Project PN0931. Defra, London.

Biggs, J., P. Williams, M. Whitfield, P. Nicolet, C. Brown, J. Hollis, D. Arnold & T. Pepper, 2007. The freshwater biota of British agricultural landscapes and their sensitivity to pesticides. Agriculture Ecosystems and Environment 122: 137–148.

Borin, M., E. Bigon, G. Zanin & L. Fava, 2004. Performance of a narrow buffer strip in abating agricultural pollutants in the shallow subsurface water flux. Environmental Pollution 131: 313–321.

Borin, M., M. Vianello, F. Morari & G. Zanin, 2005. Effectiveness of buffer strips in removing pollutants in runoff from a cultivated field in North-East Italy. Agriculture, Ecosystems and Environment 105: 101–114.

Brown, C. D., N. Turner, J. Hollis, P. Bellamy, J. Biggs, P. Williams, D. Arnold, T. Pepper & S. Maund, 2006. Morphological and physico-chemical properties of British aquatic habitats potentially exposed to pesticides. Agriculture, Ecosystems and Environment 113: 307–319.

Clay, J., 2004. World Agriculture and the Environment: A Commodity by Commodity Guide to Impacts and Practices. Island Press, Washington DC.

Collinson, N., J. Biggs, A. Corfield, M. J. Hodson, D. Walker, M. Whitfield & P. J. Williams, 1995. Temporary and permanent ponds: an assessment of the effects of drying out on the conservation value of aquatic macroinvertebrate communities. Biological Conservation 74: 125–134.

Cresser, M. S., R. Smart, M. F. Billet, C. Soulsby, C. Neal, A. Wade, S. Langan & A. C. Edwards, 2000. Modelling water chemistry for a major Scottish river from catchment attributes. Journal of Applied Ecology 37: 171–184.

Davies, B. R., J. Biggs, P. Williams, M. Whitfield, P. Nicolet, D. Sear, S. Bray & S. Maund, in press. Comparative biodiversity of aquatic habitats in the European agricultural landscape. Agriculture, Ecosystems and Environment. doi:10.1016/j.agee.2007.10.006.

De Bie, T., S. Declerck, L. De Meester, K. Martens & L. Brendonck, 2008. A comparative analysis of cladoceran communities from different water body types: patterns in community composition and diversity. Hydrobiologia. doi:10.1007/s10750-007-9222-y

Declerck, S., T. De Bie, D. Ercken, H. Hampel, S. Schrijvers, J. Van Wichelen, V. Gillard, R. Mandiki, B. Losson, D. Bauwens, S. Keijers, W. Vyverman, B. Goddeeris, L. De Meester, L. Brendonck & K. Martens, 2006. Ecological characteristics of small farmland ponds: associations with land use practices at multiple spatial scales. Biological Conservation 131: 523–532.

Donald, P. F. & A. D. Evans, 2006. Habitat connectivity and matrix restoration: the wider implications of agri-environment schemes. Journal of Applied Ecology 43: 209–218.

Donohue, I., M. L. McGarrigle & P. Mills, 2006. Linking catchment characteristics and water chemistry with the ecological status of Irish rivers. Water Research 40: 91–98.

Dosskey, M., 2002. Setting priorities for research on pollution reduction functions of agricultural buffers. Environmental Management 30: 641–650.

Environment Agency, 2006. Concise Register of Gauging Stations. Retrieved March 29th 2006 from http://www.nwl.ac.uk/ih/nrfa/station_summaries/op/EA-Thames2.html

ESRI, 2001. ESRI Support Centre: Hydro Data Model [available for download from the World Wide Web at: http://www.support.esri.com/index.cfm?fa=downloads.dataModels.filteredGateway&dmid=15]

Fitzpatrick, F. A., B. C. Scudder, B. N. Lenz & D. J. Sullivan, 2001. Effects of multi-scale environmental characteristics on agricultural stream biota in eastern Wisconsin. Journal of the American Water Resources Association 37: 1289–1507.

Foley, J. A., R. DeFries, G. P. Asner, C. Barford, G. Bonan, S. R. Carpenter, F. S. Chapin, M. T. Coe, G. C. Daily, H. K.

Gibbs, J. H. Helkowski, T. Holloway, E.A. Howard, C. J. Kucharik, C. Monfreda, J. A. Patz, C. Prentice, N. Ramankutty & P. K. Snyder, 2005. Global consequences of land use. Science 309: 570–574.
Foster, G. N., A. P. Foster, M. D. Eyre & D. T. Bilton, 1990. Classification of water beetle assemblages in arable fenland and ranking of sites in relation to conservation value. Freshwater Biology 22: 343–354.
Harding, J. S., E. F. Benfield, P. V. Bolstad, G. S. Helfman & E. B. D Jones, 1998. Stream biodiversity: the ghost of land use past. Proceeding of the National Academy of Sciences, USA 95: 14843–14847.
Host, G. E., C. Richards, L. B. Johnson & R. J. Haro, 1997. Catchment and reach-scale properties as indicators of macroinvertebrate species traits. Freshwater Biology 37: 219–230.
Hynes, H. B. N., 1975. The stream and its valley. Verhandlungen der Internationale Vereingigung Fur Limnology 19: 1–15.
Johnes, P., B. Moss & G. Phillips, 1994. Lakes – Classification and Monitoring. Environment Agency R&D Note 253. Environment Agency, Bristol.
Johnson, R. K. & W. Goedkoop, 2002. Littoral macroinvertebrate communities: spatial scale and ecological relationships. Freshwater Biology 47: 1840–1854.
Johnson, R. K., W. Goedkoop & L. Sandin, 2004. Spatial scale and ecological relationships between the macroinvertebrate communities of stony habitats of streams and lakes. Freshwater Biology 49: 1179–1194.
Quinn, J. M., 2000. Effect of pastoral development. In Collier, K. J. & M. J. Winterbourn (eds), New Zealand Stream Invertebrates: Ecology and Implications for Management. Caxton, Christchurch, New Zealand.
Lund, J. W. G. & C. S. Reynolds, 1982. The development and operation of large limnetic enclosures in Blenham Tarn, English Lake District, and their contribution to phytoplankton ecology. Progress in Phycological Research 1: 2–65.
McRae, S. E., J. D. Allan & J. B. Burch, 2004. Reach- and catchment-scale determinants of the distribution of freshwater mussels (Bivalvia: Unionidae) in south-eastern Michigan, U.S.A. Freshwater Biology 49: 127–142.
Moss, B., P. Johnes & G. Phillips, 1996. The monitoring of ecological quality and the classification of standing waters in temperate regions: a review and proposal based on a worked scheme for British Waters. Biological Reviews 71: 301–339.
Muir, R., 1999. Approaches to Landscape. Macmillan, Basingstoke.
Painter, D., 1999. Macroinvertebrate distributions and the conservation value of aquatic Coleoptera, Mollusca and Odonata in the ditches of traditionally managed and grazing fen at Wicken Fen, UK. Journal of Applied Ecology 36: 33–48.
Schmitt, T. J., M. G. Dosskey & K. D. Hoagland, 1999. Filter strip performance and processes for different vegetation, widths and contaminants. Journal of Environmental Quality 28: 1479–1489.
Tong, S. T. Y. & W. Chen, 2002. Modelling the relationship between landuse and surface water quality. Journal of Environmental Management 66: 377–393.
Wang, L. Z., J. Lyons, P. Kanehl & R. Gatti, 1997. Influences of watershed land use on habitat quality and biotic integrity in Wisconsin streams. Fisheries 22: 6–12.
Wiley, M. J., S. L. Kohler & P. W. Seelbach, 1997. Reconciling landscape and local views of aquatic communities; lessons from Michigan trout streams. Freshwater Biology 37: 133–148.
Williams, P., M. Whitfield, J. Biggs, S. Bray, G. Fox, P. Nicolet & D. Sear, 2004. Comparative biodiversity of rivers, streams, ditches and ponds in an agricultural landscape in Southern England. Biological Conservation 115: 329–341.
Williams, P., M. Whitfield & J. Biggs, 2008. How can we make new ponds biodiverse? A case study monitored over seven years. Hydrobiologia. doi:10.1007/s10750-007-9224-9
Yates, A. G., R. C. Bailey & J. A. Schwindt, 2006. No-till cultivation improves stream ecosystem quality. Journal of Soil and Water Conservation 61: 14–19.

Hydrobiologia (2008) 597:19–27
DOI 10.1007/s10750-007-9222-y

ECOLOGY OF EUROPEAN PONDS

A comparative analysis of cladoceran communities from different water body types: patterns in community composition and diversity

Tom De Bie · Steven Declerck · Koen Martens · Luc De Meester · Luc Brendonck

Abstract To develop strategies for the management and protection of aquatic biodiversity in water bodies at the landscape scale, there is a need for information on the spatial organization of diversity in different types of aquatic habitats. In this study, we compared the cladoceran composition and diversity between wheel tracks, pools, ponds, lakes, ditches, and streams, in 18 different areas of Flanders (Belgium). Multivariate analysis revealed significant differences in the composition of cladoceran communities among the different water body types. Average local and total diversity tended to be highest for lakes and lowest for streams. Despite the relatively high number of species supported by lakes, small water bodies seem to contribute considerably more to the total cladoceran richness of an average landscape in Flanders than lakes, because of their high abundance. With respect to biodiversity conservation at the landscape scale, our results point to the importance of maintaining a diversity of water body types of different size, permanence and flow regimes.

Keywords Zooplankton · Cladoceran · Ditch · Pool · Pond · Lake · Wheel track · Stream · Species richness · Biodiversity

Guest editors: R. Céréghino, J. Biggs, B. Oertli & S. Declerck
The ecology of European ponds: defining the characteristics of a neglected freshwater habitat

T. De Bie (✉) · S. Declerck · L. De Meester · L. Brendonck
Laboratory of Aquatic Ecology and Evolutionary Biology, KULeuven, Ch. Deberiotstraat 32, 3000 Leuven, Belgium
e-mail: tom.debie@bio.kuleuven.be

K. Martens
Royal Belgian Institute of Natural Sciences, Vautierstraat 29, 1000 Brussels, Belgium

Introduction

So far, most studies on biodiversity in aquatic habitats have focused on rivers, streams and lakes, and little research has been done on smaller aquatic systems. Small water bodies, such as ponds, pools, ditches, and wheel tracks are, nevertheless, ubiquitous features of the landscape and form no doubt the most common type of freshwater habitat in many areas of Europe. Despite their high abundance and the fact that they can support unique floral and faunal elements, these small ecosystems have only recently been recognized as important habitats for the maintenance of biodiversity (Armitage et al., 2003; Nicolet et al., 2004; Biggs et al., 2005; Oertli et al., 2005). In many regions these small and hence vulnerable types of water bodies are highly threatened due to eutrophication, pollution or physical destruction (Boothby, 2003). Pools, ponds, and ditches differ from lakes and rivers in many aspects (Oertli et al., 2002; Søndergaard et al., 2005) and can therefore be expected to be affected by anthropogenic stress in different ways and at different spatial scales (Allan et al., 2004; Declerck et al., 2006).

A good knowledge of patterns of species diversity within and among different types of aquatic habitat is necessary to direct strategies for the management and protection of aquatic biodiversity at the landscape scale. However, there are only few studies that have investigated aquatic biodiversity on a large catchment scale (Stendera & Johnson, 2005) or on a diverse set of water body types (Williams et al., 2004). Williams et al. (2004) compared the diversity of macroinvertebrates and macrophytes between different water body types within an entire part of a river catchment. They found that small water bodies contributed substantially to the regional biodiversity. This was to a large extent due to the high beta diversity of these systems, especially of ponds. The aim of this article is to compare the community composition and species diversity of cladocerans of different water body types and to assess the contribution of each of these water body types to average landscape cladoceran diversity.

Methods

Study area

We studied zooplankton communities in aquatic habitats from 18 different regions distributed over the major part of Flanders (Belgium) (Fig. 1). In each of the regions, we defined a circular area of 28 km^2. Within each of the areas, we sampled the zooplankton of on average 14 (range 10–21) water bodies belonging to six different water body types (see Table 1 for the definitions of water body types used in this study). In each area, the water bodies were chosen upon haphazard encounter in the field while walking in the surroundings of ponds and lakes. The number of sampled water bodies in each of the areas is presented in Table 2. Not all water body types were equally represented in each of the sampled areas. In total, we sampled ten wheel tracks, 38 pools, 151 ponds, 13 lakes, 26 ditches, and nine streams. The water bodies had no surface connections with other water bodies.

Land use in Flanders is dominated by intensive crop culture and pasture. Forest patches are scarce, fragmented, and small. The 18 areas were part of a larger survey that was carried out in the framework of the pond project MANSCAPE (Declerck et al., 2006) and each area represented a wide range in land use intensity.

Sample collection

We sampled the zooplankton of the water bodies once during the period June–August 2003. Cladoceran zooplankton samples were obtained with a conical plankton net (mesh size: 64 μm, diameter: 26 cm). Each water body was sampled for a total of 2 min with the total sampling time equally divided between the shore and the central zone of the water body. At both shore and central zone we sampled at four different spots equally distributed over the entire surface area (wheel tracks, pools, ponds, and lakes), or along a stretch of at least 10 m (ditches and streams). In very shallow habitats, water was collected with a beaker (5 l) and filtered through the

Fig. 1 Geographic location of the 18 study areas on the map of Flanders (Belgium). KN, Knokke; U, Uitkerke; BL, Diksmuide; I, Ieper; P, Ploegsteert; L, Lede; T, Temse; Ka, Kalmthout; OT, Oud Turnhout; ZE, Zemst; BE, Begijnendijk; D, Diest; HAL, Halle; TW, Tielt-Winge; Z, Zoutleeuw; HAS, Hasselt; A, Alken; BI, Bilzen

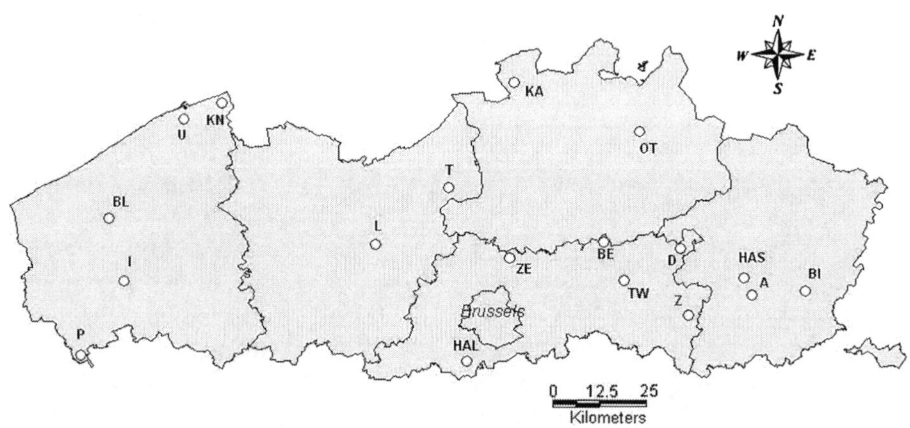

Table 1 Summary of definitions of aquatic habitats used in the survey (adapted from Williams et al., 2004)

Water body type	Definition
Wheel track (W)	Small shallow and elongated water bodies situated on sandy roads or tracks created by transport of vehicles. These aquatic habitats are temporarily and can vary strongly over time.
Pool (PL)	Temporary water bodies smaller than 25 m^2. Includes both man-made and natural water bodies.
Pond (PN)	Water bodies between 25 m^2 and 2 ha in area which may be permanent or seasonal (Collinson et al., 1995). Includes both man-made and natural water bodies.
Lake (L)	Permanent water bodies larger than 2 ha in area. Includes reservoirs and gravel pits.
Ditch (D)	Man-made channels created primarily for agricultural purposes, and which usually: (i) have a linear plan form, (ii) follow linear boundaries, often turning at right angles, and (iii) show little relationship with natural landscape contours.
Stream (S)	Small lotic water bodies created mainly by natural processes. Streams differ from ditches in (i) usually having a sinuous plan form, (ii) not following field boundaries, or if they do, pre-dating boundary creation, and (iii) showing a relationship with natural landscape contours e.g., running down valleys. All selected streams were less deep than 1 m and less width than 5 m.

Table 2 Number of water body types sampled in each of the 18 studied areas in Flanders

		Water body type					
Area	Date	W	PL	PN	L	D	S
Alken (A)	1/7/03	0	4	7	0	1	2
Begijnendijk (BE)	12/7/03	1	6	7	0	1	1
Bilzen (BI)	12/6/03	1	1	7	1	1	1
Diest (D)	20/6/03	0	1	7	0	2	0
Diksmuide (BL)	2/7/03	1	1	10	2	3	0
Halle (HAL)	4/7/03	1	3	8	1	1	1
Hasselt (HAS)	12/6/03	1	1	17	1	3	0
Ieper (I)	23/6/03	0	1	9	1	1	0
Kalmthout (KA)	15/7/03	2	1	7	2	1	0
Knokke (KN)	19/7/03	0	1	8	0	0	1
Lede (L)	14/7/03	0	2	9	0	1	0
Oud-Turnhout (OT)	6/8/03	1	3	5	2	2	1
Ploegsteert (P)	19/7/03	0	1	12	0	2	1
Temse (T)	13/7/03	0	4	7	1	1	1
Tielt-Winge (TW)	12/7/03	0	4	9	0	1	0
Uitkerke (U)	24/6/03	0	1	7	0	2	0
Zemst (ZE)	13/7/03	1	1	8	2	0	0
Zoutleeuw (Z)	27/6/03	1	2	7	0	3	0

Area, location of sampled area; date, time of sampling; W, wheel tracks; PL, pools; PN, ponds; L, lakes; D, ditches; S, streams

plankton net. Samples from deeper water were collected with a long-handled plankton net (64 μm, diameter: 26 cm). The zooplankton samples were preserved in formaldehyde (4%) saturated with sucrose. Cladocerans in the samples were analyzed up to species level following Flössner (2000). For each water body a subsample of at least 100 cladocerans was identified. Water bodies of which samples contained less than 70 cladoceran specimens in total were excluded from further data analysis (four wheel tracks, one pool, two ditches, and two streams).

Data analysis

Community composition

We formally tested for differences among the zooplankton communities of the different water body types by applying redundancy analysis (RDA) on percentage abundance data (arcsine-transformed) and canonical correspondence analysis (CCA) on species lists (presence/absence data) using the software CANOCO v5 (Lepš and Šmilauer, 2003). In these analyses, the different water body types were specified using dummy variables. We also specified the 18 studied areas as co-variables to avoid biases introduced by geographic patterns. We assessed the statistical significance of differences among water body types with random Monte Carlo permutations ($n = 999$). These permutations were restricted to areas (areas specified as 'blocks'; see Lepš and Šmilauer, 2003).

Species richness

Species richness of a water body (local species richness) was estimated as the total number of species recorded in its sample upon rarefaction to a standard number of 100 individuals with the statistical program PRIMER-5 (Clarke & Gorley, 2001). For each of the studied regions, we calculated the average local species richness for each type of water body.

We created species accumulation curves for each type of water body with the likelihood-based method developed by Mao et al. (2005). This method allowed an estimate to be made of the rate at which new species accumulate with an increasing number of samples. Through extrapolation of the available data the procedure also allowed, for each water body type, for a graphical evaluation to be made of whether the species accumulation curve approximates its asymptote and for an estimate to be made of total expected species richness with associated bootstrap confidence intervals. It should, however, be noted that extrapolations should be restricted to triple the number of samples available (Colwell et al., 2004).

Since, low sample numbers prevented us from making reliable estimates of total expected species richness for each water body type with extrapolation methods, we compared total richness among water body types with standard estimates of total richness based on a fixed number of six water bodies. We chose the number of six because this is equal to the total number of systems of the least represented water body type (wheel tracks). This procedure had the additional advantage of obtaining independent replicate estimates of total diversity for most of the water body types (except wheel tracks and streams). Replicate sets of six systems were randomly chosen within each of the water body categories without replacement.

We tested for differences in diversity among water body types with ANOVA and Tukey's post-hoc testing. All variables were log-transformed prior to analysis. ANOVA-analyses were performed with the software package STATISTICA v6. The species accumulation curves were computed with the programme R (version 2.4.0.) and algorithms were provided by C. X. Mao.

Results

Community composition

CCA and RDA revealed significant differences in both species lists ($F = 1.78$; $p < 0.01$) and proportional species composition ($F = 1.75$; $p < 0.01$) among the different water body types (Fig. 2). Pairwise post-hoc RDA tests indicated significant compositional differences between all possible pairs of water body types, except for the combinations of wheel tracks with pools and ditches with streams. The

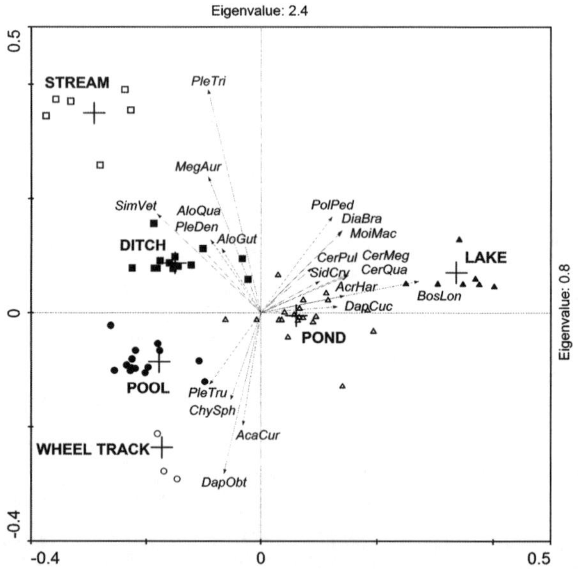

Fig. 2 Results of a redundancy analysis (RDA) showing patterns of association between species and water body type. Regional differences were partialled out by defining the 18 studied areas as co-variables. The water body types were specified as nominal variables and are represented by centroids. The samples are represented by symbols indicating the water body type: ■ = ditches, □ = streams, ▲ = lakes, △ = ponds, ● = pools and ○ = wheel tracks. AloQua, *Alona quadrangularis*; BosLon, *Bosmina longirostris*; ChySph, *Chydorus sphaericus*; DapObt, *Daphnia obtusa*; DiaBra, *Diaphanosoma brachyurum*; MegAur, *Megafenestra aurita*; MoiMac, *Moina macrocopa*; PleTri, *Pleuroxus trigonellus*; PolPed, *Polyphemus pediculus*; SimVet, *Simocephalus vetulus*; PleTru, *Pleuroxus truncatus*; AcaCur, *Acantheloberis curvirostris*; PleDen, *Pleuroxus denticulatus*; AloGut, *Alona guttata*; DapCul, *Daphnia cucullata*; CerMeg, *Ceriodaphnia megops*; Cer pul, *Ceriodaphnia puclhella*; CerQua, *Ceriodaphnia quadrangula*; SidCry, *Sida crystallina*; AcrHar, *Acroperus harpae*

most important axis of variation in the RDA analysis (Eigen value = 2.4) mainly represented the difference between lakes and other water body types, with ponds taking an intermediate position. The second axis (Eigen values = 0.8) mainly appeared to differentiate between small temporary lentic water bodies (wheel tracks and pools) and the larger systems (in terms of total surface area: lakes, ditches, and streams). The relative abundance of *Bosmina longirostris*, *Polyphemus pediculus*, *Moina macrocopa*, and *Diaphanosoma brachyurum* was higher in lakes than in the other water body types, while *Daphnia obtusa* was best represented in wheel tracks and pools (Table 3). The relative share of *Simocephalus vetulus* of total clacoderan density was highest in streams and ditches. *Pleuroxus trigonella* and *Alona quadrangularis* had their highest relative abundances in streams and ditches, respectively, whereas *Chydorus sphaericus* was most abundant in wheel tracks.

Species richness

We detected a total of 53 cladoceran species in the entire set of samples. Overall, local species richness differed significantly between water body types ($F(5) = 2.34$; $p = 0.04$). Average local species richness was highest for lakes (5.7; range: 3–11.1) (Fig. 3a). Average richness in ditches (4.1; range: 1–10.8 species) and ponds (4.2; range: 1–12.4 species) was similar and tended to be slightly higher than in pools; the difference was however not significant.

Results of the species accumulation curves (Fig. 4) show a rapid increase in total species number with an increasing number of lakes compared to pools, whereas ponds and ditches take an intermediate position. According to the curves, a total count of 20 species appears to correspond approximately with a total of six lakes, nine ditches, 11 ponds, or 18 pools. The curves also show that lakes and, to a smaller extent pools, do not approximate their asymptote within the range of samples. In contrast, the curves for ponds and ditches tend to approximate much better their asymptote. According to the likelihood-based extrapolation, total richness equals 49 species in pools (95% confidence interval: 46–51) and 40 species in ditches (95% confidence interval: 34–55). The accumulation curves of wheel tracks and streams could not be reliably characterized due to low number of available samples.

Our standardized measure of total richness showed a similar pattern as local species richness (Fig. 3b).

Table 3 Summary of all significant differences (Tukey post-hoc tests) between the studied water body types for the relative abundances of the ten most dominant species

Species		Tukey post-hoc tests					
		W	PL	PN	L	D	S
Alona quadrangularis (O.F. Müller, 1776)	D		>	>	>		
Bosmina longirostris (O.F. Müller, 1785)	L	>	>	>		>	>
	PN		>			>	
Chydorus sphaericus (O.F. Müller, 1785)	W			>	>		
Daphnia obtusa (Kurz, 1874)	W				>		
	PL			>	>	>	
Diaphanosoma brachyurum (Liévin, 1848)	L	>	>	>		>	>
Megafenestra aurita (S. Fischer, 1849)	S	>	>	>	>	>	
Moina macrocopa (Straus, 1820)	L	>	>	>		>	>
Pleuroxus trigonellus (O.F. Müller, 1776)	S	>	>	>	>	>	
Polyphemus pediculus (Linnaeus, 1761)	L	>	>	>		>	>
Simocephalus vetulus (O.F. Müller, 1776)	PL				>		
	D			>	>		
	S	>		>	>		

The symbol '>' indicates that the relative abundance of a species is significantly higher in the water body type of the first column than in the corresponding water body type of the first row (see Table 1 for the abbreviations of the water body types)

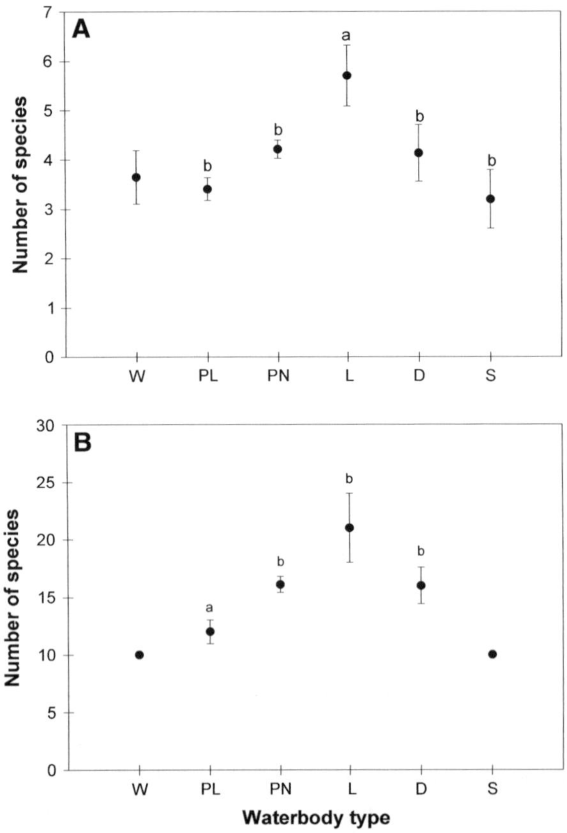

Fig. 3 Differences between water body types with respect to: (**a**) average of the mean local zooplankton diversity observed in each of the selected regions. Error bars denote SE; (**b**) average total diversity expressed as the total species richness observed in sets of six randomly selected systems. Error bars denote SE and reflect the variation among randomly chosen sets of six systems within each water body category. Significant differences between water body types according to Tukey post-hoc tests are indicated by different characters. Abbreviations of the water body types are explained in Table 1

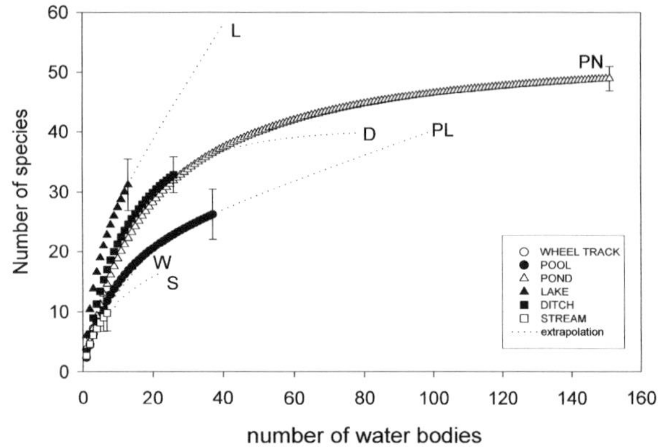

Fig. 4 Species accumulation curves for the six water body types. Expected species richness values (dotted line) were computed using the likelihood-based estimator. Extrapolations for each water body type are limited to three times the number of available samples. For reasons of clarity, 95% bootstrap confidence intervals are only presented for the total number of samples available

Lakes tended to have the highest total richness, but the difference was not significant with ponds and ditches. Pools had significantly lower total richness than ponds, lakes, and ditches (Tukey post-hoc: $p < 0.05$). Due to lack of replicate values, wheel tracks and streams could not be incorporated in this analysis.

Discussion

Community composition

Our survey revealed a high variation in cladoceran community composition among different water body types (wheel tracks, pools, ponds, lakes, ditches, and streams) in Flanders. Species composition differed considerably among the different types of systems and many species showed pronounced affinities with one or a few specific water body types. Pelagic zooplankton species, like *B. longirostris*, *D. brachyura*, *P. pediculus*, and *M. macrocopa* were clearly more associated with lakes than with other water bodies. In contrast *D. obtusa* was mainly found in small temporary systems, such as pools and wheel tracks. This species and *C. sphaericus* have been shown to be very rapid colonizers of newly created habitats (Louette & De Meester, 2005) and have often been reported for temporary waters during spring (Forró et al., 2003). Ditches and streams were quite similar in community composition and were dominated by species like *A. quadrangularis*, *P. trigonelus*, *M. aurita*, and *S. vetulus*. An important reason for the difference between the ditches and streams and the other water body types is probably that they are lotic systems. Water flow in lotic systems may act as an important source of disturbance for zooplankton communities. Cladoceran zooplankton tends to be underrepresented in flowing zones of lotic systems (Dole-Olivier et al., 2000) as complete washout of their populations can occur at water velocities higher than 3.2 cm s^{-1} (Richardson, 1992). Persistence of a zooplankton population in a lotic system is also strongly determined by the availability of flow refugia and specific characteristics of the species (e.g., strong swimming ability, use of benthic habitat and flow avoidance) (Robertson, 2000). The predominance of species of the genera *Alona*, *Pleuroxus*, and *Simocephalus* in lotic systems may at least partly be due to the fact that their species live in close association with substrate or show a strong habitat selection in favour of well-structured littoral zones, which may reduce their vulnerability to downstream washout (Richardson, 1992). Based on the relatively low species richness, lotic systems are poor habitats for cladocerans. Nevertheless, they can be important for dispersal (Havel & Shurin, 2004).

Species richness

Average local species richness in lentic systems tended to increase from small and temporary systems to larger and permanent systems. Lakes consequently had the highest local richness. This is in accordance with the literature (Fryer 1985; Collinson et al., 1995), although it is not always clear whether the degree of permanence or the size of the habitat contributed most to the increase in species richness (Bilton et al., 2001). The significant difference between the local species richness of lakes and ponds or pools and the lack of a difference between ponds and pools may, however, indicate that mainly size and not the degree of permanence is contributing most to species richness. Similarly, lakes had the highest gamma diversity. There may be several reasons why lakes are richer in species than smaller water bodies: (1) In comparison with ponds and pools, lakes are more heterogeneous in space and are less susceptible to drying out. This allows a potentially higher number of species to successfully settle in these systems; (2) The degree of connectivity may also play an important role. Lakes tend to have larger catchment areas than ponds (Davies et al., in press) and have therefore a higher chance of being colonized from neighboring water bodies; (3) Local stress events (e.g., trampling by cattle, inflow of pesticides or nutrients) have a larger impact on small water bodies than on larger-sized systems; (4) Furthermore, lakes often contain large populations that are more buffered against species loss.

Several categories of water bodies were underrepresented in our study, both in terms of absolute numbers (wheel tracks, streams, and lakes) as well as in terms of their representation in the set of samples compared to their abundance in the field. This hampers estimation of the contribution that is made by each water body type to the total cladoceran diversity in regions or to total cladoceran species richness in Flanders. The accumulation curve, nevertheless, suggests that the total number of species that is supported by lakes is larger than in ponds, pools, or ditches. Our standardized measure of total diversity also tended to be higher for lakes than for the other water bodies, suggesting that the total number of species in a set of six lakes is likely to be more species rich than a set of an equal number of other water bodies. Notwithstanding these observations, the contribution made by small water bodies to cladoceran gamma species richness in the average landscape is probably considerably higher than that of lakes. In the framework of the pond project MANSCAPE, we have estimated the number and total cumulative surface area of ponds and lakes in 26 circular regions of 28 km^2 scattered over Flanders (which include the 18 regions of this study). This was done based on aerial pictures of the National Geographic Institute (2000) through the application of the GIS software package ArcView GIS 3.2a (ESRI, Inc.). According to these estimations each region contains an average of 1.11 lakes and 80.03 ponds with an average cumulative surface area of 0.9 and 3.9 ha, respectively. This means that, in one region, there are approximately 70 times more ponds than lakes and that the cumulative surface area of ponds is 4.3 times as large as that of lakes. If both the data on pond densities and the accumulation curve are assumed to be representative for Flanders, this means that for an average surface area of 28 km^2 ponds contribute approximately 44 species to the total cladoceran richness of that region, whereas lakes only contribute five species.

It should be noted that this study is based on a single sampling campaign. As a result, temporal variation in cladoceran communities was not covered by our sampling. This may have lead to an underestimation of species numbers, especially in systems with high temporal dynamics and seasonal or annual species turnover rates (Vandekerkhove et al., 2005a, b). In addition, we may have missed rare species that typically have low population densities in habitats where they occur because we only investigated 100 specimens of cladocerans per sample.

In conclusion, we found a pronounced variation in the species lists and percentage composition of cladoceran communities between different water body types. With respect to conservation measures, these findings stress the importance of maintaining a diversity of water body types of different flow, size, and permanence regimes in the landscape. Species richness tended to be highest in lakes and lowest in lotic systems. Despite the relatively high number of species supported by lakes, small water bodies probably contribute considerably more to total cladoceran richness in the average landscape of Flanders than lakes, because of their high relative abundance.

Acknowledgments This study was financially supported by the Belgian federal government in the framework of the PODO II-program, project MANSCAPE "Integrated Management Tools for Water Bodies in Agricultural Landscapes" (EV/01/29E) and by EU IP project ALARM (GOCE-CT-2003-506675). We thank C. X. Mao for providing the algorithms to calculate the species accumulation curves. We are also grateful to J. Biggs and two anonymous referees for very useful comments on an earlier version of this text. S. Declerck is a post-doctoral fellow of the Fund for Scientific Research-Flanders.

References

Allan, J. D., 2004. Landscapes and riverscapes: the influence of land use on stream ecosystems. Annual Review of Ecology, Evolution, and Systematics 35: 257–284.

Armitage, P. D., K. Szoszkiewicz, J. H. Blackburn & I. Nesbitt, 2003. Ditch communities: a major contributor to floodplain biodiversity. Aquatic Conservation: Marine and Freshwater Ecosystems 13: 165–185.

Biggs, J., P. Williams, M. Whitfield, P. Nicolet & A. Weatherby, 2005. 15 years of pond assessment in Britain: results and lessons learned from the work of Pond Conservation. Aquatic Conservation: Marine and Freshwater Ecosystems 15: 693–714.

Bilton, D. T., A. Foggo & S. D. Rundle, 2001. Size, permanence and the proportion of predators in ponds. Archiv für Hydrobiologie 151: 451–458.

Boothby, J., 2003. Tackling degradation of a seminatural landscape: options and evaluations. Land Degradation & Development 14: 227–243.

Clarke, K. R. & R. N. Gorley, 2001. PRIMER Version 5. User Manual/Tutorial. PRIMER-E, Plymouth, UK.

Collinson, N. H., J. Biggs, A. Corfield, M. J. Hodson, D. Walker, M. Whitfield & P. J. Williams, 1995. Temporary and permanent ponds: an assessment of the effects of drying out on the conservation value of aquatic macroinvertebrate communities. Biological Conservation 74: 125–133.

Colwell, R. K., C. X. Mao & J. Chang, 2004. Interpolating, extrapolating, and comparing incidence-based species accumulation curves. Ecology 85: 2717–2727.

Davies, B. R., J. Biggs, P. J. Williams, J. T. Lee & S. Thompson, in press. A comparison of the catchment sizes of rivers, streams, ponds, ditches and lakes: implications for protecting aquatic biodiversity in an agricultural landscape. Hydrobiologia doi:10.1007/s10750-007-9227-6.

Declerck, S., T. De Bie, D. Ercken, H. Hampel, S. Schrijvers, J. VanWichelen, V. Gillardin, R. Mandiki, B. Losson, D. Bauwens, S. Keijers, W. Vyverman, B. Goddeeris, L. De Meester, L. Brendonck & K. Martens, 2006. Ecological characteristics of small farmland ponds: associations with land use practices at multiple spatial scales. Biological Conservation 131: 523–532.

Dole-Olivier, M.-J., D. M. P. Galassi, P. Marmonier & M. Creuzé des Châteliers, 2000. The biology and ecology of lotic microcrustaceans. Freshwater Biology 44: 63–91.

Flössner, D., 2000. Die Haplopoda und Cladocera (ohne Bosminidae) Mitteleuropas. Backhuys Publishers, Leiden.

Forró, L., L. De Meester, K. Cottenie & H. J. Dumont, 2003. An update on the inland cladoceran and copepod fauna of Belgium, with a note on the importance of temporary waters. Belgian Journal of Zoology 133: 31–36.

Fryer, G., 1985. Crustacean diversity in relation to the size of water bodies: some facts and problems. Freshwater Biology 15: 347–361.

Havel, J. E. & J. B. Shurin, 2004. Mechanisms, effects, and scales of dispersal in freshwater zooplankton. Limnology and Oceanography 49: 1229–1238.

Lepš, J. & T. Šmilauer, 2003. Multivariate Analysis of Ecological Data using CANOCO. Cambridge University Press, Cambridge.

Louette, G. & L. De Meester, 2005. High dispersal capacity of cladoceran zooplankton in newly founded communities. Ecology 86: 353–359.

Mao, C. X., R. K. Colwell & J. Chang, 2005. Estimating the species accumulation curve using mixtures. Biometrics 61: 433–441.

Nicolet, P., J. Biggs, G. Fox, M. J. Hodson, C. Reynolds, M. Whitfield & P. Williams, 2004. The wetland plant and macroinvertebrate assemblages of temporary ponds in England and Wales. Biological Conservation 120: 261–278.

Oertli, B., D. A. Joye, E. Castella, R. Juge, D. Cambin & J. B. Lachavanne, 2002. Does size matter? The relationship between pond area and biodiversity. Biological Conservation 104: 59–70.

Oertli, B., J. Biggs, R. Céréghino, P. Grillas, P. Joly & J. B. Lachavanne, 2005. Conservation and monitoring of pond biodiversity: introduction. Aquatic Conservation: Marine and Freshwater Ecosystems 15: 535–540.

Richardson, W. B., 1992. Microcrustaceae in flowing water: experimental analysis of washout times in a field test. Freshwater Biology 28: 217–230.

Robertson, A. L., 2000. Lotic meiofaunal community dynamics: colonisation, resilience and persistence in a spatially and temporally heterogenous environment. Freshwater Biology 135–147.

Søndergaard, M., E. Jeppesen & J. P. Jensen, 2005. Pond or lake: does it make any difference? Archiv für Hydrobiologie 162: 143–165.

Stendera, S. E. S. & R. K. Johnson, 2005. Additive partitioning of aquatic invertebrate species diversity across multiple scales. Freshwater Biology 50: 1360–1375.

Vandekerkhove, J., S. Declerck, L. Brendonck, J. M. Conde-Porcuna, E. Jeppesen, L. S. Johansson & L. De Meester, 2005a. Uncovering hidden species: hatching diapausing eggs for the analysis of cladoceran species richness. Limnology and Oceanography: Methods 3: 399–407.

Vandekerkhove, J., S. Declerck, E. Jeppesen, J. M. Conde Porcuna, L. Brendonck & L. De Meester, 2005b. Dormant propagule banks integrate spatio-temporal heterogeneity in cladoceran communities. Oecologia 142: 109–116.

Williams, P., M. Whitfield, J. Biggs, S. Bray, G. Fox, P. Nicolet & D. Sear, 2004. Comparative biodiversity of rivers, streams, ditches and ponds in an agricultural landscape in Southern England. Biological Conservation 115: 329–341.

Hydrobiologia (2008) 597:29–41
DOI 10.1007/s10750-007-9218-7

ECOLOGY OF EUROPEAN PONDS

Macroinvertebrate assemblages in 25 high alpine ponds of the Swiss National Park (Cirque of Macun) and relation to environmental variables

Beat Oertli · Nicola Indermuehle · Sandrine Angélibert · Hélène Hinden · Aurélien Stoll

Abstract High-altitude freshwater ecosystems and their biocoenosis are ideal sentinel systems to detect global change. In particular, pond communities are likely to be highly responsive to climate warming. For this reason, the Swiss National Park has included ponds as part of a long-term monitoring programme of the high-alpine Macun cirque. This cirque covers 3.6 km^2, has a mean altitude of 2,660 m a.s.l., and includes a hydrographic system composed of a stream network and more than 35 temporary and permanent ponds. The first two steps in the programme were to (i) make an inventory of the macroinvertebrates of the waterbodies in the Macun cirque, and (ii) relate the assemblages to local or regional environmental variables. Sampling was conducted in 25 ponds between 2002 and 2004. The number of taxa characterising the region (Macun cirque) was low, represented by 47 lentic taxa. None of them was endemic to the Alps, although several species were cold stenothermal. Average pond richness was low (11.3 taxa). Assemblages were dominated by Chironomidae (Diptera), and Coleoptera and Oligochaeta were also relatively well represented. Other groups, which are frequent in lowland ponds, had particularly poor species richness (Trichoptera, Heteroptera) or were absent (Gastropoda, Odonata, Ephemeroptera). Macroinvertebrate assemblages (composition, richness) were only weakly influenced by local environmental variables. The main structuring processes were those operating at regional level and, namely, the connectivity between ponds, i.e. the presence of a physical connection (tributary) and/or small geographical distance between ponds. The results suggest that during the long-term monitoring of the Macun ponds (started in 2005), two kinds of change will affect macroinvertebrate assemblages. The first change is related to the natural dynamics, with high local-scale turnover, involving the metapopulations characterising the Macun cirque. The second change is related to global warming, leading to higher local and regional richness through an increase in the number of colonisation events resulting from the upward shift of geographical ranges of species. At the same time the cold stenothermal species from Macun will be subject to extinction.

Guest editors: R. Céréghino, J. Biggs, B. Oertli & S. Declerck
The ecology of European ponds: defining the characteristics of a neglected freshwater habitat

Electronic supplementary material The online version of this article (doi:10.1007/s10750-007-9218-7) contains supplementary material, which is available to authorized users.

B. Oertli (✉) · N. Indermuehle · S. Angélibert · A. Stoll
Department of Nature Management, University of Applied Sciences of Western Switzerland - EIL, 1254 Jussy, Geneva, Switzerland
e-mail: beat.oertli@hesge.ch

H. Hinden
Laboratoire d'Ecologie et de Biologie aquatique, University of Geneva, 1206 Geneva, Switzerland

Keywords Zoobenthos · Small waterbodies · Biodiversity · Swiss Alps · Biomonitoring

Introduction

Ongoing and future global change threatens biodiversity (Thomas et al., 2004) at local, regional and global scales. In this context, it is urgent to put into action long-term programmes, to assess and monitor predicted changes. This type of monitoring has the ability to integrate sentinel ecosystems that can detect signs of stress. Mountain biocoenoses are particularly sensitive to global change (Thuillier et al., 2005), and future climatic warming in Europe is predicted to be greatest in the arctic and alpine regions (Wathne & Hansen, 1997). High-altitude aquatic ecosystems may be especially sensitive to global change (Theurillat & Guisan, 2001), and particularly to climate warming (Sommaruga-Wograth et al., 1997; Sommaruga et al. 1999; Gurung, 2005). With their small size and their relatively simple community structure, ponds constitute ideal sentinel and early warning systems (De Meester et al., 2005). This is especially true in alpine environments, where pond communities are species-poor and simpler than in lowland environments (Hinden et al., 2005). Therefore, high alpine ponds constitute ideal systems for long-term monitoring of global changes.

The Swiss National Park started in 2001 a long-term monitoring programme at the high-altitude "Macun cirque", including the ponds. The Swiss National Park is one of the 28 selected UNESCO Mountain Biosphere Reserves (Scheurer, 2004). The Macun cirque covers 3.6 km^2; it has a mean altitude of 2,660 m a.s.l., and includes a large hydrographic system composed of a stream network and more than 35 temporary and permanent ponds. Earlier scientific research in this area is limited, but includes information on diatom communities (Schanz, 1984).

One of the preliminary steps towards monitoring is to establish a large and solid baseline through intense and comprehensive field investigations. Indeed, understanding patterns of biodiversity distribution is essential to conservation strategies (Gaston, 2000), and resolving the relative contributions of local and regional processes might provide a key to understanding global patterns of biodiversity (Gaston & Spicer, 2004). As such, various abiotic and biotic investigations were conducted in the Macun cirque (starting in 2001), on all types of waterbodies (Matthaei, 2003; Hinden, 2004; Logue et al., 2004; Ruegg & Robinson, 2004; Stoll, 2005). The objectives partly followed GLOCHAMORE (Global Change and Mountain Regions) Research Strategy (Gurung, 2005), and aimed to assess the current biodiversity of the cirque, as a requisite before trying to assess or to monitor biodiversity change.

A synthesis of the investigations conducted on the ponds is presented here. The objectives of this article are to (i) present the distribution patterns of the aquatic macroinvertebrates in a network of high-alpine ponds, with a special focus on the diversity (alpha and gamma) and the composition of the taxonomic assemblages; and (ii) investigate whether observed patterns are determined by regional or local processes. As the scientific information collected is intended to constitute a baseline for the long-term monitoring of the Macun cirque, a set of questions were addressed to build a coherent monitoring programme. In particular, what taxonomic groups are dominant (density and richness) in these high altitude systems? What taxa can be used for monitoring purposes? Can flagship groups be used for monitoring, as is the case for lowland ponds (e.g. Odonata; Clausnitzer & Jödicke, 2004)? As a high level of endemism is expected (Väre et al., 2003), should there be any particular species of macroinvertebrates to monitor?

Methods

Study site

The Macun cirque was added to the Swiss National Park in 2000 and is currently designated for long-term monitoring of Alpine waterbodies. This area is situated in the high alpine zone of the Alps (>2,600 m a.s.l.) in Graubünden, Switzerland (46°44′ N, 10°08′ E). The climatic conditions are extreme, with extended ice cover (9 month). The temperature range is large (from <−15°C in winter to >20°C in summer), and precipitation is low, around 850 mm/year (Robinson & Kawecka, 2005).

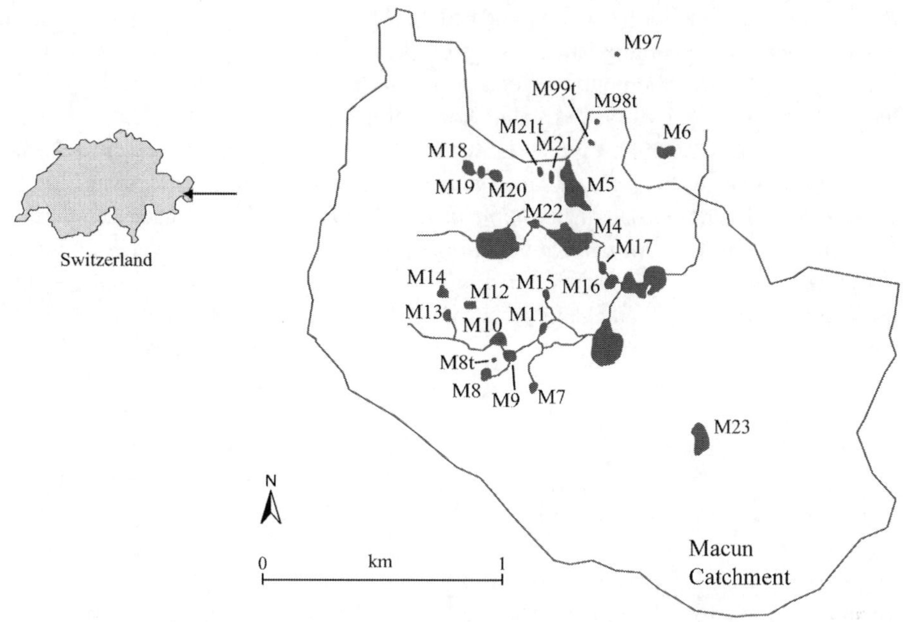

Fig. 1 The hydrological system from Macun cirque (Swiss National Park) with the position of the 25 sampled ponds

Bedrock geology is crystalline rock. The cirque covers 3.6 km^2 and includes a hydrographic system (Fig. 1) composed of a stream network, more than 35 small waterbodies (area: 24–18,000 m^2), some of the smallest being temporary. All these waterbodies can be termed "ponds", according to definition in Oertli et al. (2005b): an area less than 2 ha and maximum depth less than 8 m (except one pond, with a 10-m depth), offering water plants the potential to colonise almost the entire pond area. Some of these ponds are interconnected by streams. The hydrographic system has a natural origin with an age exceeding 4,000 year (the date of the last glacial retreat). The origin of the water differs depending on location in the Macun cirque. Robinson & Kawecka (2005) differentiated a north basin fed primarily by precipitation (mostly snow) and groundwater and a south basin fed mostly by glacial melt from rock glaciers. The different origin in each basin affects the physico-chemical characteristics of the water (Robinson & Matthaei, 2007). The south basin had temperatures on average 4°C cooler and nitrate-N levels twice the amount found in the north basin. In contrast, the north basin had higher levels (2–4 times more) of particulate-P, particulate-N, and particulate organic matter than the south basin.

The drainage area of each pond is characterized by a mixture of two types of land cover, rock and alpine grassland (typology following Delarze et al., 1998); land cover was assessed through field survey and the help of GIS (for the delimitation of the drainage area of each pond). From the 35 waterbodies, a subset of 25 was selected for sampling (Fig. 1). The choice included all types of waterbodies: different size, different duration (ponds with a low depth were temporary, dry in August), location in the North or South basin, and connected or not to streams. The pH of the sampled ponds was acid or near neutral (5.5–7.5).

Table 1 Description of the continuous environmental variables characterising the 25 sampled ponds

	Mean	SD	Median	Minimum	Maximum
Altitude (m a.s.l.)	2641	49	2653	2480	2714
Pond area (m^2)	1909	3139	607	24	12750
Mean depth (m)	0.99	1.08	0.60	0.05	4.50
Max depth (m)	2.2	2.7	1.3	0.1	10.0
Conductivity (μs/cm)	9.7	13.3	6.6	1.7	68.3
Alpine grassland in drainage area (%)	56	30	60	5	100
Total nitrogen (mg/l)[a]	0.24	0.09	0.23	0.13	0.50

[a] Measured on a subset of 16 ponds

A description of pond morphometry and other selected physico-chemical characteristics is presented in Table 1. Substrate of the ponds was relatively homogeneous (stone, gravel, sand, and occasionally bryophytes) and the habitats available to macroinvertebrates not diverse. Five of the sampled ponds (M4, M5, M14, M16, M17) had fish (*Salmo trutta fario*, *Salvelinus namaycush*, *Phoxinus phoxinus*), and were last stocked in 1993. Information of fish presence was collected by net-fishing and visual observation (Rey & Pitsch, 2004).

Macroinvertebrate sampling

The 25 ponds were sampled, either in 2002 (16–22 July) or in 2004 (27 July to 2 August). The standardised procedure "PLOCH" (Oertli et al., 2005a) was used for sampling macroinvertebrates. They were collected with a small-framed hand-net (rectangular frame 14 × 10 cm, mesh size 0.5 mm). For each sample, the net was swept through the water intensively for 30 s. The number of samples taken ranged from 2 to 20, depending on pond size. Sampling was stratified for the dominant habitats (from the land-water interface to a depth of 2 m): stones, gravel, sand and bryophytes. In all cases, the collected material was preserved in 5% formaldehyde and then comprehensively sorted in the laboratory. Specimens were identified to species level for most taxonomic groups and counted. Chironomidae and Oligochaeta were identified to species level for a subset of 20 ponds. Pupae were generally not considered; nevertheless some were identified when species level identification was impossible for larvae (see Electronic supplementary material). Collected taxa were classified either as lentic or lotic (see Electronic supplementary material). This separation was first made on the basis of ecological information available (Tachet et al., 2000; and for Chironomidae, B. Lods-Crozet personal communication). Nevertheless, some taxa known as lotic were classified as lentic if they were observed in abundance in the isolated ponds (not connected to a tributary).

Data analysis

Invertebrate data used for analysis included only the lentic taxa from the 25 sampled ponds. Lotic taxa whose presences were linked to drift from tributaries (e.g. Plecoptera and Diptera Simuliidae) were discarded (see list in Electronic supplementary material). Each analysis was conducted for both datasets; i.e. the 20 ponds where all macroinvertebrates have been identified and the 25 ponds, where Oligochaeta and Chironomidae were excluded.

Tests were conducted to check inter- and intra-year variability in the composition of the macroinvertebrate assemblage. Four ponds were sampled in both years (2002 and 2004), and one pond was sampled twice in 2004 (27 July and 2 August). Thereafter, all 30 macroinvertebrate assemblages (25 ponds, and 5 replicates) were submitted to a Ward clustering. This classification highlighted a very close grouping of the different replicates and demonstrated that differences in assemblage were minor between years (2002–2004) or within year (interval of 6 days). Therefore, for further data analyses, the ponds sampled in 2002 were added to those sampled in 2004 for constitution of a unique dataset.

The true regional diversity (gamma diversity, for the Macun cirque) was estimated through the use of the non-parametric estimator Jackknife-1 (Foggo et al., 2003; Magurran, 2003). This estimator is designed to overcome sample-size inadequacies and to estimate how many species are actually present in sampled habitats (Rosenzweig et al., 2003). Calculations were made with EstimateS software (Colwell, 2005).

Associations between seven environmental variables (mean and max depths, area, altitude, land cover, conductivity and total nitrogen) and taxonomic richness were tested by simple correlation analyses (Spearman correlation test). Three other variables are described by class (presence of tributary connected to another upstream pond, presence of fish, northern or southern position in the cirque) and their relationships with taxonomic richness were assessed by a non-parametric Mann–Whitney test.

To test the influence of environmental variables on the distribution of species, we first performed a Correspondence Analysis (CA). This analysis was realised on presence/absence data on the basis of the species list presented in Electronic supplementary material (but the rarest species, i.e. present in <5% of sites, were removed). This procedure was conducted with the whole macroinvertebrate community (20 ponds), as well as after removal of the Oligochaeta and Chironomidae (25 ponds).

Then we tested whether the environmental variables were significantly correlated to the first and/or second axis of the CA. To clarify the interpretation of the CA, we performed a cluster analysis based on Jaccard's index. As the same patterns were observed in both cases, we will present here only the results for the 25-pond dataset. These analyses were done with the Vegan Package (available at http://www.r-project.org), a contributed package for the R environment. Ward's clustering technique was used for identifying different types of assemblages based on abundances transformed into classes (detailed in Electronic supplementary material). Both datasets (20 and 25 ponds) were analysed separately.

Results

Biodiversity of the Macun ponds

The sampling of the 25 Macun cirque ponds between 2002 and 2004 yielded 57 taxa, most of which were identified to species level (Electronic supplementary material). Ten of these taxa are known as lotic (Electronic supplementary material) and were represented only in waterbodies connected to streams (i.e. drift organisms). With the removal of these lotic taxa, the regional richness (Macun cirque) was 47 lentic taxa. Taxa accumulation reached an asymptote (Fig. 2), indicating that this observed value is near the true value of richness to be expected for the cirque. Furthermore, the Jackknife-1 estimator of true richness gives a value of 57.7 taxa (±4.3), suggesting that few taxa were missed (probably about 11) and indicating that the sampling procedure used was efficient for the estimation of regional diversity.

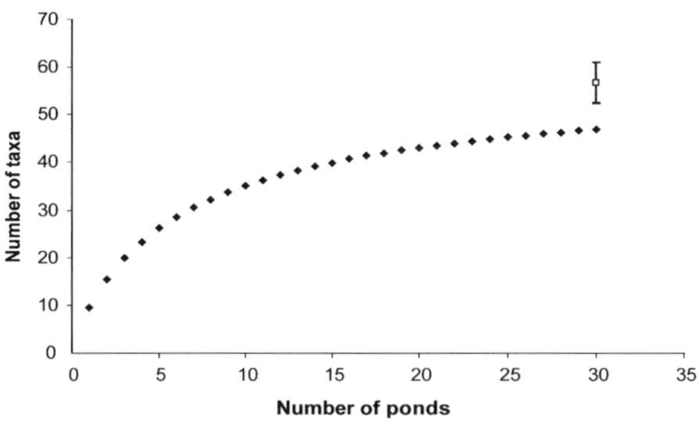

Fig. 2 Accumulation curve for observed richness (S observed) of lentic taxa of the Macun cirque ponds and estimation of true richness (S true) with estimator Jackknife-1 (±standard deviation: SD). The 30 samples are composed of 25 ponds, three being sampled twice and one being sampled three times

The richest group was the Diptera (25 Chironomidae taxa; five other families), followed by the Oligochaeta (6 taxa) and the Coleoptera (six species). Other represented groups were less diverse: Trichoptera (two species); and Heteroptera, Tricladida, and Bivalvia (each with one species). In terms of abundance (number of individuals), four groups clearly dominated (Electronic supplementary material): Chironomidae, Coleoptera, Oligochaeta and Trichoptera.

This assemblage of 47 lentic taxa is species poor compared to lowland ponds (see taxa list compiled for Swiss lowland ponds by Oertli et al., 2000). For example Hydridae, Hirudinea, Crustacea, Gastropoda, Lepidoptera, Megaloptera, Ephemeroptera, Odonata and Hydracarina are missing.

In order to assess the magnitude of the Macun cirque richness at a larger regional scale, the family Dytiscidae (Coleoptera) was chosen. This choice was motivated by the availability of relatively good knowledge of their geographical distribution in Switzerland (data bank from the Swiss Centre for Faunistic Cartography, and unpublished data from G. Carron) and in Europe (Illies, 1978). The observed species richness for lentic Dytiscidae was five species for the Macun cirque, whereas the Jackknife-1 true richness estimator indicates six species (±1). On a regional scale, the species pool for the Swiss oriental Alps (area: about 8,000 km^2) is 15 species (G. Carron, personal communication), while the pool for the whole Alps (area: about 100,000 km^2) is 22 species (data from Illies, 1978). Therefore, taking into account the small area of the Macun cirque (only 3.6 km^2), the Dytiscidae community is relatively rich.

Endemism of the Macun macroinvertebrates

Endemism of the macroinvertebrates collected in the Macun cirque was assessed for 30 lentic species based on their biogeographical distribution in Europe (Illies, 1978): none are endemic to the Alps (Fig. 3). Nevertheless, one species, the trichopteran *Acrophylax zerberus*, is restricted to only four biogeographic

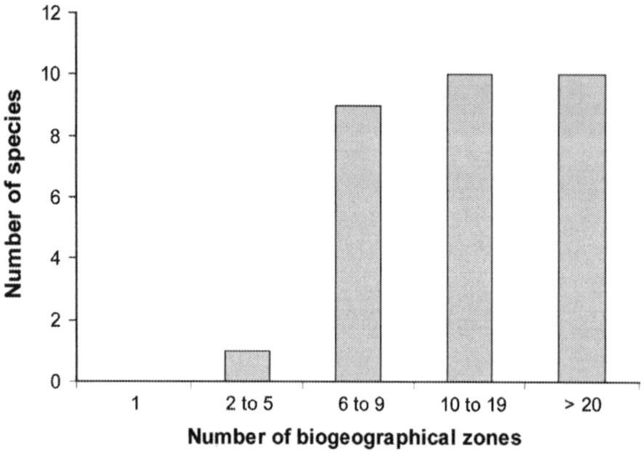

Fig. 3 Geographical distribution in Europe of 30 lentic species collected in the ponds from the Macun cirque. The number of biogeographical zones occupied by each species has been assessed through the information presented in Illies (1978), completed through the expertise of different taxonomy specialists (G. Carron, personal communication; B. Lods-Crozet, personal communication; V. Lubini, personal communication)

zones in Europe and can be considered as near-endemic. Some other species, occupying several biogeographic zones (being neither endemic nor near-endemic), are nevertheless restricted to the coldest areas of these zones. This is the case for two Dytiscidae *Hydroporus foveolatus* and *H. nivalis* and two Chironomidae *Micropsectra notescens* and *Pseudosmittia oxoniana*. These species are cold stenothermal.

Taxa richness in ponds and in samples

Richness of lentic taxa was on average 11.3 taxa per pond (min: 6; max: 24). Chironomidae was the richest group (4.8 taxa per pond), followed by the Coleoptera (2.3) and the Oligochaeta (1.9) (Fig. 4). In the case of Coleoptera, mean species richness per pond is in the range of what has been observed in ponds of the alpine belt in Switzerland (Fig. 5a). However Macun ponds appear species-poor compared to ponds situated at a lower altitude, especially to the 5×-richer ponds from the colline altitudinal belt. The mean richness per sample shows weaker differences. A sample (30 s sweeping with the handnet) taken in a Macun pond gathers a mean Coleoptera richness of 0.93 species (±0.56), a value that is higher than data observed in alpine and even in subalpine ponds (Fig. 5b). The ratio of sample

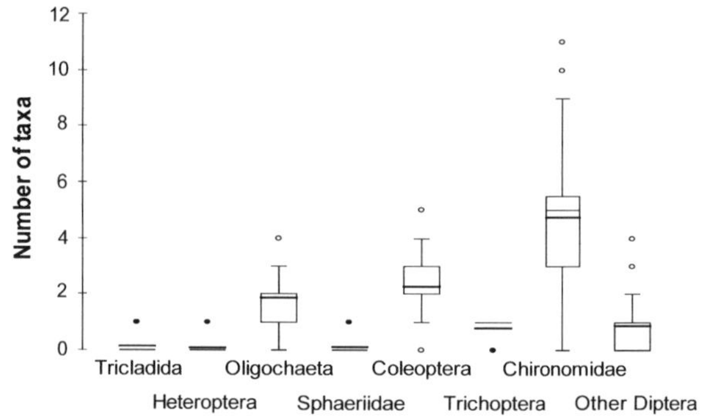

Fig. 4 Mean richness of lentic taxa per pond. Most taxa have been identified to species (see Electronic supplementary material). $n = 25$ ponds; except for Chironomidae and Oligochaeta, where $n = 20$. Each box represents the interquartile range (IQR, 25–75%) with the horizontal lines indicating the median (normal line) and the mean (bold line). Upper error bars indicate the non-outlier maximum. Lower error bars indicate the non-outlier minimum. Plain circles indicate outliers >3× IQR, empty circles indicate outliers <3× IQR

richness to pond richness illustrates the magnitude of the distribution of the species in a given pond. If species are distributed over the whole pond area (and are found in all habitats), each sample will collect all species. On the other hand, if species are restricted to some habitats, a sample will collect only a subset of the whole pond community. The ratio of 0.4 in Macun (mean from across all ponds) indicates that it is possible to catch about 40% of pond diversity with only one sample. Ratios observed elsewhere in Switzerland are usually much lower, and for all four altitudinal belts they were ~0.15 (Fig. 5c). The high ratio for Macun indicates that the species have a homogeneous distribution inside each of the Macun ponds.

Relation between taxonomic richness and environmental variables

Most of the relationships between lentic taxa richness in ponds and the environmental variables were not significant (at $P = 0.05$) (Table 2). Taxa richness is related with neither pond area nor altitude. Furthermore, temporality (expressed here by mean depth) also was not a significant variable. Nevertheless, two significant relationships were found. The clearest relationship shows that the presence of a tributary (in connection with a pond situated upstream) significantly enhances the number of lentic taxa in a pond,

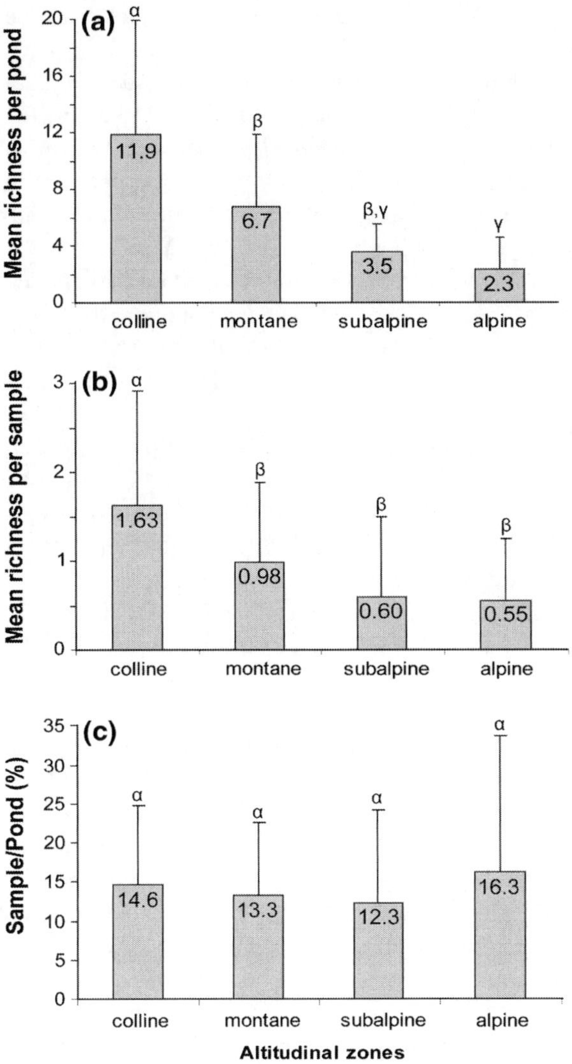

Fig. 5 Mean richness of aquatic Coleoptera per pond (**a**), per sample (**b**), and ratio sample/pond (**c**). Bars with the same letters (α, β or γ) are not significantly different (Mann–Whitney, $P > 0.05$). $n = 82$ ponds (data from Oertli et al., 2000) spread throughout the four altitudinal zones of Switzerland: colline (210–665 m a.s.l.; $n = 34$), montane (610–1400 m a.s.l.; $n = 26$), subalpine (1410–1988 m a.s.l.; $n = 10$) and alpine (1860–2650 m a.s.l.; $n = 12$) were sampled with the same standardised procedure PLOCH (Oertli et al., 2005a)

by about 50%. Mean richness increases from 9.5 (±2.5) to 14.1 (±5.9) ($P = 0.03$) when a tributary is present. When Oligochaeta and Chironomidae taxa are removed, the increase is from 3.6 (±1.4) to 5.6 (±2.4) ($P = 0.01$). When richness of active colonisers is separated from richness of passive colonisers, the relationship with the presence of a tributary is

Table 2 Relationship between the main environmental variables and the taxonomic richness in Macun ponds

	Continuous variables							Binomial variables		
	Altitude	Pond area	Mean depth	Max depth	Conductivity	Grassland in drainage area	Total nitrogen[b]	Fish presence	Tributary	North–South
S total	ns	ns	ns	ns	ns	ns	*	[a]	*	ns
($n = 20$ ponds)	$r = -0.16$	$r = 0.34$	$r = 0.28$	$r = 0.25$	$r = 0.43$	$r = 0.22$	$r = 0.68$		+; $P = 0.03$	$P = 0.49$
S without OC	ns	ns	ns	ns	ns	ns	ns	ns	**	ns
($n = 25$ ponds)	$r = -0.32$	$r = 0.04$	$r = 0.003$	$r = -0.09$	$r = 0.18$	$r = 0.15$	$r = 0.23$	$P = 0.80$	+; $P = 0.01$	$P = 0.67$

S total, all lentic taxa; S without OC, without Oligochaeta and Chironomidae

The relation is represented for continuous variables by Spearman correlation coefficient (r), or for binomial variables by the significance of a Mann–Whitney test

[a] Relation not tested (too few ponds with fish)

[b] Tested with a subset of 13 ponds

+: positive relationship

* or **: significant relationship ($P \leq 0.05$ or 0.01)

ns: non-significant relationship

significant in the case of passive colonisers ($P = 0.01$), and near significant with active colonisers ($P = 0.06$). For example, lentic Coleoptera (active colonisers) present a higher richness (2.8 ± 1.0) in ponds with a tributary than in those without a tributary (2.0 ± 1.0) ($P = 0.03$).

The other significant relationship shows the positive effect of total nitrogen on taxa richness. However, this relation is dominated by the Chironomidae and Oligochaeta and no more significant when these groups are removed.

Relation of macroinvertebrate assemblages with environmental variables

At first a CA was performed to explore the distribution of species among the ponds and to represent graphically the grouping of the 25 macroinvertebrate assemblages (Fig. 6). The first three axes of the CA explained 26%, 17% and 13% of the variance, respectively. No significant correlation was found between the measured environmental variables and the first two axes of the CA ($P > 0.05$, after 1,000 random permutations of the data).

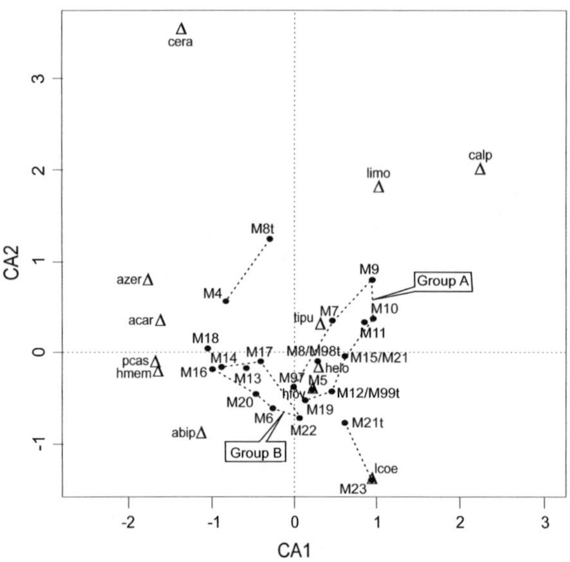

Fig. 6 Correspondence Analysis plot of the macroinvertebrate assemblages (presence/absence data) of 25 ponds from the Macun cirque. Results of clustering (using Jaccard's index; average linkage) is indicated with a dotted line. Species are represented by codes (see list in Electronic supplementary material)

Ordination of sites on the CA plot can be interpreted by the presence of species in each assemblage. Among the most common species, both Trichoptera species played a major role in the ordination of sites. *Acrophylax zerberus* (present in 28% of ponds) and *Limnephilus coenosus* (present in 52% of ponds) were never found in the same pond. Ponds which supported *A. zerberus* (M4, M8t, M13, M14 and M16 to M18) are all located on the left side of the CA plot, and ponds which supported *L. coenosus* were mostly located below in the right part of the CA plot (M6, M10 to M12, M15, M19 to M23, M97 and M99t).

The geographical distance between ponds seemed to influence the distribution of fauna. For example, among the 12 ponds of group A (Fig. 6), seven (M7–M12 and M15) are situated very close to each other in the south-western part of the cirque (see Fig. 1). In the same way, neighbour ponds M17–M16 and M13–M14 had similar faunal compositions. Results obtained using the 20-pond data set (including Oligochaeta and Chironomidae) showed the same trend (not presented here). A strong similarity among M9, M10 and M11 was also found (Fig. 6). Moreover, the similarity between M13 and M14, and M17 and M16, was stronger than in the 25-pond dataset.

An additional analysis, still with the objective of identifying groups of pond assemblages, but this time taking into account the abundances of the species, was done (Ward's clustering technique). The clustering of the 25 ponds produced three groups (Fig. 7). Taxa driving this clustering were the two species of Trichoptera (*Acrophylax zerberus, Limnephilus coenosus*) and one species of Coleoptera (*Agabus bipustulatus*). *A. zerberus* was present and abundant in all 6 ponds from group A, absent in ponds from group B, and present only once in group C. *L. coenosus* was absent from group A, present only once in group B, and present and abundant in almost all ponds from group C. *A. bipustulatus* was present in all ponds from group A but none of group B, and was present in 4 of 15 ponds from group C. None of the local environmental variables (list in Table 2) explained this clustering (Mann–Whitney tests between groups of values). Several ponds belonging to the same cluster are situated geographically close to each other in the cirque; this is the case for: (i) M8, M9 and M10, (ii) M4, M16 and M17, (iii) M13 and M14.

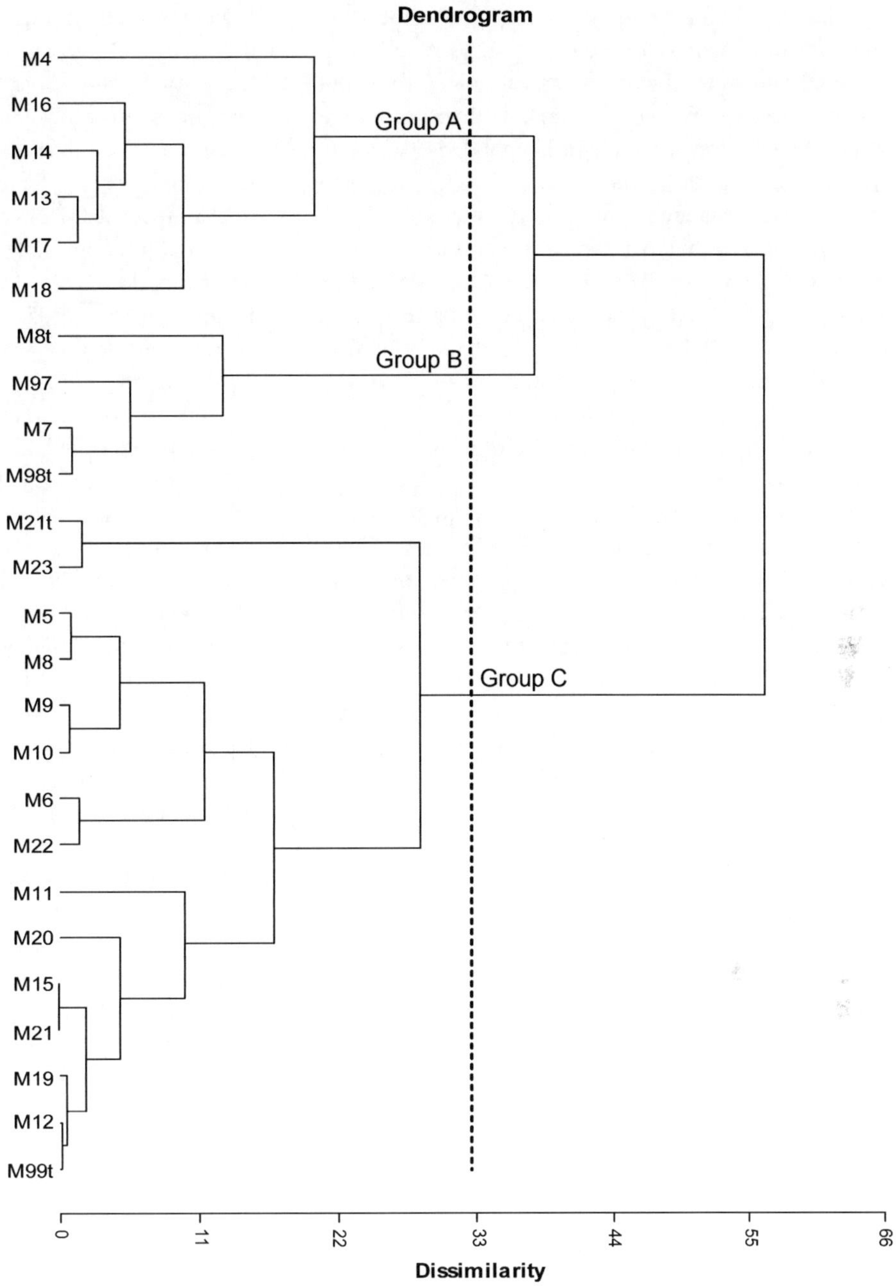

Fig. 7 Ward's Clustering of the macroinvertebrate assemblages (abundance data) of 25 ponds from the Macun cirque. These assemblages do not include Oligochaeta and Chironomidae

The clustering of the 20 ponds dataset (not presented) produced three groups. The same taxa are driving the clustering as described above. Among the environmental variables, the importance of pond morphology was highlighted, and two groups differed with regard to pond area ($P = 0.019$) and mean depth ($P = 0.06$). The other environmental variables did not influence the clustering.

Discussion

Biodiversity of the Macun cirque

The number of macroinvertebrate taxa characterising local (pond) or regional (Macun cirque) diversity investigated in this high alpine pond network was low. This was particularly obvious in comparison to lowland ponds: for example, alpine ponds hold five times fewer Coleoptera species.

Low richness in Macun has, nevertheless, to be put into perspective. When compared with other alpine ponds, Macun ponds have similar pond richness and even higher sample richness. Furthermore, regional richness (for Macun cirque) was relatively high, including nearly half of the species pool of lentic Dytiscidae from the Swiss oriental Alps. From this point of view, Macun cirque can be considered as a species rich area. This richness is perhaps related to the old age of the Macun ponds (>4,000 years), enabling multiple colonisation events. This high regional richness for the Macun cirque is in agreement with Körner (2001) who underlined the extensive biodiversity of the alpine ecosystems when related to the small surface area of the alpine zone. Väre et al. (2003) stressed the high level of endemism of plants in the Alpine life zone. From this viewpoint, the Macun macroinvertebrate community does not show the same trend, as no endemic species were identified. Nevertheless, many species were cold stenothermal. Indeed, even with a broad geographical distribution in Europe, they are restricted to the coldest areas in each country.

Relation of macroinvertebrate assemblages with environmental variables

The richness of lentic taxa from the Macun ponds was correlated with only a few environmental variables. A clear relationship was demonstrated only with the presence of a tributary (connected with an upstream pond), relating to a regional process. The presence of a tributary greatly enhanced the lentic taxa richness of ponds, by about 50%. This enrichment involved active and passive colonisers, indicating that a tributary enhances the connectivity for both types of colonisers. Most likely, a tributary ensures a continuous supply by drift of individuals living in the connected pond upstream. A tributary could also act as a corridor for migration of the adult stages of insects. Migration of lotic invertebrates in streams has already been well documented (e.g. Soderstrom, 1987; Delucchi, 1989), and lotic dispersal has also been demonstrated for lentic invertebrates (Van de Meutter et al., 2006). Our results are consistent with this last study which indicates that flying and non-flying invertebrates disperse with equal frequency. The magnitude of the enrichment through a tributary was particularly significant in the Macun ponds.

Some local variables generally recognised as important driving factors of pond biodiversity were not significant factors in the Macun ponds. This was the case for pond area, shown elsewhere as a major determinative factor (Oertli et al., 2002). However, Hinden et al. (2005) also did not find significant correlations of area with species richness for a set of alpine ponds spread throughout Switzerland. This absence of relation might be a characteristic of alpine ponds. It can reflect a weak influence of local factors, in particular biotic factors (such as competition and predation). Water chemistry (conductivity and total nitrogen) also was not a strong discriminating local variable, probably because the differences in the chemical characteristics were relatively moderate among the ponds. Nevertheless, chemical differences explain differences in the assemblages of lotic and lentic diatoms in the Macun network (Robinson & Kawecka, 2005).

As the Macun ponds with a low depth are dry during some weeks in summer, water permanence could be expected to be a key variable explaining richness or species assemblage, as this was already observed for Macun streams (Ruegg & Robinson, 2004). This was not the case for pond taxa richness. Furthermore, these ponds with a low depth (temporary ponds) did not have species assemblages which were different to those in permanent ponds. At lower altitudes, below the tree-line but still in the mountain zone, duration of water permanence seems to be a much stronger factor; e.g. in subalpine ponds in Colorado (Wissinger et al., 1999) or snowmelt ponds in Wisconsin (Schneider, 1999). A clear negative relationship was demonstrated with species richness, and hydroperiod length influenced community structure. For lowland ponds, this strong structuring effect of hydroperiod duration has already clearly been found (e.g. Wiggins et al., 1980; Williams, 1987; Collinson et al., 1995).

Ponds situated at a small geographical distance from one another tended to have similar macroinvertebrate assemblages. Such positive spatial autocorrelation of community composition is frequently observed in pond networks, at small and large spatial scales (see Briers & Biggs, 2005). As the tested local variables were of relatively weak importance for driving the species richness or the species assemblage, with probably also little pressure from predation or competition, then regional processes and particularly colonisation events should have a high chance of success, therefore playing a major role to explain the distribution of fauna. Furthermore, chance events (Jeffries, 1988) are perhaps an influential factor in these Macun ponds.

In conclusion, assemblages from ponds (composition, richness) are only weakly influenced by local environmental variables in the Macun cirque, and the main structuring processes are regional and namely the connectivity between ponds (physical connection through tributaries and/or small geographical distances among ponds). The situation is undoubtedly different from the lowlands: even in highly interconnected ponds, local environmental constraints can be strong enough to structure local communities (Cottenie et al., 2003).

Monitoring macroinvertebrates in the Macun ponds

Information collected during the sampling of the 25 Macun ponds gave some useful practical recommendations for the long-term monitoring of the area (Hinden et al., 2005; Stoll, 2005). As proposed by these authors, the PLOCH sampling methodology (Oertli et al., 2005a) must be adapted for this particular species-poor environment. First, the proposed sampling must be more intense for each pond, with a longer duration for each individual sample (double time of sweeping with handnet, from 30 to 60 s). Second, sorting of the macroinvertebrates must separate all taxonomic groups (not only Gastropoda and Coleoptera). In particular Oligochaeta and Chironomidae, often neglected, are species-rich if compared with other invertebrate groups. At 26 species, the Chironomidae represent more than half of the taxonomic richness of the lentic macroinvertebrates of the Macun cirque. During monitoring and in particular the data analyses, particular attention should be given to the cold stenothermal species, for example Dytiscidae *Hydroporus foveolatus* and *H. nivalis*, Chironomidae *Micropsectra notescens* and *Pseudosmittia oxoniana* and Trichoptera *Acrophylax zerberus*. The dynamics of the populations of such species can potentially give fingerprints of global change. Other particular attention should be given to flagship species or groups. The Odonata, often included in monitoring of lowland waterbodies, are missing in Macun. Even with rapid global warming, this group will stay species-poor for many more years, as relatively a few species comprise the alpine species pool (Maibach & Meier, 1987). As such, Coleoptera could constitute an alternative flagship group for alpine ponds. This group is well diversified in Macun, has a relatively well-known species distribution in Switzerland and is of some general interest to people.

Expected temporal changes of the species assemblages in the Macun ponds

As the ponds from Macun have low species richness, and as chance events are possibly the most influential structuring factor, each pond is the subject of colonization events that will be frequently successful. Many populations of lentic taxa from Macun vary in time and space in terms of size, and may be subject to local extinction. Therefore, they should be considered as metapopulations (see Bohonak & Jenkins, 2003) in this connected network of ponds. This would mean that species turnover is high in each Macun pond, and is consistent with Jeffries (2005) who suggests that considerable invertebrate dynamics occur in space and time, a phenomenon not adequately addressed by extensive single-year surveys.

Besides these natural changes due to metapopulation dynamics in the Macun area, other biological changes are expected to occur in the future and are related to global warming. Trends predict an upward shift of geographical ranges of species (Hodkinson & Jackson, 2005), resulting in colonisation events in the Macun ponds from species present at lower altitudes (i.e. actually present under tree-line). Many of these colonisation events will be successful in these species-poor systems, and will increase regional and local richness. Figure 5a can be used as a prediction of the magnitude of the resulting changes expected at the local scale. The alpine system will, in the future, resemble the subalpine system more and more. The remaining uncertainty in this prediction is the time needed to reach such a situation. Colonisation events take time, and global change involves not only changes in temperature, but also changes in duration of ice cover and hydrological functioning, that are much more difficult to predict and model than temperature change. Besides this increase in richness, extinction events will also occur in Macun, and the first victims will probably be the cold stenothermal species.

In conclusion, during the long-term monitoring of the Macun ponds (started in 2005), two types of change will affect the macroinvertebrate species assemblages. The first type of change is related to the natural dynamics occurring in these species-poor systems, with important local-scale turnover, involving the species pool characterising the Macun cirque. The second type of change is related to global warming and will lead to higher local and regional richness, but also to extinction of cold stenothermal species. One of the challenges of monitoring will be to identify these changes and their relationship with natural dynamics and global warming.

Acknowledgements This work was partly supported by the Research Committee from the Swiss National Park. Thanks to Thomas Scheurer and Flurin Filli for logistic support and to everyone who helped in the field—Nathalie Menetrey, Lionel Sager, Zoé Fleury and Marianna Massa. Special thanks to Chris Robinson for his helpful collaboration and also for his constructive review of the manuscript. A large part of the chemical analyses were realised by the Swiss Federal Institute of Aquatic Science and Technology. We are very grateful to the CSCF for access to the Swiss databanks on fauna. Help in identification was provided by Gilles Carron (Coleoptera), Brigitte Lods-Crozet (Chironomidae), Narcisse Giani (Oligochaeta), Verena Lubini (Trichoptera) and Nigel Thew (Sphaeriidae). Also, Jane O'Rourke and Mericia Whitfield are thanked for improving the English style. The constructive comments of two anonymous referees improved the paper.

References

Bohonak, A. J. & D. G. Jenkins, 2003. Ecological and evolutionary significance of dispersal by freshwater invertebrates. Ecology Letters 6: 783–796.

Briers, R. A. & J. Biggs, 2005. Spatial patterns in pond invertebrate communities: separating environmental and distance effects. Aquatic Conservation: Marine and Freshwater Ecosystems 15: 549–558.

Clausnitzer, V. & R. Jödicke (eds), 2004. Guardians of the watershed. Global status of dragonflies: critical species, threat and conservation. International Journal of Odonatology 7: 1–430.

Collinson, N. H., J. Biggs, A. Corfield, M. J. Hodson, D. Walker, M. Whitfield & P. J. Williams, 1995. Temporary and permanent ponds: an assessment of the effects of drying out on the conservation value of aquatic macroinvertebrate communities. Biological Conservation 74: 125–133.

Colwell, R. K., 2005. EstimateS: statistical estimation of species richness and shared species from samples. Version 7.5. User's Guide and application published at: http://purl.oclc.org/estimates.

Cottenie, K., E. Michels, N. Nuytten & L. De Meester, 2003. Zooplankton metacommunity structure: regional vs. local processes in highly interconnected ponds. Ecology 84: 991–1000.

De Meester, L., S. Declerck, R. Stoks, G. Louette, F. Van de Meutter, T. De Bie, E. Michels & L. Brendonck, 2005. Ponds and pools as model systems in conservation biology, ecology and evolutionary biology. Aquatic Conservation: Marine and Freshwater Ecosystems 15: 715–725.

Delarze, R., Y. Gonseth & P. Galland, 1998. Guide des milieux naturels de Suisse. Delachaux et Niestlé, Lausanne.

Delucchi, C. M., 1989. Movement patterns of invertebrates in temporary and permanent streams. Oecologia 78: 199–207.

Foggo, A., S. D. Rundle & D. T. Bilton, 2003. The net result: evaluating species richness extrapolation techniques for littoral pond invertebrates. Freshwater Biology 48: 1756–1764.

Gaston, K. J., 2000. Global patterns in biodiversity. Nature 405: 220–227.

Gaston, K. J. & J. I. Spicer, 2004. Biodiversity. An Introduction, 2nd edn. Blackwell Science Ltd, Malden, Oxford, Victoria.

Gurung A. B. (ed.), 2005. GLOCHAMORE Global Change and Mountain Regions. Research Strategy. Mountain Research Initiative, Bern.

Hinden, H., 2004. La biodiversité des petits plans d'eau alpins de Suisse. MS these, University of Geneva, Geneva.

Hinden, H., B. Oertli, N. Menetrey, L. Sager & J.-B. Lachavanne, 2005. Alpine pond biodiversity: what are the related environmental variables? Aquatic Conservation: Marine and Freshwater Ecosystems 15: 613–624.

Hodkinson, I. D. & J. K. Jackson, 2005. Terrestrial and aquatic invertebrates as bioindicators for environmental monitoring, with particular reference to mountain ecosystems. Environmental Management 355: 649–666.

Illies, J., 1978. Limnofauna Europaea, 2nd edn. Gustav Fischer Verlag, Stuttgart.

Jeffries, M., 1988. Measuring talling element of chance in pond populations. Freshwater Biology 20: 383–393.

Jeffries, M., 2005. Small ponds and big landscapes: the challenge of invertebrate spatial and temporal dynamics for European pond conservation. Aquatic Conservation: Marine and Freshwater Ecosystems 15: 541–548.

Körner, C., 2001. Alpine ecosystems. In Levin, S. A. (ed.), Encyclopedia of Biodiversity. Academic Press, San Diego, 133–144.

Logue, J. B., C. T. Robinson, C. Meier & J. R. Van der Meer, 2004. Relationship between sediment organic matter, bacteria composition, and the ecosystem metabolism of alpine streams. Limnology and Oceanography 49: 2001–2010.

Magurran, A. E., 2003. Measuring Biological Diversity. Blackwell Publishing, Oxford.

Maibach, A. & C. Meier, 1987. Atlas de distribution des libellules de Suisse (Odonata). Centre Suisse de

Cartographie de la Faune, Ligue suisse pour la protection de la nature, Neuchâtel.
Matthaei, S., 2003. Expansion contraction cycle of a stream/lake network in a high alpine floodplain. Diploma thesis, Swiss Federal Institute of Environmental Science and Technology, Dübendorf, Switzerland.
Oertli, B., D. Auderset Joye, E. Castella, R. Juge & J-B. Lachavanne, 2000. Diversité biologique et typologie écologique des étangs et petits lacs de Suisse. Rapport final. OFEFP et Université de Genève.
Oertli, B., D. Auderset Joye, E. Castella, R. Juge, D. Cambin & J.-B. Lachavanne, 2002. Does size matter? The relationship between pond area and biodiversity. Biological Conservation 104: 59–70.
Oertli, B., D. Auderset Joye, E. Castella, R. Juge, A. Lehmann & J.-B. Lachavanne, 2005a. PLOCH: a standardised method for sampling and assessing the biodiversity in ponds. Aquatic Conservation: Marine and Freshwater Ecosystems 15: 665–679.
Oertli, B., J. Biggs, R. Céréghino, P. Grillas, P. Joly & J.-B. Lachavanne, 2005b. Conservation and monitoring of pond biodiversity: introduction. Aquatic Conservation: Marine and Freshwater Ecosystems 15: 535–540.
Rey, P. & P. Pitsch, 2004. Die Entdeckung der Bescheidenheit: erste Einblicke in das Leben der Fische auf Macun. Cratschla 2: 23.
Robinson, C. T. & B. Kawecka, 2005. Benthic diatoms of an Alpine stream/lake network in Switzerland. Aquatic Sciences 67: 492–506.
Robinson, C. T. & S. Matthaei, 2007. Hydrological heterogeneity of an Alpine stream/lake network in Switzerland. Hydrological Processes 21: 3146–3154.
Rosenzweig, M. L., W. R. Turner, J. G. Cox & T. H. Ricketts, 2003. Estimating diversity in unsampled habitats of a biogeographical province. Conservation Biology 17: 864–874.
Ruegg, J. & C. T. Robinson, 2004. Comparison of macroinvertebrate assemblages of permanent and temporary streams in an Alpine flood plain, Switzerland. Archiv für Hydrobiologie 161: 489–510.
Schanz, F., 1984. Chemical and algological characteristics of five high mountain lakes near the Swiss National Park. Verhandlungen Internationale Vereinigung für theoretische und angewandte Limnologie 22: 1066–1070.
Scheurer, T., 2004. Global change research in mountain biosphere reserves: Swiss National Park biosphere reserve. In Lee, C. & T. Schaaf (eds), Global Change Research in Mountain Biosphere Reserves. Proceedings of the International Launching Workshop, Entlebuch Biosphere Reserve, Switzerland, 10–13 November 2003. UNESCO, Paris, 85–92.
Schneider, D. W., 1999. Influence of hydroperiod on invertebrate community structure. In Batzer, D. P., R. R. Rader & S. A. Wissinger (eds), Invertebrates in Freshwater Wetlands of North America: Ecology and Management. Wiley, New York, 299–318.
Soderstrom, O., 1987. Upstream movements of invertebrates in running waters—a review. Archiv für Hydrobiologie 111: 197–208.
Sommaruga, R., R. Psenner, E. Schafferer, K. A. Koinig & S. Sommaruga-Wograth, 1999. Dissolved organic carbon concentration and phytoplankton biomass in high-mountain lakes of the Austrian Alps: potential effect of climatic warming an UV underwater attenuation. Arctic Antarctic and Alpine Research 31: 247–253.
Sommaruga-Wograth, S., K. A. Koinig, R. Schmidt, R. Sommaruga, R. Tessadri & R. Psenner, 1997. Temperature effects on the acidity of remote alpine lakes. Nature 387: 64–67.
Stoll, A., 2005. Mise en place d'un monitoring de la biodiversité des étangs de Macun (Parc National Suisse, GR). Travail de diplôme. University of Applied Sciences of Western Switzerland, EIL, Lullier.
Tachet, H., P. Richoux, M. Bournaud & P. Usseglio-Polatera, 2000. Invertébrés d'eaux douces. Systématique, biologie, écologie. CNRS Editions, Paris.
Theurillat, J. P. & A. Guisan, 2001. Potential impact of climate change on vegetation in the European Alps: a review. Climatic Change 50: 77–109.
Thomas, C. D., A. Cameron, R. E. Green, M. Bakkenes, L. J. Beaumont, Y. C. Collingham, B. F. N. Erasmus, M. F. de Siqueira, A. Grainger, L. Hannah, L. Hughes, B. Huntley, A. S. van Jaarsveld, G. F. Midgley, L. Miles, M. A. Ortega-Huerta, A. T. Peterson, O. L. Phillips & S. E. Williams, 2004. Extinction risk from climate change. Nature 427: 145–148.
Thuillier, W., S. Lavorel, M. B. Araújo, M. T. Sykes & I. C. Prentice, 2005. Climate change threats to plant diversity in Europe. Proceedings of the National Academy of Sciences of the United States of America 102: 8245–8250.
Van de Meutter, F., R. Stoks & L. De Meester, 2006. Lotic dispersal of lentic macroinvertebrates. Ecography 29: 223–230.
Väre, H., R. Lampinen, C. Humphries & P. Williams, 2003. Taxonomic diversity of vascular plants in the European alpine areas. In Nagy, L., G. Grabherr, C. Körner & D. B. A. Thompson (eds), Alpine Biodiversity in Europe—A Europe-wide Assessment of Biological Richness and Change. Springer, 133–148.
Wathne, B. M. & H. H. Hansen, 1997. MOLAR. Measuring and modelling the dynamic response of remote mountain lake ecosystem to environmental change: a program of Mountain lake Research. MOLAR Project Manual. NIVA Report 0-96061, Oslo.
Wiggins, G. B., R. J. Mackay & I. M. Smith, 1980. Evolutionary and ecological strategies of animals in annual temporary pools. Archiv für Hydrobiologie Supplement 58: 97–206.
Williams, D. D., 1987. The Ecology of Temporary Waters. Croom Helm, London & Sydney.
Wissinger, S. A., A. J. Bohonak, H. H. Whiteman & W. S. Brown, 1999. Subalpine wetlands in central Colorado: habitat permanence, salamander predation, and invertebrate communities. In Batzer, D. P., R. R. Rader & S. A. Wissinger (eds), Invertebrates in freshwater wetlands of North America: ecology and management. Wiley, New York, 757–790.

Hydrobiologia (2008) 597:43–51
DOI 10.1007/s10750-007-9219-6

Biodiversity and distribution patterns of freshwater invertebrates in farm ponds of a south-western French agricultural landscape

R. Céréghino · A. Ruggiero · P. Marty · S. Angélibert

Abstract We assessed the importance for biodiversity of man-made farm ponds in an agricultural landscape in SW France lacking natural wetlands. The ponds were originally created to provide a variety of societal services (irrigation, visual amenity, water for cattle, etc.). We also assessed the environmental factors influencing invertebrate assemblages in these ponds. Only 18 invertebrate taxa out of 114 taxa occurring in the study area were common to ponds and rivers indicating that the contribution of farm ponds to freshwater biodiversity was potentially high. A Self-Organizing Map (SOM, neural network) was used to classify 36 farm ponds in terms of the 52 invertebrate families and genera they supported, and to specify the influence of environmental variables related to land-use and to pond characteristics on the assemblage patterns. The SOM trained with taxa occurrences showed five clusters of ponds, most taxa occurring only in 1–2 clusters of ponds. Abandoned ponds tended to support higher numbers of taxa, probably because they were allowed to undergo a natural succession. Nevertheless, abandoned ponds were also amongst the largest, so that it remained difficult to separate the effects of pond size and abandonment, although both factors were likely to interact to favour higher taxon richness. The invertebrate communities in the ponds appeared to be influenced mainly by widely acting environmental factors (e.g. area, regionalization of assemblages) with little evidence that pond use (e.g. cattle watering, amenity) generally influenced assemblage composition. Our results support the idea that agricultural landscapes containing man-made ponds make a significant contribution to freshwater biodiversity indicating that protection of farm ponds from threats such as in-filling and pollution can make a positive contribution to the maintenance of aquatic biodiversity. This added value for biodiversity should be considered when calculating the economic costs and benefits of constructing water bodies for human activities.

Keywords Agriculture · Artificial ponds · Wetlands · Macroinvertebrates · Land-use · Self-organizing maps

Guest editors: R. Céréghino, J. Biggs, B. Oertli & S. Declerck
The ecology of European ponds: defining the characteristics of a neglected freshwater habitat

Electronic supplementary material The online version of this article (doi:10.1007/s10750-007-9219-6) contains supplementary material, which is available to authorized users.

R. Céréghino (✉) · A. Ruggiero · P. Marty
EcoLab, UMR 5245, Université Paul Sabatier,
118 route de Narbonne, 31062 Toulouse cedex 9, France
e-mail: cereghin@cict.fr

S. Angélibert
Department of Nature Management,
University of Applied Sciences of Western
Switzerland – EIL, 150 route de Presinge,
1254 Jussy-Geneva, Switzerland

Introduction

Ponds are small and shallow, natural or man-made water bodies defined as wetlands by the Ramsar Convention. Ponds typically outnumber larger lakes by a ratio of about 100:1 (Oertli et al., 2005a), and recent studies have revealed their importance for the conservation of biodiversity (Pyke, 2005; Scheffer et al., 2006) because, despite their small size, they disproportionately contribute to regional diversity, e.g. when compared to streams, large rivers, or lakes (Oertli et al., 2002; Williams et al., 2004; Karaus et al., 2005). Thus, ponds challenge conventional approaches to conservation biology, where much attention has been directed towards large-scale ecosystems (Meffe & Carroll, 1997).

Natural or man-made ecosystems, such as ponds also provide a wide variety of resources that have a social and economic value (Chase & Ryberg, 2004; Hansson et al., 2005), calling for more attention to be given to their importance both for nature and people, e.g. through cost-benefit assessment of their role in promoting biodiversity in a given area at the same time as supporting human activities (Odling-Smee, 2005). For example, in a recent study, Gaston et al. (2005) showed that urban domestic gardens efficiently increased biodiversity on a local scale, while other authors have demonstrated that wetlands located in agricultural landscapes may support a diverse aquatic fauna (Hazell et al., 2004; Robson & Clay, 2005). Therefore, when ponds are artificially created to support human activities, such as recreation, ornament, agricultural practices, etc., one may consider whether, in addition to the services they offer, they also have an added value for sustaining aquatic biodiversity. The setting for our study allowed us to address such a question.

The study focused on the large numbers of ponds dug by man in an agricultural landscape in the Astarac Region of SW France, an area from which natural wetlands were largely eliminated by drainage during the 19th and 20th centuries. In this landscape we were able to assess the role of man-made ponds in sustaining aquatic life while at the same time supporting human activities. We also examined the factors influencing the composition of the invertebrate assemblage in the ponds to increase understanding of the factors influencing biodiversity in these man-made ecosystems.

In order to assess whether farm ponds contributed to the maintenance of regional freshwater invertebrate diversity, we first compared the taxonomic richness of the macroinvertebrate fauna of ponds and rivers in the study area. Then, to assess the factors influencing pond assemblage composition, we studied the distribution patterns of pond invertebrates in relation to various landscape and habitat variables. Self-Organizing Maps (SOM, neural network) were used to classify the ponds according to their macroinvertebrate assemblages, and a set of environmental variables was subsequently introduced into the SOM trained with biological variables, in an attempt to interpret the variability of the communities.

Materials and methods

Study area

The Astarac region (SW France) is a 187-km^2 area (Fig. 5) with an economy historically dominated by agriculture. The modern agricultural economy is now focussed mainly on maize and wheat growing, cattle rearing on extensive pasture, and duck farming. Water resource being naturally scarce in the Astarac (for geographic and climatic reasons, and as a result of wetland drainage), artificial channels (locally called "Nestes") were built in the 19th and early 20th centuries, to carry water from the nearby Pyrenees. At the beginning of the 20th century, each farm also had at least three ponds dug by man to hold rainwater, in order to support local activities. A preliminary inventory based on aerial photographs allowed us to locate 607 ponds (Nature Midi-Pyrénées, 2005) in the study area without a priori consideration of their origins (i.e. either natural or anthropogenic). As the region lacked natural ponds we could be confident that the contribution to biodiversity by ponds was due entirely to waterbodies of anthropogenic origin. Given this, the contribution of farm ponds to freshwater biodiversity was potentially high in the Astarac.

We studied 36 farm ponds. Twenty-seven ponds currently provided services, being used for: cattle watering (14 ponds), irrigation (7 ponds), duck farming (3), providing visual amenity (2), and baitfish culture (1). The remaining nine ponds were abandoned. Some uses (duck farming, visual amenity, baitfish culture) being under-represented, we did not attempt to define invertebrate assemblage patterns with respect to pond use. The landscape in an area 800 m around each pond was described in terms of six landscape variables: % bare area, % prairie (herbaceous areas dominated by graminaceous plants), % fields (arable crops), % bushes and % forest. Distance to the nearest pond was also calculated. Landscape variables were determined using a combination of direct field observations and examination of aerial photographs. At each pond, seven habitat variables were measured: surface area, maximum depth, water transparency (measured as Secchi depth, disc diameter = 25 cm), % cover of submerged vegetation, % cover of vegetation with aerial leaves, % cover of floating vegetation and % shade.

Invertebrate sampling

Samples were taken in each pond in March (spring) and July (summer) 2004, by intensive sweeping of a hand net (mesh size = 250 μm) through the various mesohabitats (i.e. net sweeping for a fixed time period, see Oertli et al., 2005b). The time effort in each mesohabitat (e.g. submerged vegetation, silt, *Typha* stems) was proportional to its surface area, and the total sampling time was 15 min for each pond as a whole.

Invertebrate samples were preserved in the field in 5% formalin, sorted and identified in the laboratory (see Electronic supplementary material), and then preserved in 70% ethanol. Invertebrates were keyed to the genus or family level, as for most invertebrate-based tools for the bioassessment of freshwater systems (Rosenberg & Resh, 1993).

In order to make comparisons between pond and river invertebrates in the area, a list of river invertebrates identified at a similar taxonomic level was extracted from our laboratory database. Each site was sampled in summer and winter to take into account invertebrate seasonality, and the taxa lists per site were summed. Eight samples were taken at each site and at each season from the various substratum types using a standard Surber sampler (sampling area 0.1 m^2, mesh size 0.3 mm). The pond study area surrounds site 505 in Céréghino et al. (2003). The list of river invertebrates for this study was compiled from benthic samples taken on the rivers Baïse (three sites), Arrats (four sites), Gimone (two sites), and Save (three sites), which flow through the study area from South to North. Although fewer river sites were surveyed, sampling effort in ponds and rivers was broadly equivalent.

Modelling procedures

We used Self-Organizing Maps (SOM, unsupervised neural network, Kohonen, 1982, 1995) as an analytical tool to investigate relationships between ponds and macroinvertebrate assemblages. Combining clustering and ordination approaches, this technique has been shown to have a number of advantages in the identification of assemblage patterns using presence/absence data, compared to conventional multivariate techniques (see Park et al., 2003 for theoretical considerations). The SOM Toolbox (version 2) for Matlab® was used (downloads at http://www.cis.hut.fi/projects/somtoolbox/, see also Vesanto et al., 1999). The structure of the SOM for our study consists of two layers of neurons connected by weights (i.e. connection intensities): the input layer consists of 51 neurons (one by invertebrate taxa) connected to the 36 ponds. The output layer consists of 32 neurons (visualized as hexagonal cells) organized in an array with eight rows and four columns (see results). This map size was selected based on minimum topographic error (TE). TE measures map quality (i.e. to assess whether the map has been properly trained), and is thus used for the measurement of topology preservation (see Kiviluoto, 1996; Park et al., 2003). SOM plots the data so that ponds that are similar are found together on a grid and, conversely, ponds that are very different (according to their invertebrate taxa) are far from each other. The procedure can be simply described as follows:

- The virtual samples are initialized with random samples drawn from the input data set.
- The virtual samples are updated in an iterative way:
  - A sample unit is randomly chosen as an input unit.
  - The Euclidean distance between this sample unit and every virtual sample is computed.
  - The virtual sample closest to the input unit is selected and called the 'best matching unit' (BMU).

– The BMU and its neighbours are moved gradually towards the input unit.

The training was broken down into two parts (Céréghino et al., 2001):

- Ordering phase (the 3,000 first steps): when this first phase takes place, the samples are highly modified in a region extending widely around the BMU.
- Tuning phase (7,000 steps): during this phase, there is limited modification of the virtual samples adjacent to the BMU.

At the end of the training, ponds which are neighbours on the grid are expected to represent neighbouring clusters of ponds; consequently, ponds which are further apart on the grid, (according to taxa assemblage), are expected to be distant in the feature space. Finally, a k-means algorithm (Ultsch, 1993) was applied to cluster the trained map. The SOM units (hexagons) were divided into clusters according to the weight vectors of the neurons, and subsets were justified according to the lowest Davis Bouldin Index, i.e. for a solution with low variance within clusters and high variance among clusters (Céréghino et al., 2003). Clusters of ponds were then plotted on a geographic map, to visualize further the modelled structures in a more readily interpretable manner.

In order to bring out relationships between biological and environmental variables, we introduced environmental variables into the SOM trained with biological variables, following the procedure described in Park et al. (2003). We calculated the mean value (*Ev*) of each environmental variable in each output neuron of the trained SOM. The mean value was computed as:

$$E_v = \frac{1}{n}\sum_{i=1}^{n} e_i$$

where n is the number of input vectors (ponds) assigned to each output neuron of the trained SOM, and e_i is the value of each environmental variable of input vector i. If an output neuron was not occupied by input vectors, the value was replaced with the mean value of neighbouring neurons. All mean values of environmental variables assigned on the SOM map were visualized in grey scale, and then compared with the mapping of sampling sites.

Results

Even at the genus-family level, pond and river invertebrate assemblages were very distinct. Only 18 taxa out of 114 (i.e. 16%) were common to pond and river ecosystems (Electronic supplementary material). Seventy-eight taxa occurred in rivers, of which 60 taxa (77%) were exclusive to running waters. Rivers made a higher contribution than ponds to the taxa richness of Ephemeroptera, Trichoptera, Plecoptera, and (to a lesser extent) Mollusca. On the other hand, the list of pond invertebrates derived from our samples comprised 52 taxa, among which 34 taxa (65%) were exclusively found in ponds. Specifically, farm ponds made a higher contribution to the taxonomic richness of Heteroptera, Coleoptera and Odonata in the area.

After training the SOM with invertebrate occurrences at 36 ponds, the k-means algorithm helped to derive five clusters, based on the minimum Davies Bouldin index (DBI = 0.92) (Table 1). Thus, ponds were classified into five subsets (clusters A–E) according to their macroinvertebrate assemblages (Fig. 1a), i.e. according to different spatial distribution patterns, which were characteristic of each taxon. The distribution of each taxon could be visualized on the trained SOM (using a shading scale, see examples in Fig. 1c). Distribution patterns are summarized in Electronic supplementary material. Most taxa occurred in only one (22 taxa) or two (12 taxa) clusters of ponds. They therefore had the strongest influence upon the pond classification, suggesting that they could in the future be useful as indicator taxa. Seven and eight taxa were found in three and four clusters, respectively. Finally, Diptera, Chironomidae and Oligochaeta occurred in all ponds. Any taxa association can be further examined by overlapping several taxa maps.

The upper areas of the SOM had the lowest taxon richness, whereas lower areas of the map showed higher richness (Fig. 1b). When environmental variables were introduced into the SOM trained with macroinvertebrate taxa (Fig. 2), the ordinate on the SOM showed a gradient of pond size (from small [top area of the map] to large [bottom]), whereas the abscissa of the map chiefly represented water transparency (from low [left] to high [right]). We could not bring out clear relationships between pond use and the macroinvertebrate assemblages, but

Table 1 Davies Bouldin index (DBI) of k-means clustering at different numbers of clusters on the trained self-organizing map (SOM)

	Number of clusters					
	2	3	4	5	6	7
DBI	1.1104	1.1087	1.0234	0.9251	1.1151	1.1701

The retained number of clusters (5) was justified according to the minimum DBI

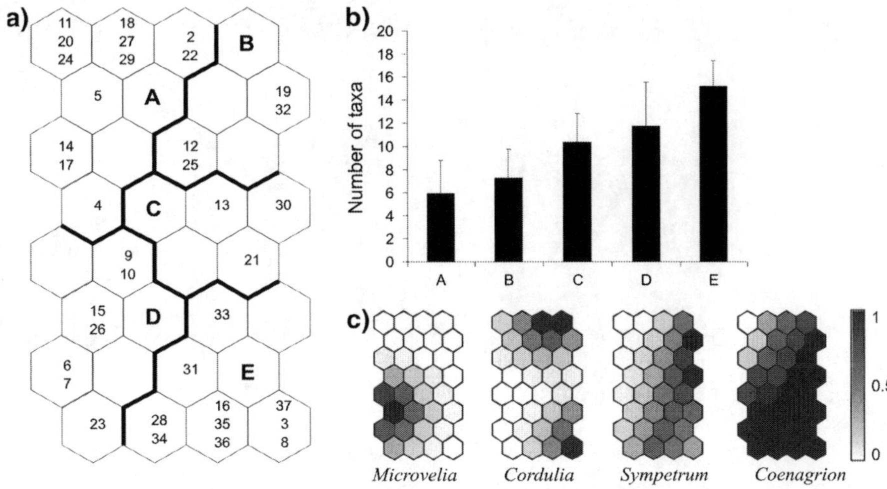

Fig. 1 (**a**) Distribution of ponds on the self-organizing map (SOM) according to the presence or absence of 51 invertebrate taxa, and clustering of the trained SOM. Ponds that are neighbours within clusters are expected to have similar invertebrate assemblages. Ponds separated by a large distance from each other, according to invertebrate taxa, are distant in the output space. Codes (11, 20, 24) correspond to ponds. Clusters A–E were derived from the k-means algorithm. (**b**) Mean number of taxa (±SE) per cluster. (**c**) Gradient analysis of the probability of occurrence of the species on the trained SOM, with visualization in shading scale (dark = high probability of occurrence, light = low probability of occurrence). Small maps (**c**) representing individual taxa should be compared with (or superimposed on) the map representing the distribution of ponds shown in **a**. One map is given as an example of the main distribution patterns

abandoned ponds (which were also among the largest ones) were rather concentrated in Cluster E (Fig. 3), i.e. where high transparency (Fig. 2) and higher taxa richness (Fig. 1b) were concentrated. Larger ponds also had the tendency to support most taxa, as suggested by the significant taxa richness-area relationship ($P < 0.01$, Fig. 4). Other variables under consideration (e.g. macrophyte cover, shading, land-use) did not show clear patterns within the SOM (Fig. 2), and did not act as structuring variables.

Clusters were finally plotted on a geographical map of the study area (Fig. 5). Small-scale autocorrelations of assemblages were suggested from the visual inspection of Fig. 5, since the taxa hosted by ponds within the same sub-areas tended to be similar (e.g. ponds 6, 7, 9, 10, 15 belonged to Cluster D; ponds 33 and 34 in Cluster E; ponds 2, 17, 18, 20, 22, 27, 29 in Cluster A; see map on Fig. 5) and those characteristics tended to differ when ponds belonged to more distinct areas.

Discussion

Much of the current research on small water bodies still aims at assessing the biodiversity that ponds and/or clusters of ponds support (Briers & Biggs, 2003), or are likely to host (Davies et al., 2004; Gaston et al., 2005). From local to regional scales, ponds also have obvious ecological functions and recognised social and economic uses (Chapman et al., 2001). Therefore, although it was conducted at a small spatial scale, and considered local services, our study provides insights into how to consider biodiversity at

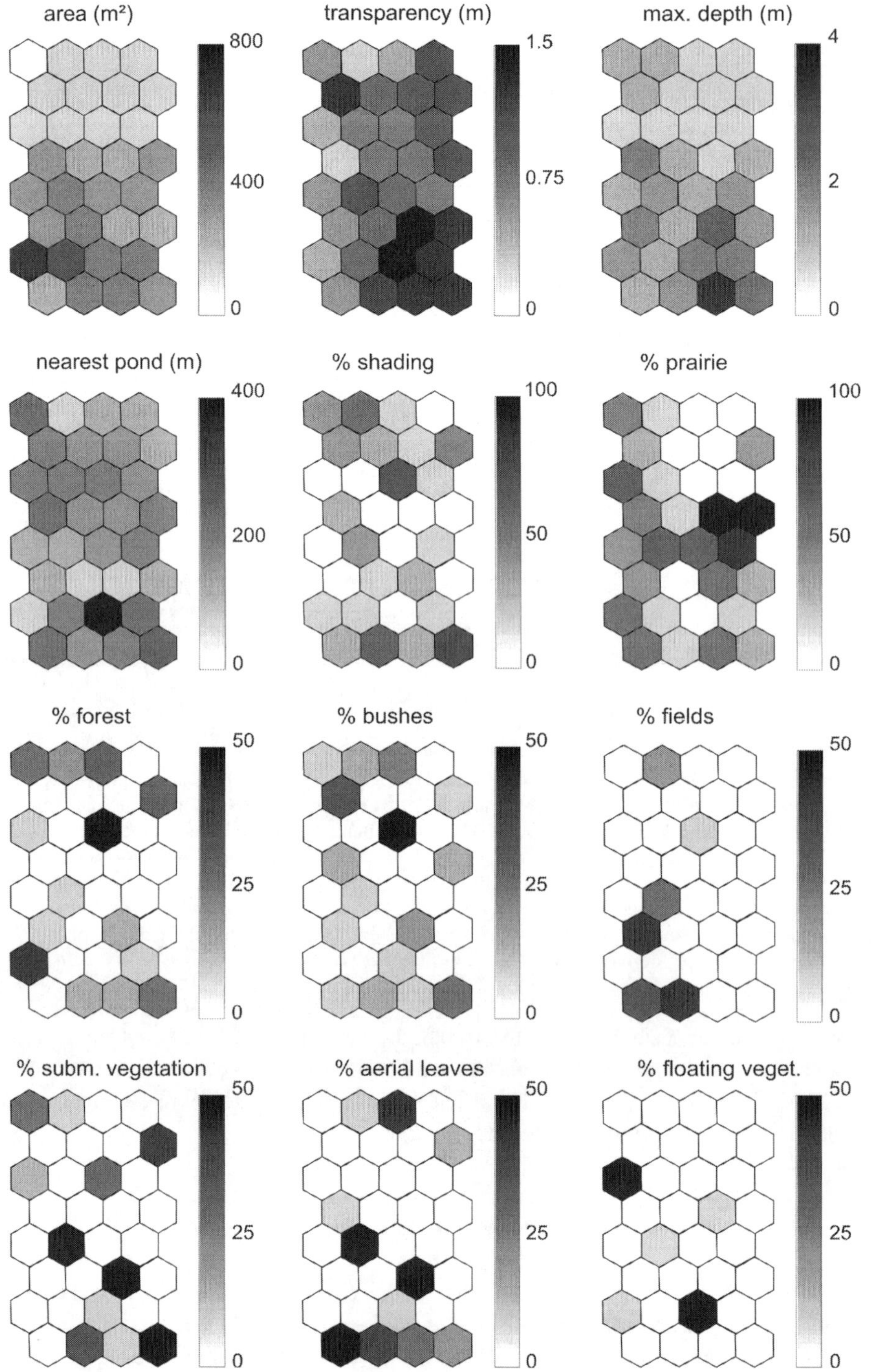

Fig. 2 Visualization of environmental variables on the SOM trained with macroinvertebrate taxa. The mean value of each variable was calculated in each output neuron of the trained SOM. Dark represents a high value and light represents low values

the same time as we promote human activities. In addition, our data set describing invertebrate taxa in farm ponds was used to assess the extent to which communities respond to environmental conditions in an agricultural landscape. Artificial, more or less intensively used ponds appeared to be efficient in sustaining biodiversity at a regional scale. All ponds, of different sizes, with different habitat features, and offering different services, contributed to the diversity of invertebrate assemblages, even if individual

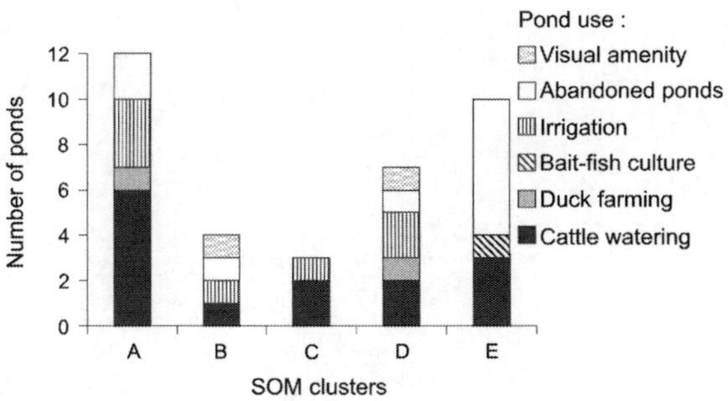

Fig. 3 Distribution of pond use type in each cluster (A–E) derived from to the Self-Organizing Map

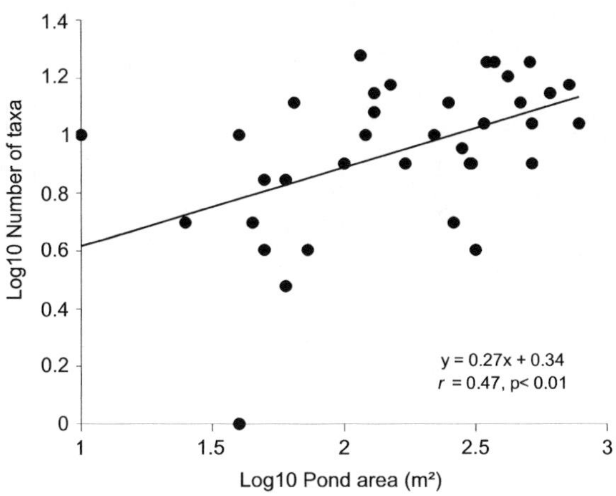

Fig. 4 Relationship between pond area (Log10 transformed) and number of taxa (Log10 transformed)

ponds hosted few taxa. The taxa we sampled in farm ponds (Ephemeroptera, Heteroptera, Odonata, Coleoptera, Mollusca, Diptera, Trichoptera, Oligochaeta) did well at representing the fauna of still waters usually found in European ponds, even under weak to moderate anthropogenic influence (see e.g. taxa lists in Angélibert et al., 2004). Given the observed differences between pond and river invertebrates, and the destruction of natural wetlands, farm ponds appeared to make a significant contribution to freshwater invertebrate diversity in our study region.

Most taxa only occurred in 1–2 clusters of ponds, and a mapping of pond clusters suggested small-scale autocorrelations of assemblages. Spatial autocorrelation of community composition is often linked to dispersal, assuming that species are more likely to reach neighbouring sites than sites far apart (Briers & Biggs, 2005). On one hand, many pond invertebrates are active dispersers (e.g. insects), whereas some taxa (e.g. molluscs) are passively transported. On the other hand, the various taxa also show different responses to a given set of environmental factors. Thus, taxa may show variable patterns of spatial autocorrelation, so that there was little chance of detecting the specific effects of dispersal vs habitat selection by invertebrates.

Although it was not possible to bring out definite patterns of invertebrate assemblages and taxa richness with respect to pond use, abandoned farm ponds tended to support higher numbers of taxa. However, abandonment also leads to the natural filling of ponds with sediment and aquatic vegetation while they undergo a natural succession which is ultimately detrimental to the aquatic fauna: Angélibert et al. (2004) proposed a conceptual scheme linking species richness to the successional stage of ponds—they proposed that aquatic invertebrate species richness progressively increases with the development of the submerged vegetation (early to mid-successional stage), then declines when the plant community is dominated by emergent species with aerial leaves, such as *Typha* spp. (late-successional stage). In our study, the fact that a higher number of taxa was often found in abandoned ponds suggests that they were close to an intermediate stage of succession, an assumption which fits with the percentages of submerged (20–50%) and aerial vegetation (<25%) calculated in Cluster E of our analysis (Fig. 2). Nevertheless, abandoned ponds were also among the largest ones, so that it remained difficult to separate the effects of pond size and absence of use, although both factors were likely to interact to increase higher taxon richness. Larger ponds were likely to show a greater habitat complexity (i.e. higher diversity of ecological niches) thus facilitating the coexistence of a larger number of taxa, while the absence of use certainly favoured successional patterns. In general, the species-area relationship (SAR) is a well-known principle (Oertli et al., 2002; Drakare et al., 2006) which can be used to address ecological mechanisms playing behind the spatial distribution of biodiversity. Even though we did not identify invertebrates to the species level, our taxa—area relationship had a slope value of 0.27, which is close to the values reported in literature for various invertebrates (around 0.20), even in study areas larger than our one. It suggests a non-homogenous distribution of taxa and a slow turnover of invertebrates among our ponds, and

Fig. 5 Distribution of the 36 study ponds in the Astarac Region, and correspondence with their location (clusters) on the SOM (see Fig. 1 and Electronic supplementary material). "Faget-Abbatial", "Tachoires", etc… correspond to municipalities, delineated with their administrative limits

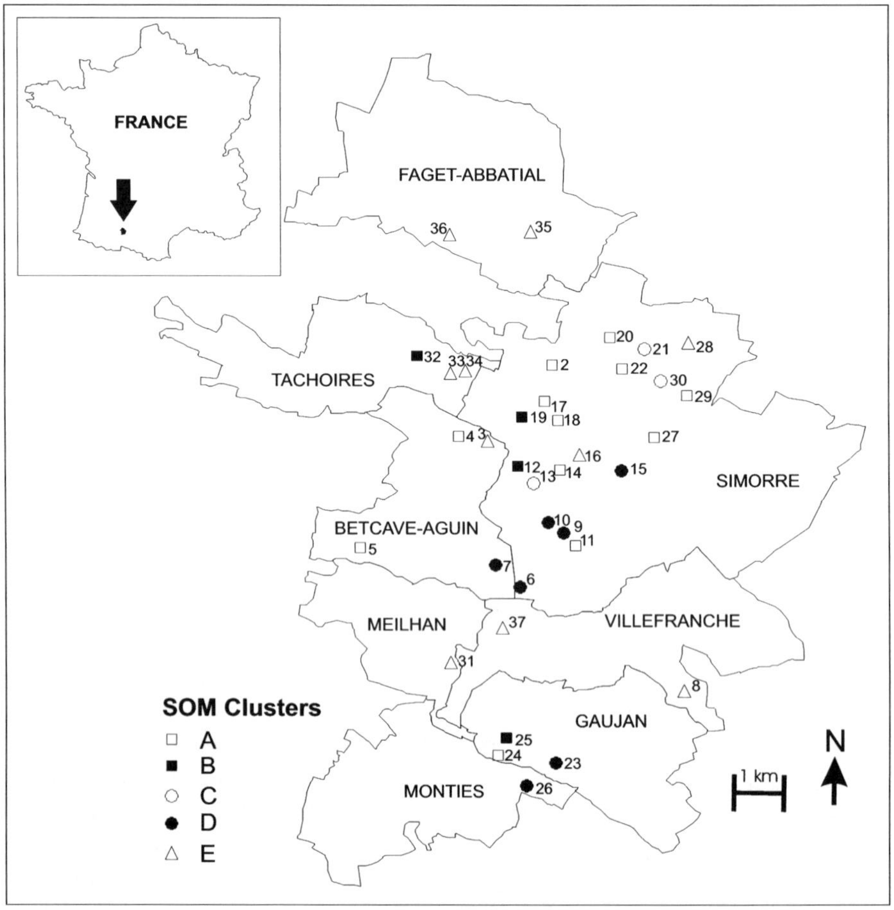

supports the idea that farm ponds provide a series of favourable habitats to the various populations which are scattered according to their different ecological requirements.

In conclusion, a striking result of our study is that the invertebrate assemblages and the taxa richness hosted by farm ponds were primarily related to general ecological patterns (regionalization of assemblages, area effect, possibly successional patterns) although the ponds were subjected to different uses. Similarly, frog assemblages in a highly anthropogenically modified agricultural landscape in south-eastern Australia were found to respond to habitat characteristics and to predation, whatever the origin (man-made or natural) of ponds (Hazell et al., 2004). Robson & Clay (2005) found that pasture wetlands efficiently contributed to biodiversity in highly managed agricultural landscapes. Such literature data combined with our own results support the ideas: (i) that agricultural landscapes containing man-made ponds make a significant contribution to freshwater biodiversity; (ii) that man-made ecosystems, for instance farm ponds, should also be protected from threats, such as natural filling at late-successional stages or pollution (Angélibert et al., 2004; Scheffer et al., 2006); and (iii) that we may consider a positive added value for biodiversity when calculating the cost/benefit ratio of constructing water bodies for human activities. Therefore, further understanding of the distribution of biological diversity in non-natural systems may facilitate the adoption of positive solutions for wildlife, with limited costs for human activities. Such an understanding is expected to help in planning management efforts, such as the creation or restoration of ecosystems (Biggs et al., 1994).

Acknowledgements M. Dessaivre and D. Hanquet (Nature Midi-Pyrénées) contributed to the study design, field work and invertebrate sorting. This study was funded by the French Water Agency (Agence de l'Eau Adour-Garonne), DIREN, Région Midi-Pyrénées, and by Nature Midi-Pyrénées. We wish to thank J. Biggs and two anonymous referees for their constructive comments on an earlier version of this article.

References

Angélibert, S., P. Marty, R. Céréghino & N. Giani, 2004. Seasonal variations in physico-chemical characteristics of ponds: implications for biodiversity conservation. Aquatic Conservation: Marine and Freshwater Ecosystems 14: 439–456.

Biggs, J., A. Corfield, D. Walker, M. Whitfield & P. Williams, 1994. New approaches to the management of ponds. British Wildlife 5: 273–287.

Briers, R. A. & J. Biggs, 2003. Indicator taxa for the conservation of pond invertebrate diversity. Aquatic Conservation: Marine and Freshwater Ecosystems 13: 323–330.

Briers, R. A. & J. Biggs, 2005. Spatial patterns in pond invertebrate communities: separating environmental and distance effects. Aquatic Conservation: Marine and Freshwater Ecosystems 15: 549–557.

Céréghino, R., J. L. Giraudel & A. Compin, 2001. Spatial analysis of stream invertebrates distribution in the Adour-Garonne drainage basin (France), using Kohonen self organising maps. Ecological Modelling 146: 167–180.

Céréghino, R., Y. S. Park, A. Compin & S. Lek, 2003. Predicting the species richness of aquatic insects in streams using a limited number of environmental variables. Journal of the North American Benthological Society 22: 442–456.

Chapman, L. J., J. Balirwa, F. W. B. Bugenyi, C. Chapman & T. L. Crisman, 2001. Wetlands of East-Africa : biodiversity, exploitation and policy perspectives. In Gopal, B., W. J. Junk & J. A. Davis (eds), Biodiversity in Wetlands: Assessment Function and Conservation, Vol. 2. Backhuys Publishers, Leiden, The Netherlands, 101–131.

Chase, J. M. & W. A. Ryberg, 2004. Connectivity, scale-dependence, and the productivity-diversity relationship. Ecology Letters 7: 676–683.

Davies, B. R., J. Biggs, J. T. Lee & S. Thompson, 2004. Identifying optimum locations for new ponds. Aquatic Conservation: Marine and Freshwater Ecosystems 14: 5–24.

Drakare, S., J. J. Lennon & H. Hillebrand, 2006. The imprint of the geographical, evolutionary and ecological context on species-area relationships. Ecology Letters 9: 215–227.

Gaston, K. J., R. M. Smith, K. Thompson & P. H. Warren, 2005. Urban domestic gardens (II): experimental tests of methods for increasing biodiversity. Biodiversity and Conservation 14: 395–413.

Hansson, L. A., C. Bronmark, P. A. Nilsson & K. Abjornsson, 2005. Conflicting demands on wetland ecosystem services: nutrient retention, biodiversity or both? Freshwater Biology 50: 705–714.

Hazell, D., J. M. Hero, D. Lindenmayer & R. Cunningham, 2004. A comparison of constructed and natural habitat for frog conservation in an Australian agricultural landscape. Biological Conservation 119: 61–71.

Karaus, U., L. Adler & K. Tockner, 2005. Concave islands: habitat heterogeneity of parafluvial ponds in a gravel-bed river. Wetlands 25: 26–37.

Kiviluoto, K., 1996. Topology preservation in self-organizing maps. In IEEE Service Center (ed.), Proceedings of ICNN'96, IEEE International Conference On Neural Networks, Piscataway, 294–299.

Kohonen, T., 1982. Self-organized formation of topologically correct feature maps. Biological Cybernetics 43: 59–69.

Kohonen, T., 1995. Self-Organizing Maps, volume 30 of Springer Series in Information Sciences. Springer, Berlin, Heidelberg.

Meffe, G. K. & C. R. Carroll, 1997. Principles of Conservation Biology, 2nd edn. Sinauer Associates, Inc., Sunderland, MA.

Nature Midi-Pyrénées, 2005. Inventaire et préservation du patrimoine des mares de l'Astarac. 114 pp. Download at http://www.premiumwanadoo.com/naturemp/.

Odling-Smee, L., 2005. Dollars and sense. Nature 437: 614–616.

Oertli, B., D. Auderset-Joye, E. Castella, R. Juge, D. Cambin & J. B. Lachavanne, 2002. Does size matter? The relationship between pond area and biodiversity. Biological Conservation 104: 59–70.

Oertli, B., J. Biggs, R. Céréghino, P. Grillas, P. Joly & J. B. Lachavanne, 2005a. Conservation and monitoring of pond biodiversity: introduction. Aquatic Conservation: Marine and Freshwater Ecosystems 15: 535–540.

Oertli, B., D. Auderset-Joye, E. Castella, R. Juge, A. Lehmann & J. B. Lachavanne, 2005b. PLOCH: a standardized method for sampling and assessing the biodiversity in ponds. Aquatic Conservation: Marine and Freshwater Ecosystems 15: 665–679.

Park, Y. S., R. Céréghino, A. Compin & S. Lek, 2003. Applications of artificial neural networks for patterning and predicting aquatic insect species richness in running waters. Ecological Modelling 160: 265–280.

Pyke, C. R., 2005. Assessing suitability for conservation action: Prioritizing interpond linkages for the California tiger salamander. Conservation Biology 19: 492–503.

Robson, B. J. & C. J. Clay, 2005. Local and regional macroinvertebrate diversity in the wetlands of a cleared agricultural landscape in south-western Victoria, Australia. Aquatic Conservation: Marine and Freshwater Ecosystems 15: 403–414.

Rosenberg, D. M. & V. H. Resh, 1993. Freshwater Biomonitoring and Benthic Macroinvertebrates. Chapman and Hall, London, UK.

Scheffer, M., G. J. van Geest, K. Zimmer, E. Jeppesen, M. Sondergaard, M. G. Butler, M. A. Hanson, S. Declerck & L. De Meester, 2006. Small habitat size and isolation can promote species richness: second-order effects on biodiversity in shallow lakes and ponds. Oikos 112: 227–231.

Ultsch, A., 1993. Self-organizing neural networks for visualization and classification. In Opitz, O., B. Lausen & R. Klar (eds), Information and Classification. Springer-Verlag, Berlin, 307–313.

Vesanto, J., J. Himberg, E. Alhoniemi & J. Parhankangas, 1999. Self-organizing map in matlab: the som toolbox. In Proceedings of the Matlab DSP Conference 1999, Comsol Oy, Espoo, Finland, 35–40.

Williams, P., M. Whitfield, J. Biggs, S. Bray, G. Fox, P. Nicolet & D. Sear, 2004. Comparative biodiversity of rivers, streams, ditches and ponds in an agricultural landscape in Southern England. Biological Conservation 115: 329–341.

Hydrobiologia (2008) 597:53–69
DOI 10.1007/s10750-007-9221-z

Patterns of composition and species richness of crustaceans and aquatic insects along environmental gradients in Mediterranean water bodies

D. Boix · S. Gascón · J. Sala · A. Badosa · S. Brucet · R. López-Flores · M. Martinoy · J. Gifre · X. D. Quintana

Abstract Differences in the dynamics of ecological processes between Mediterranean and colder temperate aquatic systems could imply different patterns in faunal communities in terms of composition and biodiversity (i.e. species richness and rarity). In order to identify some of these patterns the crustacean and aquatic insect composition and biodiversity of four water body types, classified according to their salinity and water permanence, were compared. Moreover, the relationships between species richness and water, pond and landscape variables were analysed. A total number of 91 water bodies located throughout Catalunya (NE Iberian Peninsula) were sampled. Three species assemblages were observed: one for permanent freshwaters, another for temporary freshwaters, and a third one for saline waters (SW), since permanent and temporary saline water bodies had similar composition. Differences in salinity were associated with proportion of crustaceans versus insects and with singularity. Thus, saline ponds had a higher proportion of crustaceans, and lower values of singularity. Conductivity was significantly related to total (crustaceans plus insects) richness, and also related to insect richness. The main difference between the models obtained for crustacean species richness and insect species richness is the significance of landscape variables in the latter, and this fact could be related to the different dispersion types of these two faunal groups: active for insects versus passive for crustaceans.

Keywords Mediterranean water bodies · Faunal composition · Species richness · Trophic state · Pond characteristics · Landscape variables

Electronic supplementary material The online version of this article (doi:10.1007/s10750-007-9221-z) contains supplementary material, which is available to authorized users.

Guest editors: R. Céréghino, J. Biggs, B. Oertli & S. Declerck
The ecology of European ponds: defining the characteristics of a neglected freshwater habitat

D. Boix (✉) · S. Gascón · J. Sala · A. Badosa · S. Brucet · R. López-Flores · X. D. Quintana
Faculty of Sciences, Institute of Aquatic Ecology, University of Girona, Campus Montilivi, 17071 Girona, Spain
e-mail: dani.boix@udg.edu

M. Martinoy · J. Gifre
Mosquito Control Service of Badia de Roses and Baix Ter, Girona, Spain

Introduction

In Mediterranean aquatic systems, several differences in their ecological processes have been reported when compared with the contemporary limnological paradigm, due to the fact that limnological studies have been mainly developed in colder-temperate aquatic systems (Álvarez-Cobelas et al., 2005). Despite the interest in analysing these differences, more studies have been focused on processes (e.g. Stephen et al., 2004; Romo et al., 2005), than on the structure of the Mediterranean aquatic communities. Moreover, a great number of studies analyse the aquatic community of only one water body (e.g. Bazzanti et al., 1996; Boix et al., 2004), of a few of them (e.g. Gascón et al., 2005; Serrano & Fahd, 2005), or of an area that belongs to the same wetland (e.g. Ortega-Mayagoitia et al., 2000; Martinoy et al., 2006), but studies on a non-local scale are rare (e.g., Alonso, 1998; Boronat et al., 2001). This kind of studies is needed for a general view and comprehension of the dynamics and structure of Mediterranean aquatic ecosystems (Britton & Crivelli, 1993).

The knowledge of the composition and species richness of the Mediterranean aquatic communities is also important from a management point of view. Biodiversity is one of the main criteria used in the protection of wetlands (Ramsar Convention Bureau, 2005). However, the composition of aquatic invertebrate species in wetlands is relatively poorly known, and most of the management efforts are focused on the conservation of a small number of species, mainly waterbirds (e.g. Mocci, 1983). It has already been pointed out that the biodiversity patterns of a faunal group and the main factors which determine them cannot be generalized to other faunal groups (Eitam et al., 2004a, b). Moreover, the importance of some Mediterranean aquatic ecosystems has been recognized in the European Directive that considers them priority habitats (92/43/CEE), and several studies put in evidence the value of their fauna and flora (e.g., Médail et al., 1998; Boix et al., 2001).

Salinity and water permanence are considered to be the main environmental factors that should be taken into account when classifying Mediterranean water bodies by means of water and/or substrate characteristics, fauna and flora (Britton & Podlejski, 1981; Alonso, 1998; Trobajo et al., 2002). Similarly, a recent study proposed the classification of Catalan water bodies by means of these two factors, since they appeared to be the main factors structuring aquatic communities (Boix et al., 2005). On the other hand, species richness has been related with water variables other than salinity (e.g. eutrophy; Jeppesen et al., 2000), and with pond variables other than water permanence (e.g. pond size; Oertli et al., 2002). Furthermore, new approaches put in evidence the importance of connectivity and other landscape factors for species richness (e.g. Cottenie et al., 2003).

The first objective of this study was to compare the faunal composition and species richness of four water body types: permanent fresh waters, temporary fresh waters, permanent saline waters (SW) and temporary SW. A second objective was to identify which of several water, pond and landscape factors were associated with species richness patterns in Mediterranean water bodies, and if species richness of different faunal groups, in our case crustaceans and insects, were possibly determined by similar factors.

Material and methods

Study sites

The study was carried out in 91 representative water bodies (ponds, lagoons or marshes), located throughout Catalunya (NE Iberian Peninsula). The selected water bodies are part of the official Wetland Inventory of Catalunya, that was used in order to select water bodies widely distributed on the region of Catalunya, and with a similar number of water bodies within each type. The Wetland Inventory of Catalunya is an attempt to list and characterize all the water bodies in the region with some natural heritage value. All water bodies sampled were below 800 m a.s.l. to ensure that they were influenced by Mediterranean climatic conditions, and had a maximum depth of <6 m (Fig. 1). Catalunya is a densely populated region, and a large proportion of the lowland water bodies are under several anthropogenic pressures (mainly urban and agricultural). The studied water bodies were classified by means of conductivity and water permanence (Boix et al., 2005) in permanent freshwaters (hereafter PFW; $n = 35$ water bodies), temporary freshwaters (TFW; $n = 29$ water bodies), permanent saline waters (PSW; $n = 19$ water bodies), and temporary saline waters (TSW; $n = 8$ water bodies). Two sampling periods were conducted during 2003 in order to encompass temporal variability, the first one in February and the second one in June. All water bodies were sampled twice except 17 temporary freshwater bodies that dried out before the second sampling period. Additional information on water, pond and landscape characteristics for each water body type is summarized in Table 1.

Fig. 1 Map of Catalunya (NE Iberian Peninsula) showing the 91 water bodies analysed, coded according to water body type. The shaded part of the map represents the area above 800 m a.s.l.

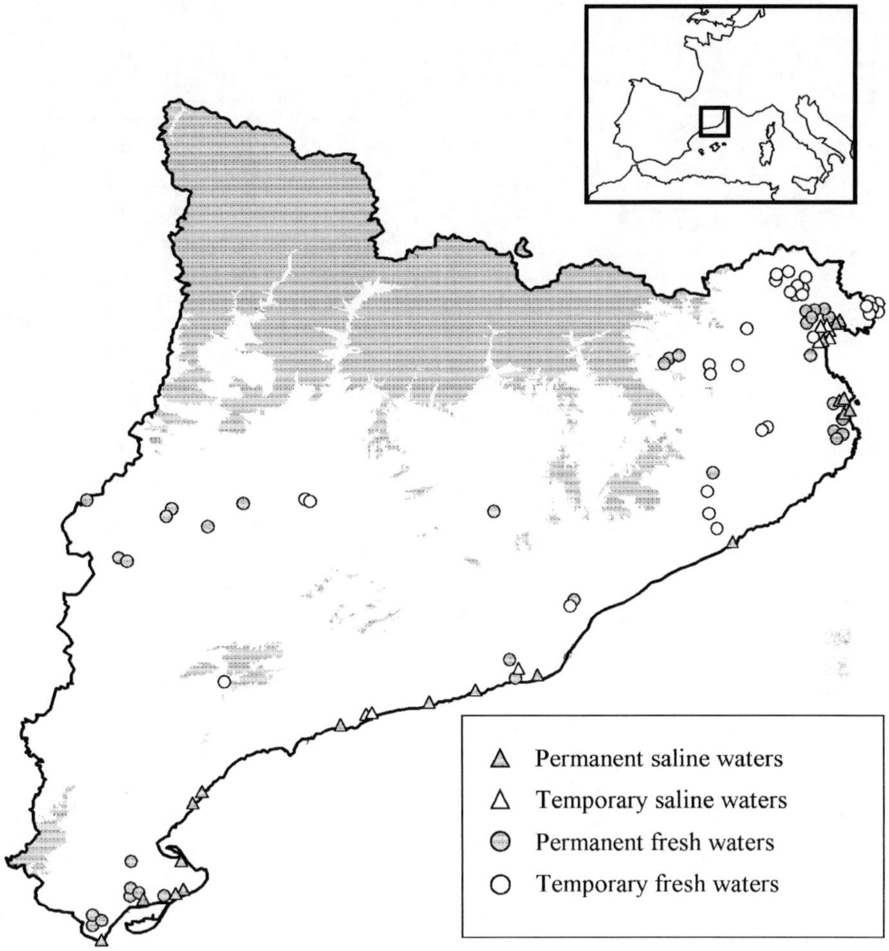

Sampling

The dataset was composed by three groups of variables: water variables (temperature, conductivity, percentage of oxygen saturation, pH, dissolved and total nutrients and chlorophyll-*a*), pond variables (pond size, depth and water permanence) and landscape variables (degree of isolation, water body density, and share of aquatic habitat). Temperature, conductivity, percentage of oxygen saturation and pH were measured in situ. Analyses of dissolved inorganic nutrients (ammonium, nitrite, nitrate and phosphate) were carried out from filtered samples, and total nutrients (nitrogen and phosphorus) from unfiltered samples, following Grasshoff et al. (1983). Chlorophyll-*a* was extracted using 80% methanol, after filtering water samples (Whatman GF/C filters), and measured following Talling & Driver (1963). Depth (three categories: water bodies shallower than 0.5 m, water bodies with a depth between 0.5 and 1.5 m, and water bodies with a depth >1.5 m) and water permanence (temporary versus permanent) were applied as categorical variables. In order to obtain pond size and landscape variables, freely available aerial photographs were used (DPTOP, 2005; MAPA, 2006). Pond size was calculated as the maximum flooded area, avoiding extreme flooding situations. Degree of isolation of the water bodies was calculated as the distance to the nearest water body, water body density as the number of water bodies within a radius of 500 m from a water body, and share of aquatic habitat as the proportion of water surface in a square kilometre centered around the water body.

Invertebrates were sampled using a 20-cm diameter dip-net (mesh size = 250 μm). At each water body, three sweeps per visit were carried out along transects. Each sweep consisted of 20 dip-net

Table 1 Characteristics of the water body types under study (PFW, permanent freshwaters; TFW, temporary freshwaters; PSW, permanent saline waters; TSW, temporary saline waters). For each variable, the median and the standard deviation (S.D.) are shown

	PFW (n = 70)		TFW (n = 41)		PSW (n = 38)		TSW (n = 16)	
	Median	S.D.	Median	S.D.	Median	S.D.	Median	S.D.
Conductivity (mS cm^{-1})	0.96	3.19	0.57	0.86	13.43	20.35	13.40	11.74
Temperature (°C)	17.3	8.3	8.9	6.1	19.2	8.7	20.8	9.7
pH	7.8	0.7	7.7	0.7	8.0	0.4	8.5	0.6
Dissolved oxygen (%)	71	40.4	75	35	95	47.0	102	70
Chlorophyll-*a* (μg l^{-1})	9.6	16.9	5.6	21.6	10.1	10.1	12.7	35.2
Phosphate (μg l^{-1})	0.4	12.2	0.2	13.9	0.3	2.8	0.5	4.6
Total phosphorus (μg l^{-1})	5.1	121.0	2.9	38.8	4.9	26.3	4.2	214.7
Ammonium (μg l^{-1})	2.4	159.6	0.7	36.0	1.32	13.6	0.7	15.4
Nitrite (μg l^{-1})	0.4	5.7	0.1	4.7	0.3	1.6	0.2	0.3
Nitrate (μg l^{-1})	23.1	206.6	0.3	339.8	7.4	98.9	0.1	7.9
Total nitrogen (μg l^{-1})	151.7	424.6	90.1	358.1	104.2	143.2	163.0	124.7
Pond size (ha)	0.81	1.19	0.23	1.20	0.63	118.93	0.37	2.32
Degree of isolation (m)	50	693	90	424	142	1042	10	151
Water body density (within 500 m radius)	4.0	4.2	4.0	3.0	5.0	8.7	8.0	4.4
Share of aquatic habitat (%)	2.2	23.3	0.8	4.1	2.6	40.2	5.6	4.0

"pushes" (each push was half a meter long) in rapid sequence, to cover all the different habitats in the littoral zone of the water body. Samples were preserved in 4% formalin. All crustaceans and several orders of insects (Ephemeroptera, Odonata, Heteroptera, Coleoptera and Trichoptera) were identified to species level.

Data analysis

Faunal composition

In order to determine the faunal composition of the different water body types, we performed two Partial Correspondence Analyses (partial-CAs), one with the species occurrence dataset, and the second with the richness of each taxonomic group. In both cases, the sampling period (February or June) was the covariable. An unimodal methodology was used because in the dataset of the first partial-CA the zeros are abundant (ter Braak & Šmilauer, 2002), while in the second partial-CA, the length of gradients, obtained by means of a DCA, were more than 3 SD (ter Braak & Prentice, 1988). The 13 taxonomic groups taken into account in this study were: large branchiopods (Anostraca and Notostraca), Cladocera Chydoridae, Cladocera non-Chydoridae, Calanoida, Cyclopoida, Harpacticoida, Ostracoda, Malacostraca, Ephemeroptera, Odonata, Heteroptera, Coleoptera and Trichoptera (see Electronic Supplementary Material). For the first partial-CA, we did not take into consideration those species with <5 occurrences (<3%), because rare species can bias the outcome of unimodal ordination methods (ter Braak & Šmilauer, 2002). This resulted in the dataset being reduced from 209 to 74 species. The number of samples was 165 for both datasets. The number of individuals and the number of species in the first and in the second partial-CA, respectively, were square root transformed. Following ter Braak & Šmilauer (2002), we downweighted for rare species in order to reduce their influence in the analyses. Significant differences among water body types in the first two partial-CA-axes were performed by means of a Welch statistic. The Welch statistic was used instead of an *F*-statistic, since scores had no variance homogeneity (Quinn & Keough, 2002). For the post hoc multiple comparisons analyses, Dunnett's T3 test was used because it does not assume homogeneity of variance (Quinn & Keough,

2002). Partial-CA analyses were performed using CANOCO 4.5 (ter Braak & Šmilauer, 2002).

Biodiversity

In order to compare the patterns of biodiversity among the different water body types, several parameters of rarity and species richness were calculated. For all the parameters, three faunal groups were analysed: crustaceans, aquatic insects, and the sum of crustacean and aquatic insect species (hereafter, total richness).

Rarity For each of the three faunal groups, we assessed rarity using the uniqueness of the set of species in each water body type and the average originality of species in each sample. Uniqueness was estimated as singularity, calculated from the entire dataset obtained for each water body type:

$$s = \frac{e}{E} \cdot 100$$

where e is the number of species only found in a given water body type and not in the other water body types, whereas E is the total number of species found in that water body type.

The average originality of species in each sample was calculated with the index of faunal originality (hereafter IFO; Puchalski, 1987):

$$\text{IFO} = \frac{\sum 1/M_i}{S}$$

where M is the total number of samples in which species i occurs (from i = 1 to S), and S is the number of species in the corresponding sample.

Species richness Three approaches were used in order to compare the species richness of the three faunal groups among water body types. The first approach was based on the sum of all species found in all samples of a given water body type. We used this cumulative species richness also to calculate the ratio between the number of crustacean species and the number of insect species.

The second approach was based on the results obtained for each sampling period in each water body, and the species richness per visit for each water body type was estimated as:

$$\text{Species richness per visit} = \frac{\sum S_i}{N}$$

where S_i is the number of species in sample i (from i = 1 to N), and N the number of samples representing the water body type.

Two-way analyses of variance (ANOVA) was used in order to study the effect of water temporality (permanent versus temporary water bodies) and salinity (saline versus freshwater water bodies) on total, crustacean and insect richness per visit. Species richness per visit values were log-transformed [$\log_{10}$(variable + 1)] in order to ensure variance homogeneity. These analyses were performed using SPSS 11.5.1 for Windows.

The third approach was performed estimating species richness using rarefaction curves. Rarefaction curves resample from a pooled group of samples to plot the expected number of species against an increasing number of samples, thus adjusting for differences in sample size. Sample-based rarefaction curves were carried out with EstimateS 7.5 software (Colwell, 2005). In order to compare species accumulation curves among water body types, samples for each water body type and sampling period (winter and spring) were randomized with replacement, and Mao Tau estimator of rarefied species richness ($\tilde{\tau}(h)$) was used, since it makes possible to perform rigorous comparisons among sample-based rarefaction curves due to the existence of legitimate confidence intervals at all points along the curve (Colwell et al., 2004). In order to obtain estimates of total richness, crustacean richness and insect richness for each water body type and sampling period, Chao2 estimator was calculated, as it is one of the most reliable predictors of total species richness when sampling effort (i.e., 20 dip-net "pushes") is comparable (Hortal et al., 2006). The bias-corrected formula of Chao2 estimator was used when Chao's estimated CV for abundance distribution was <0.5; otherwise, classic Chao2 estimator was used (Colwell, 2005). Following Colwell et al. (2004), the criteria used to determine if the results obtained with the rarefaction curves, rarefied species richness ($\tilde{\tau}(h)$) and Chao2 estimator, were significantly different ($P < 0.05$) was the absence of overlap among the 95% CI of the water body types (see values in Table 2).

Table 2 Cumulative species richness, species richness per visit (mean and range), percentage of singularity, Index of Faunal Originality (mean and range), $\tilde{\tau}(h)$ (mean and 95% CI), and total species richness (mean and 95% CI) calculated using Chao2 estimator for each water body type

	Total	Crustaceans	Insects	C/I ratio
Permanent freshwaters (PFW); n = 70				
Cumulative species richness	132	78	54	1.44
Species richness per visit (range)	8.7 (3–21)	6.4 (1–13)	3.1 (0–13)	
Water body type singularity	44.7	37.2	55.6	
Index of Faunal Originality (range)	0.14 (0.02–0.40)	0.11 (0.01–0.37)	0.22 (0.02–1.00)	
Winter $\tilde{\tau}(h)$, $n = 8$ (95% CI)	34 (26–42)	28 (21–35)	6 (3–9)	
Winter Chao2 mean (95% CI)	188 (130–317)	122 (85–219)	54 (33–110)	
Spring $\tilde{\tau}(h)$, $n = 8$ (95% CI)	46 (37–54)	29 (23–36)	16 (12–21)	
Spring Chao2 mean (95% CI)	196 (152–285)	104 (78–171)	93 (64–168)	
Temporary freshwaters (TFW); n = 41				
Cumulative species richness	100	51	49	1.05
Species richness per visit (range)	9.1 (3–27)	6.2 (3–15)	3.8 (0–18)	
Water body type singularity	45.0	33.3	57.1	
Index of Faunal Originality (range)	0.16 (0.04–0.43)	0.13 (0.04–0.38)	0.34 (0.02–1.00)	
Winter $\tilde{\tau}(h)$, $n = 8$ (95% CI)	35 (27–43)	23 (16–30)	12 (8–17)	
Winter Chao2 mean (95% CI)	108 (84–167)	49 (41–73)	51 (38–87)	
Spring $\tilde{\tau}(h)$, $n = 8$ (95% CI)	54 (43–65)	32 (23–40)	22 (15–29)	
Spring Chao2 mean (95% CI)	113 (89–170)	48 (42–67)	75 (45–171)	
Permanent saline waters (PSW); n = 38				
Cumulative species richness	65	52	13	4.00
Species richness per visit (range)	6.0 (1–14)	4.9 (0–11)	1.7 (0–6)	
Water body type singularity	15.4	13.5	23.1	
Index of Faunal Originality (range)	0.14 (0.04–0.38)	0.12 (0.03–0.29)	0.21 (0.04–0.56)	
Winter $\tilde{\tau}(h)$, $n = 8$ (95% CI)	26 (18–34)	23 (16–30)	3 (1–5)	
Winter Chao2 mean (95% CI)	91 (59–184)	73 (48–155)	11 (7–34)	
Spring $\tilde{\tau}(h)$, $n = 8$ (95% CI)	31 (23–39)	25 (18–32)	6 (3–10)	
Spring Chao2 mean (95% CI)	66 (56–95)	50 (42–73)	15 (12–30)	
Temporary saline waters (TSW); n = 16				
Cumulative species richness	39	28	11	2.55
Species richness per visit (range)	7.9 (1–12)	6.1 (1–10)	2.6 (0–7)	
Water body type singularity	5.1	3.6	9.1	
Winter $\tilde{\tau}(h)$, $n = 8$ (95% CI)	33 (23–43)	20 (12–28)	4 (1–6)	
Winter Chao2 mean (95% CI)	49 (38–85)	30 (22–65)	17 (14–31)	
Spring $\tilde{\tau}(h)$, $n = 8$ (95% CI)	22 (13–31)	19 (11–27)	3 (0–6)	
Spring Chao2 mean (95% CI)	25 (22–36)	22 (19–34)	3 (3–5)	

Note that $\tilde{\tau}(h)$ values are rarefied at a sample size of $n = 8$ (the minimum sample size) for comparison purposes. The values correspond to total species richness (crustaceans plus insects) and for insects and crustaceans separately. The ratio between crustacean and insect species (Ratio C/I) is also shown for each water body type

Environmental variables responses on species richness

Generalized Additive Models (GAMs) were used to test the importance of environmental variables for species richness. Two different datasets, one for each sampling period (winter and spring) were used in the GAM analyses, in order to avoid dependence of the datasets. Eight environmental variables were selected among the available field measurements to be used in the analyses. The selected variables included water variables (conductivity, chlorophyll-*a*, total nitrogen, and total phosphorus), pond variables (water permanence and pond size), and landscape variables (water body density, and degree of isolation). This selection was based on the necessity to reduce redundancy among the set of explanatory variables within the regression procedures. All variables, except water permanence, were log-transformed [$\log_{10}$(variable + 1)] to reduce the effect of outliers. The Pearson's correlation coefficients for the seven continuous variables are shown in Table 3. GAM calculations were carried out with S-PLUS 2000 software. A Poisson family distribution was used for the response variables (total, insect and crustacean richness), since it ensured that all fitted values were positive. The selected variables included in the final model were obtained performing an automatic stepwise selection, and the Akaike Information Criterion (AIC) was used to select the best model with increasing complexity (degrees of freedom equal to 1, 2 and 3). However, automatic stepwise procedures are rather generous leaving terms in the model. For this reason, the increase of deviance caused by each variable included in the model and obtained with the stepwise selection was tested, and only variables that caused a significant increase in the deviance were retained in the final model. When overdispersion was detected, *F*-test instead of Chi-square test was performed in order to obtain the significance of the deviance explained by each selected variable (Crawley, 2002). Drop contribution was used in order to identify the contribution of each explanatory variable, expressed as a deviance reduction associated to dropping the variable from the final model (Castella et al., 2001).

Results

In the partial-CA performed with the species occurrence dataset (Fig. 2), the first two axes explained 14.3% of the variability in the final model. The scores of both axes showed significant differences ($W_{3,48.8} = 43.5$; $P < 0.001$ for the first axis, and $W_{3,64.5} = 54.6$; $P < 0.001$ for the second axis) among water body types. Particularly, significant differences existed among PFW, TFW, and SW, while the two saline water body types were not significantly different. According to these results, we only distinguished three water body types by means of their faunal composition: PFW were characterized by a high occurrence of two cladocerans (*Chydorus sphaericus* and *Oxyurella tenuicaudis*) and two copepods (*Eucyclops serrulatus* and *Macrocyclops albidus*); TFW were characterized by a high occurrence of two copepods (*Megacyclops viridis* and *Canthocamptus staphylinus*); and SW were characterized by a high occurrence of four copepods (*Calanipeda aquaedulcis, Halicyclops rotundipes, Canuella perplexa* and *Cletocamptus confluens*), two ostracods (*Loxoconcha elliptica* and *Cyprideis torosa*) and three malacostracans (*Lekanesphaera hookeri, Gammarus aequicauda* and *Corophium orientale*).

In the second partial-CA, performed with the dataset of species richness within taxonomic groups (Fig. 3; see also Electronic Supplementary Material), the first two axes explained the 36.8% of the variability in the final model. The scores of both axes showed significant differences ($W_{3,52.4} = 26.9$; $P < 0.001$ for the first axis, and $W_{3,60.5} = 8.9$; $P < 0.001$ for the second axis) among water body

Table 3 Pearson's correlation coefficient (*r*) between the seven continuous explanatory environmental variables used in GAMs

Variable	Code	COND	CHLA	TN	TP	S	WBD
Conductivity	COND						
Chlorophyll-*a*	CHLA	0.08					
Total nitrogen	TN	0.00	0.06				
Total phosphorus	TP	0.00	0.20**	0.54**			
Size	S	0.16*	−0.04	−0.09	−0.03		
Water body density	WBD	0.18*	−0.09	−0.02	−0.14	0.71**	
Degree of isolation	I	0.04	−0.09	0.00	0.12	−0.10	−0.41**

No asterisk: *r* not significantly different from 0 at $P > 0.05$; *: *r* significantly different from 0 at $P < 0.05$; **: *r* significantly different from 0 at $P < 0.01$

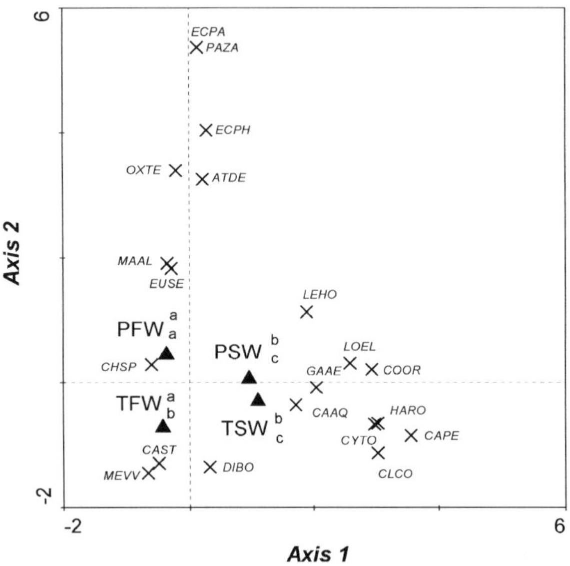

Fig. 2 Partial Correspondence Analysis: ordination of the 20 species which explain the 14.3% of the variability explained by the first two axes. Different upper and lower letters indicate significant differences ($P < 0.01$) among sample scores of the water body types on the first and second axis, respectively, after Dunnett's T3 post hoc test. ATDE: *Atyaephyra desmarestii* (Millet, 1831); CAAQ: *Calanipeda aquaedulcis* Kritschagin, 1873; CAPE: *Canuella perplexa* Scott, 1893; CAST: *Canthocamptus staphylinus* (Jurine, 1820); CHSP: *Chydorus sphaericus* (Müller, 1776); CLCO: *Cletocamptus confluens* (Schmeil, 1894); COOR: *Corophium orientale* Schellenberg, 1928; CYTO: *Cyprideis torosa* (Jones, 1850); DIBO: *Diacyclops bicuspidatus* (Schmankevitch, 1875); ECPA: *Echinogammarus pacaudi* Hubault & Ruffo, 1956; ECPH: *Ectocyclops phaleratus* (Koch, 1838); EUSE: *Eucyclops serrulatus* (Fischer, 1851); GAAE: *Gammarus aequicauda* (Martynov, 1931); HARO: *Halicyclops rotundipes* Kiefer, 1935; LEHO: *Lekanesphaera hookeri* (Leach, 1814); LOEL: *Loxoconcha elliptica* Brady, 1869; MAAL: *Macrocyclops albidus* (Jurine, 1820); MEVV: *Megacyclops viridis* (Jurine, 1820); OXTE: *Oxyurella tenuicaudis* (Sars, 1862); PAZA: *Palaemonetes zariquieyi* Sollaud, 1939

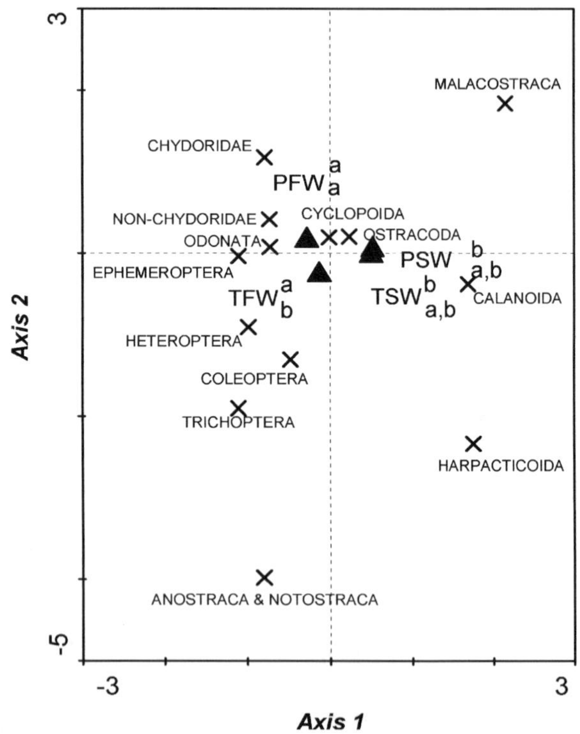

Fig. 3 Partial Correspondence Analysis: ordination of the 13 taxonomic faunal groups analysed. Different upper and lower letters indicate significant differences ($P < 0.01$) among sample scores of the water body types on the first and second axis, respectively, after Dunnett's T3 post hoc test

types. The first axis had significantly different sample scores for fresh and saline water bodies, independent of their water permanence. The second axis showed significant differences between temporary and permanent freshwater samples (Fig. 3). Again, we only observed the existence of three water body types according to the species richness of the taxonomic groups: PFW were associated with high richness of cladocerans (Chydoridae and non-Chydoridae), ephemeropterans and odonates; TFW were associated with high richness of large branchiopods (Anostraca and Notostraca), heteropterans, coleopterans and trichopterans; and SW were associated with high richness of calanoids, harpacticoids and malacostracans.

Saline water body types (TSW and PSW) had lower total cumulative species richness and singularity percentage, and higher values of crustacean/insect ratio, than freshwater types (Table 2). Cumulative richness was not a useful descriptor for comparison purposes in this study, because the number of samples for each water body type was not equal. Similarly, singularity could be influenced by rarity, since it is expected that more samples will contain more rare species and, consequently increase the singularity proportion. Nevertheless, note that TFW and PSW had similar number of samples (41 and 38, respectively), and TFW had higher values of singularity than PSW. Furthermore, PFW and TFW had similar total, crustacean and insect singularities (Table 2), although they had very different number of samples (70 and 41, respectively). Thus, the different number of water bodies was not the only explanation of the differences observed between the

Fig. 4 Sample-based rarefaction curves ($\tilde{\tau}(h)$, with 95% CI), and total species richness calculated using Chao2 estimator (a solid line for permanent water bodies and a dashed line for temporary water bodies), of total species richness (**A**, winter; **B**, spring), crustacean richness (**C**, winter; **D**, spring) and insect richness (**E**, winter; **F**, spring) for each water body type. For the curves, every second point is shown

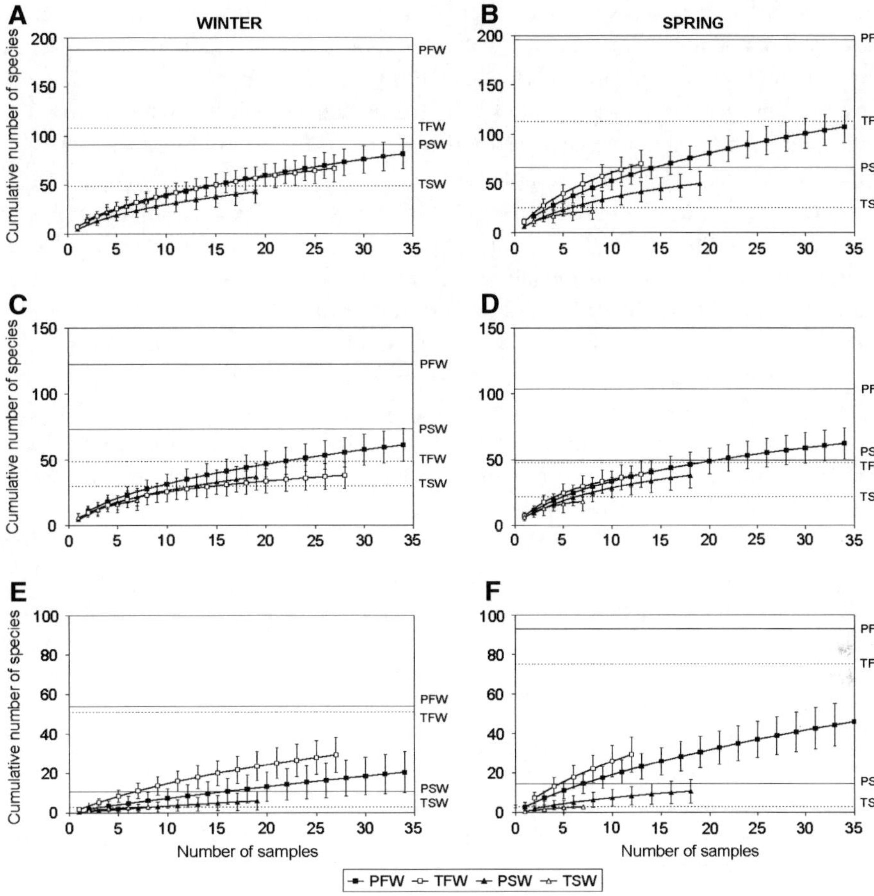

singularity of saline and freshwaters. In contrast to similarity, IFO values (calculated for each sample) showed similar values among the four water body types.

In order to compare species richness among water body types, we used species richness per visit and rarefaction curves (Table 2 and Fig. 4). Significant differences were obtained for total richness per visit and insect richness per visit in relation to salinity (ANOVA: $F = 7.87$, $P < 0.001$; ANOVA: $F = 7.74$, $P < 0.001$; respectively), but not to water permanence. Thus, SW had lower total richness per visit (mean values: 6.6 vs. 8.9) and lower insect richness per visit (mean values: 1.4 vs. 2.7) than freshwaters. In the case of crustacean richness per visit, no significant results were obtained either for salinity or for water permanence. For total species richness, spring rarefaction curves were in accordance with the results obtained with species richness per visit, whereas the confidence intervals of the winter rarefaction curves of the four water body types overlap. In spring, freshwater water bodies rose more steeply than saline water bodies, and attained significantly higher values (Table 2). However, there were no significant differences between PFW and TFW, and between PSW and TSW, due to the overlap of their confidence intervals. These results were coincident with those obtained for $\tilde{\tau}(h)$ rarefied at a sample size of $n = 8$ (Table 2), although total richness of PSW estimated by Chao2 was significantly higher than that of TSW. For crustacean richness, similar rarefaction curves were found among water body types and between sampling periods. Moreover, there were no significant differences among the four water body types, due to the overlap of the confidence intervals (Fig. 4). Similar results were obtained by $\tilde{\tau}(h)$ estimator. In winter, Chao2 estimator was significantly higher in PFW than in TFW, and, in spring, both PFW and PSW were significantly higher than TFW and TSW, respectively (Table 2). For insect richness, the curves of freshwater water bodies were steeper than those of saline water bodies (Fig. 4), and estimated insect richness by Chao2 showed significantly higher values for freshwater water bodies than for saline water bodies in spring. In winter, $\tilde{\tau}(h)$ rarefied at a sample size of $n = 8$ presented significantly higher values of insect richness for TFW than those for saline water bodies, whereas in spring, both freshwater types presented higher values of $\tilde{\tau}(h)$ than SW (Table 2). In contrast with the case of total and crustacean richness, TFW rose more steeply than the other water body types in both sampling periods, but Chao2 estimator showed similar values of insect richness for PFW and TFW (Table 2). However, some results obtained by the Chao2 estimator have to be taken carefully because they did not reach an asymptote.

For each sampling period (winter and spring), three GAMs were carried out, for total richness, insect richness and crustacean richness. For total richness, the final GAM model for the winter sampling period selected only one variable, conductivity, whereas for the spring sampling period three variables were selected: conductivity, water body density and degree of isolation (Table 4A and Fig. 5). All variables had a linear or quasi-linear negative influence on total richness. For crustacean richness, no variables were selected for the winter sampling period, and only total phosphorus was selected for the spring sampling period, showing a negative quasi-linear response (Table 4B and Fig. 6A). For insect richness, the final GAM model selected three variables for each dataset: water body density and conductivity for both sampling periods, chlorophyll-*a* for the winter sampling period and degree of isolation for the spring sampling period (Table 4C and Fig. 6B–G). Chlorophyll-*a* and degree of isolation showed a linear negative response on insect

Table 4 Significance level, degrees of freedom of the smoother, and drop contribution (contribution of each explanatory variable expressed as a deviance reduction associated to dropping the variable from the final model)

Explanatory variable		*F*-values	*P*	Smoother (df)	Drop contribution
(A) Total richness					
Water body density	(W)	–	–	–	–
	(S)	3.42	<0.05	3	19.2
Conductivity	(W)	4.14	<0.01	3	12.9
	(S)	7.10	<0.01	1	13.5
Degree of isolation	(W)	–	–	–	–
	(S)	2.95	<0.05	3	16.9
Dispersion parameter	(W)	1.06			
	(S)	2.04			
(B) Crustacean richness					
Total phosphorus	(W)	–	–	–	–
	(S)	2.79	<0.05	3	10.8
Dispersion parameter	(W)	–			
	(S)	1.37			
(C) Insect richness					
Water body density	(W)	5.18	<0.005	3	22.8
	(S)	9.12	<0.001	3	67.4
Conductivity	(W)	2.77	<0.05	3	12.2
	(S)	3.07	<0.005	3	22.3
Degree of isolation	(W)	–	–	–	–
	(S)	10.5	<0.005	1	26.6
Chlorophyll-*a*	(W)	4.06	<0.05	1	5.7
	(S)	–	–	–	–
Dispersion parameter	(W)	1.67			
	(S)	2.68			

The results obtained for total taxonomic richness (A), crustacean taxonomic richness (B), and insect taxonomic richness (C), as well as the dispersion parameter for each final model are shown. Each analysis was performed separately according to the winter (W) and spring (S) sampling periods

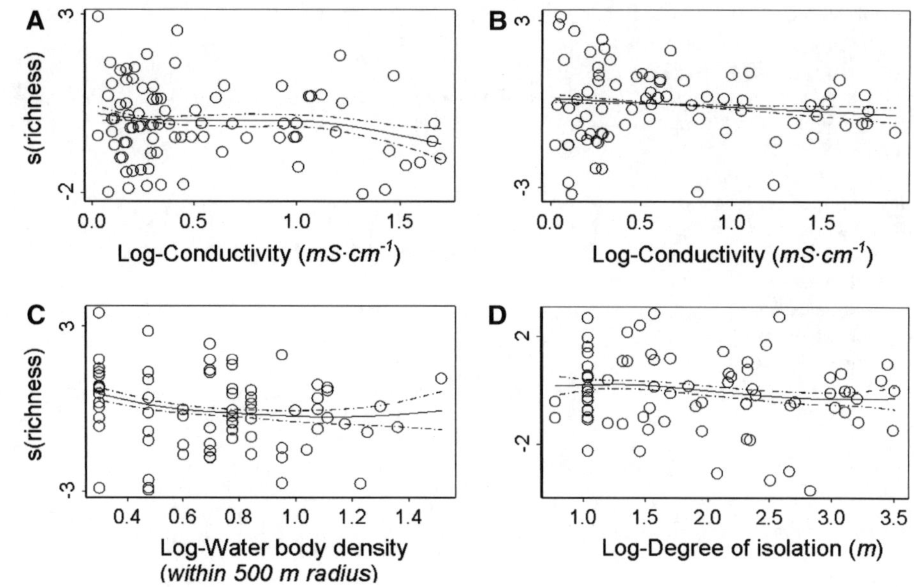

Fig. 5 Response functions for the total species richness and the variables incorporated in the GAM. (**A**) Corresponds to the winter sampling period, whereas figures **B**, **C** and **D** to the spring sampling period. The dashed lines are approximate 95% confidence intervals around the smooth function

richness, whereas for the rest of the selected variables the confidence intervals evidenced a lack of accuracy in the upper part of the gradient. Note that no pond variables (water permanence and pond size) were selected by any of the GAM models. Finally, it is also interesting to note that the GAM model for insect richness, but not the model for crustacean richness, selected landscape variables, and these variables were mainly selected for the spring sampling period dataset. However, some of the relationships established by means of the GAM analyses have to be interpreted carefully, because confidence intervals were wide, mainly at the upper limit of the gradients (e.g., water body density for insect richness; Fig. 6C, G).

Discussion

Crustacean and aquatic insect composition and rarity

The multivariate approaches (species composition and richness of taxonomic groups) showed that three different communities can be identified: PFW, TFW and SW. The absence of two community types for SW could be explained because in the studied area almost all SW are located in coastal areas, contrarily to other Mediterranean areas, where athalassohaline waters are abundant (Britton & Crivelli, 1993; Alonso, 1998). Mediterranean coastal water bodies are characterized by flooding events that connect permanent and temporary water bodies (Brucet et al., 2005; Badosa et al., 2006), whereas inland waters are more isolated.

The ratio of the number of crustacean species to insect species was higher in SW than in freshwaters. This was due to a lower-insect richness in SW rather than to a higher-crustacean richness, according to results obtained by the species richness per visit and the rarefaction curves. This pattern is also common outside the Mediterranean area (Timms, 1993; Williams & Williams, 1998). Similarly, the association between TFW and large branchiopods is expected because these animals are almost exclusive of temporary waters except a few genera which inhabit large hyperhaline lakes (Hartland-Rowe, 1972). In contrast, the other branchiopods (cladocerans) showed higher species richness in permanent waters. This difference in species richness patterns within branchiopods could be related to the different evolutive pathways of the two groups mediated by vertebrate predation (Kerfoot & Lynch, 1987). Despite the higher values of singularity obtained for freshwater water body types, a higher rarity can not be attributed to freshwater samples because IFO values were very similar among the four water body types. Thus, the low values of singularity of saline water bodies could be mainly explained by the lack of different assemblages (species composition) between the two saline water body types.

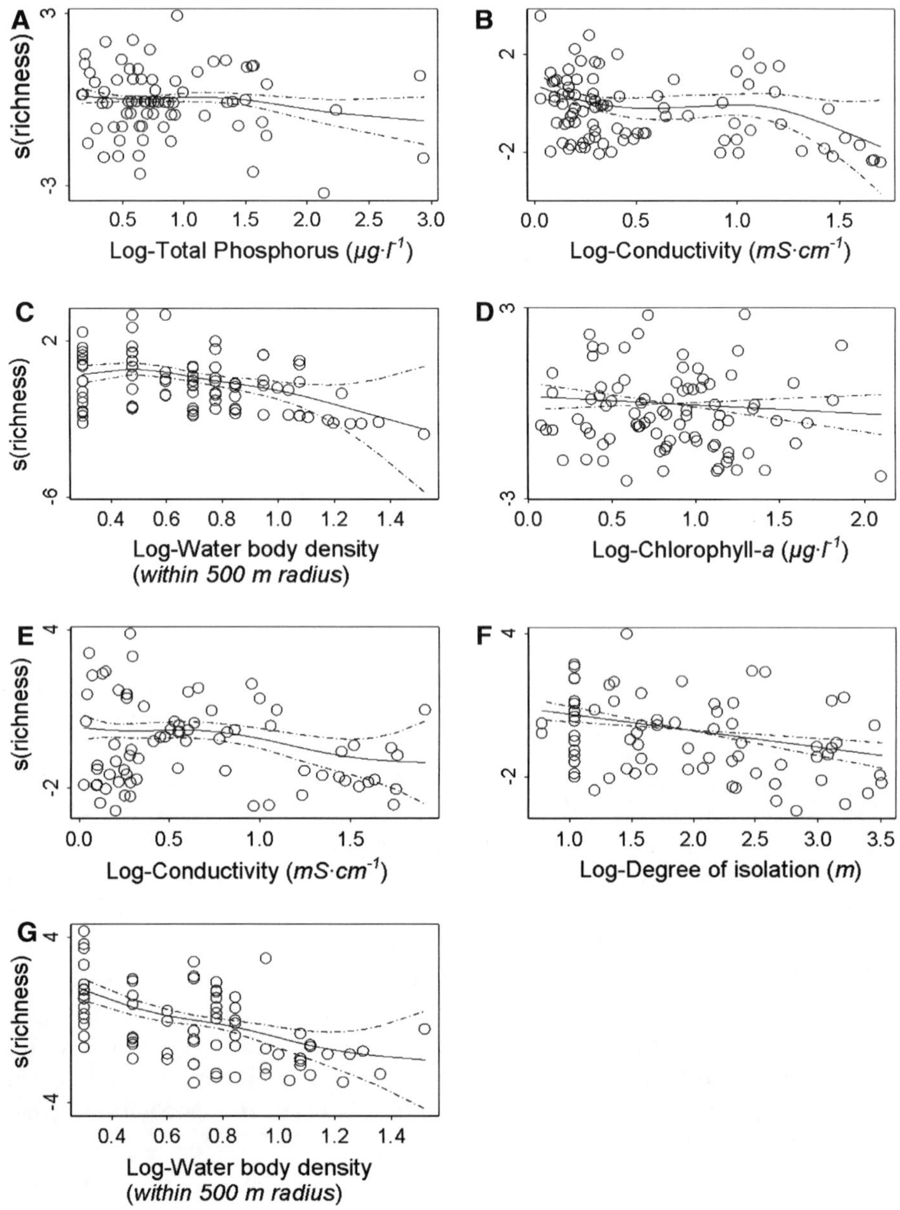

Fig. 6 Response functions for the crustacean (**A**) and insect (**B**, **C**, **D**, **E**, **F** and **G**) richness and the variables incorporated in the GAM. Figures **A**, **B**, **C** and **D** correspond to the winter sampling period, whereas figures **E**, **F** and **G** to the spring sampling period. The dashed lines are approximate 95% confidence intervals around the smooth function

In freshwaters, different assemblages of coleopterans and heteropterans between permanent and temporary systems are also known (e.g. Eyre et al., 1986; García-Avilés et al., 1996). This difference has been explained by differences in biological traits as dispersal capacity and drought tolerance (e.g. Brown, 1951; Nilsson, 1986). In the Mediterranean area, temporary ponds with high species richness of coleopterans and heteropterans are known (Ribera & Aguilera, 1996; Boix & Sala, 2002), but a positive relationship between water permanence and coleopteran and heteropteran richness has also been reported (López & Hernández, 2000; Valladares et al., 2002). These apparent contradictory results could be explained by other factors not included in these studies such as predation pressure (e.g. Wellborn et al., 1996), connectivity (e.g., Cottenie et al., 2003), pond size (e.g. Oertli et al., 2002), pond age (e.g. Fairchild et al., 2000), sampling and sample processing (e.g., Turner & Trexler, 1997; King & Richardson, 2002), etc. The relationship found between trichopterans and TFW has also been reported in colder temperate aquatic ecosystems (Wiggins, 1973). Nevertheless, this relationship has to be taken carefully, since trichopterans were scarcely captured in our study.

In our study, Calanoida were associated with SW due to their high occurrence in these systems, although we found higher species richness in freshwaters. In fact, 16 of the 24 Calanoida species known in the Iberian Peninsula are found in freshwaters (Alonso, 1998; Martinoy et al., 2006). In the case of Harpacticoida, which also were associated with SW, it is known that a decrease in the number of species takes place from the marine intertidal to low salinity areas (Huys et al., 1996). The association of Malacostraca with SW was expected because some malacostracan groups are exclusively or have high species richness in brackish waters, especially in estuaries (e.g. amphipods, isopods, mysidaceans, and decapods; Chinchilla & Comín, 1977; Cuesta et al., 2006), and some of them are poorly represented in lentic freshwaters (Karaman, 1993).

Crustacean and aquatic insect richness

A species richness decrease in the transitional zone from fresh to brackish waters has been observed in coastal wetlands (e.g. Cognetti & Maltagliati, 2000). Although in the case of insects a clear decrease of the species richness has been observed (e.g. Williams & Williams, 1998), other authors have found a lack of relationship between crustacean richness and salinity (e.g. Josefson & Hansen, 2004). Our results coincided with these studies, because a relationship between salinity and species richness was observed for insects, but not for crustaceans.

Trophic state variables, such as nutrients or chlorophyll-*a* were not selected in any of the two models for total richness. In contrast, the spring model for crustacean richness and the winter model for insect richness selected total phosphorus and chlorophyll-*a*, respectively. The relationship between species richness and trophic state variables is generally accepted (e.g. Whilh & Dorris, 1968), but contrary results have been already published. In this sense, several studies in colder temperate aquatic ecosystems did not find a significant relationship between eutrophy and zooplankton species richness (e.g. Attayde & Bozelli, 1998; Lougheed & Chow-Fraser, 2002), and in other studies in the Mediterranean area working with benthic macroinvertebrates, the relationship was weak (García-Criado et al., 2005). Decrease of zooplankton and macroinvertebrate species richness with an increase of nutrients has been also observed in Mediterranean and colder temperate areas, and a total phosphorus gradient was identified as the main factor (Jeppesen et al., 2000; Romo et al., 2005; Declerck et al., 2005). On the other hand, different relationships between species richness of different faunal groups and total phosphorus have been previously reported. Thus, Declerck et al. (2005) found a negative linear relationship for cyclopoid species richness in Spain, and a unimodal relationship for cladoceran and macroinvertebrate species richness in Denmark. In the same study, no significant relationships were found between total phosphorus and cyclopoid species richness in central and north Europe, and between total phosphorus and cladoceran and macroinvertebrate species richness in central and south Europe.

In contrast to the results obtained for salinity or trophic state, water permanence was not selected in any predictive model for the three species richness variables. Water permanence was one of the main gradients taken into account in a study of community structure of aquatic continental ecosystems (Wellborn et al., 1996), and several studies found a decrease of crustacean richness and insect richness in most temporary water bodies compared to more permanent ones (e.g. Ebert & Balko, 1987; Schneider & Frost, 1996). However, several authors already pointed out that a wide variety of freshwater species is well adapted to survive periods of drought and, even, some species are particularly associated with intermittent drying (Biggs et al., 1994; Williams, 1996). In Mediterranean water bodies, contradictory results exist. For example, higher species richness was observed in temporary ponds when planktonic crustacean richness was considered (Galindo et al., 1994), whereas lower species richness was observed in temporary ponds when benthonic macroinvertebrate richness was analysed (Della Bella et al., 2005). Furthermore, temporary sites with an intermediate hydroperiod length had higher species richness than more temporary and permanent ones (Frisch et al., 2006).

The main difference between the predictive model obtained for crustacean richness and insect richness was the inclusion of landscape variables (water body density and degree of isolation) in the latter. This relationship is in agreement with the findings of other studies. Colonisation and extinction of aquatic insects in a water body is explained not only by pond characteristics, but also by landscape characteristics (e.g. Jeffries, 2005). In contrast, the passive dispersion of crustaceans (Wiggins et al., 1980) and their high dispersion rates, especially for cladocerans (Louette & De Meester, 2005), could explain the lack of landscape variables in the final model for crustacean richness. This absence of relationship for crustacean species richness has been previously reported (Cottenie & De Meester, 2003), but studies that showed the existence of this relationship also exist (Dodson, 1992). The difference between the spatial scales employed in these studies could explain the different results. On the other hand, landscape variables were mainly selected in the spring GAM models. This result could be associated to the fact that the activity of the aerial colonisers (i.e. insects) is strongly determined by season characteristics such as air and water conditions or life cycles (e.g. Landin, 1976; Nilsson, 1986). In winter, insect populations have low densities, whereas in spring they have an important increase in numbers and activity, and colonise mainly the nearby sites. In these sense, the negative relationship between degree of isolation and insect richness could be explained. In the same way, seasonal differences of species richness in freshwaters were found for insects but not for crustaceans. Thus, insects showed higher $\tilde{\tau}(h)$ and Chao2 values observed for insect richness in spring than in winter, whereas crustaceans presented similar values. This could be related to a different dispersal activity of these two faunal groups.

Eitam et al. (2004a, b) pointed out that the species richness of different taxonomic groups is influenced by different water body characteristics. However, it is also true that the factors which explain the total species richness of a community could be not included as important factors when each group was analysed separately (e.g. Ebert & Balko, 1987; Declerck et al., 2005). In this sense, the factors involved in each taxonomic group have to be taken into account, as well as the factors involved in determining the total richness.

Acknowledgements This work was supported by a grant from the Comisión de Investigación Científica y Técnica (CICYT), Programa de Recursos Naturales (ref. CGL2004-05433/BOS). We thank Dr. R.K. Colwell for valuable advice on rarefaction curves. We are grateful to Dr. S. Declerck and to three anonymous reviewers whose comments improved the manuscript.

References

Alonso, M., 1998. Las lagunas de la España peninsular. Limnetica 15: 1–176.

Álvarez-Cobelas, M., C. Rojo & D. Angeler, 2005. Mediterranean limnology: current status, gaps and the future. Journal of Limnology 64: 13–29.

Attayde, J. L. & R. L. Bozelli, 1998. Assessing the indicator properties of zooplankton assemblages to disturbance gradients by canonical correspondence analysis. Canadian Journal of Fisheries and Aquatic Sciences 55: 1789–1797.

Badosa, A., D. Boix, S. Brucet, R. López-Flores & X. D. Quintana, 2006. Nutrients and zooplankton composition and dynamics in relation to the hydrological pattern in a confined Mediterranean salt marsh (NE Iberian Peninsula). Estuarine Coastal and Shelf Science 66: 513–522.

Bazzanti, M., S. Baldoni & M. Seminara, 1996. Invertebrate macrofauna of a temporary pond in Central Italy: composition, community parameters and temporal succession. Archiv für Hydrobiologie 137: 77–94.

Biggs, J., A. Corfield, D. Walker, M. Whitfield & P. Williams, 1994. New approaches to the management of ponds. British Wildlife 5: 273–287.

Boix, D. & J. Sala, 2002. Riqueza y rareza de los insectos acuáticos de la laguna temporal de Espolla (Pla de l'Estany, Cataluña). Boletín de la Asociación española de Entomología 26: 45–57.

Boix, D., J. Sala & R. Moreno-Amich, 2001. The faunal composition of Espolla pond (NE Iberian peninsula): the neglected biodiversity of temporary waters. Wetlands 21: 577–592.

Boix, D., J. Sala, X. D. Quintana & R. Moreno-Amich, 2004. Succession of the animal community in a mediterranean temporary pond. Journal of North American Benthological Society 23: 29–49.

Boix, D., S. Gascón, J. Sala, M. Martinoy, J. Gifre & X. D. Quintana, 2005. A new index of water quality assessment in Mediterranean wetlands based on crustacean and insect assemblages: the case of Catalunya (NE Iberian peninsula). Aquatic Conservation: Marine and Freshwater Ecosystems 15: 635–651.

Britton, R. H. & A. J. Crivelli, 1993. Wetlands of southern Europe and North Africa: Mediterranean wetlands. In Whigham, D. F., D. Dykyjová & S. Hejný (eds), Wetlands of the world I: Inventory, ecology and management. Kluwer Academic Publishers, Dordrecht, 129–194.

Boronat, L., M. R. Miracle & X. Armengol, 2001. Cladoceran assemblages in a mineralization gradient. Hydrobiologia 442: 75–88.

Britton, R. H. & V. D. Podlejski, 1981. Inventory and classification of the wetlands of the Camargue (France). Aquatic Botany 10: 195–228.

Brown, E. S., 1951. The relation between migration-rate and type of habitat in aquatic insects, with special reference to

certain species of Corixidae. Proceedings of the Zoological Society of London 121: 539–545.
Brucet, S., D. Boix, R. López-Flores, A. Badosa, R. Moreno-Amich & X. D. Quintana, 2005. Zooplankton structure and dynamics in permanent and temporary Mediterranean salt marshes: taxon-based and size-based approaches. Archiv für Hydrobiologie 162: 535–555.
Castella, E., H. Adalsteinsson, J. E. Brittain, G. M. Gislason, A. Lehmann, V. Lencioni, B. Lods-Crozet, B. Maiolini, A. M. Milner, J. S. Olafsson, S. J. Saltveit & D. L. Snook, 2001. Macrobenthic invertebrate richness and composition along a latitudinal gradient of European glacier-fed streams. Freshwater Biology 46: 1811–1831.
Chinchilla, M. & F. A. Comín, 1977. Contribució al coneixement dels crustacis del delta de l'Ebre. Treballs de la Institució Catalana d'Història Natural 8: 119–144.
Cognetti, G. & F. Maltagliati, 2000. Biodiversity and adaptative mechanisms in brackish water fauna. Marine Pollution Bulletin 40: 7–14.
Colwell, R. K., 2005. EstimateS: statistical estimation of species richness and shared species from samples. Version 7.5. User's Guide and application published at: http://purl.oclc.org/estimate
Colwell, R. K., C. X. Mao & J. Chang, 2004. Interpolating, extrapolating, and comparing incidence-based species accumulation curves. Ecology 85: 2717–2727.
Cottenie, K. & L. De Meester, 2003. Connectivity and cladoceran species richness in a metacommunity of shallow lakes. Freshwater Biology 48: 823–832.
Cottenie, K., E. Michels, N. Nuytten & L. De Meester, 2003. Zooplankton metacommunity structure: regional vs. local processes in highly interconnected ponds. Ecology 84: 991–1000.
Crawley, M. J., 2002. Statistical Computing. An Introduction to Data Analysis using S-Plus. John Wiley & Sons, Chichester.
Cuesta, J. A., E. González-Ortegón, A. Rodríguez, F. Baldó, C. Vilas & P. Drake, 2006. The decapod crustacean community of the Guadalquivir Estuary (SW Spain): seasonal and inter-year changes in community structure. Hydrobiologia 557: 85–95.
Declerck, S., J. Vandekerkhove, L. Johansson, K. Muylaert, J. M. Conde-Porcuna, K. Van der Gucht, C. Pérez-Martínez, T. Lauridsen, K. Schwenk, G. Zwart, W. Rommens, J. López-Ramos, E. Jeppesen, W. Vyverman, L. Brendonck, & L. De Meester, 2005. Multi-group biodiversity in shallow lakes along gradients of phosphorus and water plant cover. Ecology 86: 1905–1915.
Della Bella, V., M. Bazzanti & F. Chiarotti, 2005. Macroinvertebrate diversity and conservation status of Mediterranean ponds in Italy: water permanence and mesohabitat influence. Aquatic Conservation: Marine and Freshwater Ecosystems 15: 583–600.
Dodson, S., 1992. Predicting crustacean zooplankton species richness. Limnology and Oceanography 37: 848–856.
DPTOP (Departament de Política Territorial i Obres Públiques), 2005. Hipermapa. Atles electrònic de Catalunya. Available: http://www10.gencat.net/ptop/AppJava/cat/actuacions/territori/hipermapa.jsp
Ebert, T. A. & M. L. Balko, 1987. Temporary pools as islands in space and in time: the biota of vernal pools in San Diego, Southern California, USA. Archiv für Hydrobiologie 110: 101–123.
Eitam, A., L. Blaustein, K. Van Damme, H. J. Dumont & K. Martens, 2004a. Crustacean species richness in temporary pools: relationships with habitat traits. Hydrobiologia 525: 125–130.
Eitam, A., C. Noreña & L. Blaustein, 2004b. Microturbellarian species richness and community similarity among temporary pools: relationship with habitat properties. Biodiversity and Conservation 13: 2107–2117.
Eyre, M. D., S. G. Ball & G. N. Foster, 1986. An initial classification of the habitats of aquatic Coleoptera in North-East England. Journal of Applied Ecology 23: 841–852.
Fairchild, G. W., A. M. Faulds & J. F. Matta, 2000. Beetle assemblages in ponds: effects of habitat and site age. Freshwater Biology 44: 523–534.
Frisch, D., E. Moreno-Ostos & A. J. Green, 2006. Species richness and distribution of copepods and cladocerans and their relation to hydroperiod and other environmental variables in Doñana, south-west Spain. Hydrobiologia 556: 327–340.
Galindo, M. D., A. J. Mata, N. Mazuelos & L. Serrano, 1994. Microcrustacean and rotifer diversity and richness relating to water temporality in dune ponds of the Doñana National Park. Verhandlungen/Internationale Vereinigung für theoretische und angewandte Limnologie 25: 1350–1356.
García-Avilés, J., M. A. Puig & A. G. Soler, 1996. Distribution and associations of the aquatic Heteroptera of the Balearic Islands (Spain). Hydrobiologia 324: 209–217.
García-Criado, F., E. Bécares, C. Fernández-Aláez & M. Fernández-Aláez, 2005. Plant-associated invertebrates and ecological quality in some Mediterranean shallow lakes: implications for the application of the EC Water Framework Directive. Aquatic Conservation: Marine and Freshwater Ecosystems 15: 31–50.
Gascón, S., D. Boix, J. Sala & X. D. Quintana, 2005. Variability of benthic assemblages in relation to the hydrological pattern in Mediterranean salt marshes (Empordà wetlands, NE Iberian Peninsula). Archiv für Hydrobiologie 163: 163–181.
Grasshoff, K., M. Ehrhardt & K. Kremling, 1983. Methods of Seawater Analysis. Verlag Chemie, Weinheim.
Hartland-Rowe, R., 1972. The limnology of temporary waters and the ecology of Euphyllopoda. In: Clark, R. B. & R. J. Wootton (eds), Essays in Hidrobiology. University of Exeter, Exeter, 15–31.
Hortal, J., P. A. V. Borges & C. Gaspar, 2006. Evaluating the performance of species richness estimators: sensitivity to sample grain size. Journal of Animal Ecology 75: 274–287.
Huys, R., J. M. Gee, C. G. Moore & R. Hamond, 1996. Marine and Brackish Water Harpacticoid Copepods. The Linnean Society of London & The Estuarine and Coastal Sciences Association, Shrewsbury.
Jeffries, M., 2005. Local-scale turnover of pond insects: intra-pond habitat quality and inter-pond geometry are both important. Hydrobiologia 543: 207–220.
Jeppesen, E., J. P. Jensen, M. Søndergaard, T. Lauridsen & F. Landkildehus, 2000. Trophic structure, species richness

and biodiversity in Danish lakes: changes along a phosphorus gradient. Freshwater Biology 45: 201–218.
Josefson, A. B. & J. L. S. Hansen, 2004. Species richness of benthic macrofauna in Danish estuaries and coastal areas. Global Ecology and Biogeography 13: 273–288.
Karaman, G. S., 1993. Amphipoda d'acqua dolce. Edizione Calderini, Bologna.
Kerfoot, W. C. & M. Lynch, 1987. Branchiopod communities: associations with planktivorous fish in space and time. In Kerfoot, W. C. & A. Sih (eds), Predation. Direct and Indirect Impacts on Aquatic Communities. University Press of New England, Hanover, 367–378.
King, R. S. & C. J. Richardson, 2002. Evaluating subsampling approaches and macroinvertebrate taxonomic resolution for wetland bioassessment. Journal of North American Benthological Society 21: 150–171.
Landin, J., 1976. Seasonal patterns in abundance of water-beetles belonging to the Hydrophiloidea (Coleoptera). Freshwater Biology 6: 89–108.
López, T. & J. M. Hernández, 2000. Utilización de la taxocenosis de los heterópteros acuáticos (Heteroptera: Gerromorpha y Nepomorpha) en la caracterización sinecológica de las aguas epicontinentales de la provincia de Madrid (España). Boletín de la Asociación española de Entomología 24: 23–37.
Lougheed, V. L. & P. Chow-Fraser, 2002. Development and use of a zooplankton index of wetland quality in the Laurentian Great Lakes basin. Ecological Applications 12: 474–486.
Louette, G. & L. De Meester L, 2005. High dispersal capacity of cladoceran zooplankton in newly founded communities. Ecology 86: 353–359.
MAPA (Ministerio de Agricultura Pesca y Alimentación), 2006. Sistema de identificación de parcelas agrícolas. Available: http://sigpac.mapa.es/fega/visor/
Martinoy, M., D. Boix, J. Sala, S. Gascón, J. Gifre, A. Argerich, R. Barrera, S. Brucet, A. Badosa, R. López-Flores, M. Méndez, J. M. Utgé & X. D. Quintana, 2006. Crustacean and aquatic insect assemblages in the Mediterranean coastal ecosystems of Empordà Wetlands (NE Iberian Peninsula). Limnetica 25: 665–682.
Médail, F., H. Michaud, J. Molina, G. Paradis & R. Loisel, 1998. Conservation de la flore et de la végétation des mares temporaires dulçaquicoles et oligotrophes de France méditerranéenne. Ecologia Mediterranea 24: 119–134.
Mocci, A., 1983. Rapporto e risultati della "Conferenze sulla convenzione per le zone umide d'importanza internazionale" (Cagliari 24/29 novembre 1980) e riflessioni sugli stagni della Sardegna. Alytes 1: 21–40.
Nilsson, A. N., 1986. Life Cycles and habitats of the northern European Agabini (Coleoptera, Dytiscidae). Entomologica Basiliensia 11: 391–417.
Oertli, B., D. A. Joye, E. Castella, R. Juge, D. Cambin & J.-B. Lachavanne, 2002. Does size matter? The relationship between pond area and biodiversity. Biological Conservation 104: 59–70.
Ortega-Mayagoitia, E., X. Armengol & C. Rojo, 2000. Structure and dynamics of zooplankton in a semi-arid wetlands, the National Park Las Tablas de Daimiel (Spain). Wetlands 20: 629–638.
Puchalski, W., 1987. Phytoplankton assemblages in after-exploitation reservoirs. PhD Thesis, Institute of Ecology, Polish Academy of Sciences, Dziekanow Lesny.
Quinn, G. P. & M. J. Keough, 2002. Experimental Design and data analysis for biologists. Cambridge University Press, Cambridge.
Ramsar Convention Bureau, 2005. Strategic framework and guidelines for the future development of the List of Wetlands of International Importance of the Convention on Wetlands (Ramsar, Iran, 1971). Available: http://www.ramsar.org/key_guide_list2006_e.htm
Ribera, I. & P. Aguilera, 1996. Els estanys de Capmany: the missing Spanish pingo (or palsa) fens? Latissimus 7: 2–6.
Romo, S., M. J. Villena, M. Sahuquillo, J. M. Soria, M. Giménez, T. Alfonso, E. Vicente & M. R. Miracle, 2005. Response of a shallow Mediterranean lake to nutrient diversion: does it follow similar patterns as in northern shallow lakes? Freshwater Biology 50: 1706–1717.
Schneider, D. W. & T. M. Frost, 1996. Habitat duration and community structure in temporary ponds. Journal of the North American Benthological Society 15: 64–86.
Serrano, L. & K. Fahd, 2005. Zooplankton communities across a hydroperiod gradient of temporary ponds in the Doñana National Park (SW Spain). Wetlands 25: 101–111.
Stephen, D., D. M. Balayla, E. Bécares, E. Collings, C. Fernández-Aláez, M. Fernández-Aláez, C. Ferriol, P. García, J. Gomá, M. Gyllström, L.-A. Hansson, J. Hietala, T. Kairesalo, M. R. Miracle, S. Romo, J. Rueda, A. Ståhl-Delbanco, M. Svensson, K. Vakkilainen, M. Valentín, W. J. Van de Bund, E. Van Donk, E. Vicente, M. J. Villena & B. Moss, 2004. Continental-scale patterns of nutrient and fish effects on shallow lakes: introduction to a pan-European mesocosm experiment. Freshwater Biology 49: 1517–1524.
Talling, J. F. & D. Driver, 1963. Some problems in the estimation of chlorophyll a in phytoplankton. In Proceedings of the Conference on Primary Productivity Measurement, Marine and Freshwater, 1961, Hawaii: University of Hawaii, USA Atomic Energy Commission, Division of Technical Information TID-7633, 142–146.
ter Braak, C. J. F. & I. C. Prentice, 1988. A theory of gradient analysis. Advances in Ecological Research 18: 271–317.
ter Braak, C. J. F. & P. Šmilauer, 2002. CANOCO Reference manual and CanoDraw for Windows User's guide: Software for Canonical Community Ordination (version 4.5). Microcomputer Power, Ithaca.
Timms, B. V., 1993. Saline lakes of the Paroo, inland New South Wales, Australia. Hydrobiologia 267: 269–289.
Trobajo, R., X. D. Quintana & R. Moreno-Amich, 2002. Model of alternative predominance of phytoplankton-periphyton-macrophytes in lentic waters of Mediterranean coastal wetlands. Archiv für Hydrobiologie 154: 19–40.
Turner, A. M. & J. C. Trexler, 1997. Sampling aquatic invertebrates from marshes: evaluating the options. Journal of North American Benthological Society 16: 694–709.
Valladares, L. F., J. Garrido & F. García-Criado, 2002. The assemblages of aquatic Coleoptera from shallow lakes in the northern Iberian Meseta: Influence of environmental variables. European Journal of Entomology 99: 289–298.
Wellborn, G. A., D. K. Skelly & E. E. Werner, 1996. Mechanisms creating community structure across a freshwater

habitat gradient. Annual Review of Ecology and Systematics 27: 337–363.
Whilh, J. L. & T. C. Dorris, 1968. Biological parameters of water quality. Bioscience 18: 477–481.
Wiggins, G. B., 1973. A contribution to the biology of the caddisflies (Trichoptera) in temporary pools. Royal Ontario Museum Life Sciences Contributions 88: 1–28.
Wiggins, G. B., R. J. Mackay & I. M. Smith, 1980. Evolutionary and ecological strategies of animals in annual temporary pools. Archiv für Hydrobiologie Supplement 58: 97–206.
Williams, D. D., 1996. Environmental constraints in temporary fresh waters and their consequences for the insect fauna. Journal of the North American Benthological Society 15: 634–650.
Williams, D. D. & N. E. Williams, 1998. Aquatic insects in an estuarine environment: densities, distribution and salinity tolerance. Freshwater Biology 39: 411–421.

Hydrobiologia (2008) 597:71–83
DOI 10.1007/s10750-007-9215-x

Relation between macroinvertebrate life strategies and habitat traits in Mediterranean salt marsh ponds (Empordà wetlands, NE Iberian Peninsula)

Stéphanie Gascón · Dani Boix · Jordi Sala · Xavier D. Quintana

Abstract The influence of water permanence and high intra- and inter-annual hydrological variability on macrobenthos (organisms >1 mm) was studied using a taxonomical and a functional approach. The study was carried out in a Mediterranean salt marsh. Monthly samples of macrobenthic fauna were collected during two consecutive hydroperiods from six ponds with different water permanence (temporary, semi-permanent and permanent waters). Organisms were assigned to five functional response groups based on life-strategies according to their capacity to survive desiccation events, their dispersion capability and the necessity of water for their reproduction. Results from both approaches showed that the benthic community was more related to pond type than to intra- and inter-annual variability. The second aim was to analyse to which extent patterns in functional groups were determined by the existence of succession patterns or to environmental variability. In this sense, a clear succession pattern was not observed. In contrast, in most of the functional groups (4 out of 5), species within each functional group showed similar responses to water fluctuations. However, species of the fifth functional group, which comprised species without any particular adaptation to desiccation survival or avoidance, showed different responses to water level fluctuations.

Keywords Functional group · Water permanence · Hydroperiod · Habitat type · Temporal variability · Generalized Additive Model

Introduction

The relationship between species and environmental conditions has been traditionally studied using multivariate analyses (e.g. correspondence analysis, canonical analysis, multi-dimensional scaling), usually based on a taxonomical approach (e.g. ter Braak, 1987; Birks et al., 1994). The results of multivariate analyses have also been related to specific concepts of theoretical succession and predictive modelling (Jassby & Powell, 1990; Mesléard et al., 1991; Noest, 1991; Quintana, 2002a; Boix et al., 2004). The development of powerful statistic techniques has led to an expansion of predictive habitat distribution models (Guisan & Zimmermann, 2000). These are used to predict species responses to environmental fluctuations or to find similarities/generalities in these responses. For example, Generalized Additive Models (GAMs) provide an interesting extension to Generalized Linear Models (GLMs), since they allow

Guest editors: R. Céréghino, J. Biggs, B. Oertli & S. Declerck
The ecology of European ponds: defining the characteristics of a neglected freshwater habitat

S. Gascón (✉) · D. Boix · J. Sala · X. D. Quintana
Institute of Aquatic Ecology, University of Girona,
Campus de Montilivi, Facultat de Ciències, Girona 17071,
Spain
e-mail: stephanie.gascon@udg.edu

linear and non-linear response shapes, for both continuous and categorical variables, and for a combination of those within a single model (Hastie & Tibshirani, 1990). Recently, several studies have used GAMs to model macroinvertebrate species response to environmental variables (Castella et al., 2001; Horsák, 2006).

Ecosystem level processes are affected by the functional characteristics of the organisms involved, rather than by their taxonomic identity (Hooper et al., 2002). Functional groups have been defined as sets of species showing either similar responses to the environment or similar effects on major ecosystem processes (Gitay & Noble, 1997). Thus, two types of functional groups can be used: (1) functional effect groups, which are used when the goal is to investigate the effects of species on ecosystem properties (e.g. trophic groups); and (2) functional response groups, which are used when the goal is to investigate the response of species to changes in the environment, such as disturbance, resource availability or climate (e.g. life strategies). Studies of macrobenthic fauna based on functional effect groups, mainly covering feeding strategies, are very common from lotic systems (e.g. Goodman et al., 2006; Tomanova et al., 2006), while studies focusing on functional response groups are less abundant (e.g. Wiggins et al., 1980; Usseglio-Polatera et al., 2000). However, in lotic systems, some recent studies have used functional response groups to show community responses to some environmental factors (e.g. salinity; Piscart et al., 2004). On the other hand, studies on functional response groups based on life-strategies in lentic systems have been mainly used to see succession patterns in temporary waters (e.g. Williams, 1985; Bazzanti et al., 1996), and not to study community response to environmental variability.

Environmental variability is especially high in Mediterranean wetlands, where irregularity and unpredictability of hydrological patterns are well known (Quintana et al., 1998a; Álvarez-Cobelas et al., 2005). A flooding and a drying phase are distinguishable at intra-annual level, with significant effects on nutrient concentration, species composition and even in competitive and predatory interactions (Quintana et al., 1998b; Serrano et al., 1999; Boix et al., 2004). Inter-annual variability is also high and some studies have reported a higher variation of zooplankton assemblages at inter-annual than at intra-annual level (Serrano & Fahd, 2005). Inter-annual changes due to unusual drought events can have drastic effects on temperature, nutrient contents and invertebrate assemblages in Mediterranean wetlands (Angeler et al., 2002; Gascón et al., 2007). Furthermore, habitat type due to differences in water permanence has been shown to determine benthic species composition, and differences in benthos (Gascón et al., 2005) and zooplankton (Brucet et al., 2005; Serrano & Fahd, 2005) assemblages in these Mediterranean environments. Using a taxonomical approach, differences in benthic communities have already been found to be mainly related to water level fluctuations of the ponds (Gascón et al., 2005), but it remains to be established whether these hydrological patterns determine the functional response of the species involved.

Our objectives were to investigate if macrobenthic communities are influenced by pond type (classified according to water permanence), and high inter- and intra-annual hydrological variability, and to compare if observed effects are similar depending on whether a taxonomical or a functional approach is used. Secondly, we will analyse to which extent patterns in functional response groups, related to life-strategies, are determined by the existence of succession patterns or by environmental variability, as a function of water level fluctuations. Water level was chosen as target factor since differences in water level are related to water permanence, and also to (intra- and inter-annual) temporal variability.

Material and methods

Study site

Empordà wetlands salt marshes are located close to the Mediterranean Sea, in the northeast of the Iberian Peninsula, and are free from tidal influence. The hydrology is characterized by sudden and irregular flooding (caused by rainfall, inputs from rivers or channels and sea storms), followed by dry periods, when most of the ponds become isolated and gradually dry out (Quintana, 2002b). Thus, two phases can be differentiated: a flooding phase from autumn to winter, and a drying phase from spring to summer. Note that the study site is characterized by a high connectivity, since flooding events usually connect all ponds (Quintana, 2002b).

The ponds under study are depressions occurring between sand bars in salt marshes, where water accumulates. Their depth varies greatly depending on the flooding regime, but they are usually less than 1 m deep. The ponds are flat, not vegetated and they do not communicate directly with the sea. They differ in water permanence, from temporary to permanent ponds. Following the classification of temporary ponds made by Wiggins (1973), the studied ponds which dry out completely are similar to temporary autumnal ponds, since they have a wet phase of approximately 9 months and a dry phase of 3 months, the wet phase starting in autumn (Brucet et al., 2005). However, during the study the wet phase was unusually short, due to intrinsic interannual variability of the Mediterranean climate (Table 1).

Sampling procedure and processing

The study was undertaken from November 1997 until July 1999 during two hydroperiods (from November 1997 to July 1998, and from November 1998 to July 1999) in six ponds of Empordà wetlands salt marshes (Fig. 1). The ponds were grouped into pond types according to their water permanence, size and landscape variables. To obtain landscape variables (isolation, waterbodies density and predominance of aquatic habitat), freely available aerial photographs were used (DPTOP, 2005; MAPA, 2006). Isolation of the wetlands was calculated as the distance (m) to the nearest pond, the waterbodies density as the number of waterbodies within a radius of 500 m from the waterbody and the predominance of aquatic habitat as the proportion of water surface in a square kilometre centred in the waterbody. Ponds which were connected during most of the hydroperiod, were considered in the same pond type. Thus, three pond

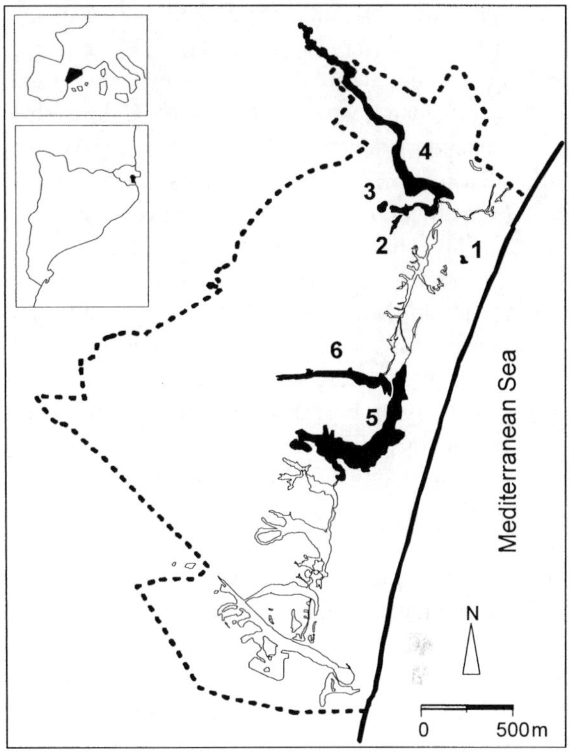

Fig. 1 Map of the study site, indicating the basins studied in black: 1 and 3 are temporary ponds, 2 and 4 semi-permanent ponds and 5 and 6 permanent ponds. The discontinuous line indicates the limit of the integral reserve of the Empordà Wetlands Natural Park

Table 1 Water and landscape characteristics of the ponds during the study (mean; range in brackets and number of samples in italics)

	Temporary ponds	Semi-permanent ponds	Permanent ponds
Pond	1 and 3	2 and 4	5 and 6
Surface (m^2)	(983–1,649)	(3,236–55,837)	(87,369–19,151)
Isolation (m)	(90–10)	(10–35)	(10–10)
Waterbodies density (n° of waterbodies/500 m radius)	(11–11)	(11–11)	(5–5)
Predominance of aquatic habitat (%)	(5.3–6.3)	(5.9–6.3)	(13.2–15.4)
Hydroperiod mean length (month/year)	7	7	12
Water level (cm)	41.8 (0–108.6), *40*	45.8 (0–111.8), *37*	54.3 (21.7–109.4), *37*
Water temperature (°C)	16.0 (5.9–27.5), *38*	17.5 (4.9–26.8), *35*	15.3 (7.2–24.0), *36*
Water conductivity (mS cm^{-1})	35.3 (20.8–73.2), *38*	30.3 (13.0–41.3), *35*	22.1 (14.0–38.8), *36*

Hydroperiod mean length was observed during the study period types were differentiated by water permanence, size and landscape variables (Table 1): (1) temporary ponds (T) dry out completely every year, have small sizes and the landscape is characterized by a low predominance of aquatic habitat, and a high waterbodies density (ponds 1 and 3); (2) semi-permanent ponds (SP) do not dry out completely every year, and have similar characteristics of predominance of aquatic habitat and waterbodies density than those observed in temporary ponds, but they have intermediate sizes (ponds 2 and 4); (3) permanent ponds (P) never dry out completely, and in contrast with the rest of the ponds, have bigger sizes, a higher predominance of aquatic habitat and the lowest waterbodies density (ponds 5 and 6). Water level was measured monthly using a graduated gauge firmly fixed on the bottom of the ponds. Water temperature and conductivity were measured monthly in situ.

Monthly samples of macroinvertebrates (organisms >1 mm) were obtained using an Ekman grab (225 cm^2), taking two replicates for the smallest ponds (1 and 3) and four replicates for the bigger ponds (2, 4, 5 and 6). Sampling was carried out until the sampling site dried out. In the case of permanent ponds, sampling sites were located at levels which also dried out due to high water level fluctuation. Organisms were separated alive from the sediment using a 1 mm mesh-size sieve and preserved in 4% formalin until taxonomic identification. All individuals were identified to species whenever it was possible. Abundances were estimated by counting all individuals retained on the sieve. Following Wiggins et al.'s (1980) classification and according to the literature (Wiggins et al., 1980; Takeda & Nagata, 1998; Herbst, 1999; Tachet et al., 2002), taxa were grouped in four life strategies. A new group different from those established was added, comprising species without any particular adaptation to avoid or survive desiccation (Table 2).

Table 2 Characteristics of each life-strategy group

	Type of dispersion	Need of water to reproduce	Can survive desiccation in the basin
G1	Passive	Yes	Yes
G2	Active	Yes	Yes
G3	Active	No	Yes
G4	Active	Yes	No
G5	Passive	Yes	No

Groups 1, 2, 3 and 4 correspond to the classification made by Wiggins et al. (1980); group 5 was created to include organisms with passive dispersion and without adaptations to desiccation

Statistical analyses

Environmental data

Differences in water level, temperature and conductivity among different pond types (temporary, semi-permanent and permanent ponds), between phases (flooding vs. drying phases) and hydroperiods (1997–1998 vs. 1998–1999) were analysed using a multivariate analysis of variance (MANOVA). To ensure homogeneity of the variances, water level was log transformed [log (variable + 1)]. Initially, MANOVA was run with the most complex model, introducing all possible interactions. Then, to increase statistical power, the model was simplified by removing non-significant interactions, identified using Pillai's statistic ($P > 0.05$). The relationship between species richness and hydroperiod length was calculated using Spearman's correlation. Species richness values were obtained for each hydroperiod and pond, and were related to the number of weeks in which ponds remained flooded for each hydroperiod. The relationship between species richness and pond surface was analysed using Pearson's correlation after log transforming all variables [log (variable + 1)]. Species richness values corresponded to accumulative richness from this study and were related to the surface of each pond. All statistical analyses were performed using SPSS 11.5.1 for Windows.

Pond type and temporal variability

For the taxonomical approach, a canonical correspondence analysis (CCA) using CANOCO 4.5 (ter Braak & Šmilauer, 2002) was performed to study temporal variability (between phases and hydroperiods) and pond type (water permanence) influence on macrobenthic community. Water permanence has been chosen to summarize pond type characteristics, since all the variables that describe pond types are highly related. The rest of pond type descriptors have been included as supplementary variables. Only taxa which appeared in more than one sample were considered in the analysis (Table 3). The species-abundance matrix was squareroot transformed. Following ter Braak & Šmilauer (2002), we downweighted rare species to reduce their influence in the analysis. Forward-selection procedure available in CANOCO 4.5 was used to obtain the conditional effect for each variable, and the significance of the explanatory effect for each variable was evaluated using a Monte Carlo permutation test (ter Braak & Šmilauer, 2002).

For the functional approach, pond type and temporal (phases and hydroperiods) variability for each functional group were analysed using generalized linear models (GLMs), instead of ANOVA, which it is not appropriate for count data. GLM were calculated for each functional group using a Poisson distribution as error and the log link function, which ensures that all the fitted values are positive. The significance of the different explanatory variables

Table 3 Abundance (ind · 10 cm^{-2}) of the 21 most frequent taxa (occurrence higher than 1%)

Taxa	Life strategy group	Temporary ponds	Semi-permanent ponds	Permanent ponds
Gastropoda				
Hydrobia acuta (Draparnaud, 1805)	5	0.01 (0.00–0.06)	0.01 (0.00–0.10)	<0.01 (0.00–0.03)
Polychaeta				
Nereis diversicolor Müller, 1776	5	0.24 (0.00–2.46)	0.11 (0.00–0.80)	0.20 (0.00–1.36)
Streblospio shrubsolii (Buchanan, 1890)	5	–	–	0.05 (0.00–0.22)
Oligochaeta				
Paranais sp.	1[a]	–	<0.01 (0.00–0.02)	0.01 (0.00–0.17)
Nais sp.	1[a]	–	–	<0.01 (0.00–0.03)
Tubificidae undet.	1[a]	<0.01 (0.00–0.09)	<0.01 (0.00–0.02)	<0.01 (0.00–0.01)
Malacostraca				
Leptocheirus pilosus Zaddach, 1844	5	–	–	<0.01 (0.00–0.02)
Corophium orientale Schellenberg, 1928	5	–	<0.01 (0.00–0.01)	0.63 (0.00–7.34)
Gammarus aequicauda (Martynov, 1931)	5	0.01 (0.00–0.29)	0.09 (0.00–1.56)	<0.01 (0.00–0.03)
Orchestia sp.	5	<0.01 (0.00–0.02)	<0.01 (0.00–0.01)	–
Lekanesphaera hookeri (Leach, 1814)	5	–	<0.01 (0.00–0.02)	0.01 (0.00–0.06)
Coleoptera				
Hydroporus planus (Fabricius, 1781)	2[a]	0.01 (0.00–0.11)	<0.01 (0.00–0.02)	–
Enochrus bicolor (Fabricius, 1792)	4[a]	<0.01 (0.00–0.06)	–	–
Diptera				
Culicoides sp.	2[a]	<0.01 (0.00–0.06)	<0.01 (0.00–0.02)	<0.01 (0.00–0.04)
Halocladius varians (Staeger, 1839)	2[a]	0.13 (0.00–2.70)	0.01 (0.00–0.15)	<0.01 (0.00–0.05)
Chironomus gr. salinarius Kieffer, 1915	2[a]	2.29 (0.00–8.62)	0.35 (0.00–1.82)	0.05 (0.00–0.82)
Dolichopodidae undet.	3[b]	<0.01 (0.00–0.02)	<0.01 (0.00–0.02)	<0.01 (0.00–0.04)
Chloropidae undet.	3[c]	–	<0.01 (0.00–0.04)	<0.01 (0.00–0.02)
Ephydra sp.	4[d]	0.04 (0.00–0.86)	<0.01 (0.00–0.04)	–
Scatella sp1	4[d]	<0.01 (0.00–0.06)	–	0.01 (0.00–0.20)
Scatella sp2	4[d]	–	–	<0.01 (0.00–0.02)
Richness without rare taxa		13	15	17
Total richness		(18)	(17)	(21)

Richness without rare taxa (occurrence <1%) and, in brackets, total richness are also shown. References are indicated by superscript: (a) Wiggins et al., 1980; (b) Tachet et al., 2002; (c) Takeda & Nagata, 1998 and (d) Herbst, 1999 was checked through comparisons of changes in deviance and degrees of freedom with a Chi-square distribution (Crawley, 2002). GLM analyses were carried out with S-PLUS 2000.

Species responses

Generalized Additive Models (GAMs) were used to analyse to which extent patterns in functional response groups were determined by the existence of succession patterns related to life-strategies (functional response groups), or to environmental variability summarized by water level fluctuations. We used the log link function and Poisson distribution as error. Two degrees of freedom were specified to obtain a low complexity model (i.e. a more general pattern). A stepwise selection using the Akaike Information Criterion (AIC) was used to select the best model with increasing complexity (degrees of freedom equal to 1 and 2). GAM analyses were performed with CANOCO 4.5 (ter Braak & Šmilauer, 2002).

Results

Environmental variability

During the study, semi-permanent ponds dried out completely (Fig. 2). Thus, temporary and semi-permanent ponds had similar hydroperiod length. There were no significant differences in water level ($F_{2,100} = 1.04$; $P = 0.357$) and temperature values ($F_{2,100} = 7.06$; $P = 0.407$) among different pond types. In contrast, significant differences were found in conductivity values ($F_{2,100} = 18.27$; $P < 0.001$): temporary ponds had higher values than semi-permanent ponds, and semi-permanent ponds had higher values than permanent ponds (Table 1, Fig. 3).

Significant temporal variability was detected for water level, conductivity and temperature values. At intra-annual level (between phases), significant lower water level ($F_{1,100} = 8.78$; $P < 0.05$) and higher temperature values ($F_{1,100} = 172.92$; $P < 0.001$) were observed during the drying phase (Fig. 3). At inter-annual level (between hydroperiods), significant differences were also found. Water level values ($F_{1,100} = 14.79$; $P < 0.001$) and temperature values ($F_{1,100} = 16.19$; $P < 0.05$) were significantly higher during the first hydroperiod. However, the temporal variability of conductivity values was more complex, due to the significant interactions found between phases and hydroperiods ($F_{1,100} = 18.46$; $P < 0.001$). Thus, while conductivity values remained similar during the flooding phases for both hydroperiods, an increase in conductivity values was observed during the drying phase only in the second hydroperiod (Fig. 3).

Pond type and temporal variability

Taxonomic approach

Eight of the 21 most frequent taxa (occurrence >1) were found in all pond types (Table 3). In contrast,

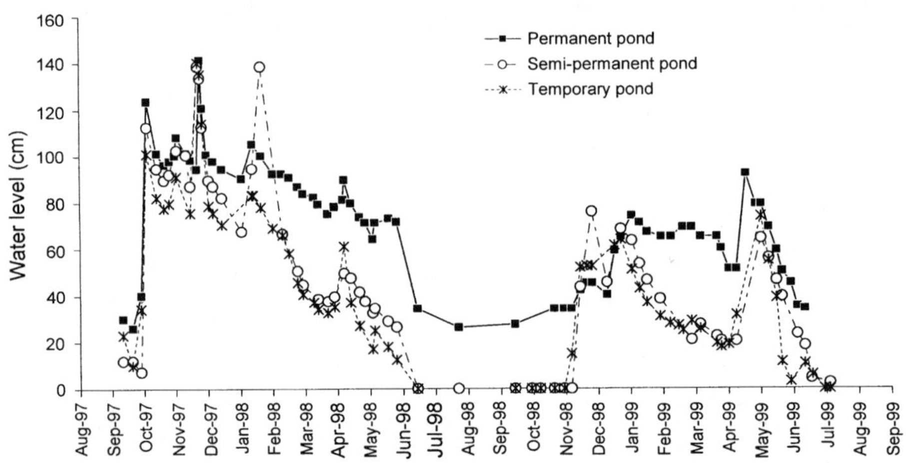

Fig. 2 Water level fluctuation (cm) in the three pond types. The values correspond to examples of each pond type: pond 1 for temporary, pond 4 for semi-permanent and pond 5 for permanent pond

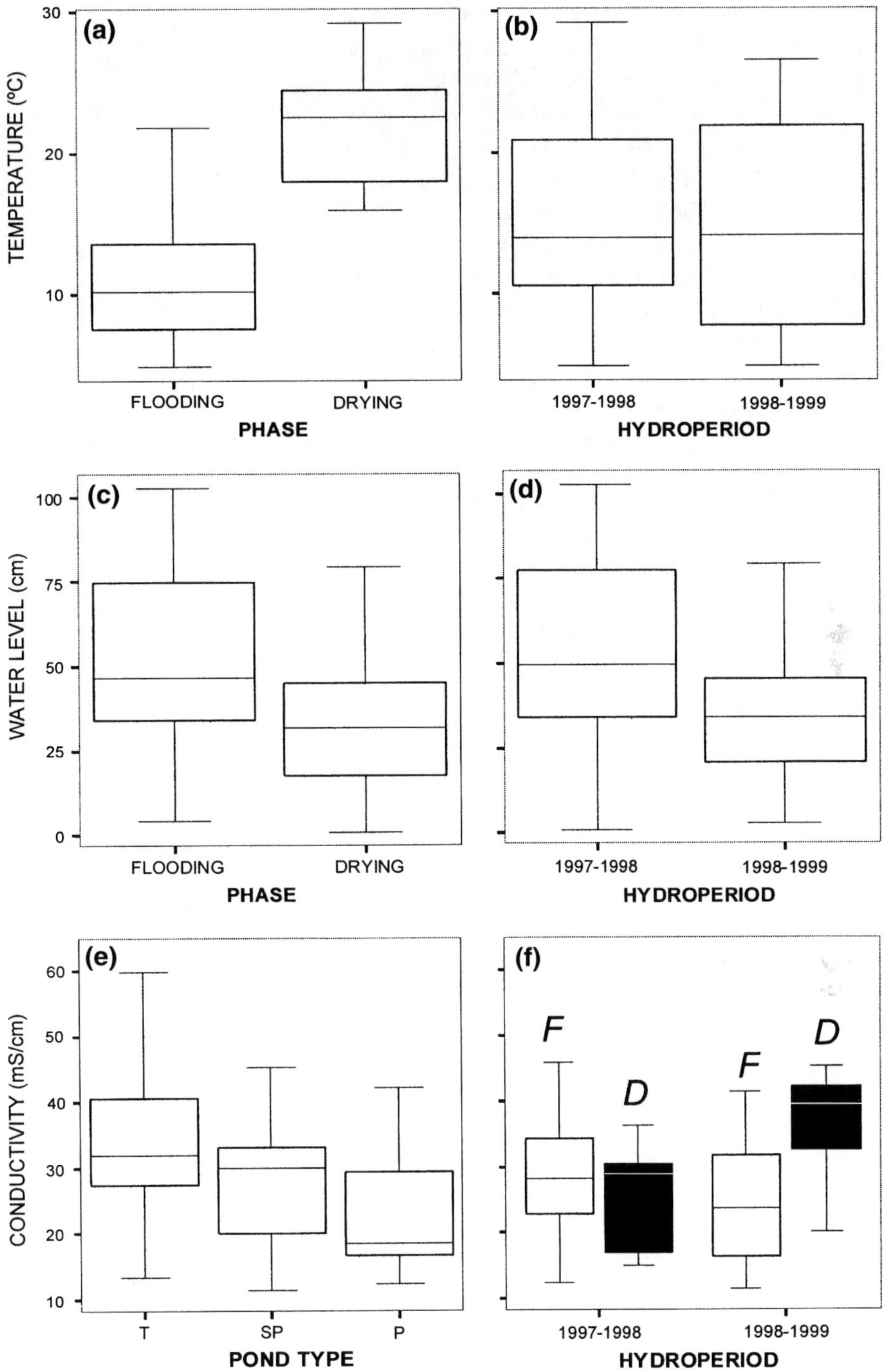

Fig. 3 Box plots showing significant differences of environmental variables among factors after a MANOVA. (**a**) Temperature vs. phases, (**b**) Temperature vs. hydroperiod, (**c**) Water level vs. phases, (**d**) Water level vs. hydroperiod, (**e**) Conductivity vs. pond type (T—temporary, SP—semi-permanent, P—permanent waters) and (**f**) Conductivity vs. phases (*F*—flooding, *D*—drying) within each hydroperiod

only 5 taxa were exclusive of one pond type: 1 taxa was exclusive of temporary ponds (the coleopteran *Enochrus bicolor*), and the other 4 taxa were exclusive of permanent ponds (the polychaeta *Streblospio shrubsolii*, the oligochaeta *Nais* sp., the amphipod *Leptocheirus pilosus* and the dipteran *Scatella* sp2). No taxon was found to be exclusive of semi-permanent ponds. Species richness was similar among pond types, for both frequent and all taxa (Table 3). Additionally, no significant relationship was found between hydroperiod length and species richness. Similarly, no significant relationship was found between pond surface and species richness.

Twenty-one taxa (those with occurrence >1) were used in the CCA. The first two axes of the CCA explained 16.2% of the total variability observed in the species dataset. The first axis explained 12.4% of the total variance, while the second axis explained 3.8% of the total variance. The explanatory variable which fitted best the species variability was the water permanence which is a characteristic of the different pond types (conditional effect = 0.46; $F = 11.94$; $P = 0.001$). Hydroperiod was the second variable which also significantly fitted the variability observed in the species data set (conditional effect = 0.14; $F = 3.68$; $P = 0.001$), whereas phases did not improve significantly the explanation of the species variability, as it is shown by the non-significant result of the permutation test (conditional effect = 0.04; $F = 1.05$; $P = 0.407$). Samples of permanent ponds were mostly found in positive coordinates of the first axis, while samples of temporary and semi-permanent ponds were found in negative coordinates (Fig. 4). The relation among landscape variables and water permanence is also shown in Fig. 4. Differences between hydroperiods (inter-annual variability) were related to the second canonical axis. Thus, the centroid for the second hydroperiod is found in positive coordinates of this axis, while the centroid for the first hydroperiod is found in negative coordinates (Fig. 4).

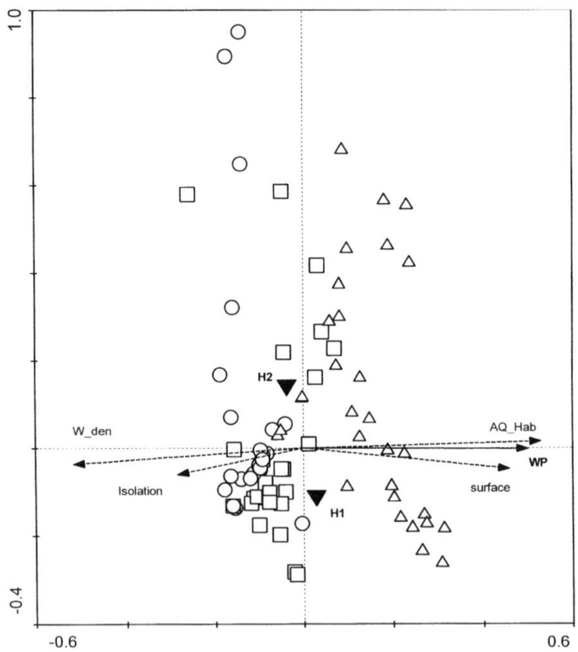

Fig. 4 Weighted average scores of the samples for the different pond types (up-triangles stand for permanent ponds; squares for semi-permanent ponds and circles for temporary ponds) in the first two canonical axes. Only variables with significant relation to species dataset are shown in bold: down-triangles showed centroid position for the hydroperiods (H1: 1997–1998; H2: 1998–1999), solid arrow indicates water permanence (WP). Dashed arrows indicate the position of landscape variables included as supplementary variables: isolation (isolation), Waterbodies density (W_den), surface (surface), predominance of aquatic habitat (AQ_Hab)

Functional approach

Species of the five functional groups were observed during the study, but some groups were better represented than others. Groups 2 and 5 had higher abundances (mean ± standard deviation: 0.904 ± 1.899 and 0.464 ± 1.045 ind $\cdot$ 10 cm^{-2}, respectively), and different species richness, with higher values in group 5 (4 and 8 taxa, respectively). Groups 1 and 4 had lower abundances (mean ± standard deviation: 0.007 ± 0.032 and 0.018 ± 0.101 ind $\cdot$ 10 cm^{-2}, respectively), and similar richness (3 and 4 taxa, respectively). Finally, group 3 had the lowest abundance (mean ± standard deviation: 0.003 ± 0.008 ind $\cdot$ 10 cm^{-2}) and richness (2 taxa).

Significant differences among functional groups, and pond type and temporal (phase and hydroperiod) factors were only found for those groups that had higher abundances (Table 4). Pond type explained the highest proportion of variation in both groups (48.8% and 17.4% for group 2 and 5, respectively). However, they showed a different pattern: while group 2 had higher abundances in temporary ponds, group 5 had higher abundances in permanent ponds (Fig. 5). Differences in temporal variability were also found, but they were less important (less proportion of variation). Group 2 showed significant differences between phases (accounting for 9.5% of the variation), and group 5 had significant differences between hydroperiods (accounting for 4.5% of the variation; Table 4). Group 2 had higher abundances during the drying phase than the flooding one, and group 5 had higher abundances during the first hydroperiod (Fig. 5). Significant interactions were found between phase and hydroperiod (accounting for 6% of the variation) for group 2, and between hydroperiod and

Table 4 Results of the GLM on the influence of phase (flooding, drying), hydroperiod (1997–1998, 1998–1999) and pond type (temporary, semi-permanent, permanent ponds) in the abundance of each life-strategy group studied

Factor	df	G1	G2	G3	G4	G5
Phase (P)	1	0.2	**9.5*****	4.4	6.8	0.7
Hydroperiod (H)	1	8.7	0.4	0.1	11.6	**4.5***
Pond type (Pt)	2	6.7	**48.8*****	6.3	16.5	**17.4****
P × H	1	1.2	**6.0****	1.4	0.0	1.5
P × Pt	2	0.0	1.3	5.8	18.3	1.7
H × Pt	2	10.1	2.9	2.4	2.5	**8.3***
P × H × Pt	2	1.5	0.9	9.3	0.0	2.3
Error	74					
Total % deviance		28.4	69.7	29.6	55.8	36.4
Dispersion parameter		0.001	0.833	0.010	0.058	0.800

Values are the percentage of the total deviance explained by each factor. Significant results appear in bold. Asterisks show the significance level (*$P < 0.05$, **$P < 0.001$, ***$P < 0.0001$)

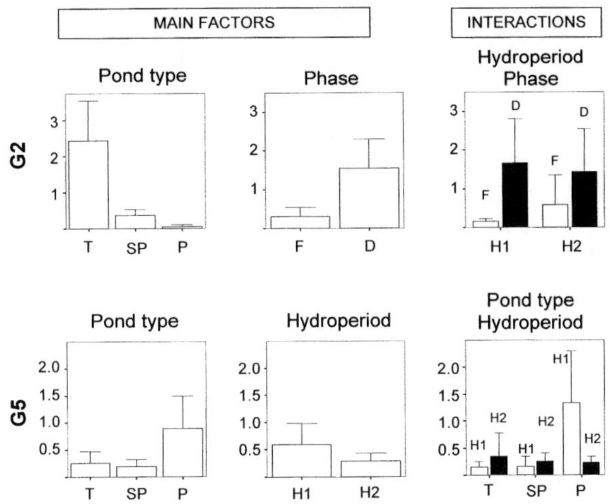

Fig. 5 Abundances of life-strategy groups 2 and 5 related to their significant factors after a GLM analysis: pond types (T: temporary, SP: semi-permanent and P: permanent ponds), hydroperiod phases (F: flooding and D: drying phases) and hydroperiod (H1: 1997–1998 and H2: 1998–1999). Significant interactions are also shown: in group 2, white bars correspond to flooding phase and black bars correspond to drying phase; in group 5, white bars correspond to the first hydroperiod (1997–1998) and black bars correspond to the second hydroperiod (1998–1999)

pond type (accounting for 8.3% of the variation) for group 5. In this sense, the differences observed between the two phases during the second hydroperiod in the abundances of group 2 were lower than between the two phases of the first hydroperiod (Fig. 5). Group 5 showed differences between hydroperiods among pond types: for more temporary waters (temporary and semi-permanent ponds), group 5 was more abundant during the second hydroperiod, while the opposite pattern was observed in permanent ponds, where it was more abundant during the first hydroperiod (Fig. 5).

Species responses

All functional groups had species that presented a significant response to the GAMs (Figs. 6, 7). The species responses to time (days after flooding) were not similar within each functional group (Fig. 6). Thus, there was not a similar succession pattern for taxa within the same functional group.

In contrast, the GAM results for the first four functional groups showed that species of the same functional group had similar responses to water level variation (Fig. 7). Therefore, taxa from group 1 had a unimodal response to water level variation, and their maximum abundances were expected at high water level values. Taxa of group 2 had their maximum abundances in low water level values, but not as low as those observed to be expected for taxa of groups 3 and 4. Finally, different taxa in group 5 had differing response curves. For example, the snail *Hydrobia acuta* showed a unimodal response with higher abundances at intermediate values of water level, while the amphipod *Corophium orientale* reached its maximum abundance at the highest water level values.

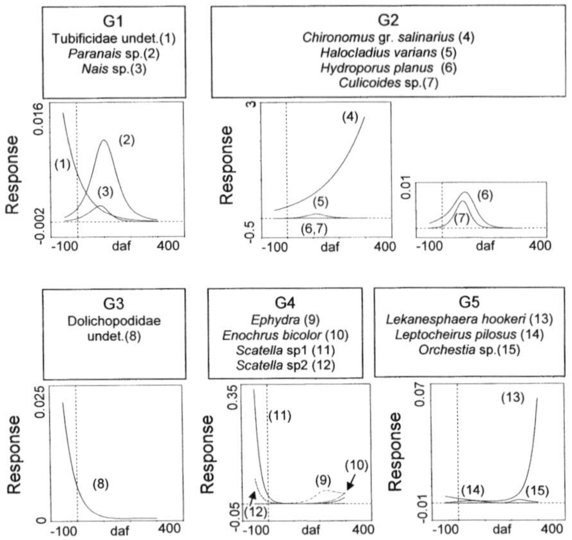

Fig. 6 Species responses to time (days after flooding (daf)) fitted with the Generalized Additive Model

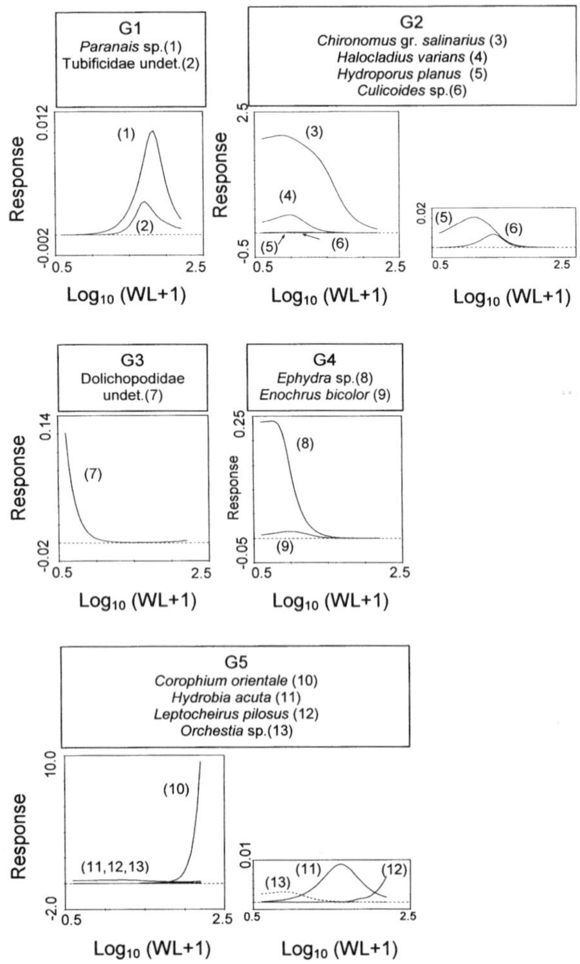

Fig. 7 Species responses to water level (WL) fitted with the Generalized Additive Model

Discussion

The functional and the taxonomical approaches showed similar results in relation to pond types, showing that permanent ponds had a different benthic composition than more temporary ones. Previous studies developed in the Empordà salt marshes already noted the importance of water permanence for benthic assemblages (Gascón et al., 2005). Similarly, it has been shown that water permanence also plays an important role in the structure and composition of benthic assemblages in other aquatic systems (Schneider, 1999; Schwartz & Jenkins, 2000). However, as water permanence covaried with other pond characteristics, it is not possible to ascertain that water permanence was only responsible to the observed differences.

Life histories of species have been frequently related to environmental characteristics in both lentic and lotic systems (e.g. Williams, 1985; Richards et al., 1997). In Empordà salt marshes, life history traits determine species preferences for ponds with a certain water permanence. Thus, organisms adapted to avoid desiccation, with active dispersion and needing water for reproduction (group 2), seemed to select temporary ponds, since they always present higher abundances in these environments. In contrast, taxa which are less adapted to desiccation (group 5), without active dispersion and lacking strategies to endure dry periods, select permanent ponds to ensure their survival. In this sense, Wiggins et al. (1980) described that taxa belonging to group 2 (mainly chironomids, which were the most abundant taxa of this group in our study site) can successfully inhabit ponds which start their inundation in autumn and remain with water until summer, since these organisms spend the winter under water, and hence the winter stress is reduced. Wiggins et al. (1980) also described that in fluctuating waters which never dry out, species of all functional response groups are expected to be found. However, in Empordà salt marshes group 5 species were the dominant taxa in permanent water ponds.

The lack of differences between semi-permanent and temporary waters could be explained by the fact that during the study these ponds had a similar hydroperiod length. This is in accordance with other studies, which have shown that wetland hydroperiod length influences invertebrate community composition and structure (e.g. Schneider & Frost, 1996; Wellborn et al., 1996). Other authors pointed out the effects of hydroperiod length on species richness (e.g. Brooks, 2000; Tarr et al., 2005), but our study showed no such relation. Furthermore, the lack of a species richness-area relationship could be explained by the dry out constraint, since in our study site, short hydroperiod length also occurred in large ponds (e.g. basin 4). This is in agreement with other studies where water permanence has been shown to be more important for species richness than pond surface area (e.g. Schneider & Frost, 1996; Della Bella et al., 2005), although results stating otherwise also exist (March & Bass, 1995).

Inter-annual variability was observed using both approaches. However, only one functional group, comprising organisms without any particular adaptation to survive or avoid desiccation (group 5), presented significant differences between hydroperiods. This was even clearer in permanent ponds, which have a characteristic benthic community not found in more stressed environmental conditions (Gascón et al., 2005). Group 5 had lower abundance during the second hydroperiod which was characterized by more extreme environmental conditions (the temperature during the flooding phase was significantly cooler; and water level values were significantly lower during the second hydroperiod, indicating less water inputs). Thus, the decrease in abundance of group 5 may respond to the higher environmental stress occurring during the second hydroperiod.

Non-taxonomic aggregations based on life history are a better approach to determine seasonal variability, since seasonal patterns of occurrence and abundance in invertebrates depend on their life history and behavioural characteristics (Wolda, 1988; Bêche et al., 2006). Supporting this idea, we observed intra-annual variability using functional groups, whereas differences were not observed using a taxonomical approach. However, this temporal pattern seems to be more related to water level fluctuations than to a succession pattern. Bazzanti et al. (1996) found the highest importance of group 2 taxa at lower water level values. Similarly, during our study group 2 taxa had significantly higher abundances during the drying phase, which was characterized by significantly lower water level values.

Macroinvertebrate assemblages in Empordà wetlands were highly dominated by few species, which is a typical situation in Mediterranean salt marshes due to their marked daily and seasonal variations in physical and chemical parameters (Guelorget & Perthuisot, 1983; Victor & Victor, 1997; Mistri et al., 2001; Reizopoulou & Nicolaidou, 2004; Álvarez-Cobelas et al., 2005). Thus, our results might be conditioned by particular life history traits of the dominant species. This possible artefact may be easily discarded, since similar response patterns of species within each functional group were observed in four of the five life-strategy groups studied. Only group 5 presented different responses for each species. This group is composed of species with a complete dependence on water permanence. Thus, their different responses could be explained by their preference for other factors which covary with water level, such as water conductivity (Pearson correlation; $r = -0.249$, $P < 0.05$) or by a different degree of desiccation tolerance.

On the other hand, the functional response groups 1, 2, 3 and 4 described by Wiggins et al. (1980) were created to group organisms with similar life history traits, but all share some independence of water presence, as they are adapted to avoid or survive desiccation. Hence, the response curves indicate their preference for a specific water level. In this sense, groups with active dispersion colonize faster new ponds and can remain in the pond until the end of the wet phase (Barnes, 1983; Ribera et al., 1994). Our results support this idea, since all groups with active dispersion are expected to be abundant with low water levels, and these levels occurred at the start and end of the hydroperiods. Moreover, our study revealed that organisms adapted to avoid desiccation, with active dispersion and needing water for reproduction (group 2) preferred temporary ponds, since they always presented higher abundances in this environment.

In summary, the abundance of species with a particular life-strategy was influenced by environmental variability such as water fluctuations. Moreover, species without any adaptation to survive or avoid desiccation did not show similar responses to water fluctuations. Thus, the inter-annual differences observed in this functional response group are more determined by the life-strategy of the dominant taxon (the amphipod *Corophium orientale*) than by a general functional pattern extended to the rest of the species of this group.

Acknowledgements This work was supported by a grant from the Comisión de Investigación Científica y Técnica (CICYT), Programa de Recursos Naturales (ref. CGL2004-05433/BOS).

References

Álvarez-Cobelas, M., C. Rojo & D. G. Angeler, 2005. Mediterranean limnology: current status, gaps and the future. Journal of Limnology 64: 13–29.

Angeler D. G., M. Alvarez-Cobelas, S. Sánchez-Carrillo & M. A. Rodrigo, 2002. Assessment of exotic fish impacts on water-quality and zooplankton in a degraded semi-arid floodplain wetland. Aquatic Science 64: 76–86.

Barnes, L. E., 1983. The colonization of ball-clay ponds by macroinvertebrates and macrophytes. Freshwater Biology 13: 561–578.

Bazzanti, M., S. Baldoni & M. Seminara, 1996. Invertebrate macrofauna of a temporary pond in Central Italy: composition, community parameters and temporal succession. Archiv für Hydrobiologie 137: 77–94.

Bêche, L. A., E. P. McElravy & V. H. Resh, 2006. Long-term seasonal variation in the biological traits of benthic-macroinvertebrates in two Mediterranean-climate streams in California, U.S.A. Freshwater Biology 51: 56–75.

Birks, H. J. B., S. M. Peglar & H. A. Austin, 1994. An Annotated Bibliography of Canonical Correspondence Analysis and Related Constrained Ordination Methods 1986–1993. Botanical Institute, University of Bergen, Bergen.

Boix, D., J. Sala, X. D. Quintana & R. Moreno-Amich, 2004. Succession of the animal community in a Mediterranean temporary pond. Journal of the North American Benthological Society 23: 29–49.

Brooks, R. T., 2000. Annual and seasonal variation and the effects of hydroperiod on benthic macroinvertebrates of seasonal forest ("vernal") ponds in central Massachusetts, U.S.A. Wetlands 20: 707–715.

Brucet, S., D. Boix, R. López-Flores, A. Badosa, R. Moreno-Amich & X. D. Quintana, 2005. Zooplankton structure and dynamics in permanent and temporary Mediterranean salt marshes: taxon-based and size-based approaches. Archiv für Hydrobiologie 162: 535–555.

Castella, E., H. Adalsteinsson, J. E. Brittain, G. M. Gislason, A. Lehmann, V. Lencioni, B. Lods-Crozet, B. Maiolini, A. M. Milner, J. S. Olafsson, S. J. Saltveit & D. L. Snook, 2001. Macrobenthic invertebrates richness and composition along a latitudinal gradient of European glacier-fed streams. Freshwater Biology 46: 1811–1831.

Crawley, M. J., 2002. Statistical Computing. An Introduction to Data Analysis using S-Plus. John Wiley & Sons, Chichester.

Della Bella, V., M. Bazzanti & F. Chiarotti, 2005. Macroinvertebrate diversity and conservation status of Mediterranean ponds in Italy: water permanence and mesohabitat influence. Aquatic Conservation: Marine and Freshwater Ecosystems 15: 583–600.

DPTOP (Departament de Política Territorial i Obres Públiques), 2005. Hipermapa. Atles electrònic de Catalunya. Available: http://www10.gencat.net/ptop/AppJava/cat/actuacions/territori/hipermapa.jsp

Gascón, S., D. Boix, J. Sala & X. D. Quintana, 2005. Variability of benthic assemblages in relation to the hydrological pattern in Mediterranean salt marshes (Empordà wetlands, NE Iberian Peninsula). Archiv für Hydrobiologie 163: 163–181.

Gascón, S., D. Boix, J. Sala & X. D. Quintana, 2007. Changes on macrobenthic fauna of a Mediterranean salt marsh (Empordà Wetlands, NE Iberian Peninsula) after a severe drought, with special emphasis on the *Corophium orientale* population. Vie et Milieu 57: 3–12.

Gitay, H. & I. R. Noble, 1997. What are functional types and how should we seek them? In Smith, T. M., H. H. Shugart & F. I. Woodward (eds), Plant Functional Types. University Press, Cambridge, 3–19.

Goodman, K. J., A. E. Hershey & K. Fortino, 2006. The effect of forest type on benthic macroinvertebrate structure and ecological function in a pine plantation in the North Carolina Piedmont. Hydrobiologia 559: 305–318.

Guisan, A. & N. E. Zimmermann, 2000. Predictive habitat distribution models in ecology. Ecological Modelling 135: 147–186.

Guelorget, O. & J. P. Perthuisot, 1983. Le domaine paralique: Expressions géologiques, biologiques et économiques du confinement. Travaux du laboratoire de Géologie 16. Presses de l'École Normale Supérieure, París.

Hastie T. J. & R. J. Tibshirani, 1990. Generalized Additive Models. Chapman & Hall, London.

Herbst, D. B., 1999. Biogeography and physiological adaptations of the brine fly genus *Ephydra* (Diptera: Ephydridae) in saline waters of the Great Basin. Great Basin Naturalist 59: 127–135.

Hooper, D. U., M. Solan, A. Symstad, S. Díaz, M. O. Gessner, N. Buchmann, V. Degrange, P. Grime, F. Hulot, F. Mermillod-Blondin, J. Roy, E. Spehn & L. van Peer, 2002. Species diversity, functional diversity, and ecosystem functioning. In Loreau, M., S. Naeem & P. Inchausti (eds), Biodiversity and Ecosystem Functioning: Synthesis and Prespectives. Oxford University Press, Oxford, 195–281.

Horsák, M., 2006. Mollusc community patterns and species response curves along a mineral richness gradient: a case study in fens. Journal of Biogeography 33: 98–107.

Jassby, A. D. & T. M. Powell, 1990. Detecting changes in ecological time series. Ecology 71: 2044–2052.

MAPA (Ministerio de Agricultura Pesca y Alimentación), 2006. Sistema de identificación de parcelas agrícolas. Available: http://www.sigpac.mapa.es/fega/visor/.

March, F. & D. Bass, 1995. Application of island biogeography theory to temporary pools. Journal of Freshwater Ecology 10: 83–85.

Mistri, M., R. Rossi & E. A. Fano, 2001. Structure and secondary production of a soft bottom macrobenthic community in a brackish lagoon (Sacca di Goro, north-

eastern Italy). Estuarine, Coastal and Shelf Science 52: 605–616.
Mesléard, F., P. Grillas & J. Lepart, 1991. Plant community succession in a coastal wetland after abandonment of cultivation: the example of the Rhone delta. Vegetatio 94: 35–45.
Noest, V., 1991. Simulated impact of sea level rise on phreatical level and vegetation of dune slacks in the Voorne dune area (The Netherlands). Landscape Ecology 6: 89–97.
Piscart, C., A. Lecerf, P. Usseglio-Polatera, J. C. Moreteau & J. N. Beisel, 2004. Biodiversity patterns along a salinity gradient: the case of net-spinning caddisflies. Biodiversity and Conservation 14: 2235–2249.
Quintana, X. D., 2002a. Measuring the intensity of disturbance in zooplankton communities of Mediterranean salt marshes using multivariate analysis. Journal of Plankton Research 24: 255–265.
Quintana, X. D., 2002b. Estimation of water circulation in a Mediterranean salt marsh and its relationship with Flooding causes. Limnética 21: 25–35.
Quintana, X. D., R. Moreno-Amich & F. A. Comín, 1998a. Nutrient and plankton dynamics in a Mediterranean salt marsh dominated by incidents of flooding. Part1: differential confinement of nutrients. Journal of Plankton Research 20: 2089–2107.
Quintana, X. D., F. A. Comín & R. Moreno-Amich, 1998b. Nutrient and plankton dynamics in a Mediterranean salt marsh dominated by incidents of flooding. Part2: response of the zooplankton community to disturbances. Journal of Plankton Research 20: 2109–2127.
Ribera, I., J. Isart & J. A. Régil, 1994. Coleópteros acuáticos de los estanys de Capmany (Girona): Hydradephaga. Scientia Gerundensis, 20: 17–34.
Richards, C., R. Haro, L. Johnson & G. Host, 1997. Catchment and reach-scale properties as indicators of macroinvertebrate species traits. Freshwater Biology 37: 219–230.
Reizopoulou, S. & A. Nicolaidou, 2004. Benthic diversity of coastal brackish-water lagoons in western Greece. Aquatic Conservation: Marine and Freshwater Ecosystems 14: 93–102.
Schneider, D. W., 1999. Influence of hydroperiod on invertebrate community structure. In Batzer, D., R. B. Rader & S. A. Wissinger (eds), Invertebrates in Freshwater Wetlands of North America. John Wiley & Sons, New York, 299–318.
Schneider, D. W. & T. M. Frost, 1996. Habitat duration and community structure in temporary ponds. Journal of the North American Benthological Society 15: 64–86.
Schwartz, S. S. & D. G. Jenkins, 2000. Temporary aquatic habitats: constraints and opportunities. Aquatic Ecology 34: 3–8.
Serrano, L. & K. Fahd, 2005. Zooplankton communities across a hydroperiod gradient of temporary ponds in the Doñana National Park (SW Spain). Wetlands 25: 101–111.
Serrano, L., M. D. Burgos, A. Díaz-Espejo & J. Toja, 1999. Phosphorus inputs to wetlands following storm events after drought. Wetlands 9: 318–326.
Tachet, H., P. Richoux, M. Bournaud & P. Usseclio-Polatera, 2002. Invertébrés d'eau douce: systématique, biologie, écologie. CNRS Éditions, Paris.
Takeda, M. & T. Nagata, 1998. Effects of temperature and winter diapause on survival and development in bivoltine and trivoltine ecotypes of rice stem maggot, *Chlorops oryzae* Matsumura (Diptera: Chloropidae), reared on a winter host. Applied Entomology and Zoology 33: 85–96.
Tarr, T. L., M. J. Baber & K. J. Babbitt, 2005. Macroinvertebrate community structure across a wetland hydroperiod gradient in southern New Hampshire, USA. Wetlands Ecology and Management 13: 321–334.
ter Braak, C. J. F., 1987. Unimodal models to relate species to environment. PhD Thesis, University of Wageningen, Wageningen.
ter Braak, C. J. F. & P. Šmilauer, 2002. CANOCO Reference manual and CanoDraw for Windows User's guide: Software for Canonical Community Ordination (version 4.5). Microcomputer Power, Ithaca.
Tomanova, S., E. Goitia & J. Helešic, 2006. Trophic levels and functional feeding groups of macroinvertebrates in neotropical streams. Hydrobiologia 556: 251–264.
Usseglio-Polatera, P., M. Bournaud, P. Richoux & H. Tachet, 2000. Biomonitoring through biological traits of benthic macroinvertebrates: how to use species trait databases? Hydrobiologia 422/423: 153–162.
Victor, R. & J. R. Victor, 1997. Some aspects of the ecology of littoral invertebrates in a coastal lagoon of southern Oman. Journal of Arid Environments 37: 33–44.
Wellborn, G. A., D. K. Skelly & E. E. Werner, 1996. Mechanisms creating community structure across a freshwater habitat gradient. Annual Review of Ecology and Systematics 27: 337–363.
Wiggins, G. B., 1973. A contribution to the biology of caddisflies (Trichoptera) in temporary pools. Royal Ontario Museum of Life Sciences Contributions 8: 1–28.
Wiggins, G. B., R. J. Macklay & I. M. Smith, 1980. Evolutionary and ecological strategies of animals in annual temporary pools. Archiv für Hydrobiologie supplement 58: 97–206.
Williams, W.D., 1985. Biotic adaptations in temporary lentic waters, with special reference to those in semi-arid and arid regions. Hydrobiologia 125: 85–110.
Wolda, H., 1988. Insect seasonality: why? Annual Review of Ecology and Systematics 19: 1–18.

Hydrobiologia (2008) 597:85–95
DOI 10.1007/s10750-007-9216-9

ECOLOGY OF EUROPEAN PONDS

Macrophyte diversity and physico-chemical characteristics of Tyrrhenian coast ponds in central Italy: implications for conservation

Valentina Della Bella · Marcello Bazzanti · Maria Giuseppina Dowgiallo · Mauro Iberite

Abstract Awareness of pond conservation value is growing all over Europe. Ponds are recognized as important ecosystems supporting large numbers of species and several rare and threatened aquatic plants, macroinvertebrates and amphibians. Notwithstanding ponds, particularly temporary ones, are still neglected in Italy. There are some gaps in our understanding of the macrophyte ecology and the conservation value of Mediterranean small still waters. Therefore, this study investigated the macrophyte communities and physico-chemical characteristics of 8 permanent and 13 temporary ponds along the Tyrrhenian coast near Rome, with the aim to relate the distribution of aquatic plants to environmental variables, and to define the botanical conservation value of ponds. Throughout the study period (Spring 2002), Principal Component Analysis performed on abiotic variables clearly discriminated temporary ponds, smaller and more eutrophic, from permanent ponds, larger and with higher pH and oxygen concentration. A total of 73 macrophyte taxa were collected in the study ponds. Temporary waters hosted a smaller number of plant species than permanent ones. Besides hydroperiod length, the environmental factors related to plant richness were maximum depth, surface area, dissolved oxygen and nitrogen concentration in the water. Moreover, the Non-metric Multidimensional Scaling showed a high dissimilarity in the taxonomic composition of aquatic plants between temporary and permanent ponds. The former contained more annual fast-growing species (*Callitriche* sp. pl. and *Ranunculus* sp. pl.), while in the latter species with long life-cycles (i.e. *Potamogeton* sp. pl.) were more abundant. Our results highlighted that temporary and permanent ponds in central Italy have different macrophyte assemblages, with aquatic species (including some of conservation interest at regional scale) exclusively found in each pond type. This suggested that both type of ponds could give an irreplaceable contribution to the conservation of aquatic plant diversity of these freshwater ecosystems.

Keywords Hydroperiod · Species richness · Mediterranean temporary and permanent ponds · Wetland plants

Guest editors: R. Céréghino, J. Biggs, B. Oertli & S. Declerck
The ecology of European ponds: defining the characteristics of a neglected freshwater habitat

Electronic supplementary material The online version of this article (doi:10.1007/s10750-007-9216-9) contains supplementary material, which is available to authorized users.

V. Della Bella (✉) · M. Bazzanti
Department of Animal and Human Biology, University of Rome La Sapienza, Viale dell'Università 32, 00185 Rome, Italy
e-mail: valentina.dellabella@uniroma1.it

M. G. Dowgiallo · M. Iberite
Department of Plant Biology, University of Rome La Sapienza, P.le Aldo Moro 5, 00185 Rome, Italy

Introduction

As it has been demonstrated by the contributions of two recent European Ponds Workshops hosted in Geneva, Switzerland (2004) and in Toulouse, France (2006), the awareness of pond conservation value as biodiversity resource is growing all over Europe. Ponds are recognized as particularly important for amphibian (Beebee, 1997; Beja & Alcazar, 2003; Hazell et al., 2004), macroinvertebrate (Collinson et al., 1995; Oertli et al., 2000, 2002; Nicolet, 2001; Nicolet et al., 2004; Della Bella et al., 2005), and aquatic plant conservation (Grillas & Roché, 1997; Linton & Goulder, 2000, Oertli et al., 2000, 2002; Nicolet, 2001; Nicolet et al., 2004). They support large numbers of species and several rare and threatened species of all of these groups (Grillas et al., 2004a, b), and they strongly contribute to freshwater biodiversity at a regional level (Williams et al., 2004).

Ponds, particularly temporary ones, are aquatic habitats with multiple constraints due to their great abiotic variability, but this offers to species with particular adaptations many opportunities to succeed (Schwartz & Jenkins, 2000). For macrophytes inhabiting temporary waters, drought is the principal constraint. This constraint is even greater because of its unpredictability, and alternation of dry and wet-phase varies from year-to-year, especially in Mediterranean regions (Grillas & Roché, 1997). The survival strategies of macrophytes to fluctuations of environmental conditions involve resistant spores, seeds, dormant vegetative parts and flexibility of life cycles (Williams, 1985; Brock, 1988; Grillas & Roché, 1997; Grillas et al., 2004a; Nicolet et al., 2004; Cherry & Gough, 2006). The development of life cycles with different length, the dominance of one form of reproduction (sexual or vegetative), the major or minor investment in seed production and the germination patterns, might concur in structuring the macrophyte assemblages in wetlands with different hydroperiod length (Casanova & Brock, 1996; Grillas & Roché, 1997; Fernández-Aláez et al., 1999; Capon, 2003; Warwick & Brock, 2003; Grillas et al., 2004a). Seasonal wetland vegetation is mostly linked to semi-arid conditions and is widespread in regions with Mediterranean climate, such as the Mediterranean Basin, California, West Africa and Australia, where it has been thoroughly described (Grillas & Roché, 1997; Grillas et al., 2004a, b; Bagella et al., 2005; Barbour et al., 2005; Molina, 2005; Müller & Deil, 2005; Pignatti & Pignatti, 2005; Rudner, 2005). In Mediterranean temporary habitats, most plants are short-lived species with rapid life-cycles and with the ability to rapidly exploit periods of favourable conditions for germination and growth. Sexual reproduction is the dominant form of reproduction and therefore plants make a great investment in seed production to withstand alternate periods of flooding and desiccation (Grillas & Roché, 1997; Bissels et al., 2005; Pignatti & Pignatti, 2005; Rhazi et al., 2005). On the contrary, plants with long life-cycles have an advantage in permanent and more stable aquatic habitats and the vegetative reproduction is generally the commonest form of reproduction (Grillas & Roché, 1997). While the above studies have shown the influence of water regime on composition and distribution of wetland vegetation, comparative studies on the plant community characteristics in temporary and permanent ponds are still limited (Grillas, 1990; Williams et al., 1998; Bianco et al., 2001; Nicolet, 2001).

Macrophyte assemblages, species richness and botanical conservation value of ponds were investigated in some regions of Europe, such as UK (Jeffries, 1998; Williams et al., 1998; Linton & Goulder, 2000; Nicolet, 2001; Nicolet et al., 2004), Germany (Brose, 2001), Switzerland (Oertli et al., 2000, 2002) and France (Grillas, 1990; Grillas & Roché, 1997; Grillas et al., 2004a, b). In Italy the diversity of pond macrophytes has been little recognized. To date, the studies concerning pond vegetation in Italy are very limited (Bianco et al., 2001; Bagella et al., 2005) and there are some gaps in our understanding of macrophyte ecology in small still waters, and in their conservation value. The aims of this study were (i) to investigate the distribution of aquatic plants in some temporary and permanent ponds in central Italy, (ii) to determine the relationships between macrophyte richness and physico-chemical variables, and (iii) to define the botanical conservation value of the study ponds. Such results should allow us to provide useful advises on the management of these aquatic environments.

Methods

Study area

We conducted our study on 8 permanent and 13 temporary ponds located in four protected areas in central Italy, along the Tyrrhenian coast near Rome: WWF Oasis of Palo Laziale, Litorale Romano Natural State Reserve, Decima Malafede Natural Reserve and Presidential Estate of Castelporziano (Fig. 1). These protected areas include the last residues of the original Mediterranean plain forest formerly covering the Latium coast, now surrounded by an urban and agricultural landscape. All these four sites were proposed under the Birds Directive (CEC, 1979) and Habitats Directive (CEC, 1992) as part of the Natura 2000 network (IT6030022/5/7/8, IT6030053, IT6030084; Regione Lazio, 2004). Most of the permanent ponds are ground water fed. In spite of this, water level widely fluctuated during the study year (2002). The sampled temporary ponds may be considered as autumnal ponds (*sensu* Wiggins et al., 1980) and the length of their hydroperiod depends on rainfall, which usually peaks in autumn and spring. In the study year one temporary pond holded water for 60 days, three had a wet-phase duration between 100 and 200 days, and nine between 200 and 300 days.

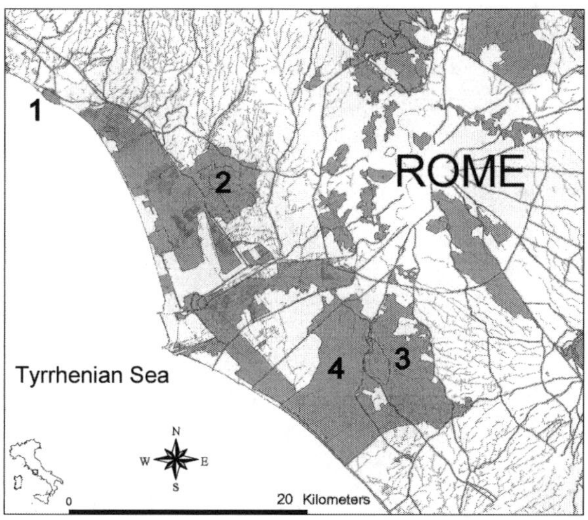

Fig. 1 Map of the sampling area. 1. WWF Oasis of Palo Laziale; 2. Litorale Romano Natural State Reserve; 3. Decima Malafede Natural Reserve; 4. Presidential Estate of Castelporziano

Sampling and laboratory methods

Macrophytes

Macrophyte algae (or macroalgae) and vascular plants (submerged, floating and emergent) were collected by walking around and throughout ponds. In order to collect the highest number of species we repeated the surveys of ponds in March, May and June 2002, covering the flowering period of many plants to facilitate the identification. At the start of the study, the spring level of the studied ponds was similar to the winter level. We recorded all plants present in each sampling date within the perimeter of the pond, as defined by the water's edge, and within 1 m of drawdown zone around it. Most taxa were identified to species level (see Electronic supplementary material), sometimes to genera, and rarely to family (Characeae and some species belonging to Gramineae and Umbelliferae). Species were assigned a distribution and conservation status according to the Regional Red List of Italian plants (Conti et al., 1997; Anzalone et al., in press).

Moreover, at the time of sampling, we also visually estimated macrophyte covers for each pond and the percentage of water surface covered by macrophytes was grouped in five class (0 = 0%, 1 = 1–25%, 2 = 26–50%, 3 = 51–75% and 4 = 76–100%), following the phytosociologic approach (Pignatti & Mengarda, 1962; Braun-Blanquet, 1976).

Physical and chemical data

The study ponds were characterised using variables describing morphology, water and sediment chemistry. At each visit, we measured the maximum depth and area of ponds as reported by Bazzanti et al. (1996). Conductivity, pH and dissolved oxygen were recorded by field metres. Water transparency was measured by visual judgement (1 = clear water, 2 = intermediate, 3 = turbid water). We determined organic matter, organic carbon, total phosphorus and total nitrogen contents and granulometric composition of the sediments, according to methods reported in Cummins (1962), Gaudette et al. (1974), Marengo & Baudo (1988), Bremner (1965), respectively. Nitrogen and phosphorus concentrations in waters were also measured following standard methods reported in IRSA (1994) and Wetzel & Likens (2000).

Data analysis

We employed Principal Component Analysis (PCA), based on morphology, water and sediment characteristics, to summarize variations among ponds and to highlight environmental gradients. Before the analyses, all variables were standardized following

$$X_{st} = (X - \overline{X})/SD.$$

Relationships between numbers of species [$\log_{10}(x + 1)$ transformed] and environmental variables were explored using stepwise multiple regressions. Since physico-chemical data were intercorrelated, the first two axes extracted by PCA (PC factors) were used as independent variables in the subsequent regression analysis to determine how much variation in species richness could be accounted for by the environmental variables. Variables with factor loadings $\geq|0.60|$ on an axis were considered important for that particular PC factor. Durbin–Watson test (Durbin & Watson, 1951; Olsen, 1995) was used to control autocorrelation of residuals due to temporal pseudoreplication (Hurlbert, 1984). Values of the test's parameter (d) near 2 indicate absence of autocorrelation.

In order to measure similarity among pond macrophyte communities, we performed a 2-d Non metric Multidimensional Scaling (N-MDS) on the similarity matrix based on the Bray–Curtis similarity coefficient (Bray & Curtis, 1957; Clarke & Warwik, 1994) which was calculated on presence/absence data of all macrophyte taxa collected in the ponds (in the analysis three ponds were excluded because they hosted only one macrophyte species or were without aquatic vegetation). In order to identify which species were "typical" (found with consistent high frequencies in most of samples) of temporary or permanent pond, we used the Similarity Percentage analysis (SIMPER; Clarke & Warwick, 1994). This procedure decomposes the similarities of all within-pond type comparisons into their contributions from each species and lists species in decreasing order of their importance in typifying the two types of ponds.

Spearman rank coefficient of correlation (r_s) was adopted to discover relationships between the environmental variables and PCA factor scores of ponds, and also their NMDS configuration. The Mann–Whitney U-test was used to highlight any significant differences between variables of temporary and permanent ponds.

We conducted our statistical analyses with Statistica (version 5) and PRIMER 5 (version 5.2.0) software.

Results

Physical and chemical characteristics of ponds

The first two components extracted in the PCA accounted for 57.6% of variance in the original data (Table 1 and Fig. 2). The strongest variations were in the sediment variables, with the contents of phosphorus, nitrogen, carbon and organic matter increasing on the first PC factor, along with silt and clay. The second factor defined a gradient from permanent ponds, with higher values of depth, surface area, pH and higher concentrations of dissolved oxygen in the water, to more temporary ponds, with higher concentrations of phosphorus and nitrogen in the water. Temporary ponds having shorter wet-phase duration (<200 days) are plotted on the most negative side of the second factor because of extreme values of these variables. There were significant differences in the values of these variables between temporary and permanent ponds at least for one sampling occasion, whereas granulometric analysis showed no significant differences in the fraction texture between the two types of ponds and the sediments resulted to be composed predominantly of silt and clay (Table 2).

Macrophyte species richness and assemblages

A total of 73 taxa (88% identified to species level) were collected from 21 study ponds (Electronic supplementary material). Fifty-three (more than 70% of the total) were typical or exclusive species of wetland habitat and represent 13% of the aquatic species of Latium Region. During the entire study, 20 ponds hosted aquatic vegetation and the species richness of sites ranged between 1 and 26 (mean = 9

Table 1 Factor loadings of physico-chemical variables used in the PCA analysis and their respective codes

Variables	Code	PC 1	PC 2
Maximum depth (cm)	Depth	−0.01	**0.87**
Surface area (m^2)	Area	0.20	**0.82**
Total Phophorus in the water ($mg\ l^{-1}$)	TP water	−0.02	−0.57
Total Nitrogen in the water ($mg\ l^{-1}$)	TN water	−0.10	**−0.84**
Dissolved oxygen ($mg\ l^{-1}$)	DO	0.07	**0.79**
Conductivity (μS/cm)	Cond	0.36	0.18
pH	pH	0.09	**0.67**
Transparency (class)	Transp	0.45	0.41
Total Phophorus in the sediments (‰)	TP sediment	**0.64**	0.28
Total Nitrogen in the sediments (%)	TN sediment	**0.74**	−0.23
Organic Carbon in the sediments (%)	OC sediment	**0.73**	−0.30
Organic Matter in the sediments (%)	OM sediment	**0.89**	0.20
Coarse sand (%)	C sand	**−0.81**	−0.02
Medium sand (%)	M sand	**−0.91**	−0.13
Fine sand (%)	F sand	**−0.71**	−0.15
Silt (%)	Silt	**0.76**	0.27
Clay (%)	Clay	**0.64**	0.03
Eigenvalues		5.67	4.12
Percentage of variance explained		36.5	21.1

The contributions >|0.60| are reported in bold

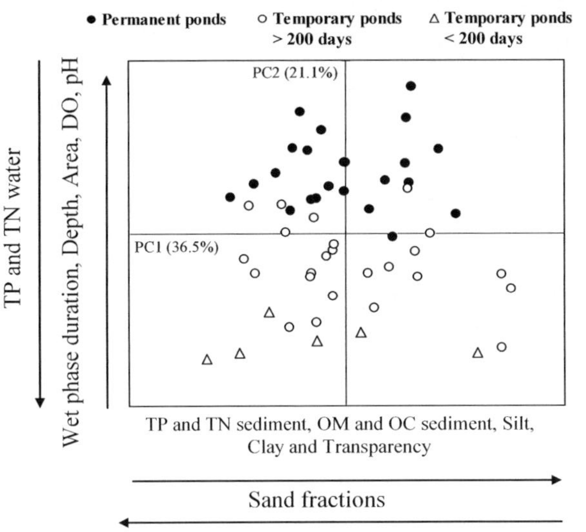

Fig. 2 Principal Component Analysis performed on physico-chemical and morphological variables (the percentage of variance explained by two first axes is reported in brackets and wet-phase duration of temporary ponds is indicated in the legend). Arrows indicate the correlation (significance at least $P < 0.05$) between axis pond scores and environmental variables

± 1.6 S.E.; median = 6). The overall macrophyte richness was significantly higher (Mann–Whitney U-test; $P < 0.01$) in permanent ponds than in temporary ones (Fig. 3) although the maximum number of species was found in a pond with temporary character. On the contrary, the percentages of pond surface area covered by macrophytes did not result significantly different in two pond typologies. Multiple Regression Analysis (Durbin–Watson test for autocorrelation: $d = 1.8$) found significant relationships between PC factors and species richness of macrophytes. Both axes explained 44% of the variation in macrophyte richness ($0.63 + 0.09$ PC1 + 0.18 PC2; $P < 0.001$); but PC2 ($P < 0.001$) seemed to be more important than PC1 ($P < 0.03$). Besides hydroperiod length, the environmental factors related to plant richness were maximum depth, surface area, dissolved oxygen and nitrogen concentration in the water (Tables 1 and 3).

Non-metric Multidimensional scaling, performed on presence/absence of all macrophyte taxa collected within the ponds during the study year, showed a clear dissimilarity in the taxonomic composition of aquatic vegetation between temporary and permanent ponds (Fig. 4) along the first axis according to increasing values of nitrogen content in the water and decreasing values of wet-phase duration, surface area, depth, conductivity, pH, transparency, oxygen and phosphorus contents in the sediments. The

Table 2 Physico-chemical features of permanent and temporary ponds and their significant differences for each sampling period

Variable code	Pond type		U-test results			
			Significance for sampling period			
	Permanent	Temporary	March	May	June	
Morphology						
Depth	150	50	***	***	***	P
Area	10000	530	***	***	***	P
Water variables						
TP water	0.21	0.32		*		T
TN water	1.53	3.66	**	***	*	T
DO	10.77	5.60	*	***	*	P
Cond	883.60	704.07	*			P
pH	8.65	7.43		****	***	P
Transp	1	2		*		T
Sediment variables						
TP sediment	0.52	0.40				
TN sediment	0.15	0.18				
OC sediment	1.16	1.53				
OM sediment	10.87	10.51				
C sand	3.11	2.73				
M sand	14.16	15.29				
F sand	11.45	15.04				
Silt	44.44	35.02				
Clay	27.39	31.77				

Mann–Whiteny U-test: $*P < 0.05$; $**P < 0.01$; $***P < 0.005$; $****P < 0.001$

Means (permanent pond: $N = 8$; temporary ponds: $N = 13$) are reported except for Depth and Area (maxima), and water transparency (median). The last column indicates whether values are significantly higher in permanent (P) or temporary (T) ponds. For code of variables see Table 1

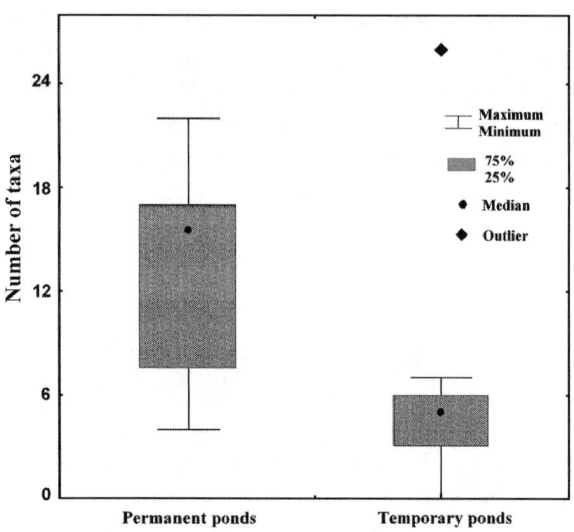

Fig. 3 Total richness of macrophyte species in the two types of pond (permanent or temporary)

Similarity Percentages Analysis (SIMPER) showed that temporary and permanent ponds had a Bray–Curtis dissimilarity of 90.6%. In Table 4 the species were listed in decreasing order of their importance in typifying the two groups of ponds. Permanent ponds were characterised by an exclusive presence of *Veronica anagallis-aquatica, Mentha aquatica*, Characeae, *Myriophyllum spicatum* and all species belonging to Potamogetonaceae (*Potamogeton crispus, P. natans, P. nodosus, P. trichoides*), and also by a high occurrence of *Lythrum junceum, Sparganium erectum* and *Rumex conglomeratus*. On the other hand, a lot of species belonging to Callitrichaceae (*Callitriche truncata, C. stagnalis, C. hamulata*) and Ranunculaceae family (*Ranunculus ophioglossifolius, R. aquatilis, R. peltatus, R. trichophyllus*) were exclusively collected in temporary ponds and *Damasonium alisma* and *Lythrum portula* were found

Table 3 Multiple regression model relating macrophyte species richness [$\log_{10}(x + 1)$ transformed] and the two orthogonal factors extracted using PCA in order to reduce the number of environmental variables

	Estimate	S.E.	Coefficient	S.E.	t (50)	P
Intercept			0.63	0.04	16.38	<0.001
PC1	0.26	0.12	0.09	0.04	2.22	0.03
PC2	0.52	0.12	0.18	0.04	4.50	<0.001

$r = 0.66$, $r^2 = 0.44$; $F_{2,50} = 19.65$, $P < 0.001$

with higher occurrence in this type of ponds than in permanent ones.

Finally, in order to assess the botanical conservation value of studied ponds, the number of species of conservation concern was determined at regional level. Out of 53 aquatic species evaluated, five were of conservation interest. One is Vulnerable and exclusively found in a temporary pond, and four were Lower Risk (IUCN, 1994), of which one was exclusively found in permanent ponds and two in temporary ones (Electronic supplementary material).

Discussion

Besides hydroperiod length, the size (maximum depth and surface area), pH, oxygen and nitrogen contents in the water seem to be the main physico-chemical variables which determine the separation between temporary and permanent ponds in central Italy. Principal Component Analysis discriminated temporary ponds, smaller and more eutrophic, from permanent ponds, larger and with higher pH and oxygen concentration along the second axis. However the analysis showed that most of variance among ponds seems to be explained by some sediment variables not related with wet-phase duration of ponds. Likely geology of the area and soil type are factors of primary importance in pond classification. To date there are very few studies on sediment characteristics of ponds, both temporary and permanent, to compare our findings that highlighted the role played by pond sediments. In freshwater ecosystems, sediments represent both nutrient accumulation level and release zone of nutrients (Häkanson, 1984; Salomons, 1985; Chapman, 1989) and, therefore, the sediments can provide an evaluation of the "history" of pond, in spite of the strong variability of their water characteristics.

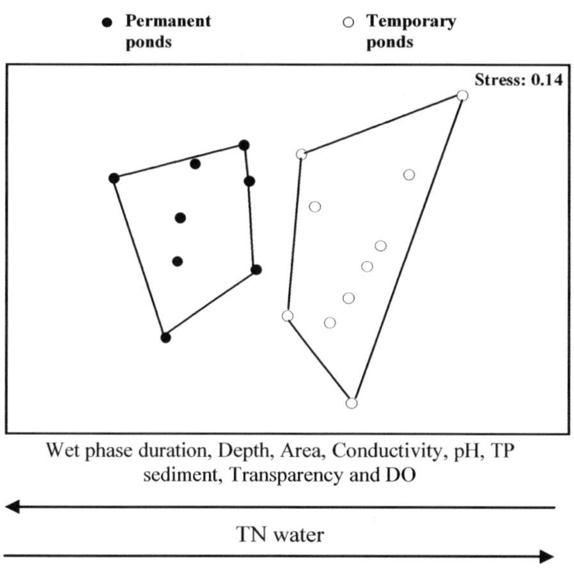

Fig. 4 Non-metric Multidimensional Scaling performed on presence/absence of aquatic vegetation species. Arrows indicate the correlation (significance at least $P < 0.05$) between axis pond scores and environmental variables

This study also showed that ponds are valuable for wetland plant biodiversity. In the 21 studied ponds a high number of aquatic plants was found according to the most recent works in Europe. Similarly to our investigation, Williams et al. (1998) in a study of lowland ponds of Great Britain recorded a mean number of six plant species per pond with a range 0–25 species in temporary ponds and a mean of 11 with a range 0–35 in permanent ones. In another study on temporary ponds in Germany, Brose (2001) found a similar number of species per pond (9) with a range 1–24 species. In further investigations in Great Britain, some authors (Nicolet, 2001; Nicolet et al., 2004) recorded ponds, minimally impaired by anthropogenic activities, with an average of seventeen species per temporary pond (range 0–37) and 23 in permanent ponds (range 3–56). All these studies maintain that temporary ponds, although they

Table 4 List of macrophyte species in decreasing order of their importance in typifying permanent and temporary ponds identified by SIMPER analysis performed on presence/absence data of all macrophyte taxa

Permanent ponds		Temporary ponds	
Similarity: 23.43%		Similarity: 18.89%	
Taxa	Contribution (%)	Taxa	Contribution (%)
Lythrum junceum	12.53	Gramineae indet.	37.57
Gramineae indet.	10.89	*Callitriche truncata*	15.74
Potamogeton natans	10.3	*Ranunculus ophioglossifolius*	11.41
Cynodon dactylon	8.5	*Damasonium alisma*	8.06
Potamogeton trichoides	7.64	*Lythrum portula*	6.82
Potamogeton crispus	6.27	*Ranunculus sardous*	4.95
Veronica anagallis-aquatica	6.22		
Characeae indet.	6.01		
Mentha aquatica	5.19		
Sparganium erectum	5.11		
Rumex conglomeratus	3.5		

Percentage of similarity within the two types of ponds and percentage contribution of each macrophyte species to similarity are reported. Indet = Indetermined

generally have a few species when compared with permanent ones, could potentially host-plant communities rich in species. This assertion is confirmed by our investigation where the maximum number of species (26) was registered in a temporary pond. This can be attributable to many pioneer species from humid meadows colonising drawdown zone, free from hydrophyte competitors. Yet, median richness was higher in permanent ponds.

A part of wet-phase duration, we found that plant species richness seems mainly depend on pond surface area, depth and nitrogen in the water. In the ponds described here area, depth and nitrogen are highly correlated with hydroperiod because large and less eutrophic ponds are generally permanent. Therefore, it is difficult to discern the effect of hydroperiod, size, nitrogen enrichment on species richness variation as described in a previous work on macroinvertebrate communities hosted in these ponds (Della Bella et al., 2005). However, the relationship between area and macrophyte richness is well documented in aquatic systems (Rørslett, 1991; Vestergaard & Sand-Jensen, 2000; Oertli et al., 2000, 2002; Murphy, 2002; Jones et al., 2003; Rolon & Maltchik, 2006). For aquatic plants, the positive relationship between pond size and biodiversity can be considered a valid generalization for many cases (Gee et al., 1997; Jeffries, 1998; Oertli et al., 2000, 2002; Brose, 2001) although some authors found some controversial results (Friday, 1987; Linton & Goulder, 2000). The influence of water chemistry on aquatic plant richness was analysed in several studies. Generally, nutrient availability was described as major predictor of species distributions and the highest macrophyte diversity was observed in mesotrophic or slightly eutrophic ecosystems (Rørslett, 1991; Vestergaard & Sand-Jensen, 2000, Heegaard et al., 2001; Murphy, 2002). Consistently with our results, Oertli et al. (2000) found a negative relationship between nitrate concentration and the diversity of aquatic plants in ponds, while other authors failed to find this relationship in lakes (Jones et al., 2003) and wetlands (Rolon & Maltchik, 2006).

Moreover, our study highlighted a high dissimilarity in plant species composition between temporary and permanent ponds as shown in the ordination analysis where a clear separation is pointed out between the two types of ponds. The studied temporary waters were characterised by exclusive presence of many species of *Callitriche* and *Ranunculus*. These taxa are fast-growing species with an annual life cycle (Pignatti, 1982), and are capable to avoid the pond dry phase through a coincidence between the start of their life cycle and the filling of the basin with the autumn rainfall. Seed production occurs before summer then the plant dies when the pond dries up. Other species without this surviving strategy collected in our temporary ponds, like *Damasonium alisma* and *Alisma lanceolatum*, can survive if the moisture of soil is high or the wet-phase is long enough. These findings concur with other studies on plant community composition of temporary ponds and marshes (Grillas & Roché, 1997; Grillas et al., 2004a; Nicolet et al., 2004). On the contrary, the water presence all year round in permanent ponds allows the development of species with perennial life cycles that need more time for growth and flowering (Pignatti, 1982). In fact, in our permanent ponds we found hydrophytes, such as *Potamogeton* sp. pl., *Myriophyllum spicatum* and *Veronica anagallis-aquatica*, with long life-cycles that need to be continuously submerged and that do not tolerate dry soil, thus absent from temporary waters. Hydrological regime is recognized as one of the main factors determining the distribution and characteristics of aquatic vegetation in wetlands with different hydroperiod length (Capon, 2003; Warwick & Brock, 2003). In Mediterranean temporary waters most of plants are annuals, and the seed bank in the soil is of great importance for their survival. In these habitats, aquatic plants made major investment in seed production, thus sexual reproduction is the commonest form of reproduction (Grillas & Roché, 1997). Perennial species and vegetative reproduction dominate in more stable habitats, such as permanent waters, where this kind of plants are advantaged in competition for space, light and nutrients by their life form (Grillas & Roché, 1997). Comparative studies on the plant community composition between temporary and permanent ponds are still limited but they seem to confirm our findings. Bianco et al. (2001) in a previous study on some ponds in the Castelporziano Reserve (Italy) found plant communities dominated by *Callitriche* sp. pl. and *Ranunculus aquatilis* in temporary ponds, with turbid and eutrophic waters, and communities rich in *Potamogeton* species in permanent ones, with more transparent and oxygenated waters. Grillas (1990) investigated submerged macrophyte assemblages in the marshes of the Camargue (France) and found that *Callitriche* sp. pl., *Ranunculus* sp. pl. and other species (i.e. *Tolypella* sp.pl.) dominated communities in temporarily flooded oligohaline marshes whereas permanently flooded marshes are dominated by *Potamogeton* sp pl. and *Myriophyllum spicatum*. In the latter, they maintained an important cover and biomass in winter thus preventing the growth of annual species.

In conclusion, this study highlighted that temporary and permanent ponds in central Italy have different macrophyte species composition, with aquatic species exclusively found in each pond type. Permanent ponds are strictly aquatic habitats dominated by hydrophytes. Lowland temporary ponds, while capable of hosting some hydrophytes, are tightly linked to humid meadows. They have similar environmental conditions (temporariness of flooding period, low depth, hydromorphic soils) and they potentially can share similar species, such as *Ranunculus ophioglossifolius*, *R. sardous* and *Lythrum portula*. Our results also showed that the relationship between macrophyte species and pond wet-phase duration depends on pond size and some physicochemical variables of water. For conservation purposes, the studied temporary and permanent ponds hosted some species of conservation interest at regional scale. Among these, *Potamogeton trichoides* and *Callitriche truncata* (Lower Risk Category) are exclusively found more than one time in only one type of pond (see Electronic supplementary material). Therefore, the two types of pond should be preserved or created because they both could give an irreplaceable contribution to the conservation of aquatic plant diversity of small still water bodies in Italy.

Acknowledgements The study was carried out as part of the PhD of V.D.B., and supported by a MURST 60% grant to M.B. We thank F. Grezzi for his valuable help in the field work. We wish to thank also the Presidential Estate of Castelporziano, WWF Lazio, Councils of Roma and Fiumicino, and Roma Natura for granting us permission to conduct our research within their natural reserves. We are also grateful to two anonymous referees for their suggestions and to B. Oertli for his editing care that improved the earlier version of the article.

References

Anzalone, B., M. Iberite & E. Lattanzi. La flora del Lazio, in press.

Bagella, S., E. Farris, S. Pisanu & R. Filigheddu, 2005. Ricchezza floristica e diversità degli habitat umidi temporanei nella Sardegna Nord-Occidentale. Atti 100° Congresso della Società Botanica Italiana (Roma). Informatore Botanico Italiano 37: 112–113.

Barbour, M. G., A. I. Solomeshch, R. F. Holland, C. W. Witham, R. L. Macdonald, S. S. Cilliers, J. A. Molina, J.

J. Buck & J. M. Hillman, 2005. Vernal pool vegetation of California: communities of long-inundated deep habitats. Phytocoenologia 35: 177–200.

Bazzanti, M., S. Baldoni & M. Seminara, 1996. Invertebrate macrofauna of a temporary ponds in Central Italy: composition, community parameters and temporal succession. Archiv für Hydrobiologie 137: 77–94.

Beebee, T. J. C., 1997. Changes in dewpond numbers and amphibian diversity over 20 years on Chalk Downland in Sussex, England. Biological Conservation 81: 215–219.

Beja, P. & R. Alcazar, 2003. Conservation of Mediterranean temporary ponds under agricultural intensification: an evaluation using amphibians. Biological Conservation 114: 317–326.

Bianco, P. M., M. De Lillis & A. Tinelli, 2001. Biodiversità in aree umide di recente realizzazione. In Accademia Nazionale delle Scienze detta dei XL (ed.), "Scritti e documenti" XXVI. 2. Il sistema ambientale della Tenuta Presidenziale di Castelporziano. Ricerche sulla complessità di un ecosistema forestale costiero mediterraneo. Segretariato Generale della Presidenza della Repubblica Italiana, 835–842.

Bissels, S., T. W. Donath, N. Hölzel & A. Otte, 2005. Ephemeral wetland vegetation in irregularly flooded arable fields along the northern Upper Rhine: the importance of persistent seedbanks. Phytocoenologia 35: 469–488.

Braun-Blanquet, J., 1976. Pflanzensoziologie. 3 Aufl. Springer, Vienna.

Bray, J. R. & J. T. Curtis, 1957. An ordination of the upland forest communities of Southern Wisconsin. Ecological Monograph 27: 325–349.

Bremner, J. M., 1965. Total nitrogen. In Black, C. A (ed.), Methods of soil analysis. 2. American Society of Agronomic, Inc., Vol. 9. Madison, 1149–1178.

Brock, M. A., 1988. Flexibility of life cycle pattern as a mechanism for tolerance of fluctuations of environmental conditions by aquatic plants. Verhandlungen der Internationale Vereinigung für theoretische und angewandte Limnologie 23: 1949–1953.

Brose, U., 2001. Relative importance of isolation, area and habitat heterogeneity for vascular plant species richness of temporary wetlands in east-German farmland. Ecography 24: 722–730.

Capon, S. J., 2003. Plant community responses to wetting and drying in a large arid floodplain. River Research and Applications 19: 509–520.

Casanova, M. T. & M. A. Brock, 1996. Can oospore germination patterns explain charophyte distribution in permanent and temporary wetland? Aquatic Botany 54: 297–312.

CEC, 1979. Council of European Communities Directive 79/409/EEC on the conservation of wild birds. Official Journal of European Communities, C103.

CEC, 1992. Council of European Communities Directive 92/43/EEC on the conservation of natural habitats and of wild fauna and flora. Official Journal of European Communities, L206.

Chapman, P. M., 1989. Current approaches to developing sediment quality criteria. Environment Toxicological Chemistry 8: 589–599.

Cherry, J. A. & L. Gough, 2006. Temporary floating island formation maintains wetland plant species richness: The role of the seed bank. Aquatic Botany 85: 29–36.

Clarke, K. R. & R. M. Warwick, 1994. Change in Marine Communities: An Approach to Statistical Analysis and Interpretation. Natural Environment Research Council, UK.

Collinson, N. H., J. Biggs, A. Corfield, M. J. Hodson, D. Walker, M. Whitfield & P. J. Williams, 1995. Temporary and permanent ponds: an assessment of the effects of drying out on the conservation value of aquatic macroinvertebrate communities. Biological Conservation 74: 125–133.

Conti, F., A. Manzi & F. Pedrotti, 1997. Liste Rosse Regionali delle Piante d'Italia. WWF Italia, Società Botanica Italiana, Camerino.

Cummins, K. W., 1962. An evaluation of some technique for the collection and analysis of benthic sample with special emphasis on lotic water. American Midland Naturalist 67: 447–504.

Della Bella, V., M. Bazzanti & F. Chiarotti, 2005. Macroinvertebrate diversity and conservation status of Mediterranean ponds in Italy: water permancence and mesohabitat influence. Aquatic Conservation: Marine and Freshwater Ecosystems 15: 583–600.

Durbin, J. & G. S. Watson, 1951. Testing for serial correlation in least-squares regression. II Biometrika 38: 159–178.

Fernández-Aláez, C., M. Fernández-Aláez & E. Bécares, 1999. Influence of water level fluctuation on the structure and composition of the macrophyte vegetation in two small temporary lakes in the northwest of Spain. Hydrobiologia 415: 155–162.

Friday, L. E., 1987. The diversity of macroinvertebrate and macrophyte communities in ponds. Freshwater Biology 18: 87–104.

Gaudette, H. E., W. R. Flight, L. Toner & D. W. Folger, 1974. An inexpensive titration method for the determination of organic carbon in recent sediments. Journal of Sediment Petrology 44: 249–253.

Gee, J. H. R., B. D. Smith, K. M. Lee & S. W. Griffiths, 1997. The ecological basis of freshwater pond management for biodiversity. Aquatic Conservation: Marine and Freshwater Ecosystems 7: 91–104.

Grillas, P., 1990. Distribution of submerged macrophyte in the Camargue in relation to environmental factors. Journal of Vegetation Science 1: 393–402.

Grillas, P. & J. Roché, 1997. Vegetation of Temporary Marshes. Ecology and Management. Station Biologique de la Tour du Valat, Arles.

Grillas, P., P. Gauthier, N. Yavercovski & C. Perennou, 2004a. Mediterranean Temporary Pools, Vol. 1. Issues Relating to Conservation, Functioning and Management. Station Biologique de la Tour du Valat, Arles.

Grillas, P., P. Gauthier, N. Yavercovski & C. Perennou, 2004b. Mediterranean Temporary Pools, Vol. 2. Species Information Sheets. Station biologique de la Tour du Valat, Arles.

Häkanson, L., 1984. On the relationship between lake trophic level and lake sediments. Water Research 18: 303–314.

Hazell, D., J. M. Hero, D. Lindenmayer & R. Cunningham, 2004. A comparison of constructed and natural habitat for

frog conservation in an Australian agricultural landscape. Biological Conservation 119: 61–71.
Heegaard, E., H. H. Birks, C. E. Gibson, S. J. Smith & S. Wolfe-Murphy, 2001. Species-environmental relationships of aquatic macrophtes in Northern Ireland. Aquatic Botany 70: 175–223.
Hurlbert, S. J., 1984. Pseudoreplication and the design of ecological field experiments. Ecological Monographs 54: 187–211.
IRSA, CNR, 1994. Metodi analitici per le acque. Istituto poligrafico e zecca dello stato, Roma.
IUCN, 1994. IUCN Red List Categories. Prepared by the IUCN Species Survival Commission. IUCN, Gland, Switzerland.
Jeffries, M. J., 1998. Pond macrophyte assemblages, biodisparity and spatial distribution of ponds in the Northumberland coastal plain, UK. Aquatic Conservation: Marine and Freshwater Ecosystems 8: 657–667.
Jones, J. I., W. Li & S. C. Maberly, 2003. Area, altitude aquatic plant diversity. Ecography 26: 411–420.
Linton, S. & R. Goulder, 2000. Botanical conservation value related to origin and management of ponds. Aquatic Conservation: Marine and Freshwater Ecosystems 10: 77–91.
Marengo, G. & R. Baudo, 1988. Forme del fosforo nei sedimenti lacustri. Acqua Aria 6: 717–721.
Molina, J. A., 2005. The vegetation of temporary ponds with Isoetes in the Iberian Peninsula. Phytocoenologia 35: 219–230.
Müller, J. V. & U. Deil, 2005. The ephemeral vegetation of seasonal and semipermanent ponds in tropical West Africa. Phytocoenologia 35: 327–388.
Murphy, K. J., 2002. Plant communities and plant diversity in softwater lakes of northern Europe. Aquatic Botany 73: 287–324.
Nicolet, P., 2001. Temporary ponds in the UK: a critical biodiversity resource for freshwater plants and animals. Freshwater Forum 17: 16–25.
Nicolet, P., J. Biggs, G. Fox, M. J. Hodson, C. Reynolds, M. Withfield & P. Williams, 2004. The wetland plant and macroinvertebrate assemblages of temporary ponds in England and Wales. Biological Conservation 120: 265–282.
Oertli, B., D. Auderset Joye, E. Castella, R. Juge & J. B. Lachavanne, 2000. Diversité biologique et typologie écologique des étangs et petits lacs de Suisse. Université de Genève, Office Federal de l'Environnement, des Forets et du Paysage (OFEFP), Genève.
Oertli, B., D. Auderset Joye, E. Castella, R. Juge, D. Cambin & J. B. Lachavanne, 2002. Does size matter? The relationship between pond area and biodiversity. Biological Conservation 104: 59–70.
Olsen, W., 1995. Regression Techniques. SAGE publishing inc., New York.
Pignatti, S., 1982. Flora d'Italia. 3 voll. Edagricole, Bologna.
Pignatti, S. & F. Mengarda, 1962. Un nuovo procedimento per l'elaborazione delle tabelle fitosociologiche. Accademia Nazionale dei Lincei. Rendiconti Lincei, Scienze Fisiche e Naturali, Serie 8, 32: 215–222.
Pignatti, E. & S. Pignatti, 2005. Ephemeral wetland vegetation of Western Australia. Phytocoenologia 35: 201–218.
Regione Lazio, 2004. La Rete Natura 2000 nel Lazio. Caratterizzazione dei Siti di Importanza Comunitaria e delle Zone di Protezione per l'attuazione della sottomisura I.1.2. Seconda edizione. Assessorato Ambiente, Dipartimento Territorio, Direzione Regionale Ambiente e Protezione Civile, Roma.
Rhazi, M., P. Grillas, F. Médail & L. Rhazi, 2005. Consequences of shrub clearing on the richness of aquatic vegetation in oligotrophic seasonal pools in Southern France. Phytocoenologia 35: 489–510.
Rolon, A. S. & L. Maltchik, 2006. Environmental factors as predictors of aquatic macrophyte and composition in wetlands of southern Brazil. Hydrobiologia 556: 221–231.
Rørslett, B., 1991. Principal determinants of aquatic macrophyte richness in northern European lakes. Aquatic Botany 39: 173–193.
Rudner, M., 2005. Seasonal and interannual dynamics in dwarf rush vegetation in the Southwestern Iberian Peninsula. Phytocoenologia 35: 403–420.
Salomons, W., 1985. Sediments and water quality. Environmental Technology Letters 6: 315–326.
Schwartz, S. S. & J. D. Genkins, 2000. Temporary aquatic habitats: constraints and opportunities. Aquatic Ecology 34: 3–8.
Vestergaard, O. & K. Sand-Jensen, 2000. Aquatic macrophyte richness in Danish lakes in relation to alkalinity, transparency, and lake area. Canadian Journal of Fisheries and Aquatic Science 57: 2022–2031.
Warwick, N. W. M. & M. A. Brock, 2003. Plant reproduction in temporary wetlands: the effects of seasonal timing, depth, and duration of flooding. Aquatic Botany 77: 153–167.
Wetzel, R. G. & G. E. Likens, 2000. Limnological Analyses. 3rd edn. Springer-Verlag, New York.
Wiggins, G. B., R. J. Mckay & I. M. Smith, 1980. Evolutionary and ecological strategies of animals in annual temporary pools. Archiv für Hydrobiologie Suppl. 58: 97–206.
Williams, W. D., 1985. Biotic adaptation in temporary lentic waters, with special reference to those in semi-arid and arid regions. Hydrobiologia 125: 85–110.
Williams, P., J. Biggs, C. J. Barr, C. P. Cummins, M. K. Gillespie, T. C. G. Rich, A. Baker, J. Baker, J. Beesley, A. Corfield, D. Dobson, A. S. Culling, G. Fox, D. C. Howard, K. Luursema, M. Rich, D. Samson, W. A. Scott, R. White & M. Whitfield, 1998. Lowland Pond Survey 1996. Department of the Environment, Trade and Regions, London.
Williams, P., M. Whitfield, J. Biggs, S. Bray, G. Fox, P. Nicolet & D. Sear, 2004. Comparative biodiversity of rivers, streams, ditches and ponds in an agricultural landscape in Southern England. Biological Conservation 115: 329–341.

Hydrobiologia (2008) 597:97–107
DOI 10.1007/s10750-007-9217-8

ECOLOGY OF EUROPEAN PONDS

Evaluation of sampling methods for macroinvertebrate biodiversity estimation in heavily vegetated ponds

G. Becerra Jurado · M. Masterson · R. Harrington · M. Kelly-Quinn

Abstract This article presents an evaluation of two sampling methods for assessing the biodiversity of heavily vegetated wetlands. The aim was to establish an effective sampling regime to maximise total taxon richness and minimise sampling effort. Three Integrated Constructed Wetland (ICW) systems in Annetown Valley, Co. Waterford, SE of the Republic of Ireland, were sampled during spring and summer 2005. The two methods that were evaluated were pond netting and two types of horizontal activity traps, namely "horizontal activity traps" (HATs) and modified "horizontal activity traps" (modified HATs). The activity traps provided a one-way funnel system and were constructed from 2 l plastic bottles, allowing for the passive collection of taxa. HATs were designed to capture macroinvertebrates in open water and modified HATs, which were designed specifically for this study, were used to sample within stands of dense emergent vegetation. Results show that a combination of pond netting and activity traps will yield a more complete estimate of taxon richness. The performance of Modified HATs was not significantly different from that of the HATs in dense vegetation. Tests on the sampling effort required for each method are also discussed.

Keywords Macroinvertebrates · Sampling methods · Pond · Wetland · Biodiversity · Activity traps

Guest editors: R. Céréghino, J. Biggs, B. Oertli and S. Declerck
The ecology of European ponds: defining the characteristics of a neglected freshwater habitat

Electronic supplementary material The online version of this article (doi:10.1007/s10750-007-9217-8) contains supplementary material, which is available to authorized users.

G. Becerra Jurado (✉) · R. Harrington · M. Kelly-Quinn
Freshwater Ecology Research Group, University College Dublin, Science Education and Research Centre West, Belfield Campus, Dublin 4, Ireland
e-mail: gustavo.becerra@ucdconnect.ie

M. Masterson
Department of Environment, Heritage and Local Government, Integrated Constructed Wetlands Initiative, Old Custom House, 106 The Quay, Waterford, Ireland

Introduction

Human alteration of the environment seems to be the single most important factor causing the decrease of global biodiversity (Chapin et al., 2000). Aquatic habitats are particularly vulnerable to impacts from anthropogenic inputs and physical habitat destruction. Wetlands and small water bodies such as ponds have been severely degraded or destroyed. Despite the high potential macroinvertebrate diversity of freshwater ponds (Oertli et al., 2002; Williams et al., 2004), research work has not historically been focused on them. Instead, larger water bodies such as streams, rivers or lakes, have received more attention in limnology literature (SIL, 2004) and especially so in recent years with the adoption of the Water Framework Directive. Consequently, there have been few efforts to develop methodologies for the assessment of aquatic macroinvertebrate biodiversity in ponds. The present study addressed this knowledge gap as part of a study examining the biodiversity potential of constructed wetlands. Essentially, these consist of a series of heavily vegetated ponds with variable amounts of open water.

To date, pond netting has been the most commonly used method employed in various pond studies (Biggs et al., 1998; Nicolet et al., 2004; Oertli et al., 2005). Despite the widespread use of pond netting (Indermuehle et al., 2004), few articles have tested its performance (Muzaffar & Colbo, 2002; O'Connor et al., 2004; García-Criado & Trigal, 2005). Even though the effectiveness of netting in open water habitats seems to be high, its performance in other mesohabitats such as within dense stands of emergent vegetation has not been widely evaluated. Hydrophyte habitats constitute an essential part of wetlands (Mitsch & Gosselink, 2000) and a rigorous sampling methodology is needed in order to avoid underestimation of the biodiversity of these environments. It is highly likely that different mesohabitats and groups of macroinvertebrates require different methods. Consequently, the use of a single method for all mesohabitats could yield misleading conclusions on biodiversity assessment.

Aquatic activity traps constitute another option to incorporate into the standard sampling of the epifauna in pond habitats. They seem to be effective for catching fast moving macroinvertebrates, which are normally top predators in the trophic web of these ecosystems. Traps have been found to be quite convenient in terms of deployment in the field and easily standardised between field workers (Pieczynski, 1961; Murkin et al., 1983; Elmberg et al., 1992; Hyvönen & Nummi, 2000). Furthermore, they are relatively simple to make and should be adaptable to the mesohabitat in which they are used. For this reason bottle activity traps were developed. This article presents the application and effectiveness of both netting and activity trapping and, in doing so, aims to shed some light on the difficult task of evaluating the macroinvertebrate biodiversity of freshwater wetland habitats. The following hypotheses were proposed: (1) sampling with pond netting and traps increases the taxon richness count when compared to only using pond netting and (2) the percentage of taxa caught exclusively with traps increases with vegetation cover.

Material and methods

Study area

The study area comprised 25 km^2 within the catchment area of Annetown River, Co. Waterford, Ireland. Thirteen ICWs were developed in that area. ICWs are a series of interconnected ponds (normally four) with a gradient of increasing water depth (from 10 cm to more than 1.5 m) and decreasing emergent vegetation cover (from 95% to 2%), as a consequence of the effluent cleansing processes. Emergent vegetation constitutes more than 80% of the vegetation cover of the ponds. These systems are used to treat organic wastes. The incorporation of a diversity of preferably autochthon macrophytes and local soils facilitate the ICW's capacity to mimic natural conditions.

Three of these wetland systems were selected for this study. They were: Dunhill Village (S06870 2690), Castle Farm (S50397 0649) and Milo's Farm (S5305 04087), treating both domestic effluent and farmyard dirty water. Their physico-chemical characteristics are summarised in Table 1. Dunhill Village ICW treats the domestic wastewater from circa 100 houses and has been operational since 1996. The influent passes through a primary sedimentation tank before entering the first of the four man-made ponds. The effluent is discharged into the adjacent Annetown River. In the case of the Castle Farm ICW, the nature of the waste is agricultural run-off and this pond system became operational in 1999. After passing through a collecting pond, the contaminated waters slowly flow through the wetland system, comprising four ponds, before being released into the Annetown River. Milo's Farm ICW consists of four ponds and also treats agricultural wastes.

Sampling methods

In this study, two methods were evaluated: pond netting and two types of activity traps, namely "horizontal activity traps" (HATs) and modified

Table 1 Physico-chemical characteristics of study ponds (MRP, molibdate reactive phosphorus; n/a, not applicable since only one measurement was taken)

Pond	pH	Ammonium (mg/l N)	MRP (mg/l P)	Nitrate (mg/l N)	Chloride (mg/l Cl)	Area (m^2)	Average depth (m)	Length (m)	Width (m)	Vegetation cover (%)
M1	7.57	25.79	5.99	1.17	51.84	1,906	0.25	100	18	95
	Stdv: n/a	Stdv: 11.92	Stdv: 2.59	Stdv: 1.91	Stdv: 11.46					
M2	7.58	8.96	5.34	0.89	43.08	2,126	0.1	130	15	95
	Stdv: n/a	Stdv: 6.13	Stdv: 2.95	Stdv: 1.06	Stdv: 15.35					
M3	7.58	8.96	5.34	0.89	43.08	2,435	1.8	250	10	2
	Stdv: n/a	Stdv: 6.13	Stdv: 2.95	Stdv: 1.06	Stdv: 15.35					
D1	6.62	50.73	7.97	2.30	97.10	120	0.4	20	6	75
	Stdv: 0.32	Stdv: 17.70	Stdv: 1.68	Stdv: 1.77	Stdv: 37.50					
D2	6.88	49.52	8.76	1.16	75.04	375	0.75	25	15	75
	Stdv: 0.19	Stdv: 10.36	Stdv: 1.56	Stdv: 0.86	Stdv: 15.15					
D3	6.67	32.65	7.72	1.55	75.46	352	0.75	22	16	75
	Stdv: 0.30	Stdv: 8.17	Stdv: 0.80	Stdv: 1.03	Stdv: 10.85					
C1	7.04	19.96	8.58	1.51	70.02	6,161	0.75	135	15	75
	Stdv: 1.13	Stdv: 12.39	Stdv: 3.10	Stdv: 1.01	Stdv: 23.68					
C2	7.69	6.97	5.00	2.67	61.33	1,907	1.5	120	15	10
	Stdv: 0.37	Stdv: 4.06	Stdv: 1.64	Stdv: 0.26	Stdv: 1.84					
C3	7.70	0.11	3.32	1.23	68.12	2,245	1.5	170	30	5
	Stdv: 0.37	Stdv: 0.06	Stdv: 0.13	Stdv: 0.17	Stdv: 1.52					

"horizontal activity traps" (modified HATs). A standard 1 mm pond net (frame size 20 × 25 cm) was used. Pond netting consisted of vigorous sweeping through the upper part of the water column, following the multihabitat technique recommended by the UK Pond Survey (Biggs et al., 1998). In each 3-min sampling period all mesohabitats, as defined in Table 2, were sampled in proportion to the occurrence. The traps were constructed from 2 l plastic bottles (32 cm high and 10 cm in diameter) and provided a one-way funnel system, allowing for the passive collection of biota. HATs were placed in both open water areas and dense stands of emergent vegetation. HATs were used in open water areas because of their adequate design. Modified HATs were designed to allow the traps to be submerged in areas with dense, heavy emergent vegetation. They consisted of plastic bottles placed in vertical position and fitted with randomly allocated tubes (5 cm in length, 3 cm in diameter) in order to allow the one-way funnel effect to take place. Both traps were held in position by the same support mechanism. They were fixed to a 1 m rod by means of a brace. The brace consisted of two rings cut from a four-inch drainage pipe and screwed together to form a figure of eight. The traps were supported by one ring and the second ring had two holes drilled in it for attachment to the support rod. The position of the traps on the rod could be varied, depending on depth of water (Fig. 1).

Sampling

Sampling was conducted on two separate occasions, March–April and July–August 2005. All ponds, except the first one, in each of the three pond systems were sampled. The first pond in each system had heavily polluted conditions that would have compromised health and safety standards due to the presence of untreated sewage and animal slurry. Generally, three, 3-min multihabitat samples were collected in each pond, but for the generation of the species accumulation curves five samples were taken in two ponds. The same applied to the traps. In general, 10 activity traps were placed in each mesohabitat for two consecutive nights. However, 20 traps were taken in two of the ponds. The entire HAT was submerged in the water to a water depth of 10 cm or the deepest possible without resting on the sediments if water

Table 2 Mesohabitats, plant species and ranges of water depth found within ponds of the study

Mesohabitats	Plant species	Water depth (cm)
Matted vegetation	Decayed *Carex* spp.	0 to10
	Lemna minor	
Emergent vegetation and open water	*Typha latifolia*	20 to 50
Open water		20 to >100
Matted and emergent vegetation		
Matted	*Lemna minor*	20 to 30
Emergent	*Potamogeton natans*	
	Alisma plantago-aquatica	
	Apium nodiflorum	
	Carex spp.	
	Eleocharis sp.	
	Epilobium sp.	
	Equisetum sp.	
	Filipendula ulmaria	
	Glyceria maxima	
	Myriophyllum verticillatum	
	Iris sp.	
	Juncus effusus	
	Phalaris sp.	
	Phragmites australis	
	Polygonum hydropiper	
	Stachys sp.	
	Rorippa nasturtium-aquaticum	
	Solanum dulcamara	
	Sparganium sp.	
	Typha latifolia	
	Ranunculus sceleratus	
Dense emergent vegetation	*Carex* spp.	0 to 30
Emergent vegetation	As "emergent" in "matted and emergent vegetation" above	20 to 30

Proportions of plant species of each mesohabitat varied seasonally and between ponds

depth was shallow. Before retrieval, the bottle was adjusted to a vertical position. The characteristics of the mesohabitat dictated the depth at which the modified HAT could be set and, therefore, not all entrance tubes were necessarily submerged. On some

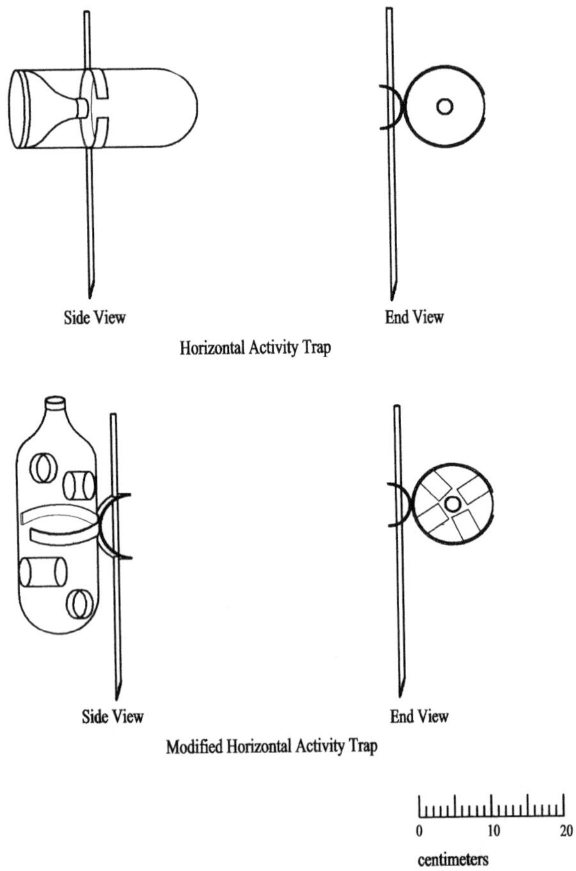

Fig. 1 Lateral and end view of horizontal activity trap and modified horizontal activity trap

occasions vegetation and sediment were moved to allow for deeper application of the trap. For retrieval, a cover made from a plastic bottle was placed over the modified HATs to prevent loss of the sample. Traps and support mechanism were carried to the bank and the entire contents, including the outer surfaces of the traps, were decanted into a 250 μm sieve. The contents of the sieve were then backwashed with 70% IMS solution into an appropriately labelled plastic bag. Once in the laboratory, all macroinvertebrates were sorted from the samples, except for *Asellus aquaticus* which was subsampled due to the high number of specimens present. Macroinvertebrates were identified to species level where possible using the keys Macan & Cooper (1977), Elliott & Mann (1979), Elliott et al. (1988), Friday (1988), Savage (1989), Wallace et al. (1990), Edington & Hildrew (1995) and Nilsson (1996, 1997). Oligochaetes, water mites and dipteran larvae were not identified further than family.

Statistical analysis

Unless stated differently, the data used for the analyses were derived from the first and last sampled ponds of Castle and Dunhill constructed wetlands, that is, C1, C3, D1 and D3. The data from spring and summer for each particular pond were pooled. SPSS (Version 11) was the statistical package used. The assumption of normality was tested by applying the Kolmogorov–Smirnov test. Differences between taxon richness count for multihabitat net sampling and activity traps were tested by two level nested design ANOVA. The comparison between HATs and Modified HATs was conducted using *t*-tests on samples collected in summer from dense stands of emergent vegetation in two randomly selected ponds. The data from the two seasons were combined for each sampling method to allow estimates of the numbers of taxa common and exclusive to each method. This was performed separately for each pond because of the statistical differences detected using the ANOVA analysis. Pond M2 was not included in the analysis because it had dried out during the summer.

The accumulation curve analyses were carried out taking into account the difference between the observed pond richness (S_{obs}), that is Mao Tau estimator and various estimators of the pond richness (S_{true}), Jackknife 1 (Jack 1), Chao 1, Chao 2 and ICE. The selection of the estimators was carried out following the recommendations by Brose et al. (2003), Heltshe & Forrester (1983), Hortal et al. (2006) and Foggo et al. (2003). For these analyses, 20 trap samples from both open water and dense stands of emergent vegetation in two randomly selected ponds were collected. Also collected in these ponds were five, 3-min multihabitat net samples. These analyses were computed using EstimateS (Version 7.5) with 1,000 randomisations.

Results

Common and exclusive taxa

There were significant differences in the taxon richness count between ponds both for multihabitat net sampling ($F_{1,2} = 77.796$, $P < 0.01$) and activity traps ($F_{1,2} = 4.606$, $P < 0.05$). In terms of their community composition, the coefficient of association T also revealed significant differences between ponds (Table 3).

The modified HATs within dense stands of emergent vegetation seemed to collect an average of 8% more taxa than the HATS. However, the difference was not significant ($P > 0.05$). When the exclusive taxa for HATs and modified HATs were analysed and compared to the netting of that mesohabitat, two taxa were found in modified HATs that were not present in HATs: *Hydrometra stagnorum* (L.) and Syrphidae.

There was no significant relationship between the number of exclusive taxa and percentage vegetation cover for the activity traps (Pearson correlations, $P > 0.05$). However, the relationship between exclusive taxa caught by netting and vegetation cover was significant (Pearson correlation $r = -0.871$, $P < 0.01$, Fig. 2). Fewer exclusive taxa were caught by netting in the heavily vegetated ponds.

Comparisons were made between the taxa caught by each method to establish which taxa were exclusive to a particular sampling method. For D1, 20% of the taxa were exclusive to traps, 25% were exclusive to nets and 55% were common to both methods. For D3, 23.8% of the taxa were exclusive to traps, 16.7% were exclusive to nets and 59.5% were common to both methods. In the case of C1, 25% of the taxa were exclusive to traps, 25% were exclusive to nets and 50% were common to both methods. Last but not least, for C3, 10.9% of the taxa were exclusive to traps, 38.2% were exclusive to nets and 50.9% were common to both methods. On average, 19.9% of the taxa were exclusive to traps, 26.2% were exclusive to nets and 53.9% were common to both methods. Highly mobile macroinvertebrates such as *Dytiscus marginalis* (L.) or *Acilius sulcatus* (L.) were generally exclusive to traps (Appendix 1).

Accumulation curves

Randomised species accumulation curves and various estimators, computed using EstimateS, were employed to evaluate the sampling effort required for 70%, 75% and 80% capture efficiency. This calculation was carried out taking into account the difference between the observed pond richness (S_{obs})

Table 3 Coefficient of association T comparing ponds D1, D3, C1 and C3 using the combination of 10 activity traps and three, 3 min multihabitat net sampling

Between ponds 1 and 2	A (pond 1)	B (pond 2)	C	S	U	V	Y	X	W	T	E	Significance
D1 compared to D3	41	43	30	52	11	9	13	11.0	−2	−0.49	0.273	NO
D1 compared to C1	41	45	33	53	12	8	12	8.0	0	−0.23	0.271	NO
D1 compared to C3	41	56	32	65	24	9	24	9.0	0	−0.31	0.244	NO
D3 compared to C1	43	45	34	54	11	9	11	9.0	0	−0.23	0.268	NO
D3 compared to C3	43	56	36	63	20	7	20	7.0	0	−0.24	0.248	NO
C1 compared to C3	45	56	40	61	16	5	16	5.0	0	−0.18	0.252	NO

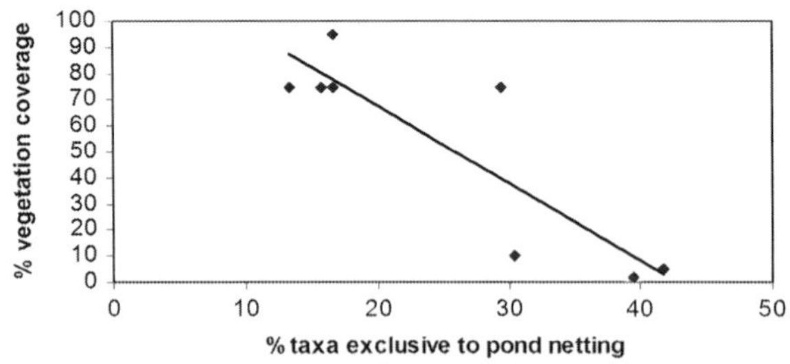

Fig. 2 Relationship between percentage of taxa exclusive to pond netting and percentage of vegetation cover (Pearson correlation, $r = -0.871$, $P < 0.01$)

and the estimated pond richness (S_{true}), Jack 1, Chao 1, Chao 2 and ICE estimators in this case (Figs. 3–5). For traps, depending on the estimator used, between 7 and 9 samples were required for an efficiency of 70%, between 9 and 11 for 75% efficiency and between 11 and 14 for 80% efficiency. In the case of the net sampling, the result also depended on the estimator used but the numbers were lower. Between 2 and 3 samples were required for an efficiency of 70%, between 3 and 4 for 75% and between 4 and 5 for 80% efficiency (Table 4).

Discussion

The variably vegetated nature of the ponds in ICWs makes them an ideal environment for testing the performance of activity traps and pond netting for the evaluation of aquatic macroinvertebrate biodiversity. This study showed variable efficiency of pond netting and activity traps in terms of the capture of exclusive taxa and that a high percentage of taxa collected were common to both methods. Therefore, as hypothesised, a combination of pond netting and activity traps will yield a more complete estimate of taxon richness. Hilsenhoff (1991) and Turner & Trexler (1997) achieved similar findings in comparative studies of netting and activity traps. Furthermore, Helgen et al. (1993) recognised that netting did not capture highly mobile macroinvertebrates. However, it is not possible to state with certainty that the taxa exclusive to traps of this study would not have been caught by netting if the sampling effort had been increased.

There is a need to distinguish between different mesohabitats in order to assess pond netting effectiveness. García-Criado & Trigal (2005) found that sweep netting performed generally well within dense stands of submerged vegetation. On the other hand, O'Connor et al. (2004) found that pond netting was not a suitable method among mosses when compared to the box method. This study also showed that the performance of netting was reduced in the heavily vegetated areas presumably because the vegetation can impede the free movement of the pond net and thus the probability of highly mobile macroinvertebrate capture decreases. Oertli et al. (2005) highlighted the difficulties of catching mobile adult Coleoptera. Even though a hand net of small rectangular frame (14 × 10 cm) was used to "facilitate movement within dense aquatic vegetation", its performance was limited. Another factor that is worth mentioning is the substrate. Muzaffar & Colbo (2002) showed that sweep netting was not effective when compared to rock bags for coarse rocks. Even though no precise description of substrate of the ponds of this study was feasible due to the low water visibility, it became apparent that coarse rocky material was not present.

The lack of significant difference in the performance of the HATs and modified HATs within dense stands of emergent vegetation indicates that different

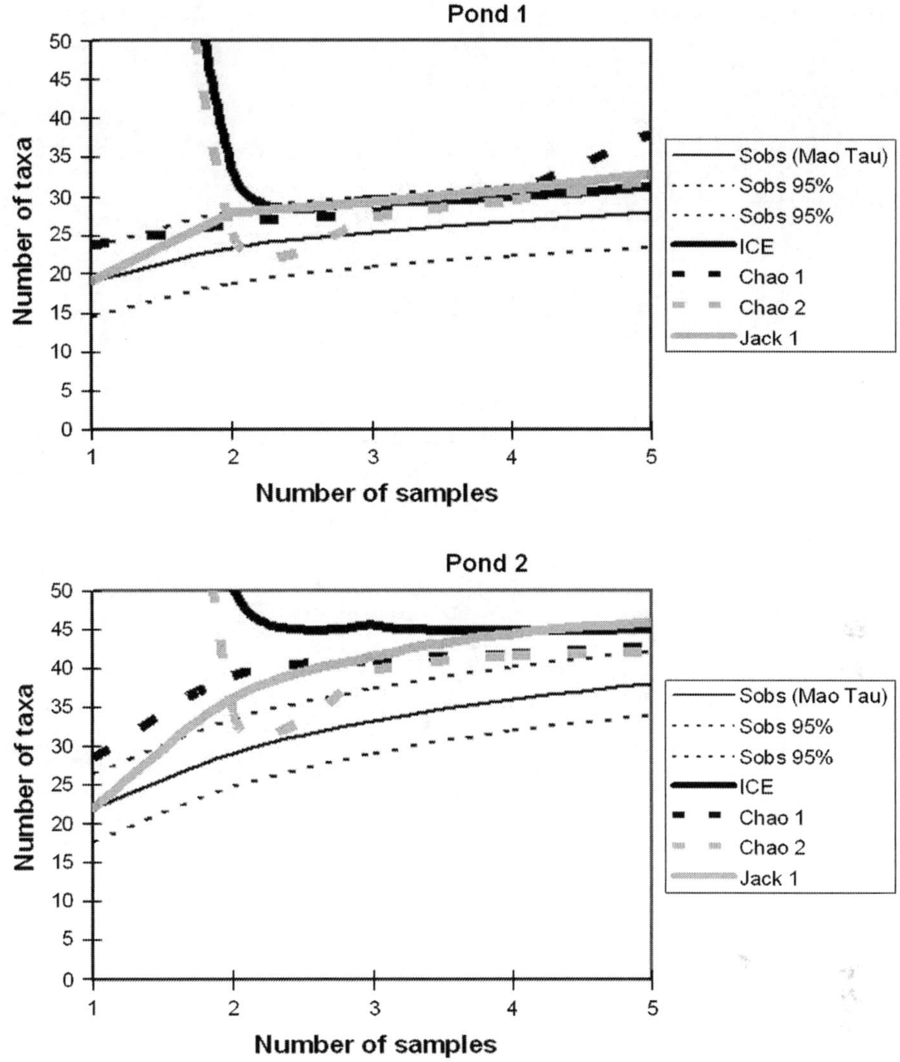

Fig. 3 Randomised species accumulation curve for five, 3 min multihabitat net sampling in two randomly selected ponds and ICE, Chao 1, Chao 2 and Jack 1 estimators (S_{obs} = observed number of taxa)

designs for water activity traps should be considered to maximise the success of capture. Both are equally efficient in vegetated areas and the design most appropriate for the conditions can be applied. However, modified HATs collected two exclusive taxa that were not collected by HATs: *Hydrometra stagnorum* and Syrphidae. The reason for capturing Syrphidae with modified HATs might have been that on some occasions, modified HATs rested on the sediments facilitating their entrance. In the case of *Hydrometra stagnorum*, the presence of the top tubes of the modified HATS at the surface level could allow surface dwellers to enter the traps. This corresponds with the type of taxa caught by Hanson et al. (2000) in specially designed surface-associated activity traps.

The results suggest that three samples for 3-min multihabitat netting and nine samples for activity traps would yield 70% efficiency. However, if higher efficiency is required a more rigorous testing of the method will be necessary.

Although there were no extreme weather fluctuations that would differ from average local conditions during the sampling periods, it cannot be stated with certainty how water temperature varied as no measurements were taken. Water temperature may indeed affect the performance of activity traps since temperature affects activity of invertebrates (Henrikson & Oscarson, 1978). On the other hand, Murkin et al. (1983) suggested that water temperature was not significantly correlated with the abundance of invertebrates captured with activity traps. The direction of

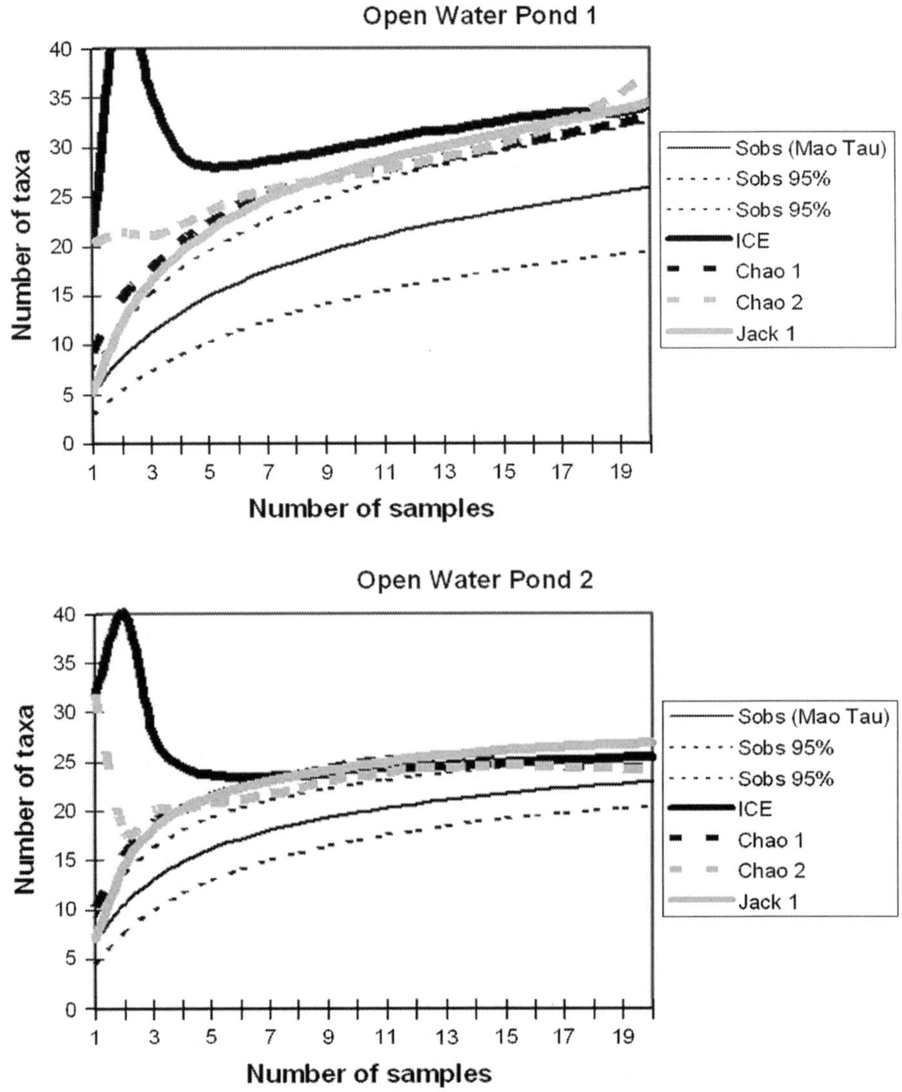

Fig. 4 Randomised species accumulation curve for a set of 20 activity traps in open water of two randomly selected ponds and ICE, Chao 1, Chao 2 and Jack 1 estimators (S_{obs} = observed number of taxa)

deployment of activity traps (horizontal or vertical) may also affect their effectiveness. Horizontally deployed activity traps were used for this study to minimise the effect of pond water depth. Vertically deployed activity traps are apparently not feasible for shallow ponds as observed by Muscha et al. (2001). Furthermore, the direction of the funnel in horizontally deployed activity traps may also be a potential factor to consider since steady prevailing winds may have an effect on pond water circulation which, in turn, may alter macroinvertebrate movements.

During this study, it became apparent that sampling with HATs is not as time-consuming as sampling using pond netting since samples are relatively clean. Sorting pond net samples can become a tedious process, especially in heavily vegetated ponds. The use of HATs is also beneficial because it is easy to standardise between field workers (Murkin et al., 1983). A disadvantage we have encountered is that these traps improve in efficiency when deployed for two consecutive nights, making a second visit to the pond a must. This disadvantage was also observed by Kenttämies et al. (1985) and Hyvönen & Nummi (2000). Another disadvantage is the effect of predation within the traps by fish, amphibians or invertebrate predators since this might alter the performance of activity traps. Elmberg et al. (1992) suggested that fish but

Fig. 5 Randomised species accumulation curve for a set of 20 activity traps in dense stands of emergent vegetation in two randomly selected ponds and ICE, Chao 1, Chao 2 and Jack 1 estimators (S_{obs} = observed number of taxa)

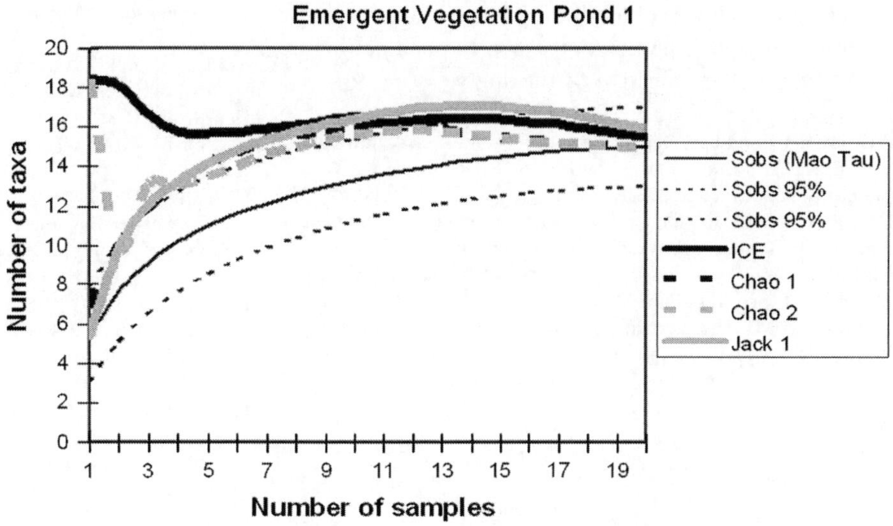

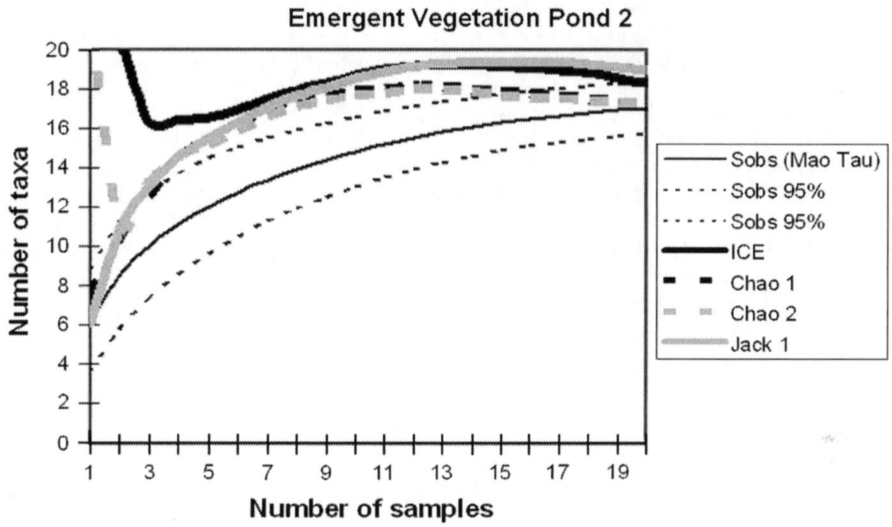

Table 4 Sampling effort required for an efficiency of 70%, 75% and 80% using the estimators ICE, Chao 1, Chao 2 and Jack 1 for both activity traps and three, 3 min multihabitat net sampling

	Estimated efficiency (%)	ICE	Chao 1	Chao 2	Jack 1
Number of activity trap samples	70	8	7	7	9
	75	10	9	9	11
	80	13	11	12	14
Number of 3 min multihabitat net samples	70	3	3	2	3
	75	3	4	3	4
	80	4	5	4	5

not invertebrate predators could affect the effectiveness of activity traps in terms of the invertebrate richness. In the present study, small *Gasterosteus aculeatus* L. specimens (<5 cm) were caught in a small number (<10%) of the traps. Overall, activity traps seem to be an effective macroinvertebrate capture method to be used in ponds, alone or in combination with other methods such as pond netting depending on the objective of the study.

Acknowledgements This project has been funded through the European Union programme INTERREG IIIA. We gratefully acknowledge the help and enthusiasm of Waterford County Council and the constant support of Prof. Thomas Bolger, Dr. Tasman Crowe, Dr. Jan-Robert Baars, Dr. Ronan Matson, Ciara Smith B.Sc., Dr. Robert Cruishanks, Maria Callanan, B.Sc., Siobhán McCarthy, B.Sc., Ms. Katharine Maurer and Ms. Pascale Morisset with various aspects of the project. We would also like to thank two anonymous referees for their valuable comments on a draft of this manuscript.

References

Biggs, J., G. Fox, P. Nicolet, D. Walker, M. M. Whitfield & P. Williams, 1998. A Guide to the Methods of the National Pond Survey. Pond Action, Oxford.

Brose, U., N. D. Martínez & R. J. Williams, 2003. Estimating species richness: sensitivity to sample coverage and insensitivity to spatial patterns. Ecology 84: 2364–2377.

Chapin, F. S., E. S. Zavaleta, V. T. Eviner, R. L. Naylor, P. M. Vitousek, H. L. Reynolds, D. U. Hooper, S. Lavorel, E. O. Sala, S. E. Hobbie, M. C. Mack & S. Díaz, 2000. Consequences of changing biodiversity. Nature 405: 234–242.

Edington, J. M. & A. G. Hildrew, 1995. A Revised Key to the Caseless Caddis Larvae of the British Isles with Notes on Their Ecology. Freshwater Biological Association, Scientific Publication No. 53, Ambleside.

Elliott, J. M. & K. H. Mann, 1979. A Key to the British Freshwater Leeches. Freshwater Biological Association, Scientific Publication No. 40, Ambleside.

Elliott, J. M., U. H. Humpesch & T. T. Macan, 1988. Larvae of the British Ephemeroptera. Freshwater Biological Association, Scientific Publication No. 49, Ambleside.

Elmberg, J., P. Nummi, H. Pöysä & K. Sjöberg, 1992. Do introducing predators affect the reliability of catches in activity traps? Hydrobiologia 239: 187–193.

Friday, L. E., 1988. A Key to Adults of British Water Beetles. Field Studies Council Publication 189.

Foggo, A., S. D. Rundle & D. T. Bilton, 2003. The net result: evaluating species richness extrapolation techniques for littoral pond invertebrates. Freshwater Biology 48: 1756–1764.

García-Criado, F. & C. Trigal, 2005. Comparison of several techniques for sampling macroinvertebrates in different habitats of a North Iberian pond. Hydrobiologia 545: 103–115.

Hanson, M. A., C. C. Roy, N. H. Euliss Jr., K. D. Zimmer, M. R. Riggs & M. G. Butler, 2000. A surface-associated activity trap for capturing water-surface and aquatic invertebrates in wetlands. Wetlands 20: 205–212.

Helgen, J. H., K. Thompson, J. P. Gathman M. Gernes, L. C. Ferrington & C. Wright, 1993. Developing an Index of Biological Integrity for 33 Depressional Wetlands in Minnesota. Minnesota Pollution Control Agency, Minnesota, USA.

Heltshe, J. & N. E. Forrester, 1983. Estimating species richness using the jackknife procedure. Biometrics 39: 1–11.

Henrikson, L. & H. Oscarson, 1978. A quantitative sampler for air-breathing aquatic insects. Freshwater Biology 8: 73–77.

Hilsenhoff, W. L., 1991. Comparison of bottle traps with a D-frame net for collecting adults and larvae of Dytiscidae and Hydrophilidae (Coleoptera). Coleopterists Bulletin 45: 143–146.

Hortal, J., P. A. V. Borges & C. Gaspar, 2006. Evaluating the performance of species richness estimators: sensitivity to sample grain size. Journal of Animal Ecology 75: 274–287.

Hyvönen, T. & P. Nummi, 2000. Activity traps and the corer: complementary methods for sampling aquatic invertebrates. Hydrobiologia 432: 121–125.

Indermuehle, N., B. Oertli, N. Menetrey & L. Sager, 2004. An overview of methods potentially suitable for pond biodiversity assessment. Archives des Sciences 57: 131–140.

Kenttämies, K., S. Haapaniemi, J. Hynynen, P. Joki-Heiskala & J. Kämäri, 1985. Biological characteristics of small acidic lakes in southern Finland. Aqua Fennica 15: 21–33.

Macan, T. T. & R. D. Cooper, 1977. A Key to the British Fresh- and Brackish-Water Gastropods. Freshwater Biological Association, Scientific Publication No. 13, Ambleside.

Mitsch, J. W. & J. G. Gosselink, 2000. Wetlands, 3rd edn. John Wiley, New York.

Murkin, H. R., P. G. Abbott & J. A. Kadlec, 1983. A comparison of activity traps and sweep nets for sampling nektonic invertebrates in wetlands. Freshwater Invertebrate Biology 2: 99–106.

Muscha, M. J., K. D. Zimmer, M. G. Butler & M. A. Hanson, 2001. A comparison of horizontally and vertically deployed aquatic invertebrate activity traps. Wetlands 21: 301–307.

Muzaffar, S. B. & M. H. Colbo, 2002. The effects of sampling technique on the ecological characterization of shallow, benthic macroinvertebrate communities in two Newfoundland ponds. Hydrobiologia 477: 31–39.

Nicolet, P., J. Biggs, G. Fox, M. J. Hodson, C. Reynolds, M. Whitfield & P. Williams, 2004. The wetland plant and macroinvertebrate assemblages of temporary ponds in England and Wales. Biological Conservation 120: 261–278.

Nilsson, A. N., 1996. Aquatic Insects of Northern Europe. A Taxonomic Handbook, Vol. 1. Apollo Books, Stenstrup.

Nilsson, A. N., 1997. Aquatic Insects of Northern Europe. A Taxonomic Handbook, Vol. 2. Apollo Books, Stenstrup.

O'Connor, A., S. Bradish, T. Reed, J. Moran, E. Regan, M. Visser, M. Gormally & M. Sheehy Skeffington, 2004. A comparison of the efficacy of pond-net and box sampling methods in turloughs—Irish ephemeral aquatic systems. Hydrobiologia 524: 133–144.

Oertli, B., J. D. Auderset, E. Castella, R. Juge, D. Cambin & J. B. Lachavanne, 2002. Does size matter? The relationship between pond area and biodiversity. Biological Conservation 104: 59–70.

Oertli, B., D. A. Joye, E. Castella, R. Juge, A. Lehmann & J. B. Lachavanne, 2005. PLOCH: a standardized method for sampling and assessing the biodiversity in ponds. Aquatic Conservation 15: 665–679.

Pieczynski, E. 1961. The trap method for capturing water mites (Hydracarina). Ekologia Polska Serie B: 111–115.

Savage, A. A., 1989. Adults of the British Aquatic Hemiptera Heteroptera: A Key with Ecological Notes. Freshwater Biological Association, Scientific Publication No. 50, Ambleside.

SIL, 2004. Conference programme of XXIX Societas Internationalis Limnologiae Congress. Lahti, Finland.

Turner, A. M. & J. C. Trexler, 1997. Sampling aquatic invertebrates from marshes: evaluating the option. Journal of the North American Benthological Society 16: 694–709.

Wallace, I. D., B. Wallace & G. N. Philipson, 1990. A Key to the Case-Bearing Caddis Larvae of Britain and Ireland. Freshwater Biological Association, Scientific Publication No. 51, Ambleside.

Williams, P., M. Whitfield, J. Biggs, S. Bray, G. Fox, P. Nicolet & D. Sear, 2004. Comparative biodiversity of rivers, streams, ditches and ponds in an agricultural landscape in Southern England. Biological Conservation 115: 329–341.

Hydrobiologia (2008) 597:109–123
DOI 10.1007/s10750-007-9226-7

ECOLOGY OF EUROPEAN PONDS

Developing a multimetric index of ecological integrity based on macroinvertebrates of mountain ponds in central Italy

Angelo G. Solimini · Marcello Bazzanti · Antonio Ruggiero · Gianmaria Carchini

Abstract The lack of biological systems for the assessment of ecological quality specific to mountain ponds prevents the effective management of these natural resources. In this article we develop an index based on macroinvertebrates sensitive to the gradient of nutrient enrichment. With this aim, we sampled 31 ponds along a gradient of trophy and with similar geomorphological characteristics and watershed use in protected areas of the central Apennines. A bioassessment protocol was adopted to collect and process benthic samples and key-associated physical, chemical, and biological variables during the summer growth season of 1998. We collected 61 genera of macroinvertebrates belonging to 31 families. We calculated 31 macroinvertebrate metrics based on selected and total taxa richness, richness of some key groups, abundance, functional groups and tolerance to organic pollution. The gradient of trophy was quantified with summer concentrations of chlorophyll *a*. We followed a stepwise procedure to evaluate the effectiveness of a given metric for use in the multimetric index. Those were the pollution tolerance metric ASPT, three metrics based on taxonomic richness (the richness of macroinvertebrate genera, the richness of chironomid taxa, and the percentage of total richness composed by Ephemeroptera, Odonata, and Trichoptera), two metrics based on FFG attributes (richness of collector gatherer taxa and richness of scraper taxa) and the habit-based metric richness of burrowers. The 95th percentile of each metric distribution among all ponds was trisected for metric scoring. The final Pond Macroinvertebrate Integrity Index ranged from 7 to 35 and had a good correlation ($R^2 = 0.71$) with the original gradient of environmental degradation.

Keywords Apennines · Ecological integrity · Macroinvertebrates · Mountain ponds · Bioassessment

Guest editors: R. Céréghino, J. Biggs, B. Oertli & S. Declerck
The ecology of European ponds: defining the characteristics of a neglected freshwater habitat

A. G. Solimini (✉)
European Commission, Joint Research Centre, Institute for Environment and Sustainability, TP 290, 21027 Ispra, Italy
e-mail: angelo.solimini@jrc.it

M. Bazzanti
Department of Animal and Human Biology, University 'La Sapienza', Viale dell'Università 32, 00185 Rome, Italy

A. Ruggiero
Laboratoire d'Ecologie des Hydrosystèmes, UMR 5177, Université Paul Sabatier, 118 route de Narbonne, 31062 Toulouse cedex 9, France

G. Carchini
Department of Biology, University 'Tor Vergata', Via della Ricerca Scientifica, 00133 Rome, Italy

Introduction

Small lentic freshwater bodies like wetlands, ponds, ditches, and ephemeral pools are increasingly threatened at the European scale as a result of the human-induced changes in the landscape. Evidence of a decrease in the number of ponds in several European countries has appeared in the last decade in the scientific literature (see for example the recent articles by Biggs et al., 2005; Sondergaard et al., 2005). At the same time, there has been a remarkable increase in awareness of the importance of ponds and wetlands as habitats for a variety of (unique) flora and fauna, initiating urgent scientific research on those ecosystems. While, on the one hand, research efforts were directed to increase our knowledge of basic pond ecology (De Meester et al., 2005), on the another hand, practical assessment tools that can be used by land managers were proposed (Biggs et al., 2000; Hicks & Nedean, 2000; Boix et al., 2005; Oertli et al., 2005a).

Most of the mountain areas in central Italy (central Apennines) are protected by local and/or national authorities. Here, the calcareous soil and karstic process result in a rapid loss of surface water. Therefore, the few natural and artificial ponds, often placed in natural parks, represent valuable freshwater ecosystems. Although being of high conservation status, the ponds of those areas are often affected by nutrient enrichment coming from livestock reared in the watershed during snow-free months. Livestock-derived excreta enter lakes and ponds either directly or through watershed drainage and represent a high potential source of nutrients (Steinman et al., 2003) and a contributory factor to the eutrophication of these ecosystems (Ruggiero et al., 2004). Park managers are faced with a range of problems related to eutrophication, such as blooms of algae, turbidity of waters and fish kills, and ask for reliable mitigating and preventative management strategies. The lack of a system for the assessment of ecological quality specific for these pond types prevents an effective management of these natural resources.

In shallow lentic ecosystems, nutrient enrichment can cause a shift from a clear water state with a dominance of aquatic macrophytes to a turbid water state dominated by phytoplankton (Scheffer et al., 1993). A previous study has already shown that patterns of nutrients, phytoplankton biomass, macrophyte cover, and water transparency among Appennine ponds are consistent with the existence of two alternative stable states (Ruggiero et al., 2005). These two stable states are the result of complex interactions, occurring in the pelagic and the benthic habitats, between the biota and some physical and chemical variables, such as water depth, extension of macrophyte coverage, and resuspension of sediments (Scheffer et al., 1993). Since the presence of submerged macrophytes has a strong influence on pond invertebrates (Diehl & Kornijòw, 1998; Solimini et al., 2000; Bazzanti et al., 2003; Della Bella et al., 2005), the clear and the turbid states may result in substantial differences in macroinvertebrate community structure. Therefore, the establishment of an assessment method based on macroinvertebrates is likely to prove an effective means of pond assessment.

In this article we develop a multimetric index of pond ecological degradation based on macroinvertebrates, sensitive to the gradient of nutrient enrichment (e.g., nutrient pressure). The index is specific to this pond type and nutrient pressure, and it is developed using the dataset gathered from a survey of water quality and macroinvertebrates targeting Appenninic mountain ponds more than 1,000 m above see level (asl), most of them impacted solely by cattle density (see Ruggiero et al., 2005).

Material and methods

Study ponds type description

The 31 study sites are representative of mountain ponds in the central Apennines. Watersheds (in general smaller than 300 ha) are mainly characterized by a large percentage of bare rocks with vegetation being represented only by meadows and bushes. Ponds are limited in depth and size: only six ponds are larger than 1 ha while only four ponds are deeper than 3 m (see Ruggiero et al., 2005; Carchini et al., 2005 for pond general morphological attributes). Their water originates from local precipitation and snow melting and the water level shows substantial variation over the year. During the winter, these ponds are ice-covered for a period ranging from 3 to 5 months.

Environmental variables

We sampled ponds twice during the summer of 1998. Recorded environmental variables included chlorophyll *a*, soluble reactive phosphorus (SRP, as P–PO_4), dissolved inorganic nitrogen (DIN, as N–NO_3 + N–NH_4), Secchi disk transparency, pH, and conductivity. Additional variables recorded were pond maximum depth, aquatic macrophyte coverage extension, pond area and altitude. Sampling dates, field, and laboratory methods, as well as water chemistry data for each pond are reported in detail elsewhere (Carchini et al., 2005; Ruggiero et al., 2005) and are not repeated here. Among the sampled ponds, three ponds were excluded and are not considered further in this article. They were (see Ruggiero et al., 2005 for pond location and acronyms): AQP because of exceptional water level fluctuation through the study period, NUR because of temporary attributes and PTZ because of its exceptionally high levels of humic substances.

Pond characterization

A previous detailed description of nutrients and chlorophyll *a* temporal patterns, based on an intensive sampling campaign (every 15 days during snow-free months) from a subset of Apennine ponds, indicated that summer water samples (e.g., June–September) are representative of the nutrient condition of a given pond (Ruggiero et al., 2003). The nutrient input in the watershed comes from cattle reared in the watersheds in summer (Solimini et al., 2000; Ruggiero et al., 2003). Additional variables linked to cattle presence were found to be water level variation and the extent of macrophyte beds (see Solimini et al., 2000; Ruggiero et al., 2005). For the purpose of this article (see below—metric evaluation; Table 1), ponds near pristine ("reference") condition were defined as those having low summer chlorophyll *a* (average summer chlorophyll *a* $<10\ \mu g\ l^{-1}$), extensive macrophyte beds (summer coverage >40% of pond area) and high transparency (high Secchi disk depth). Impaired ponds were defined as those having average summer chlorophyll *a* over $100\ \mu g\ l^{-1}$, very low water transparency and limited macrophyte cover. All the others were placed in an intermediate group. Applying those criteria, of the 28 ponds considered in this study, 4 ponds were in the reference group and 10 ponds in the impaired group.

Macroinvertebrate survey

The methods and timing of the macroinvertebrate survey was decided after a pilot study undertaken in 1997 in a subset of the ponds. The main outcomes of this pilot study were: (a) sample collection needs to include each macrophyte taxon because some macroinvertebrate taxa are only found associated with specific macrophyte taxa (if macrophytes are present; see also Solimini et al., 2003); (b) the early summer samples provide higher density and diversity of macroinvertebrates, but in late summer some additional taxa appear; (c) macroinvertebrate presence–absence data are enough to discriminate between impacted and not impacted by cattle (see Solimini et al., 2000).

We collected macroinvertebrate samples during summer 1998 (see Ruggiero et al., 2005 for sampling dates), visiting ponds twice (i.e., at the start and at the end of the summer season) with a hand net (dimension 30 × 30 cm; mesh size 0.5 mm). The chosen sampling period correspond with the time window, when macrophyte beds are well developed and when macroinvertebrates exhibit the higher density and diversity. At each pond, a total time of 3 min was proportionally allocated to the area of each pond mesohabitat, including aquatic vegetation (when present), sand sediments, and muddy bottom, following the Pond Action methods (1989). Special care was taken to sample diverse patches within the pond as shoreline, small beds of distinct macrophyte species, etc. Each vegetated mesohabitat was vigorously sampled, while less effort was used over soft sediment to avoid excessive clogging of the net. The samples were washed in the field to reduce the amount of material, composited in a single plastic bag and taken to the laboratory. Here, samples were sieved (1 mm mesh) to reduce further the amount of organic and inorganic material associated with macroinvertebrates and preserved in 70% ethanol.

Macroinvertebrate sorting and identification

In order to shorten the sorting time and because temporal variation is not of interest in this study, for

Table 1 Median, minimum, maximum and percentiles of the distribution of pond environmental variables in reference, intermediate and impaired groups (see text for definitions)

Group	Variable (units)	Median	Min	Max	25th perc	75th perc
Reference	Macrophyte cover (%)	72.5	40.0	100.0	45.0	97.5
	Chlorophyll *a* ($\mu g\ l^{-1}$)	4.3	1.8	8.2	2.0	7.3
	DIN ($mg\ l^{-1}$)	2.6	2.2	3.5	2.4	3.1
	Altitude (m asl)	1466	1350	1548	1393	1522
	Area (ha)	3729	1236	5311	1992	5010
	Maximum depth (m)	1.2	0.5	4.5	0.5	3.2
	Conductivity ($\mu s\ cm^{-1}$)	221	153	432	153	360
	pH	9.0	7.9	10.3	8.4	9.8
	SRP ($mg\ l^{-1}$)	0.08	0.06	0.11	0.07	0.10
	Secchi depth (m)	0.86	0.34	2.10	0.40	1.68
Impaired	Macrophyte cover (%)	0.0	0.0	25.0	0.0	5.0
	Chlorophyll *a* ($\mu g\ l^{-1}$)	389.9	123.1	1704.1	258.1	1137.3
	DIN ($mg\ l^{-1}$)	5.7	3.2	10.7	4.4	7.3
	Altitude (m asl)	1588	1115	2005	1225	1780
	Area (ha)	6360	352	12262	1030	9012
	Maximum depth (m)	1.0	0.5	3.5	0.7	2.1
	Conductivity ($\mu g\ cm^{-1}$)	181	72	464	143	216
	pH	9.0	7.6	9.7	8.4	9.4
	SRP ($mg\ l^{-1}$)	0.08	0.05	0.31	0.06	0.10
	Secchi depth (m)	0.14	0.03	0.54	0.07	0.14
Intermediate	Macrophyte cover (%)	67.5	5.0	100.0	15.0	95.0
	Chlorophyll *a* ($\mu g\ l^{-1}$)	24.1	1.6	80.8	5.3	55.1
	DIN ($mg\ l^{-1}$)	3.6	2.4	9.6	3.2	4.5
	Altitude (m asl)	1409	1014	1788	1243	1568
	Area (ha)	3116	120	27475	1649	7536
	Maximum depth (m)	1.3	0.3	7.5	0.4	1.7
	Conductivity ($\mu s\ cm^{-1}$)	224	134	438	189	259
	pH	8.5	7.2	9.6	8.0	8.9
	SRP ($mg\ l^{-1}$)	0.09	0.03	0.82	0.06	0.15
	Secchi depth (m)	0.37	0.22	0.78	0.26	0.58

each pond the samples coming from the two sampling dates were pooled and completely mixed. After pooling, a portion of the sample was placed in a white tray, diluted with water as needed and sorted by eye for macroinvertebrates. This procedure was repeated until the sample was completely sorted. For Oligochaeta and Chironomidae and for a few other groups (*Cloeon*, *Gyraulus*, Ceratopogonidae) depending on the specific pond, we stopped collection when ca. 100 individuals were taken up because of their very high number in the samples. When possible, macroinvertebrates were identified to the genus level, with the exception of Oligochaeta (family), Chironomidae (subfamily and tribe) and Ceratopogonidae (subfamily).

Multivariate analysis

A prerequisite in developing a biotic index is the response of the chosen group to the variables indicative of environmental degradation, after taken into account other potential covariables. In order to evaluate the relationships among macroinvertebrate assemblages (presence–absence) and environmental variables we ran a Canonical Correspondence

Analysis (CCA), using the Vegan package for R (Oksanen, 2005; available at http://www.cran.r-project.org/). Some of the environmental variables were correlated with one another and we used stepwise selection procedure to select a subset of them, after examination of ordination plots and calculation of Akaike information criterion (AIC). The selected model was chosen so that the variance inflation factor (VIF) of selected variables had VIF <10, giving a low AIC (see Oksanen, 2005 for details).

Macroinvertebrate metrics

We calculated 31 metrics based on taxa richness, pollution tolerance, habit traits and functional feeding groups (Table 2). Candidate metrics were chosen by reviewing the literature (US EPA, 1998; Biggs et al., 2000; Blocksom et al., 2002, Oertli et al., 2005b; Menetrey et al., 2005) or by our professional judgment.

Taxa richness was calculated both as total number of genera and as total number of families present in a sample, with the exceptions of Oligochaeta (family), Ceratopogonidae (subfamily) and Chironomidae (subfamily and tribe). In addition, we calculated a variety of richness measures for specific taxonomic groups that are known to respond to environmental degradation (number of genera of Ephemeroptera, Odonata, Mollusca, Coleoptera, and the number of tribes/subfamilies of Chironomidae). Those metrics were also calculated as a percentage of total assemblage richness composed by a specific taxonomic group.

As a metric of pollution tolerance, we considered only Average Score Per Taxon (ASPT), which is already included in another pond assessment protocol (Biggs et al., 2000). For the ASPT calculus, a score from 1 to 10 was assigned to each macroinvertebrate family based on the known tolerance to organic pollution (higher score means lower tolerance). Scores in a given sample were then averaged.

Metrics based on taxon habit and FFG (following Merritt & Cummins, 1984; Moog et al., 2002) were expressed both as simple richness of a given habit or FFG (e.g., total number of taxa of a given habit or FFG) and as richness of a given habit or FFG relative to total assemblage richness (e.g., the proportion of total richness of a given habit or FFG relative to overall taxa richness of the pond).

Metric evaluation

We followed a stepwise procedure to evaluate the effectiveness of a given metric for use in the multimetric index as described in US EPA (1998) and Blocksom et al. (2002). Several metrics were immediately eliminated because of poor applicability (low applicability is when the data for a given metric are too sparse, as most ponds have zero values). Four characteristics were then evaluated for each metric: (1) discriminatory power, (2) relative scope of impairment (RSI), (3) relationship with stressors, and (4) redundancy.

The discriminatory power is the metric ability to discriminate among reference and impaired lakes. A given metrics was judged of low discriminatory power when the interquartile (IQ) ranges of the metric distribution in reference and impaired sites were overlapping (see US EPA, 1998 for details). RSI is an indication of the inherent variability of a given metric because metrics too variable even in reference sites are unlikely to be effective for the assessment. Using only values from reference ponds RSI was calculated for each metric as $RSI = (IQ)/(q^a - q^b)$; where IQ is the interquartile range, q^a and q^b are the percentiles of the metric distribution. For metrics that decrease with environmental degradation $a = 25$ and $b = 0$ (minimum), whereas for metrics that increase with degradation $a =$ maximum and $b = 75$ (Barbour et al., 1996; US EPA, 1998). Metrics with an RSI >1 were excluded from further consideration (Blocksom et al., 2002).

The remaining metrics were checked against the trophic gradient (e.g., chlorophyll *a*, see below) by calculating the Spearman rank correlation coefficient. Only metrics having significant coefficients were retained. Finally, the Spearman rank correlations among the remaining metrics were calculated. If pairs of metrics were highly correlated (Spearman rank correlation >0.9), we retained the one having the highest correlation with the stressor.

Multimetric index development

The remaining metrics were scored based on expected values determined from the metric

Table 2 Definitions of tested macroinvertebrate metrics. All the metrics are expected to decrease in response to stress with the exception of pr_rich and pr_rich_p (see text)

Metric type	Metric	Definition
Richness measures	richgen	Number of genera
	richfam	Number of families
	eotr	Number of genera of Ephemeroptera + Odonata + Trichoptera
	odonor	Number of Odonata genera
	coleopr	Number of Coleoptera genera
	chr	Number of Chironomidae taxa
	molr	Number of Mollusca genera
	eotr_p	eotr/richgen
	odonor_p	odonor/richgen
	coleopr_p	coleopr/richgen
	chr_p	chr/richgen
	molr_p	molr/richgen
Pollution tolerance	ASPT	Average score per taxon
Functional feeding/ trophic group	pr_rich	Number of predator genera
	pr_rich_p	pr_rich/richgen
	cofi_rich	Number of collector filterer genera
	cofi_rich_p	cofi_rich/richgen
	coga_rich	Number of collector gatherer genera
	coga_rich_p	coga_rich/richgen
	sc_rich	Number of scraper genera
	sc_rich_p	sc_rich/richgen
	sh_rich	Number of shredder genera
	sh_rich_p	sh_rich/richgen
	omn_rich	Number of omnivorous genera
	omn_rich_p	omn_rich/richgen
Habit	burrow_rich	Number of burrower genera
	burrow_r_p	burrow_rich/richgen
	swim_rich	Number of swimmer genera
	swim_r_p	swim_rich/richgen
	climb_rich	Number of climber genera
	climb_rich_p	climb_rich/richgen

distribution across all sites. This method is preferable over one using the metric distribution from reference sites only when true reference conditions are difficult to determine (US EPA, 1998). The range between the minimum value and the 95th percentile of frequency distributions of metrics were trisected in equal slices and values of 1 (lower third part of frequency distribution), 3 (middle third), and 5 (top-one third) were assigned to each metric score (US EPA, 1998). The 95th percentile was chosen to avoid using anomalously high outliers as best expected values. Finally, these scores were summed into a single multimetric index (Pond Macroinvertebrate Integrity Index, PMII).

Results

Relationships between environmental variables and macroinvertebrate taxa

Some of the environmental variables were cross-correlated (Table 3). In particular, chlorophyll *a* was negatively correlated with macrophyte coverage and

Table 3 List of pairwise Spearman correlations among environmental variables (*$P < 0.05$; **$P < 0.01$)

	Macrophyte cover	Altitude	Area	Maximum depth	Chlorophyll *a*	SRP	Conductivity	pH	DIN
Altitude	−0.14								
Area	−0.09	0.07							
Maximum depth	−0.01	−0.06	0.43*						
Chlorophyll *a*	−0.79**	0.23	0.22	0.06					
SRP	0.17	−0.22	−0.28	−0.39*	0.01				
Conductivity	0.09	0.02	−0.51*	−0.27	−0.18	0.26			
pH	−0.17	−0.05	0.33	0.34	0.10	−0.08	−0.41*		
DIN	−0.44*	0.27	0.39*	0.04	0.61**	−0.21	−0.25	−0.01	
Secchi depth	0.57**	−0.20	0.22	0.48*	−0.64**	−0.14	0.11	0.15	−0.51*

SRP, soluble reactive phosphorus; DIN, dissolved inorganic nitrogen

Secchi depth and positively correlated with DIN. As expected, Secchi depth was correlated positively also with pond depth and macrophyte coverage. None of the variables were correlated with altitude.

We collected 61 genera of macroinvertebrates belonging to 31 families (Table 4). Canonical correspondence analysis was run excluding those taxa that occur in one pond only, and was used to analyze the environmental–macroinvertebrates relationships. The stepwise CCA selected a non-redundant subset of variables important in explaining the distribution of macroinvertebrates among ponds. The selected model was significant (Monte Carlo test $P < 0.05$ after 1,000 permutations under the full model) and the first three axes explained 19% of cumulative variance in macroinvertebrate data. On the first ordination axis, chlorophyll *a*, transparency and aquatic macrophyte coverage were the most important environmental variables (Fig. 1). The second axis of the CCA was defined by pH and conductivity. The macroinvertebrates characterizing the most eutrophic ponds were the hemipterans *Hesperocorixa* and *Sigara* and the chironomid *Chironomus*, whereas the taxa characterizing less eutrophic ponds were Enchytraeidae, the chironomids Pentaneurini, Corynoneurini and Tanytarsini, the ephemeropteran *Cloeon* and some odonate and coleopteran genera (Fig. 1).

Multimetric index development

The 31 metrics tested are showed in Table 5. Eight of the metrics were not applicable, because most of the ponds had zero or none values. Many of the metrics were highly correlated with one another and with total invertebrate richness. Comparing the frequency distributions of all metrics in reference and impaired lakes, we eliminated 10 metrics with overlapping IQ ranges (Table 6). Of the remaining metrics, four had an IQ coefficient greater than 1, indicating poor RSI. The remaining 9 metrics were checked against the environmental degradation gradient by calculating the Spearman rank correlation coefficient with chlorophyll *a*. All the metrics were significantly correlated with chlorophyll *a* and were retained. Finally, 2 metrics (omn_rich and richfam) were highly correlated with richgen (Spearman rank correlation >0.9) and were eliminated at this stage. The final selection included 7 metrics. Those were the pollution tolerance metric ASPT, 3 metrics based on taxonomic richness (the richness of macroinvertebrate genera, the richness of chironomid taxa, and the percentage of total richness composed by Ephemeroptera, Odonata and Trichoptera), 2 metrics based on FFG attributes (richness of collector-gatherer taxa and richness of scraper taxa) and the habit-based metric richness of burrowers.

The 95th percentile distribution of values of the selected metrics among all ponds (Table 5) was divided into three equal portions and each pond received a value of 1, 3, or 5 accordingly. The thresholds among classes resulted as follow (moderate, best): ASPT (3.4, 4.1); chr (2.9, 4.7); coga_rich (4.0, 5.5); eotr_p (0.1, 0.2); rich_gen (10.7, 17.4); burrow_rich (3.5, 5.0); sc_rich (1.6, 2.7). For each pond, the final multimetric index (Pond

Table 4 List of taxa and number of ponds in which macroinvertebrates were collected

Major taxonomic group	Family	Genus/tribe/ subfamily	Label	Number of ponds
Coleoptera	Dryopidae	*Dryops*	Dry	2
	Dytiscidae	*Acilius*	Aci	3
		Agabus	Aga	4
		Colymbetes	Col	4
		Dytiscus	Dyt	8
		Hydroporus	Hydr	2
		Hygrotus	Hyg	4
		Ilybius	Ilb	1
		Laccophilus	Lac	6
		Porhydrus	Por	1
		Potamonectes	Pot	1
	Haliplidae	*Haliplus*	Hal	7
	Helophoridae	*Helophorus*	Helo	4
	Hydrophilidae	*Enochrus*	Eno	3
		Helochares	Heo	1
		Hydrobius	Hyd	3
	Noteridae	*Noterus*	Not	3
Diptera	Ceratopogonidae	Heleinae	Hele	1
	Chironomidae	Chironomini	Chi	20
		Chironomus	Chir	21
		Corynoneurini	Cory	3
		Orthocladiinae	Ort	23
		Pentaneurini	Pen	4
		Tanypodini	Tan	16
		Tanytarsini	Tany	13
	Culicidae	*Culex*	Cul	1
Ephemeroptera	Baetidae	*Cloeon*	Cle	13
Heteroptera	Corixidae	*Corixa*	Cor	15
		Hesperocorixa	Hes	8
		Micronecta	Mic	3
		Sigara	Sig	11
	Naucoridae	*Ilyocoris*	Ily	5
	Notonectidae	*Anisops*	Ani	1
		Notonecta	Noto	17
	Pleidae	*Plea*	Ple	1
Hirudinea	Erpobdellidae	*Erpobdella*	Erp	3
	Glossiphonidae	*Helobdella*	Hel	6
	Hirudinidae	*Hirudo*	Hir	3
Isopoda	Asellidae	*Proasellus*	Pro	1
Mollusca	Lymnaeidae	*Lymnaea*	Lym	10
	Pisidiidae	*Pisidium*	Pis	7
	Planorbidae	*Gyraulus*	Gyr	1
	Sphaeriidae	*Musculium*	Mus	4

Table 4 continued

Major taxonomic group	Family	Genus/tribe/ subfamily	Label	Number of ponds
Odonata	Aeshnidae	*Aeshna*	Aes	1
		Anax	Ana	7
	Coenagrionidae	*Cercion*	Cer	2
		Coenagrion	Coe	9
		Enallagma	Ena	16
		Ischnura	Isc	6
	Lestidae	*Lestes*	Les	1
		Sympecma	Syp	1
	Libellulidae	*Crocothemis*	Cro	1
		Libellula	Lib	8
		Orthetrum	Orth	1
		Sympetrum	Sym	4
Oligochaeta	Enchytraeidae		Enc	2
	Naididae		Nai	16
	Tubificidae		Tub	28
Plecoptera	Nemouridae	*Nemoura*	Nem	1
Seriata	Planaridae	*Planaria*	Pla	4
Trichoptera	Limnephilidae	*Limnephilus*	Lim	13

Labels are used in Fig. 1

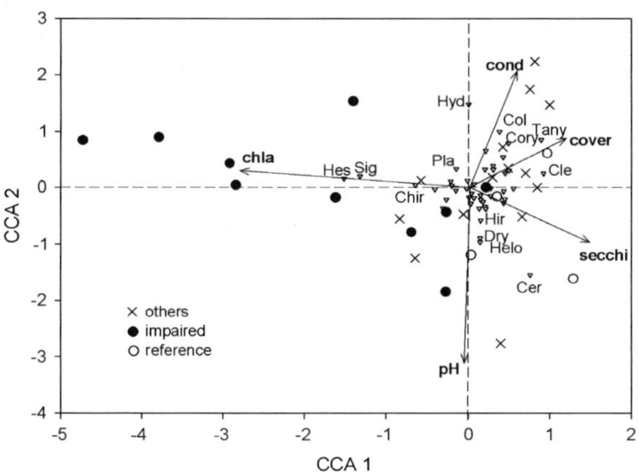

Fig. 1 Canonical Correspondence Analysis biplot of Apennine ponds (*open circles* for reference ponds, *closed circles* for impaired ponds and *x* for other ponds) and selected macroinvertebrate taxa (*triangles*). The environmental variables were selected with a stepwise procedure (chl-a = chlorophyll a; secchi = Secchi depth; cover = macrophyte coverage; cond = conductivity; see text). Macroinvertebrate taxa acronyms are in Table 4

Macroinvertebrate Integrity Index, PMII) summed the 7 scores obtained by each index. The PMII ranged from 7 to 35 and showed a good correlation with the original gradient of environmental degradation represented by the chlorophyll a (Pearson $R^2 = 0.71$, $N = 28$, $P < 0.05$; Fig. 2).

Discussion

Trophic gradient and reference ponds

When developing a pressure specific bioassessment system, the choice of the environmental variable representative of the pressure gradient is critical for the final result. Here, we quantified the gradient of trophy using mean summer chlorophyll a. As ice deeply covers ponds during winter (approximately from late November to March), the temporal window for plant growth is shorter than in lowland ponds. The maximum values of chlorophyll a are reached between June and July and are maintained until the temperature drops in late August (Ruggiero et al., 2003). This temporal pattern is also consistent among subsequent years (Ruggiero et al., 2004).

As a quantitative indicator of pond trophy, we preferred the concentration of chlorophyll a over measures of nutrient concentrations. Since ponds with low chlorophyll a, high transparency, and abundant macrophytes can also exist at relatively

Table 5 Median, maximum and percentiles of metric distributions for Apennine ponds

Metric	All ponds						Reference					Impaired				
	Median	Min	Max	25th perc	75th perc	95th perc	Median	Min	Max	25th perc	75th perc	Median	Min	Max	25th perc	75th perc
ASPT	4.0	2.7	4.8	3.4	4.3	4.8	4.2	3.8	4.8	3.9	4.4	3.6	2.7	4.3	2.9	4.2
burrow_r_p	0.3	0.2	0.8	0.3	0.4	0.7	0.3	0.3	0.8	0.3	0.4	0.4	0.2	0.8	0.3	0.6
burrow_rich	4.5	2.0	6.5	3.1	5.5	6.5	5.5	4.5	6.5	4.6	6.4	3.0	2.0	6.5	2.0	3.6
chr	4.0	1.0	7.0	2.3	4.8	6.6	5.0	4.0	7.0	4.3	5.8	2.0	1.0	4.0	1.8	3.0
chr_p	0.3	0.1	0.5	0.2	0.3	0.5	0.3	0.3	0.5	0.3	0.4	0.3	0.1	0.5	0.2	0.4
climb_rich	3.0	0.5	7.0	1.5	4.4	6.6	3.8	1.5	7.0	2.0	4.8	1.8	0.5	3.5	1.0	2.6
climb_rich_p	0.2	0.1	0.4	0.2	0.3	0.4	0.2	0.1	0.4	0.1	0.2	0.2	0.1	0.3	0.2	0.3
cofi_rich	0.0	0.0	2.0	0.0	0.8	2.0	0.0	0.0	2.0	0.0	0.8	0.0	0.0	2.0	0.0	0.0
cofi_rich_p	0.0	0.0	0.1	0.0	0.0	0.1	0.0	0.0	0.1	0.0	0.0	0.0	0.0	0.1	0.0	0.0
coga_rich	4.7	2.5	7.2	3.5	5.8	7.0	5.3	4.8	7.2	4.9	6.5	3.5	2.5	5.0	3.0	4.4
coga_rich_p	0.4	0.2	0.8	0.3	0.5	0.7	0.3	0.2	0.8	0.3	0.4	0.5	0.2	0.8	0.4	0.6
coleopr	1.5	0.0	6.0	0.0	3.0	6.0	1.5	0.0	6.0	0.0	3.8	0.0	0.0	2.0	0.0	1.3
coleopr_p	0.1	0.0	0.3	0.0	0.2	0.3	0.1	0.0	0.3	0.0	0.2	0.0	0.0	0.2	0.0	0.1
eotr	2.5	0.0	8.0	1.0	4.0	7.6	4.5	2.0	8.0	2.3	6.8	1.0	0.0	2.0	0.0	2.0
eotr_p	0.2	0.0	0.4	0.1	0.3	0.4	0.3	0.1	0.4	0.2	0.3	0.1	0.0	0.3	0.0	0.2
molr	0.5	0.0	3.0	0.0	1.0	3.0	0.5	0.0	3.0	0.0	1.8	0.0	0.0	2.0	0.0	0.3
molr_p	0.02	0.0	0.2	0.0	0.1	0.2	0.0	0.0	0.2	0.0	0.1	0.0	0.0	0.1	0.0	0.0
odonor	2.0	0.0	6.0	1.0	3.0	5.6	2.5	1.0	6.0	1.0	4.8	0.5	0.0	2.0	0.0	1.3
odonor_p	0.1	0.0	0.3	0.1	0.2	0.3	0.1	0.1	0.3	0.1	0.2	0.0	0.0	0.3	0.0	0.2
omn_rich	3.8	0.5	7.5	2.0	5.5	6.8	5.0	4.0	7.5	4.1	5.9	1.8	0.5	2.5	0.9	2.1
omn_rich_p	0.3	0.1	0.4	0.2	0.3	0.4	0.3	0.3	0.4	0.3	0.3	0.2	0.1	0.4	0.1	0.3
pre_rich	5.0	0.0	12.5	2.3	7.8	11.8	7.3	3.0	12.5	3.5	10.6	2.0	0.0	5.0	0.0	4.3
pre_rich_p	0.4	0.0	0.6	0.3	0.4	0.5	0.4	0.2	0.6	0.3	0.5	0.3	0.0	0.4	0.0	0.4
richfam	9.0	3.0	17.0	5.5	13.0	16.1	12.0	9.0	17.0	9.5	13.0	4.5	3.0	9.0	3.8	8.0
richgen	14.0	4.0	25.0	8.5	19.5	24.1	17.0	13.0	25.0	13.3	20.8	7.5	4.0	14.0	4.8	11.3
sc_rich	2.0	0.5	4.2	1.0	2.3	3.8	2.2	2.0	4.2	2.0	2.3	1.0	0.5	1.5	0.5	1.4
sc_rich_p	0.1	0.0	0.2	0.0	0.1	0.2	0.1	0.0	0.2	0.0	0.1	0.1	0.0	0.2	0.0	0.1
sh_rich	1.2	0.0	2.7	0.5	1.7	2.6	1.9	1.5	2.7	1.6	2.4	0.7	0.0	1.8	0.5	1.5
sh_rich_p	0.05	0.0	0.2	0.0	0.1	0.1	0.0	0.0	0.3	0.0	0.1	0.1	0.0	0.2	0.0	0.1
Swim_r_p	0.2	0.0	0.4	0.1	0.2	0.4	0.2	0.1	0.4	0.1	0.2	0.2	0.0	0.4	0.1	0.3
Swim_rich	3.0	0.0	5.5	1.1	4.0	5.5	2.8	2.0	5.5	2.1	3.8	1.8	0.0	4.0	0.9	3.1

Values are showed separately for all ponds together ($N = 28$) and for reference ($N = 4$) and impaired ($N = 10$) groups. Metrics acronyms as in Table 2

high nutrient concentrations (see Ruggiero et al., 2003 for details), these concentrations alone cannot predict effectively the status of the ponds. Moreover, a single nutrient (either phosphorus or nitrogen) is unlikely to limit phytoplankton growth in vegetated ponds as macrophytes control algal biomass via several mechanisms (reviewed in Scheffer, 1998). Alternatively, we could have used the extension of macrophyte coverage that was negatively correlated with chlorophyll *a*. However, since macrophyte area extension is often visually estimated (as it was in this present study), it is influenced by the operator who makes the assessment.

The identification of reference condition is a complex task in modern multimetric bioassessment systems (Cardoso et al., 2007; see also Oertli et al., 2005b). We could not find in the literature any numerical criteria for defining a pond in near pristine

Table 6 Metric evaluation and stepwise selection procedure (explanations in the text) Metric acronyms like in Table 2. Spearman correlation coefficient between each metric and chlorophyll *a* (**$P < 0.001$; *$P < 0.05$; ns = not significant). Applicability of metrics is low when data are too sparse. Discriminatory power is low when the interquartile ranges of impaired and reference groups are overlapping (see Table 5). Relative scope of impairment (RSI) is the metric ability to discriminate among reference and impaired (worst when >1; see Methods for explanation on RSI calculus). Metrics are redundant if highly correlated with other metrics selected at this stage

Metric	Spearman	Applicability	Discriminatory power	Relative scope of impairment	Redundant	Selected
ASPT	−0.56**			0.1		Yes
burrow_r_p	0.38^{ns}		Low	0.2		
burrow_rich	−0.77**			0.4		Yes
chr	−0.69**			0.4		Yes
chr_p	0.20^{ns}		Low	0.4		
climb_rich	−0.56**			1.4		
climb_rich_p	0.16^{ns}		Low	0.6		
cofi_rich	0.20^{ns}	Low				
cofi_rich_p	0.17^{ns}	Low				
coga_rich	−0.67**			0.3		Yes
coga_rich_p	−0.59**		Low	0.4		
coleopr	−0.54**	Low				
coleopr_p	−0.42*	Low				
eotr	−0.81**			2.0		
eotr_p	−0.61**			0.9		Yes
molr	−0.39*	Low				
molr_p	0.32^{ns}	Low				
odonor	0.75**			3.8		
odonor_p	−0.51**		Low	2.0		
omn_rich	−0.70**			0.4	Yes[a]	
omn_rich_p	−0.67**	Low		0.1		
pre_rich	−0.72**			2.0		
pre_rich_p	−0.52**		Low	0.9		
richfam	−0.79**			0.4	Yes[b]	
richgen	−0.79**			0.6		Yes
sc_rich	−0.80**			0.2		Yes
sc_rich_p	−0.52**		Low	1.1		
sh_rich	−0.44*	Low		0.5		
sh_rich_p	−0.53**		Low	1.3		
swim_r_p	0.30^{ns}		Low	0.3		
swim_rich	0.33^{ns}		Low	0.8		

Notes: [a] Correlation with richgen (Spearman rho = 0.95; $P < 0.01$); [b] Correlation with richgen (Spearman rho = 0.92; $P < 0.01$)

condition. Recently Carvalho & Moe (2005) reported summer chlorophyll *a* concentrations between 5 and 8 μg l^{-1} for a large number of European shallow lakes (e.g., average depth <3 m) as being in reference condition. As small shallow lakes and ponds share some limnological features (see Fairchild et al., 2005; Sondegard et al., 2005), for the purpose of this article, we used a similar threshold (chlorophyll *a* <10 μg l^{-1}) to identify ponds considered to be in reference conditions. Although we did not define numerical criteria for nutrients (see the above argument on trophic gradient), we checked to ensure if in this group there were no obviously high concentrations of dissolved nutrients. In addition, we excluded ponds not having extended macrophyte beds (less than 40% of the pond area) and with turbid waters (low Secchi depths). We also excluded a few ponds having low chlorophyll *a* but showing large water level lowering during summer because of the critical effect of this variable on macrophytes (Chow-Fraser, 2005; Van Geest et al., 2005). Altogether those five conditions should ensure a correct application of the reference concept, at least for the Apennine pond type.

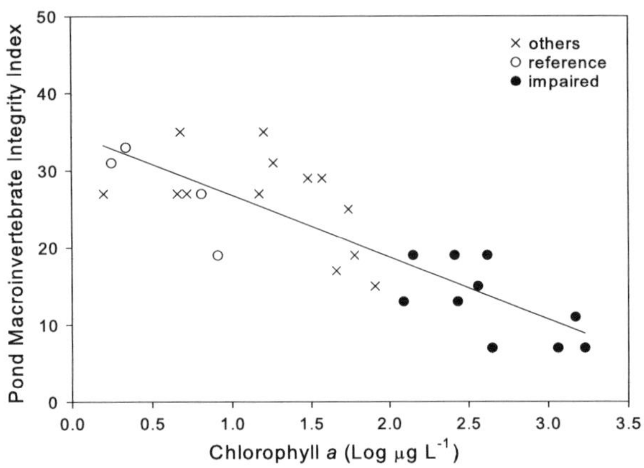

Fig. 2 Relationship between the Pond Macroinvertebrate Integrity Index (PMII) and the trophic gradient in Apennine ponds. The regression was significant ($R^2 = 0.71$, $N = 28$, $P < 0.05$). Chlorophyll *a* ($\mu g\ l^{-1}$) was log transformed

Macroinvertebrate response to trophic gradient

In the CCA ordination, very eutrophic ponds were grouped away from those with higher aquatic vegetation coverage, pointing out the close relationships between macroinvertebrates and macrophytes. Therefore, the macroinvertebrate assemblage probably responds to modification in the extension of pond macrophyte coverage. For example, the most eutrophic and muddy ponds were characterized by the presence of two hemipterans Corixidae (*Hesperocorixa* and *Sigara*) and the chironomid *Chironomus*. These taxa are known as tolerant to deoxygenation of water (Saether, 1979; Wiederholm, 1980; Rosemberg & Resh, 1993) and transitory anoxic conditions are likely to occur in those ponds (Ruggiero et al., 2004; Fairchild et al., 2005). In contrast, the chironomids Pentaneurini, Corynoneurini, and Tanytarsini are tribes often linked to less eutrophic and more vegetation-rich ponds (Wiederholm, 1983; Coffman & Ferrington, 1984).

Although significant, the CCA analysis based on presence/absence data showed a low variance explained by first canonical axes, reflecting the fact that most of the taxa were grouped near the origin of the ordination. This limited variability and low taxa richness were already reported from observations on a specific group (Odonata, Carchini et al., 2005) and may be a general feature of macroinvertebrate assemblages of Apennine ponds. Organisms living in Apennine ponds must cope with a short activity period when ponds are ice free, large daily temperature and wind fluctuations, a variable regime of rain, and nutrient enrichment coming from cattle.

Comparisons of taxonomic richness among different studies are difficult because of different methodological approaches, but it is instructive to compare the family richness of the Apennine ponds with similar environments in the Swiss Alps. In the most impacted ponds we recorded only three macroinvertebrate families, whereas the overall maximum was 17. A low-macroinvertebrate family richness (ranging between 2 and 11) is also reported for Swiss Alpine ponds over 1,800 m asl. (Hinden et al., 2005), but here the low level of nutrients, in addition to climate related factors, explained the pattern. Menetrey et al. (2005) found a median family richness ranging from 25 for eutrophic to 20 for hypertrophic ponds of the Swiss colline belt, closer to our findings. Another reason of the low family richness that we recorded in the Apennines probably reflects the fact that most of the ponds selected were somehow impacted by nutrient enrichment. Extending the dataset to other pristine ponds might undoubtedly give more insights on this issue.

Almost all the richness-based metrics were negatively correlated with the chlorophyll *a* gradient. The decrease of pond macroinvertebrate species richness along a trophic gradient was already reported for Swiss mountain ponds by Menetrey et al. (2005) and for a set of Italian lowland ponds by Della Bella et al. (2005). In our study, this negative pattern was robust regardless of the taxonomic level used (genus or family). On the one hand, increasing the taxonomic level from genus to family would decrease the time needed for processing the biological samples; however, on the another hand, functional attributes are usually different among genera within a family (Merritt & Cummins, 1984). Therefore, for correct FFG attribution of taxa, the genus level is needed.

Also most of the metrics based on functional attributes showed a diminishing pattern with chlorophyll *a*. Proportion indexes based on FFG (richness of a given FFG over total richness) significantly correlated with chlorophyll *a* were predator richness (negatively) and the collector gatherer richness (positively). Also the proportion of omnivorous taxa richness over total richness and the simple burrower richness were negatively correlated with chlorophyll *a*. Therefore, the macroinvertebrate assemblage of more eutrophic ponds is largely composed of collector–gatherer and burrower taxa that can switch to different food sources and of few (tolerant) predators and omnivores. This suggests that an increase in pond algal biomass and related increasing stress tend to reduce the diversity of all FFG and habit groups, according to the pattern of species richness.

Multimetric index development

Low variability in reference conditions and high discriminatory power between impacted and reference ponds are highly desirable characteristics of biological metrics. In the stepwise procedure of metric selection that we followed here, most of the macroinvertebrate metrics were eliminated because they were unable to discriminate among reference and impacted ponds or because they were too variable in reference ponds.

The seven metrics selected to form the multimetric index reflect different features of the invertebrate assemblage. Three are based on taxa richness. Of those, two are based on simple taxa richness (total genera richness and Chironomidae richness) and another is the total richness composed by Ephemeroptera, Odonata, and Trichoptera. High richness generally indicates undisturbed or unpolluted conditions (Barbour et al., 1996). In our study, richness-related metrics were among the highest correlated with the trophic gradient. Ponds with higher taxa richness were those having a large macrophyte cover, which increases invertebrate habitat diversity by providing food, substrate, and refuge (Diehl & Kornijow, 1998). The inclusion of chironomid richness in the index may pose serious problems when processing samples because of the difficulties of larval identification for genera and tribes. However, chironomids are a major numerical constituent of pond macroinvertebarte assemblages and are widely used for biological assessment of lentic waters (cf. Saether, 1979; Wiederholm, 1980; Rosenberg & Resh, 1993). Therefore, we advocate the need to include this group also for pond bioassessment. The identification of chironomidae genera or even species, although time demanding, may provide a valuable information on the response to environmental stress.

Three other metrics selected for the index reflect functional attributes and are based on functional feeding groups (richness of collector–gatherers and richness of scrapers) and on ecological habits (richness of burrowers). Functional feeding group metrics have a long history of application in streams and wadeable rivers (Plafkin et al., 1989) and were applied recently to pond macroinvertebrates analysis (Merritt et al., 2002; Bazzanti & Della Bella, 2004). However, the use of these functional aspects in the biological assessment of ponds is still lagging behind. In this study, the diminishing of richness of scraper taxa is linked with the progressive disappearance of macrophyte beds along the gradient of trophy. Submerged macrophytes, although of little nutritive value when alive (Newman, 1991), greatly increase the ecological opportunities for macroinvertebrates, adding a third dimension in the pond ecosystem (Varga, 2003). For example, a larger number of invertebrate predators (like zygopterans, some dipterans and coleopterans) are associated with dense vegetation cover, while few predator taxa (like anisopterans and other dipterans) are found in macrophyte-free habitats. In eutrophic ponds, even small increases in the portion covered by submerged macrophytes can have a positive effect on the invertebrate assemblage diversity (see for example Solimini et al., 2003; Della Bella et al., 2005). Large-sized scrapers, like snails and ephemeropterans, consume periphyton that can grow on floating and submerged plants and dominate the assemblage. However, when aquatic vegetation is not present, gathering collectors, like oligochaetes and *Chironomus*, feed on the organic matter which falls down from the productive pelagic habitat and are numerically predominant even if they exhibit a low diversity (Bazzanti & Della Bella, 2004).

Finally, the pollution tolerant metric ASPT is already part of the British PSYM system (Biggs et al., 2000) for pond ecological assessment and indicates the average tolerance of the macrobenthic assemblage. The selection of such a metric in the final index is remarkable, as it was the only metric directly related to organic pollution that we tested. Although the tolerance scores for some taxa probably need refinement and adaptation for use in Italian ponds (e.g., those computed for the Odonata genera), the ASPT showed a high correlation with the trophic gradient and a good capacity of discrimination among reference and impacted ponds.

Conclusion

The final PMII is the summation of seven metrics reflecting different aspects of the structure and function of the macroinvertebrate assemblage. The index range (7–35) can be subdivided into any number of categories (e.g., classes of ecological quality) corresponding to various level of impairment. However, if five categories are needed, as for an assessment compliant with the Water Framework Directive (see Heiskanen et al., 2004), careful calibration of class boundary values is needed against an independent dataset. As regards the scoring of metrics, the trisection choice (US EPA, 1998) was forced by the relatively low variation of each metric. Additionally, we decided to use the 95th percentile of the whole population of ponds as upper limit, because of the limited number of reference ponds that we could include in the study.

The very good correlation of the index PMII is a promising result. The type and pressure specific assessment are no doubt key to the performance of the index. However, the limited variation of many metrics and the low number of reference ponds included in the study stress the need for further refinement and validation of the index. Validation on an independent dataset and the assessment of temporal variability of the index are important issues and efforts should be directed to collect data from additional ponds. In particular, testing its performance for biological assessment of lowland ponds is recommended.

Acknowledgments We thank G. Free and two anonymous reviewers for their constructive comments on an early draft of the manuscript. Assistance to the field and lab work was provided by M. Anello, A. Mutschlechner and L. Proia. We are particularly grateful to G. Nardi, M. De Cicco, F. Cianficconi, B. Todini, M. Rizzotti-Vlach, F. Baldari, V. Della Bella for identification of some zoological groups. This work was funded by CNR grants to A. Ruggiero (Bando n. 201.16.15 Codice n. 01.16.18; Bando n. 201.16.13 Codice n. 21.16.06), by a grant of Tor Vergata University to G. Carchini, by a MURST grant to M. Bazzanti and by a grant of Istituto Nazionale della Montagna (IMONT) to A.G. Solimini (Agenzia 2002, gruppo 3 cod. 191: Uso sostenibile della risorsa acqua nei parchi naturali montani dell'Appennino Centrale, studio pilota sulla relazione tra attività agropastorali e la diversità degli invertebrati delle pozze d'alpeggio).

References

Barbour, M. T., J. Gerritsen, G. E. Griffith, R. Frydenborg, E. McCarron, J. S. White & M. L. Bastian, 1996. A framework for biological criteria for Florida streams using benthic macroinvertebrates. Journal of North American Benthological Society 15: 185–211.

Bazzanti, M. & V. Della Bella, 2004. Functional feeding and habit organization of macroinvertebrate communities in permanent and temporary ponds in central Italy. Journal of Freshwater Ecology 19: 493–497.

Bazzanti, M., V. Della Bella & M. Seminara, 2003. Factors affecting macroinvertebrate communities in astatic ponds in central Italy. Journal of Freshwater Ecology 18: 537–548.

Biggs, J., P. Williams, M. Whitfield, G. Fox & P. Nicolet, 2000. Biological techniques of still water quality assessment. Phase 3. Method development. R&D Technical Report E110, Environment Agency, Bristol.

Biggs, J., P. Williams, M. Whitfield, P. Nicolet & A. Weatherby, 2005. 15 years of pond assessment in Britain: results and lessons learned from the work of Pond Conservation. Aquatic Conservation: Marine and Freshwater Ecosystems 15: 693–714.

Blocksom, K. A., J. P. Kurtenbach, D. J. Klemm, F. A. Fulk & S. M. Cormier, 2002. Development and evaluation of the lake macroinvertebrate integrity index (LMII) for New Jersey lakes and reservoirs. Environmental Monitoring and Assessment 77: 311–333.

Boix, D., S. Gascon, J. Sala, M. Martinoy, J. Gifre & X. D. Quintana, 2005. A new index of water quality assessment in Mediterranean wetlands based on crustacean and insect assemblages: the case of Catalonya (NE Iberian peninsula). Aquatic Conservation: Marine and Freshwater Ecosystems 15: 635–651.

Carchini, G., A. G. Solimini & A. Ruggiero, 2005. Habitat characteristics and odonate diversity in mountain ponds of central Italy. Aquatic Conservation: Marine and Freshwater Ecosystems 15: 573–581.

Cardoso, A. C., A. G. Solimini, G. Premazzi, L. Carvalho, A. Lyche & S. Relolainen, 2007. Phosphorus reference concentrations in European lakes. Hydrobiologia 584: 3–12.

Carvalho, L. & J. Moe, 2005. In A. Lyche-Solheim (ed.), Reference Conditions of European Lakes. Indicators and Methods for the Water Framework Directive Assessment of Reference conditions. EC-FP6 project Rebecca Report D7 (Contract number SSPI-CT-2003-502158). Available at http://www.rbm-toolbox.net.

Chow-Fraser, P., 2005. Ecosystem response to changes in water level of Lake Ontario marshes: lessons from the restoration of Cootes Paradise Marsh. Hydrobiologia 539: 189–204.

Coffman, W. P. & L. C. Ferrington, 1984. Chironomidae. In Merrit R.W. & K.W Cummins (eds), An Introduction to the Aquatic Insects of North America, 2nd edn. Kendall/ Hunt, Dubuque, 551–652.

Della Bella, V., M. Bazzanti & F. Chiarotti, 2005. Macroinvertebrate diversity and conservation status of Mediterranean ponds in Italy: water permanence and mesohabitat influence. Aquatic Conservation: Marine and Freshwater Ecosystems 15: 583–600.

De Meester, L., S. Declerck, R. Stoks, G. Louette, F. Van de Meutter, T. De Bie, E. Michels & L. Brendonck, 2005. Ponds and pools as model systems in conservation biology, ecology and evolutionary biology. Aquatic Conservation: Marine and Freshwater Ecosystems 15: 715–725.

Diehl, S. & R. Kornijòw, 1998. Influence of submerged macrophytes on trophic interaction among fish and macroinvertebrates. In Jeppesen E., M. Søndergaard, M. Søndergaard & K. Christoffrsen (eds), The Structuring Role of Submerged Macrophytes in Lakes. Springer, New York, 24–46.

Fairchild, G. W., J. N. Anderson & D. J. Velinsky, 2005. The trophic state 'chain of relationships' in ponds: does size matter? Hydrobiologia 539: 35–46.

Heiskanen, A. S., W. van de Bund, A. C. Cardoso & P. Noges, 2004. Towards good ecological status of surface waters in Europe—interpretation and harmonization of the concept. Water Science and Technology 49: 169–177.

Hick, A. L. & E. J. Nedean, 2000. New England Freshwater Wetlands Invertebrate Biomonitoring Protocol. (NEF-WIBP) UMass Extension, NREC & Massachusetts CZM.

Hinden, H., B. Oertli, N. Menetrey, L. Sager & J. B. Lachavanne, 2005. Alpine pond biodiversity: what are the related environmental variables? Aquatic Conservation: Marine and Freshwater Ecosystems 15: 613–624.

Menetrey, N., L. Sager, B. Oertli & J. B. Lachavanne, 2005. Looking for metrics to assess the trophic state of ponds. Macroinvertebrates and amphibians. Aquatic Conservation: Marine and Freshwater Ecosystems 15: 653–664.

Merritt, R. W. & K. W. Cummins, 1984. An Introduction to the Aquatic Insects of North America. Kendall/Hunt, Dubuque, USA.

Merritt, R. W., K. W. Cummins, M. B. Berg, J. A. Novak, M. J. Higgins, K. J. Wessel & J. L. Lessard, 2002. Development and application of a macroinvertebrate functional-group approach in the bioassessment of remnant river oxbows in southwest Florida. Journal of North American Benthological Society 21: 290–310.

Moog, O. (ed.), 2002. Fauna Aquatica Austriaca, 2nd edn. Wasserwirtschaftskataster, Bundesministerium für Land- und Forstwirtschaft, Umwelt und Wasserwirtschaft, Wien.

Newman, R. M., 1991. Herbivory and detritivory on freshwater macrophytes by invertebrates: A review. Journal of North American Bentholological Society 10: 89–114.

Oertli, B., J. Biggs, R. Cereghino, P. Grillas, P. Joly & J.-B. Lachavanne, 2005a. Conservation and monitoring of pond biodiversity: introduction. Aquatic Conservation: Marine and Freshwater Ecosystems 15: 535–540.

Oertli, B., D. Auderset Joye, E. Castella, R. Juge, A. Lehmann & J. B. Lachavanne, 2005b. PLOCH: a standardized method for sampling and assessing the biodiversity in ponds. Aquatic Conservation: Marine and Freshwater Ecosystems 15: 665–679.

Oksanen, J., 2005. The Vegan Package, A Community Ecology Package. Version 1.6-7. (available at http://www.cran.r-project.org/).

Plafkin, J. L., M. T. Barbour, K. D. Porter, S. K. Gross & R. M. Hughes, 1989. Rapid Bioassessment Protocols for Use in Streams and Rivers: Benthic Macroinvertebrates and Fish. EPA 444/4-89-001, US Environmental Protection Agency, Washington, DC.

Pond Action, 1989. National Pond Survey, Methods Booklet. National Pond Survey Group, WWF, Oxford Polytechnic, Oxford, UK.

Rosenberg, D. M. & V. H. Resh, 1993. Freshwater Biomonitoring and Benthic Macroinvertebrates. Chapman and Hall, New York.

Ruggiero, A., A. G. Solimini & G. Carchini, 2003. Nutrient and chlorophyll *a* temporal patterns in eutrophic mountain ponds with contrasting macrophyte coverage. Hydrobiologia 506–509: 657–663.

Ruggiero, A., A. G. Solimini & G. Carchini, 2004. Limnological study of a mountain shallow lake in the Italian Apennines. International Journal of Limnology 40: 89–99.

Ruggiero, A., A. G. Solimini & G. Carchini, 2005. The alternative stable state concept and the management of Apennine mountain ponds. Aquatic Conservation: Marine and Freshwater Ecosystems 15: 625–634.

Saether, O. A., 1979. Chironomid communities as water quality indicators. Holarctic Ecology 2: 65–74.

Scheffer, M., 1998. Ecology of Shallow Lakes. Chapman and Hall, London.

Scheffer, M., S. H. Hosper, M. L. Meijer, B. Moss & E. Jeppesen, 1993. Alternative equilibria in shallow lakes. Trends in Ecology and Evolution 8: 275–279.

Solimini, A. G., A. Ruggiero, M. Anello, A. Mutschlechner & G. Carchini, 2000. The benthic community structure in mountain ponds affected by livestock watering in natural reserves of central Italy. Verhlungen of Internationale Vereinigung fur Limnologie 27: 501–505.

Solimini, A. G., A. Ruggiero, V. Bernardini & G. Carchini, 2003. Temporal pattern of macroinvertebrate diversity and production in a new man made shallow lake. Hydrobiologia 506–509: 373–379.

Søndergaard, M., E. Jeppesen & J. P. Jensen, 2005. Pond or lake: does it make any difference? Archiv fur Hydrobiologia 162: 143–165.

Steinman, A. D., J. Conklin, P. J. Bohlen & D. G. Uzarski, 2003. Influence of cattle grazing and pasture land use on macroinvertebrate communities in freshwater wetlands. Wetlands 23: 877–889.

US EPA, 1998. Lake and Reservoir Bioassessment and Biocriteria: Technical Guidance Document, EPA 841-B-98-007, U.S. EPA, Washington, D.C.

Van Geest, G. J., H. Wolters, F. C. J. M. Roozen, H. Coops, R. M. M. Roijackers, A. D. Buijse & M. Scheffer, 2005. Water-level fluctuations affect macrophyte richness in floodplain lakes. Hydrobiologia 539: 239–248.

Varga, I., 2003. Structure and changes of macroinvertebrate community colonizing decomposing rhizome litter of common reed at Lake Fertö/Neusiedler See (Hungary). Hydrobiologia 506–509: 413–420.

Wiederholm, T., 1980. Use of benthos in lake monitoring. Journal of the Water Pollution Contontrol Federation 52: 537–547.

Wiederholm, T., (ed.) 1983. Chironomidae of Holarctic region. Keys and diagnosis. Part 1. Larvae. Entomologica Scandinavica (suppl) 19: 1–457.

Hydrobiologia (2008) 597:125–135
DOI 10.1007/s10750-007-9223-x

Eutrophication: are mayflies (Ephemeroptera) good bioindicators for ponds?

N. Menetrey · B. Oertli · M. Sartori · A. Wagner · J. B. Lachavanne

Abstract Ephemeroptera larvae are recognized worldwide for their sensitivity to oxygen depletion in running waters, and are therefore commonly used as bioindicators in many monitoring programmes. Mayflies inhabiting lentic waters, like lakes and ponds, in contrary have been poorly prospected in biomonitoring. For this purpose, a better understanding of their distribution in lentic habitats and of the relations of species presence with environmental conditions are needed. Within this framework, 104 ponds were sampled in Switzerland. The Ephemeroptera are found to be an insect order particularly well represented in the ponds studied here (93% of the lowland ponds). Nevertheless, in terms of diversity, they are relatively poorly represented (mean species number = 1.9). Two species dominated: *Cloeon dipterum* (Baetidae) and *Caenis horaria* (Caenidae). The investigations contributed to the updating of the geographical distribution of the species in Switzerland, as many of the observations appear to be from new localities. The trophic state of ponds appears here to be important for Ephemeroptera communities. First, there is a negative relationship between total phosphorus (TP) concentrations and species richness. Second, the presence of *Caenis horaria* or *Cloeon dipterum* is dependent on the trophic state. *Caenis horaria* is most closely associated with low levels of TP concentrations, while *Cloeon dipterum* appears to be less sensitive, and is most frequently found in hypertrophic conditions. A probable consequence of these relations, is that Baetidae are always present when Caenidae are also present. Contrastingly, Baetidae is observed as the only mayflies family present in several ponds.

Guest editors: R. Céréghino, J. Biggs, B. Oertli & S. Declerck
The ecology of European ponds: defining the characteristics of a neglected freshwater habitat

N. Menetrey (✉) · J. B. Lachavanne
Laboratory of Ecology and Aquatic Biology,
University of Geneva, Ch. des Clochettes 18,
1206 Geneva, Switzerland
e-mail: nathalie.menetrey@leba.unige.ch

B. Oertli
Ecole d'Ingénieurs de Lullier, University of Applied Sciences of Western Switzerland, 150 rte de Presinge,
1254 Jussy, Geneva, Switzerland

M. Sartori · A. Wagner
Museum of Zoology, Place Riponne 6, 1014 Lausanne,
Switzerland

Keywords Ephemeroptera larvae · Aquatic macroinvertebrates · Small waterbodies · Eutrophication · Water quality · Biomonitoring

Introduction

Mayflies are considered as "keystone" species and their presence is believed to be an important environmental indicator of oligotrophic to mesotrophic (i.e. low to moderately productive) conditions in running waters (Barbour et al., 1999; Bauernfeind & Moog, 2000). A high sensitivity of mayfly taxa to oxygen depletion, acidification, and various contaminants including metals, ammonia and other chemicals was demonstrated in both observational and experimental studies (Hubbard & Peters, 1978; Resh & Jackson, 1993; Moog et al., 1997; Hickey & Clements, 1998). Various Biological Indices including mayflies to assess water quality have been developed over the years (Lenat, 1988; Metcalfe, 1989; Kerans & Karr, 1994). Subsequently, many of the biological water quality assessment methods for streams include Ephemeroptera, as for example the EPT (Ephemeroptera + Plecoptera + Trichoptera) taxa richness (Lenat & Penrose, 1996) which has been incorporated into studies in the United States and in many other countries. Other examples include the River InVertebrate Prediction and Classification System (RIVPACS) for the UK (Wright et al., 1998) and the Indice Biologique Global Normalisé (IBGN) for France (AFNOR, 1992). A major EU project with 14 participating member states entitled STAndardisation of River Classifications (STAR) has now been established, which will calibrate different biological survey results against ecological quality classifications that have to be developed for the Water Framework Directive of 2000 (Furse et al., 2006).

On the contrary, mayflies inhabiting lentic waters (e.g. lakes and ponds), have been poorly used in biomonitoring programmes (see however Madenjian et al., 1998). Nevertheless, in such environments, we could expect that mayflies also adequately integrate some aspects of water quality. Ephemeroptera have also other advantages for monitoring: they are highly visible, relatively easy to sample and are represented by only a few species in such habitats, which makes identification easier. In Lake Erie, Ephemeroptera are successfully used in biomonitoring, following the example of a recent study that showed burrowing mayfly nymphs (*Hexagenia* spp.) to be associated with an improvement of the ecosystem health (Schloesser & Nalepa, 2002). In smaller waterbodies like ponds, the water quality is rarely assessed. Nevertheless, with the implementation of the directive, such procedures will be developed. This is already the case in some European states (UK, see Biggs et al., 2000; Catalonia, see Boix et al., 2005; Switzerland, see Menetrey et al., 2005).

For the purpose of better understanding the importance of Ephemeroptera in the assessment of water quality in lentic habitats and especially in ponds, a better understanding is needed of: (i) the distribution of mayflies in such habitats and (ii) the relations of species presence with environmental conditions. In this study, the distribution of mayflies is investigated for 104 ponds from Switzerland. In a second step, their presence is assessed in relation to environmental variables, particularly the trophic state indicators (total phosphorus (TP), total nitrogen (TN) and conductivity). Finally, we will examine whether a new metric using Ephemeroptera can be proposed for inclusion in rapid bioassessments methods for swiss ponds.

Materials and methods

Study area

Table 1 shows the location of the 104 permanent small water bodies sampled within the following four altitudinal vegetation belts in Switzerland: colline, montane, subalpine, and alpine. They vary in size from 5 m^2 to 10 ha (Table 2), with a mean depth comprising between 15 and 910 cm. We will further refer to these small water bodies as "ponds", since most of them correspond to the criteria of the definition of a pond presented by Oertli et al. (2005a). Only one third of these ponds are known to have a natural origin with an age exceeding 4,000 years (last glacial retreat). The others, with

Table 1 Number of sampled ponds per altitudinal vegetation belt (colline (200–800 m), montane (600–1,400 m), subalpine (1,300–2,000 m), alpine (>1,800 m)) and trophic state (based on the concentration of total phosphorus (TP) and total nitrogen (TN) as described by OECD (1982) and Wetzel (1983)

	Colline	Montane	Subalpine	Alpine	n = total of ponds
Oligotrophic	1 (1)	1 (1)	1 (1)	11(2)	14 (5)
Mesotrophic	4 (4)	7 (7)	9 (4)	6 (1)	26 (16)
Eutrophic	19 (19)	12 (11)	0 (0)	1 (0)	32 (30)
Hypertrophic	20 (17)	7 (5)	4 (3)	1 (0)	32 (25)
n = total of ponds	44 (41)	27 (24)	14 (8)	19 (3)	104 (76)

In brackets: number of ponds of each type containing Ephemeroptera various ages (1–900 years), are artificial, linked to past or present human activities (gravel or clay extraction, fish production, nature conservation, etc.). The range of altitude is from 210 to 2,757 m. The trophic state varies between oligotrophic and hypertrophic (Table 1). Additionally, each pond was characterised with environmental and geo-morphological data (Table 2) (site details are available on request).

Sampling

Each pond was sampled once during the summer months (June to early August) from 1996 to 2005 following the PLOCH method (Oertli et al., 2005b). Mayflies were collected using a small hand-net (rectangular frame 14 × 10 cm, mesh size 0.5 mm). For each sample, the net was swept intensively through the pre-selected dominant habitats for 30 s. In all cases, the collected material was preserved in either 4% formaldehyde or 70% alcohol solutions and then sorted in the laboratory.

Table 2 Mean values and ranges of selected variables characterising the 104 ponds

	Mean	Median	Minimum	Maximum
Altitude (m a.s.l.)	1069	733	210	2757
Area (m^2)	8619	2328	6	96200
Mean depth (cm)	175.5	113	15	910
Maximal depth (cm)	343	210	40	2400
Age (years)	1258	68	1	4000
Total nitrogen (TN) (mg N/l)	1.07	0.55	0.04	8.79
Total phosphorus (TP) (μg P/l)	65	26	1	611
Conductivity (μS/cm)	350	360	3	1367
Hardness ($CaCo_3$ mg/l)	174	175	0.8	884
Transparence (Snellen, cm)	44	54	3	60
Number of habitats sampled	4	4	1	9
Sinuosity of the shoreline	1.5	1.3	1	3.3
Macrophyte species richness	11	10	0	34
Macroinvertebrate family richness	19	18	3	44

The physico-chemistry of the water was measured during winter and summer months, as described by Oertli et al. (2000), by establishing a profile using WTW field probes down to the deepest point of the pond (to measure conductivity, pH and oxygen concentration). The transparency was additionally recorded from a surface water sample using a Snellen tube. Laboratory analyses of the content of TP and TN were made with winter water samples. TP concentrations and TN concentrations were then used to classify each pond into one of the four following trophic categories: oligotrophic, mesotrophic, eutrophic or hypertrophic, as described by the Organization for Economic Cooperation and Development (1982) and Wetzel (1983).

Statistical analyses

Statistical analyses were performed exclusively on 71 out of the 104 ponds from the colline and montane vegetation belts. The remaining 33 ponds from the subalpine and alpine belts were excluded from this dataset because of the particularity of their mayfly assemblages: only 11 ponds contained Ephemeroptera (Table 1). In addition, *Cloeon dipterum* and *Caenis horaria*, the two most abundant species present in many lowland ponds, were much less common at these altitudes. Indeed, most of the mayflies that are present in the subalpine and alpine belts were rare species.

A between-class Principal Component Analysis (PCA) was performed to test if there was an overall difference between the ponds containing Caenidae + Baetidae (33 ponds) and those with Baetidae only (31 ponds) for 12 relevant selected environmental and physico-chemical variables. Three of these variables were log-transformed: area, mean depth and sinuosity of the shoreline; five were transformed in categories: TP, TN, conductivity, transparency and altitudinal vegetation belt; and the last four were not transformed: presence versus absence of fishes, % of natural zone surrounding the waterbody, % of catchment area and macrophytes species richness.

A non-parametric Mann–Whitney *U* test was conducted to test if there was a significant difference for three trophic state variables considered separately (concentrations of TP, TN or conductivity) between ponds where *Cloeon dipterum* or *Caenis horaria* were present or absent, respectively. In addition, a non-parametric Mann–Whitney *U* test was performed to analyse the differences in the mayfly species richness between groups of ponds based upon their trophic state (being defined separately by TP, TN or conductivity values).

Furthermore, Generalized Additive Models (GAMs) were used to model the occurrence of *Cloeon dipterum*, or of *Caenis horaria* with the purpose of (i) identifying the physico-chemical and environmental variables explaining the presence of these species in the ponds, and (ii) building predictive models of their occurrence. GAMs are nonparametric regressions that lead to complex response curves, which differ from the linear and parabolic responses; therefore, non-normally distributed data (including binomial distributions) can be modelled. GAMs were carried out with S-PLUS software using a set of functions developed to perform generalized regression analyses and spatial predictions (GRASP) (Lehmann et al., 2002). After an exploratory stepwise procedure of the same twelve selected variables as the ones taken for PCA, the least contributive were discarded to avoid an over-parameterization of the models. This means that the final model was built around the five most relevant variables: altitude, log of area, TP (expressed as four trophic categories), log of mean depth, and macrophytes species richness.

The diagnostic procedure for the GAMs included: (1) the most relevant variables retained in the two final regression models at $P = 0.05$ level, (2) the contributions of each explanatory variable expressed as a deviance reduction associated to dropping the variable from the model, (3) the percentage of the deviance explained by the models, (4) a linear correlation ratio (*r*) between observed and predictive values derived from a cross-validation procedure.

Results

Ephemeroptera species distribution in ponds

Mayflies were found to be present in 76 of the 104 sampled ponds. Of the 85 species (and 11 families) of Ephemeroptera present in Switzerland, 12 species from five families were identified (Table 3). This list included logically a majority of lentic species; however, lotic species were also found to be present (i.e. *Baetis rhodani*, *Centroptilum luteolum*, *Ephemera danica* and *Siphlonurus aestivalis*), which could be explained by the presence of tributaries. The lotic species *Baetis alpinus* was additionally found in one alpine pond, and this independently of the presence of a tributary. An explanation for the presence of lotic species in lentic ecosystem is that in alpine ponds, the physico-chemical conditions (oxygen, nutrient content, T°C) are similar to those observed in streams (Hieber et al., 2005).

Amongst the 12 identified species, four are mentioned in the red list of threathened species for Switzerland (Sartori et al., 1994): *Centroptilum luteolum*, *Cloeon simile*, *Ephemera danica* (all three potentially endangered) *and Siphlonurus aestivalis* (endangered). The finding of *Habrophlebia fusca* was a first for Switzerland, while *Habrophlebia lauta* was observed for the first time in the Canton of Graubünden.

The 28 ponds where Ephemeroptera were absent included a set of ponds situated at an altitude over 1,410 m (22 ponds) or another set with hypertrophic conditions (six ponds). However, mayflies were not always absent from ponds with hypertrophic conditions. Baetidae were observed in 26 hypertrophic ponds, and of these, 15 ponds also contained Caenidae. Likewise, mayflies were not always absent from ponds over an altitude of 1,410 m: nine ponds over 1,410 m contained mayflies, mostly from the Baetidae or Caenidae families.

When present in a pond, the Ephemeroptera community diversity was low (see Table 3 and Fig. 1) and composed of only a few taxa (mean species number = 1.9 and mean family number = 1.6). The dominant lentic species were *Cloeon dipterum* (in 93% of the ponds containing Ephemeroptera) and *Caenis horaria* (in 45%). In 43% of the cases, ponds included only one family, generally the Baetidae with, in most of these cases, *Cloeon dipterum* being found alone. Otherwise, there was one case each where *Cloeon simile* was found alone or both together with *Cloeon dipterum*. For 55% of the ponds containing Ephemeroptera, two families were recorded, with Baetidae (*Cloeon dipterum*) present in all cases. One pond included three families. Therefore, Baetidae appeared as the most common mayfly family to be found in Swiss ponds. An interesting observation was that Caenidae were

Table 3 List of the 12 Ephemeroptera species sampled in 104 ponds from Switzerland, with frequency of observation and altitudinal range

Families	Species	Current (preference)	nb of ponds	Red list	Altitudinal range (m) in Switzerland	
					Known (Sartori & Landolt, 1999)	Observed (our study)
Baetidae	*Baetis rhodani* (Pictet, 1843)	lo	2	nd	200–1900	458–910
	Baetis alpinus (Pictet, 1843)	lo	1	nd	200–2600	2191
	Centroptilum luteolum (Müller, 1776)	lo	1	4	300–1100	910
	Cloeon dipterum (Linné, 1761)	le	71	nd	300–1500	210–1855
	Cloeon simile (Eaton, 1870)	le	9	4	300–1000	350–1813
Caenidae	*Caenis horaria* (Linné, 1758)	le	34	nd	200–1200	210–1813
	Caenis luctuosa (Burmeister, 1839)	le	6	nd	300–600	350–725
	Caenis robusta (Eaton, 1884)	le	12	nd	300–500	419–1685
Ephemeridae	*Ephemera danica* (Müller, 1764)	lo	1	4	200–1200	838
Leptophlebiidae	*Habrophlebia fusca* (Curtis, 1834)	lo	2	nd	–	425–910
	Habrophlebia lauta (Eaton, 1884)	lo	2	nd	200–1200	930–1907
Siphlonuridae	*Siphlonurus aestivalis* (Eaton, 1903)	lo	1	3	200–800	665

Red list for Switzerland (Sartori et al., 1994): nd, status not defined; 3 = endangered; 4 = potentially endangered. le, lentic taxa; lo, lotic taxa

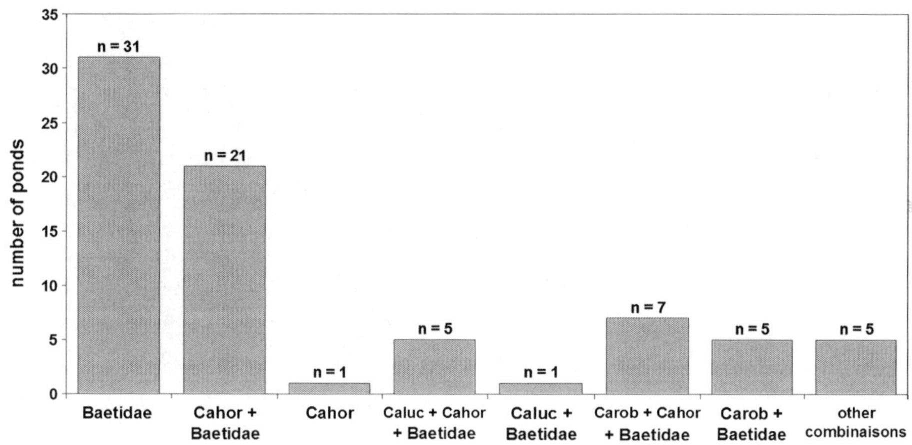

Fig. 1 Distribution of the mayflies among the 76 ponds containing Ephemeroptera. n = number of ponds. cahor = *Caenis horaria*; caluc = *Caenis luctuosa*; carob = *Caenis robusta*. The case of "other combinaisons" comprises: three ponds with Baetidae + another family than Caenidae, one pond with Leptophlebiidae only and one pond with three families: Baetidae, Caenidae and Leptophlebiidae

only present when Baetidae were present (with one exception). However, considering the selected environmental and physico-chemical variables, these were found to have no relevance in differentiating the ponds between sites with the presence of both Baetidae and Caenidae and sites with Baetidae alone. Only 3% of the variability given by the between-class PCA could be explained by environmental and physico-chemical variables. The Monte Carlo *P*-value was not significant for the parameters tested ($P = 0.654$).

Out of the 40 ponds containing Caenidae, *Caenis horaria* was found once alone, while in most cases (52%, Fig. 1) its presence was associated with the

Baetidae family. Otherwise, with the presence of Baetidae in most cases, *C. luctuosa* was observed together with *Caenis horaria* (12% of cases, only once alone). *Caenis robusta* was associated with *Caenis horaria* (17%) or without (13%). *Cloeon simile* could be observed alone, or in combination with *Cloeon dipterum*, and/or *Caenis horaria*, and/or *Caenis robusta* and/or even *Caenis luctuosa*. But *Caenis luctuosa* and *Caenis robusta* were never found in the same pond together.

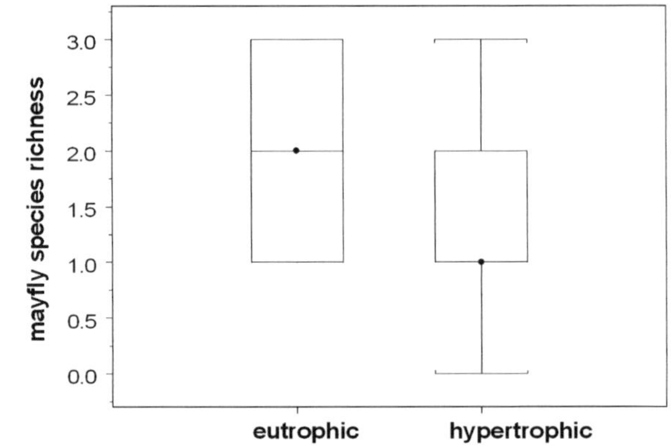

Fig. 2 Differences near level of significance (Mann–Whitney *U* test; $P = 0.096$) in mayfly species richness (box-plots) between groups of ponds based upon the trophic state of TP. Oligotrophic and mesotrophic box-plots are not showed, thus they were no significant difference for these trophic states. $n = 71$ ponds (from colline and montane vegetation belts). Each box represents the interquartile distance (25–75%) with the horizontal lines indicating the median. Upper error bars indicate the non-outlier maximum. Lower error bars indicate the non-outlier minimum

Mayflies and eutrophication

There was no significant difference in the values of the trophic state variables (concentration of TP, TN or conductivity) between the group of ponds with *Cloeon dipterum* (or *Caenis horaria*) present and the group of ponds without the species (Mann–Whitney *U* test; $P > 0.05$, see Table 4). Nevertheless the relationship between TP and *Cloeon dipterum* was near to being significant ($P = 0.085$).

There was also no significant difference of the mayfly species richness present between the groups of ponds based upon their trophic state (TP, TN or conductivity). However, the relationship between eutrophic and hypertrophic ponds for TP (Fig. 2) was also almost significant (Mann–Whitney *U* test; $P = 0.096$).

Generalised Additive Model regressions were calculated for the two most frequent taxa, *Caenis horaria* and *Cloeon dipterum*. Table 5 presents the most relevant variables (two variables for *Cloeon dipterum*, four for *Caenis horaria*) retained in the two final regression models at $P = 0.05$ level and their relative contributions. Cross-validation ratios were high for both models, with r above 0.7. Consequently, regarding r^2 values, more than 50% of the species' distribution could be explained by the two, respectively the four variables retained in the models. The models explained between 14.6% and 32.1% of the deviance for *Cloeon dipterum* and *Caenis horaria* respectively. The response curves for the variables retained in the models are presented in Fig. 3. Confidence intervals were usually wider at both ends of all gradients where there were fewer observations. For both species, the regression models showed one similar trend: the linear positive influence of area. This finding indicates that the two species were more frequently associated with larger sized ponds than with smaller ones. Also for both species, the trophic state of the pond was a significant variable, although the shape of the response curve was different for each species explaining that *Caenis horaria* was mostly present in oligotrophic ponds, while *Cloeon dipterum* was associated mainly with eutrophic ponds. The model for *Caenis horaria* incorporated two more variables: mean depth which showed a complex response curve that seemed incoherent; and macrophytes species richness which showed a bell-shaped response curve: *Caenis horaria* seemed therefore to be associated with species-rich ponds.

Table 4 Signification (*P*-values) of the differences for three trophic state variables between the group of ponds with *Cloeon dipterum* (or *Caenis horaria*) present and the group of ponds without the species (Mann–Whitney *U* test)

Trophic state variables	*Cloeon dipterum*	*Caenis horaria*
Total phosphorus (TP) (μg P/l)	0.085	n.s.
Total nitrogen (TN) (mg N/l)	n.s.	n.s.
Conductivity (μS/cm)	n.s.	n.s.

n.s. = not statistically significant ($P > 0.05$). Value of *P* is indicate if near to significance ($0.05 < P < 0.10$). $n = 71$ ponds (from colline and montane vegetation belts)

Table 5 Contributions of the explanatory variables and diagnostic parameters for the GAM of *Cloeon dipterum* and *Caenis horaria* at $P = 0.05$ level. GAMs included 71 ponds from colline and montane vegetation belts

Taxon	*n*	Explanatory variables				Diagnostic parameters	
		Area	Total phosphorus (TP)	Mean depth	Macrophyte species richness	Explained deviance (%)	Linear correlation ratio (r^2)
Cloeon dipterum	63	5.4	5.5	–	–	14.6	0.748
Caenis horaria	29	4.5	2.5	2.5	2.7	32.1	0.716

n, number of ponds where the species was present. area = $\log_e$ transformed m^2; total phosphorus (TP) = transformed into one of the four trophic categories as described by OECD (1982); mean depth = $\log_e$ transformed cm; macrophyte species richness = not transformed

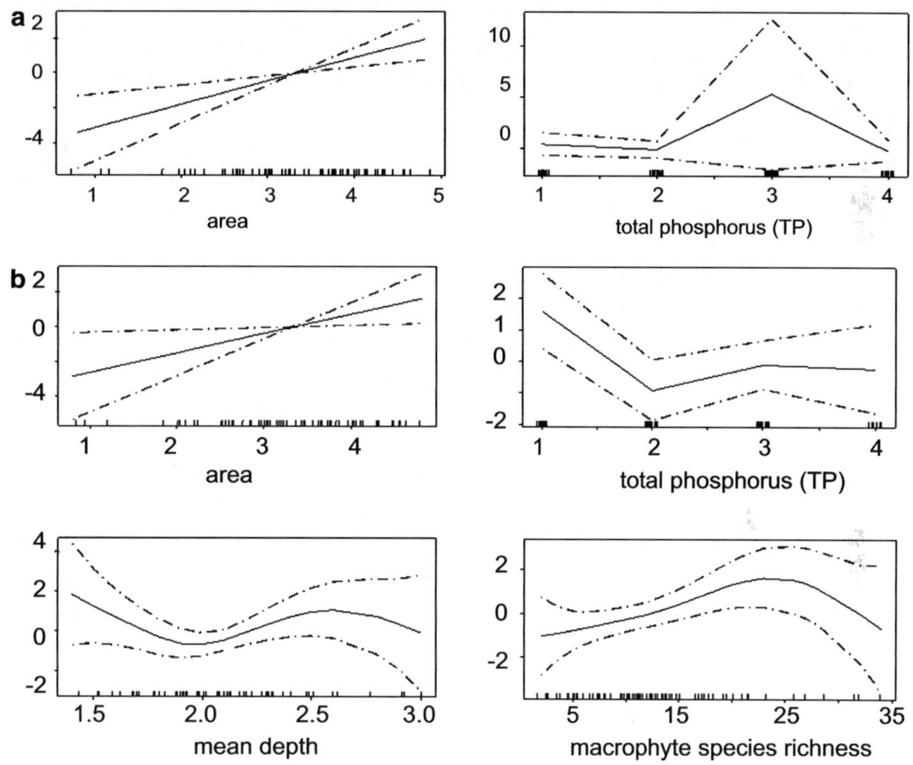

Fig. 3 Response curves for the variables incorporated in the Generalized Additive Models (GAMs) calculated for the presence of (**a**) *Cloeon dipterum* and (**b**) *Caenis horaria*. The dashed lines are approximate 95% confidence intervals around the smooth function lines. area = $\log_e$ transformed m^2; total phosphorus (TP) = transformed into one of the four trophic categories as described by OECD (1982); mean depth = $\log_e$ transformed cm; macrophyte species richness = not transformed. Vertical axes are scaled according to the dimensionless linear predictor

Discussion

Ephemeroptera species distribution in ponds

The Ephemeroptera are particularly well represented, being observed in 93% of the lowland ponds (colline and montane). Nevertheless, in terms of species richness, they are relatively poorly represented (mean species number = 1.9). This is mainly due to the fact that most mayflies are adapted to living in running waters where the environmental conditions are drastically different from those of standing waters. Both Caenidae and Baetidae families contain some of the most resistant species to organic pollution and to low levels of oxygen (Macan, 1973; Brönmark & Hansson, 2000). Two species dominate our data group: *Cloeon dipterum* and *Caenis horaria*, both of which are known to be very resistant to eutrophic conditions (group 6 in Soldán et al., 1998, or groups E-G in Kelly-Quinn & Bracken, 2000).

Only a few additional mayfly species could potentially be observed in Swiss ponds. These are:

Leptophlebia marginata, *Leptophlebia verspertina*, *Rhithrogena loyolaea* (above altitude of 2,800 m), *Ephemera glaucops* (recently discovered in one location in eastern Switzerland); *Ecdyonurus* sp. (in drift conditions or at altitude), *Paraleptophlebia werneri* (elsewhere) and perhaps *Arthroplea congener* (although at present only found in Germany and Austria). Interestingly, all these species mentioned are rare species and found in special conditions.

Many of our observations include new localities for Switzerland's Ephemeroptera, and therefore will contribute to the updating of the geographical distribution of the species presented in Sartori and Landolt (1999). Furthermore, altitudinal ranges presented by these authors will be largely revised, with new data for 7 out of the 12 species found in the sampled ponds (see Table 3).

Mayflies and eutrophication

Ephemeroptera species richness has a negative relationship with an increase of eutrophication (based on TP). Nevertheless, the presence of Ephemeroptera species in the studied ponds cannot be explained by the trophic state alone, since all simple direct relationships between the presence of *Caenis horaria* and *Cloeon dipterum* and the trophic state of water are not significant. However, in the model, taking into account the other predominant environmental variables (i.e. altitude, area, mean depth and macrophyte species richness), trophic state, based on TP, is significant. The two species appear to avoid hypertrophic ponds. Their optimum conditions are oligotrophic for *Caenis horaria* and eutrophic for *Cloeon dipterum*. This relationship with trophic conditions has already often been demonstrated in running water studies. For example, in Tachet et al. (2000), the biological traits for *Caenis* sp. indicate that mesotrophic conditions are optimal for this genera. Contrastingly, *Cloeon* sp. could be found in either mesotrophic or eutrophic habitats. Furthermore, Baetidae appears as one of the Ephemeropteran families the most tolerant to organic pollution. For example, *Cloeon dipterum* is the European species that exhibits the greatest saprobic index among mayflies (SI = 2.6) making it a characteristic element of β-α mesosaprobic conditions. *Caenis horaria* and *C. robusta* are ranked as less tolerant (SI = 2.2) and more confined to β mesosaprobic environments (Moog, 1995; Moog et al., 1997).

Following these relations, a really interesting observation was made on the association of the two most common families observed in the ponds sampled. Baetidae is observed as the only Ephemeroptera family present in several ponds. This is not the case for Caenidae, which are found only if Baetidae are already present. This particularity has also been observed for several sets of water-bodies in the French Rhône and Ain floodplains (Castella et al., 1984, 1991; Castella, 1987).

The presence of Caenidae alongside Baetidae, could be important for bioassessment work. As the potential number of mayfly species in such environment is normally low (five in general), the occurrence of an additional species from the Caenidae family may have some significance with regard to environmental conditions. On the contrary, the presence of rare species in ponds seems to depend more on special conditions than on trophic states. This finding represents an important development for the use of Ephemeroptera as bioindicators in Switzerland. Therefore, frequently observed species like *Caenis* sp. and *Cloeon* sp. seem to be more suitable to assess the trophic state of ponds (as demonstrated in this study).

The fact that Caenidae are more often absent from sampled ponds than Baetidae could be a discrepancy in the sampling dates among ponds. One hypothesis could be that Caenidae were not present in ponds sampled in late summer because the adult emergence occurs earlier in the season and before the sampling session. *Caenis* sp. shows large variations in their life history patterns as demonstrated for *Caenis luctuosa* by Cayrou & Cereghino (2003) and for *Caenis horaria* and *Caenis luctuosa* by Oertli (1992) and Bänziger (2000). Nevertheless, the population dynamics presented by these authors demonstrate that individuals of *Caenis* sp. are present in the water throughout the year (even if their repartition in size classes largely varies); This being the case, the time of sampling is probably not an explanation for the absence of this taxa. An alternative explanation could be the capacity of dispersion and colonisation since it is known that this capacity is greater for Baetidae than for Caenidae. Furthermore, *Cloeon dipterum* is relatively well known as a pioneering coloniser of new waterbodies and of temporary habitats (Sartori &

Landolt, 1999). The explanation of *Cloeon dipterum's* "success story" is to be found in its peculiar biology, ecology, and physiology. It is one of the rare ovoviviparous species in Europe, with females having an unusual life span of about 2 weeks during which the whole embryonic development takes place in the genital ducts (Degrange, 1959; Soldán, 1979). Females are often found quite far from the waterbody where they were born and disperse actively towards new habitats, making it a true colonizer species. Finally nymphs are detritivorous (Brown, 1961; Cianciara, 1980) and can afford very low levels of oxygen concentration, even anoxia in some conditions (Nagell, 1977a, b; Nagell & Fagerström, 1978) and seem tolerant to rapid temperature changes (McKee & Atkinson, 2000). These traits enable *C. dipterum* to be very successful in small ponds where it encounters few competitors. In fact, this species is known to be relatively independent of environmental factors. Another hypothesis is that ponds where only Baetidae are present are young ponds or temporary ponds. However, this is not supported by our data since only 12 of the 71 ponds containing Baetidae are younger than 25 years old. Furthermore they are all permanent ponds. Therefore, the most likely explanation of this singularity is the trophic state of ponds as discussed in the previous section.

In conclusion, our study has demonstrated that there is a great potential in using mayflies as bioindicators for the management of the water quality of ponds. Indeed, a relationship with the trophic state of ponds is hereby revealed for Ephemeroptera species richness and for the presence of *Caenis horaria* and *Cloeon dipterum*. These findings allow us to propose two new metrics for the water assessment of Swiss ponds: first, the Ephemeroptera species richness and second the presence of Caenidae associated with Baetidae. Nevertheless, these metrics need to be tested before being integrated into routine monitoring. Furthermore, other investigations must be made to confirm the suitability of these pond bioindicators for areas outside of Switzerland. Moreover, as these two metrics are only based on a small number of species, it would be necessary to use them in conjunction with other metrics to enable accurate assessments. Other such metrics, based on species or families richness (from macroinvertebrates and macrophytes assemblages), are currently in development (Hering et al., 2004; Menetrey et al., 2005; Furse et al., 2006). These should help to put into practice a scientific-based management of water quality in ponds as required for other waterbodies by the European Water Framework Directive.

Acknowledgements We are grateful to our collaborators Christiane Ilg, Hélène Hinden, Marc Pellaud, Gilles Carron, Amaël Paillex and Dominique Auderset-Joye for their help in the field and laboratory work. The authors appreciate also the great help and constructive comments from Emmanuel Castella. Thanks to the CSCF for the access to the Swiss databanks on fauna, Swiss National Park for collaboration studies, Nathalie Rimann and Emilie Hafner for access to a subset of the pond data. Thanks to David McCrae for his great help in improving the English style. Previous help in identification was provided by Diana Cambin. We are also grateful to the Bourse Augustin Lombard of the "*Société de physique et d'histoire naturelle de Genève* (SPHN)" and the "*Office of Environment and Energy*" of the canton of Luzern for their financial support. Finally, thanks to the water agencies of the cantons of Geneva and Vaud for completing some of the chemical analyses of the water samples. The database was in great part provided by the previous PLOCH study, financially supported by the "*Swiss Agency for the Environment, Forests and Landscape*".

References

AFNOR, 1992. Détermination de l'Indice Biologique Global Normalisé (IBGN). Norme française T 90-350. Association française de normalization, Paris.

Bänziger, R., 2000. Spatio-temporal distribution of size classes and larval instars of aquatic insects (Ephemeroptera, Trichoptera and Lepidoptera) in a *Potamogeton pectinatus* L. bed (Lake Geneva, Switzerland). Revue Suisse de Zoologie 107: 139–151.

Barbour, M. T., J. Gerritsen, B. D. Snyder & J. B. Stribling, 1999. Rapid Bioassessment Protocols for Use in Streams and Wadeable Rivers: Periphyton, Benthic Macroinvertebrates and Fish, 2nd edn. EPA 841-B-99-002, Office of Water, U.S. Environmental Protection Agency, Washington.

Bauernfeind, E. & O. Moog, 2000. Mayflies (Insecta: Ephemeroptera) and the assessment of ecological integrity: a methodological approach. Hydrobiologia 422: 71–83.

Biggs, J., P. Williams, M. Whitfield, G. Fox & P. Nicolet, 2000. Biological techniques of still water quality assessment: phase 3. Method development. P. Action. R&D Technical Report E56. Environment Agency, Bristol.

Boix, D., S. Gascon, J. Sala, M. Martinoy, J. Gifre & X. D. Quintana, 2005. A new index of water quality assessment in Mediterranean wetlands based on crustacean and insect assemblages: the case of Catalunya (NE Iberian peninsula). Aquatic Conservation: Marine and Freshwater Ecosystems 15: 635–651.

Brönmark, C. & L. A. Hansson, 2000. The biology of Lakes and Ponds. In Crawley, M., C. Little, T. R. E. Southwood, S. Ulfstrand (eds), Biology of Habitats, 285 Sidor. Oxford University Press, New York.

Brown, D. S., 1961. The food of the larvae of *Cloeon dipterum* L. and *Baetis rhodani* (Pictet) (Insecta, Ephemeroptera). Journal of Animal Ecology 30: 55–75.

Castella, E., 1987. Apport des macroinvertébrés aquatiques au diagnostic écologique des écosystèmes abandonnés par les fleuves. Recherches methodologiques sur le Haut-Rhône français. Université Claude Bernard – Lyon 1, Lyon.

Castella, E., M. Richardot-Coulet, C. Roux & P. Richoux, 1984. Macroinvertebrates as describers of morphological and hydrological types of aquatic ecosystems abandoned by the Rhone River. Hydrobiologia 119: 219–226.

Castella, E., M. Richardot-Coulet, C. Roux & P. Richoux, 1991. Aquatic macroinvertebrate assemblages of two contrasting floodplains: the Rhône and Ain rivers, France. Regulated Rivers: Research & Management 6: 289–300.

Cayrou, J. & R. Cereghino, 2003. Life history, growth and secondary production of *Caenis luctuosa* and *Cloeon simile* (Ephemeroptera) in a small pond, SW France. Aquatic Insects 25: 191–201.

Cianciara, S., 1980. Food preference of *Cloeon dipterum* (L. larvae and dependence of their development and growth on the type of food. Polskie Archiwum Hydrobiologii 27: 143–160.

Degrange, C., 1959. L'ovolarviparité de *Cloeon dipterum* (L.) (Ephemereptera, Baetidae). Bulletin de la Société Zoologique de France 64: 94–100.

Furse, M., D. Hering, O. Moog, P. Verdonschot, L. Sandin, K. Brabec, K. Gritzalis, A. Buffagni, P. Pinto, N. Friberg, J. Murray-Bligh, J. Kokes, R. Alber, P. Usseglio-Polatera, P. Haase, R. Sweeting, B. Bis, K. Szoszkiewicz, H. Soszka, G. Springe, F. Sporka & I. Krno, 2006. The STAR project: context, objectives and approaches. Hydrobiologia 566: 3–29.

Hering, D., O. Moog, L. Sandin & P. F. M. Verdonschot, 2004. Overview and application of the AQEM assessment system. Hydrobiologia 516: 1–20.

Hickey, C. W. & W. H. Clements, 1998. Effects of heavy metals on benthic macroinvertebrate communities in New Zealand streams. Environmental Toxicology and Chemistry 17: 2338–2346.

Hieber, M., C. T. Robinson, U. Uehlinger & J. V. Ward, 2005. A comparison of benthic macroinvertebrate assemblages among different types of alpine streams. Freshwater Biology 50: 2087–2100.

Hubbard, M. D. & W. L. Peters, 1978. Environmental requirements and pollution tolerance of Ephemeroptera. EPA 600-4-78-061 U.S. Environmental Protection Agency, Cincinnati, OH, VI.

Kelly-Quinn, M. & J. J. Bracken, 2000. Ephemeropteran assemblages in Ireland. Verhandlungen Internationale Vereinigung für Theoretische und Angewandte Limnologie 27: 963–969.

Kerans, B. L. & J. R. Karr, 1994. A benthic index of biotic integrity (B-Ibi) for rivers of the Tennessee valley. Ecological Applications 4: 768–785.

Lehmann, A., J. M. Overton & J. R. Leathwick, 2002. GRASP: generalized regression analysis and spatial prediction. Ecological Modelling 157: 189–207.

Lenat, D. R., 1988. Water quality assessment of streams using a qualitative collection method for benthic macroinvertebrates. Journal of the North American Benthological Society 7: 222–233.

Lenat, D. R. & D. L. Penrose, 1996. History of the EPT taxa richness metric. Bulletin of the North American Benthological Society 12: 305–306.

Macan, T. T., 1973. Ponds and Lakes. G. A. U. Ltd., London.

Madenjian, C. P., D. W. Schloesser & K. A. Krieger, 1998. Population models of burrowing mayfly recolonization in western Lake Erie. Ecological Applications 8: 1206–1212.

McKee, D. & D. Atkinson, 2000. The influence of climate change scenarios on populations of the mayfly *Cloeon dipterum*. Hydrobiologia 441: 55–62.

Menetrey, N., L. Sager, B. Oertli & J.-B. Lachavanne, 2005. Looking for metrics to assess the trophic state of ponds. Macroinvertebrates and amphibians. Aquatic Conservation: Marine and Freshwater Ecosystems 15: 653–664.

Metcalfe, J. L., 1989. Biological water quality assessment of running waters based on macroinvertebrate communities: history and present status in Europe. Environmental Pollution 60: 101–139.

Moog, O., 1995. Fauna aquatica Austriaca. Katalog zur autökologischen Einstufung Aquatischer Organismen Osterreichs. O. Moog. Fauna aquatica Austriaca. Abteilung für Hydrobiologie, Fischereiwirtschaft und Aquakultur des Universität für Bodenkultur, Wien.

Moog, O., E. Bauernfeind & P. Weichselbaumer, 1997. The use of Ephemeroptera as saprobic indicators in Austria. In Landolt, P. & M. Sartori Fribourg, (eds), Ephemeroptera & Plecoptera. Biology-Ecology-Systematics, Mauron, Tinguely & Lachat SA, 254–260.

Nagell, B., 1977a. Phototactic and thermotactic responses facilitating survival of *Cloeon dipterum* (Ephemeroptera) larvae under winter anoxia. Oikos 29: 342–247.

Nagell, B., 1977b. Survival of *Cloeon dipterum* (Ephemeroptera) larvae under anoxic conditions in winter. Oikos 29: 161–165.

Nagell, B. & T. Fagerström, 1978. Adaptations and resistance to anoxia in *Cloeon dipterum* (Ephemeroptera) and *Nemoura cinerea* (Plecoptera). Oikos 30: 95–99.

OECD, 1982. Eutrophication of Waters: Monitoring, Assessment and Control. Organization of Economic Cooperation and Development (OECD), Paris.

Oertli, B., 1992. L'influence de trois substrats (Typha, Chara, feuilles mortes) d'un étang forestier sur la densité, la biomasse et la production des macroinvertébrés aquatiques. Faculté des Sciences, LEBA, Université de Genève.

Oertli, B., D. Auderset Joye, E. Castella, R. Juge & J.-B. Lachavanne, 2000. Diversité biologique et typologie écologique des étangs et petits lacs de Suisse. OFEFP. LEBA, Université de Genève, Genève.

Oertli, B., J. Biggs, R. Cereghino, P. Grillas, A. Jolly & J.-B. Lachavanne, 2005a. Conservation and monitoring of pond biodiversity: introduction. Aquatic Conservation: Marine and Freshwater Ecosystems 15: 535–540.

Oertli, B., D. Auderset Joye, C. Castella, R. Juge, A. Lehmann & J.-B. Lachavanne, 2005b. PLOCH: a standardized method for sampling and assessing the biodiversity in ponds. Aquatic Conservation: Marine and Freshwater Ecosystems 15: 665–679.

Resh, V. H. & J. K. Jackson, 1993. Rapid assessment approaches to biomonitoring using benthic macroinvertebrates. In Rosenberg, D. M. & V. H. Resh (eds), Freshwater Biomonitoring and Benthic Macroinvertebrates. Chapmann and Hall, V. H. New York: 159–194.

Sartori, M. & P. Landolt, 1999. Atlas de distribution des éphémères de Suisse (Insecta, Ephemeroptera). Neuchâtel, Centre suisse de cartographie de la faune (CSCF/SZKF), Fauna Helvetica.

Sartori, M., P. Landolt & P. A. Zurwerra, 1994. Liste rouge des éphémères de Suisse (Ephemeroptera). In Duelli, P. (ed.), Liste rouge des espèces animales menacées de Suisse. Office fédéral de l'environnement, des forêts et du paysage (OFEFP), Berne, 72–74.

Schloesser, D. W. & T. F. Nalepa, 2002. Comparison of 5 benthic samplers to collect burrowing mayfly nymphs (*Hexagenia* spp: Ephemeroptera: Ephemeridae) in sediments of the Laurentian Great Lakes. Journal of the North American Benthological Society 21: 487–501.

Sladecek, V., 1973. System of water quality from the biological point of view. Archiv fur Hydrobiologie, Beiheft Ergebnisse der Limnologie 7: 1–218.

Soldán, T., 1979. The structure and development of the female internal reproductive system in six European species of Ephemeroptera. Acta Entomologica Bohemoslovaca 76: 353–365.

Soldán, T., S. Zahradkova, J. Helesic, L. Dusek & V. Landa, 1998. Distributional and quantitative patterns of Ephemeroptera and Plecoptera in the Czech republic: a possibility of detection of long-term environmental changes of aquatic biotopes. Folia Facultatis Scientiarum Naturalium Universitatis Masarykianae Brunensis 98: 1–305.

Tachet, H., P. Richoux, M. Bournaud & P. Usseglio-Polatera, 2000. Invertébrés d'eaux douces. Systématique, biologie, écologie. CNRS Editions, Paris.

Wetzel, R. G., 1983. Limnology, 2nd edn. CBS College Publishing, New York.

Wright, J. F., M. T. Furse & D. Moss, 1998. River classification using invertebrates: RIVPACS applications. Aquatic Conservation: Marine and Freshwater Ecosystems 8: 617–631.

Hydrobiologia (2008) 597:137–148
DOI 10.1007/s10750-007-9224-9

ECOLOGY OF EUROPEAN PONDS

How can we make new ponds biodiverse? A case study monitored over 7 years

P. Williams · M. Whitfield · J. Biggs

Abstract A new pond complex, designed to enhance aquatic biodiversity, was monitored over a 7-year period. The Pinkhill Meadow site, located in grassland adjacent to the R. Thames, proved unusually rich in terms of its macrophyte, aquatic macroinvertebrate and wetland bird assemblages. In total, the 3.2 ha mosaic of ca. 40 permanent, semi-permanent and seasonal ponds and pools was colonized by approximately 20% of all UK wetland plant and macroinvertebrate species over the 7-year survey period. This included eight invertebrate species that are Nationally Scarce in the UK. The site supported three breeding species of wading bird and was used by an additional 54 species of waders, waterfowl and other wetland birds. The results from four monitoring ponds investigated in more detail showed that these ponds significantly supported more plant and macroinvertebrate species than both minimally impaired UK reference ponds, and other new ponds for which compatible data were available. Comparisons of the physico-chemical, hydrological and land-use characteristics of the Pinkhill pools with those of other new ponds showed that the site was unusual in having a high proportion of wetlands in the near surrounds. It also had significantly lower water conductivity than other ponds and a higher proportion of (non-woodland) semi-natural land in its surroundings. Given that ponds are known to contribute significantly to UK biodiversity at a landscape level, and that several thousand new ponds are created each year in the UK alone, the findings suggest that well designed and located pond complexes could be used to significantly enhance freshwater biodiversity within catchments.

Keywords Constructed ponds · Compensatory wetlands · Pond age · Colonisation · Hydroperiod

Guest editors: R. Céréghino, J. Biggs, B. Oertli & S. Declerck
The ecology of European ponds: defining the characteristics of a neglected freshwater habitat

Electronic supplementary material The online version of this article (doi:10.1007/s10750-007-9224-9) contains supplementary material, which is available to authorized users.

P. Williams (✉) · M. Whitfield · J. Biggs
Pond Conservation: The Water Habitats Trust, Oxford Brookes University, Gipsy Lane, Headington, Oxford OX3 0BP, UK
e-mail: pwilliams@brookes.ac.uk

Introduction

Ponds are an important freshwater habitat in Britain. They are species-rich, supporting populations of at least two-thirds of Britain's freshwater plant and animal species (Williams et al., 1999) and, in terms of both species richness and rarity, the biodiversity of ponds appears to compare well with that of other freshwater ecosystems, such as lakes, rivers, streams and ditches (Godreau et al., 1999; Williams et al., 2004; Davies, 2005). A considerable number of new ponds are created in Britain each year, with around 2000 excavated annually in the lowlands alone (Williams et al., 1998). Yet, despite the popularity of pond creation, a little is known of the ecological value or characteristics of new pond sites. Across Europe as a whole in the last decade, a mere handful of papers describe the value of new ponds for amphibian and macrophyte assemblages (e.g., Gee et al., 1997; Stumpel & van der Voet, 1998; Baker & Halliday, 1999; Fleury & Strehler Perrin, 2004; Hansson et al., 2005), and fewer still the biodiversity value of new ponds for other groups, such as macroinvertebrates, or wetland birds (Gee et al., 1997; Fairchild et al., 2000; Hansson et al., 2005).

Perhaps most significantly, there has been exceptionally little research into the factors that drive new pond quality. In particular we know little of the key locational and physico-chemical characteristics likely to promote development of high biodiversity in new ponds. This is an important omission. If ponds both support a large proportion of freshwater biodiversity in catchments and, are continually being created in large numbers, then, through the application of good design principles, there is considerable potential to use the creation of highly biodiverse pond sites as a tool for enhancing catchment biodiversity.

This article describes the biodiversity value of a new pond complex, created in the early 1990s in southern England and subsequently monitored over a 7-year period. The richness of the wetland plant, aquatic macroinvertebrate and wetland bird assemblages is compared with other available datasets, and the factors likely to promote the development of biodiverse new ponds are evaluated.

Methods

Site description

Pinkhill Meadow in Oxfordshire, England (UK national grid coordinates: SP 439 067), lies in an area of floodplain grassland surrounded on two sides by a meander of the upper River Thames and on the third side by Farmoor Reservoir, which is the largest area of standing open water in the county. Pinkhill Meadow as a whole is small (approximately 4.5 ha), and relatively disturbed by the public using perimeter footpaths that surround the site.

Soil cores through the meadow substrata show that the site is underlain by clayey alluvium (0.8 to >4 m thick). This, in turn, overlies Quaternary gravels that support a shallow confined aquifer. Excavations into the gravel are filled rapidly by groundwater that fluctuates by approximately 0.4 m during the year. Excavations into the alluvial layer alone fill with surfacewater that fluctuates by ca. 1 m during the year. The area is subject to inundation from 1 in 15 year floods from the River Thames. During the monitoring period described here, the new pond complex was flooded twice: in the winters of 1991/2 and 1992/3.

Construction of the pond complex

The Pinkhill Wetland Enhancement Project was conceived in 1990 as a joint initiative by Thames Water Utilities Ltd (TWUL) and the Environment Agency (Thames Region). Pond Conservation provided ecological guidance for the design of the site and on-site supervision during construction work. A principle objective of the Pinkhill scheme was to provide complementary breeding and feeding habitats for wetland birds that would extend the existing feeding and roosting areas provided by the concrete-rimmed reservoir. Associated objectives were to provide habitats for a diverse range of wetland plant and aquatic macroinvertebrate species.

The final design of the wetland comprised a complex of approximately 40 permanent, semi-permanent and seasonal pools, sited within a low wetland area ca. 3.2 ha in total. The ponds ranged in area from the largest waterbody (the Main Pond) which is about 0.75 ha and has a number of mud and gravel small islands, to many tiny permanent and seasonal pools ca. 1–2 m^2 in area. Most ponds were dug into groundwater, but some were surfacewater fed. Winter pond depths ranged from a few centimetres to 2.5 m. Part of the site perimeter, adjacent to footpaths, was planted-up with a narrow reed bed and willow hedge to act as a screen.

Excavation of the new wetland occurred in two stages. Phase 1 excavation was undertaken in June and July 1990 and involved the creation of the four waterbodies (the Main Pond, Scrape, Groundwater Pond and Surfacewater Pond), which were subsequently the main focus of site monitoring. Phase 2, undertaken in winter 1991/92, extended the areas of shallow water, wet meadow, mudflats and temporary pool habitats.

The ponds were allowed to colonise naturally, so that, with the exception of *Phragmites australis* (Cav.) Trin. ex Steud which was planted around the edge of the site as a screen, no plants were deliberately introduced to the site.

Survey methods

Water chemistry samples were collected to provide background data describing the site's water quality. Water sampling focused on the four waterbodies created during Phase 1 of the project (above). For these ponds, replicate samples were taken monthly from April 1991 to March 1992 and then bimonthly from July 1992 to July 1993 except during the bird-breeding season. Water samples and meter readings were taken from the centre of each pond at mid column depth. Samples were analysed at an accredited laboratory for the following determinands: pH, conductivity, total oxidised nitrogen, ionised ammoniacal nitrogen ($NH_4^+ \cdot N$), unionised ammoniacal nitrogen (NH_3N), soluble reactive phosphorus (SRP) and biochemical oxygen demand (BOD). Water temperature was recorded on site on each sampling occasion.

Physico-chemical data were collected from each of the four main monitoring ponds. Details of the methods used to collect data are given in Pond Action (1998). However, in summary: pond area was measured from a 1:500 scale map of the site levelled after construction. Land-use cover was estimated in the field within the 100 m zone supplemented by map evidence where necessary. Water and sediment depth measurements were an average from 5 measurements taken along two perpendicular transects. Organic matter and sediment particle size categories were estimated in the field. Inflow velocity was estimated as the number of seconds for a floating object to travel 1 m downstream × average water width × average water depth.

Wetland macrophyte species were recorded in August 1991–1996. On each occasion, plant species lists were compiled for the site as a whole and individually for each of the four monitoring ponds. Plants were surveyed while walking and wading the margin and shallow water areas of the waterbodies. In the deep Main Pond, submerged macrophytes were surveyed using a grapnel thrown from the bank or islands. 'Wetland macrophytes' were defined as plants listed in the National Pond Survey methods guide (Pond Action, 1998), which comprises a standard list of the ca. 400 submerged, floating-leaved and marginal wetland plants recorded in the UK.

Aquatic macroinvertebrate species lists were compiled annually for the four monitoring ponds between 1990 and 1997. These surveys were undertaken in July, except in 1991 (May) and 1995 (August). Water areas were sampled using a standard 1 mm mesh hand-net, frame-size 0.26 × 0.30 m. The sampling method followed the National Pond Survey protocol (Pond Action, 1998), with each site sampled for 3 min and the sampling time divided equally between major mesohabitats identified at the pond. In addition, for the years 1994–1997, the samples from the available mesohabitats (either 2, 4 or 8, depending on the pond in that year) were sub-divided, as appropriate, to give a total of 16 sub-samples from each pond. In the current article, these data were used to provide a means of comparing Pinkhill results with other survey data collected using shorter sampling durations (see below), but for the majority of the analyses the data were collated to give a single 3-min sample for each pond. The samples collected were exhaustively live-sorted in the laboratory to remove all individual macroinvertebrates, with the exception of very abundant taxa (>100 individuals), which were sub-sampled. In 4 years (1992–1994 and 1997) a more comprehensive survey of the site as a whole was carried out in autumn. This used a field sorting approach and covered all parts of the site. In order to standardise this field survey, the site was divided into 14 specified water areas of similar size or potential. Each of these areas was searched for macroinvertebrate species (using a pond-net and large white sorting tray), for 1 h per water area on each occasion. Most species were identified on site. Taxa requiring microscopic identification were preserved in 70% ethanol for return to the laboratory. In order to reduce the possibility of recorder bias, the same two surveyors carried out the survey on each occasion. Macroinvertebrate taxa were identified to species level in the groups for which reliable UK distribution data and Red Data Book information is available. These were: Tricladida (flatworms), Hirudinea (leeches), Mollusca (snails and bivalves, but excluding *Pisidium* species), Malacostraca (shrimps and slaters), Ephemeroptera (mayflies), Odonata (dragonflies and damselflies), Plecoptera (stoneflies), Heteroptera (bugs), Coleoptera (water beetles), Neuroptera (alderflies and spongeflies) and Trichoptera (caddis flies). Other taxa (mainly Diptera larvae and Oligochaeta) were noted at family or genus level, but were not included in the analysis of species richness.

Data describing wader and waterfowl use of the site as a whole (but not of the individual waterbodies) were derived from records entered in the log-books located in the hide overlooking Pinkhill and from the adjacent Farmoor Reservoir. In 1991, during the Phase I monitoring programme, J. Biggs evaluated the accuracy of the log-book data compared to a complete survey (i.e., full-day observations) of Pinkhill (Pond Action, 1992). The results indicated that for recording breeding species, birdwatchers visited the site sufficiently regularly to provide accurate and near daily summaries of the progress of breeding species. Records of the daily peak count of individuals were also sufficiently reliable for waders and scarce species (e.g., Shoveler *Anas clypeata* L., Redshank *Tringa totanus* (L.). and Water Rail *Rallus aquaticus* L.). Peak numbers of common waterfowl (e.g., Tufted Duck *Aythya fuligula*, (L.) and Mallard *Anas platyrhynchos* L.) were more likely to be underestimated, since many observers ignored these birds. However, the observations still provided an indication of the relative abundance of the common species across the site as a whole.

Analysis of biodiversity

Macrophyte and macroinvertebrate data were analysed to assess the biodiversity value of the different waterbody types in terms of: (i) species richness and (ii) species rarity. Richness was measured as the number of species, or distinctive taxa, recorded. Wetland plant richness was further subdivided into aquatic species (i.e., the total number of submerged and floating-leaved taxa) and marginal emergents. Measures of plant richness excluded *Phragmites australis*, which was deliberately introduced using plants purchased from commercial suppliers, for use as a screening reed bed around the periphery of the site. Two other common ornamental species that appeared in this reed bed the year after planting (*Mimulus guttatus* DC and *Bolboschoenus maritimus* (L.) Palla), were also excluded because of the likelihood that they were introduced with the *Phragmites* plants. For both plant and macroinvertebrate assemblages, comparisons of species rarity were made within the following two categories: (a) "Local", defined for invertebrates as species either confined to certain limited geographical areas, where populations may be common, or of widespread distribution but with few populations, and for plants as those species recorded from fewer than 25% of the 10 × 10 km grid squares ($n = 2823$) in Britain (Preston et al., 2002) and (b) "Nationally Scarce" (both invertebrates and plants), species recorded from 15 to 100 10 × 10 km grid squares in Britain.

Comparison with other data

The Pinkhill plant and macroinvertebrate data were compared with results from a range of national pond surveys, and surveys of new ponds. The national surveys were (i) the National Pond Survey (NPS): a survey of high quality reference sites located in areas of semi-natural land use across the UK (Biggs et al., 2005) (ii) the Impacted Ponds Database (IPD): a stratified random survey of ponds in England and Wales excluding ponds in semi-natural landscapes (Biggs et al., 2005), and (iii) Lowland Pond Survey (LPS96): a plant only survey of ponds representative of lowland countryside areas of the UK (Williams et al., 1998). Both the IPD and LPS96 datasets were, predominantly composed of ponds exposed to a wide range of anthropogenic stresses including urban and road runoff, agricultural runoff, organic pollution, hydrological stresses and overstocking with fish and waterfowl. Sub-sets of data specifically describing new ponds were extracted from these three datasets. For the NPS and IPD datasets this included ponds <10-years-old. LPS ponds were all less than 12 years old. In addition, the Pinkhill data were compared with results from the Welsh Farm Pond Survey (Gee et al., 1994, Gee & Smith, 1995), a study of new and renovated ponds located in mid and west Wales. For the current assessment, a subset of the Welsh data was used which included ponds 3–10 years in age (mean 5.2 years) and which excluded renovated sites.

For plants, the field methods used for data collection were, in all cases, directly comparable with those used at Pinkhill (see above). Macroinvertebrate data were collected using directly compatible 3-min survey data for (i) the NPS (summer season data) and (ii) the IPD survey. Gee et al.'s survey of farm ponds in Wales provides useful invertebrate data from new ponds, but was based on surveys using only a 1-min sample. In order to enable a direct comparison with the Pinkhill ponds, sub-sample data from Pinkhill were analysed (see above). In order to provide these data five of the 16 sub-samples (each 11 s duration) were selected at random for each survey pond and the data collated to give a 55 s sample from each pond. The Pinkhill dataset spans a 7-year period. In order to provide the fairest comparison with the other new pond datasets (for which the average or median pond age was 5 years), the 1995 Pinkhill data was used, when the four monitoring ponds were also 5-years-old. Physico-chemical data compatible with the Pinkhill survey results were available from all three national pond surveys (NPS, IPD, LPS96), although LPS96 lacked a full suite of chemical determinands (limited to pH, calcium, conductivity).

Statistical analysis

Differences between (i) the four Pinkhill monitoring ponds and (ii) the monitoring ponds and other new ponds, were analysed in terms of species richness and physico-chemical characteristics. These data differed in the extent to which they approached normality and were, therefore, compared using non-parametric methods. The significance of differences was tested using two-tailed Mann–Whitney U, Kruskal–Wallis or Friedman tests.

Results

Physico-chemical data

The four Pinkhill monitoring ponds were un-shaded permanent pools between 0.75 ha and 0.02 ha in area, with a mean water depth of 0.2–1.5 m. In general, their water chemistry profiles were typical of calcium-rich ponds in lowland Oxfordshire with a mean pH of 7.9. Conductivity was low with a mean of 208 μS. Total oxidised nitrogen (TON), Soluble reactive phosphorus (SRP) and ammoniacal nitrogen ($NH_4^+ \cdot N$ and $NH_3 \cdot N$) levels were also low: falling below the analysis detection limits on most occasions (<0.2, <0.06 and <0.05 mg/l respectively). The exception was winter TON levels where there was a maximum of 0.3–6.05 mg/l: almost certainly the result of inputs of groundwater from the adjacent River Thames.

Plant species

The Pinkhill site was colonised rapidly by both marginal and aquatic plants. Within 6 months of the site's creation the new pool complex as a whole had been colonised by 34 species of wetland plant, and richness in the four main monitoring ponds ranged from 9 to 19 species (Fig. 1a). By 1997 the complex as a whole supported at least 67 plant species (Appendix 1). Four early colonist plant species had been lost by this time, but overall the site was still accumulating taxa (Fig. 1b). Plant richness in the four main monitoring ponds after 7 years varied from 27 to 50 species (mean 36 species).

During its first 7 years the Pinkhill site supported between three and nine locally uncommon plant species. Two early colonising local aquatics (*Potamogeton obtusifolius* Mert. & W.D.J. Koch and *Potamogeton perfoliatus* L.) disappeared co-incidentally with the invasion of two non-native taxa: *Elodea nuttallii* (Planch.) H. St. John and *Lagarosiphon major* (Ridl.) Moss.

Macroinvertebrate species

The four monitoring ponds were relatively slowly colonised by macroinvertebrate species. Four to eight species (mainly Coleoptera and Hemiptera) were recorded from the waterbodies in summer 1990, a few months after they were dug. However, in the 3 years after this, species richness increased rapidly so that in 1993, invertebrate richness in the four ponds averaged 52 species (range 46–57). After 3 years, average invertebrate richness plateauxed (Fig. 1c), although there were considerable differences between individual ponds in different years. In the 4 years when the Pinkhill site as a whole was surveyed, the field survey recorded between 87

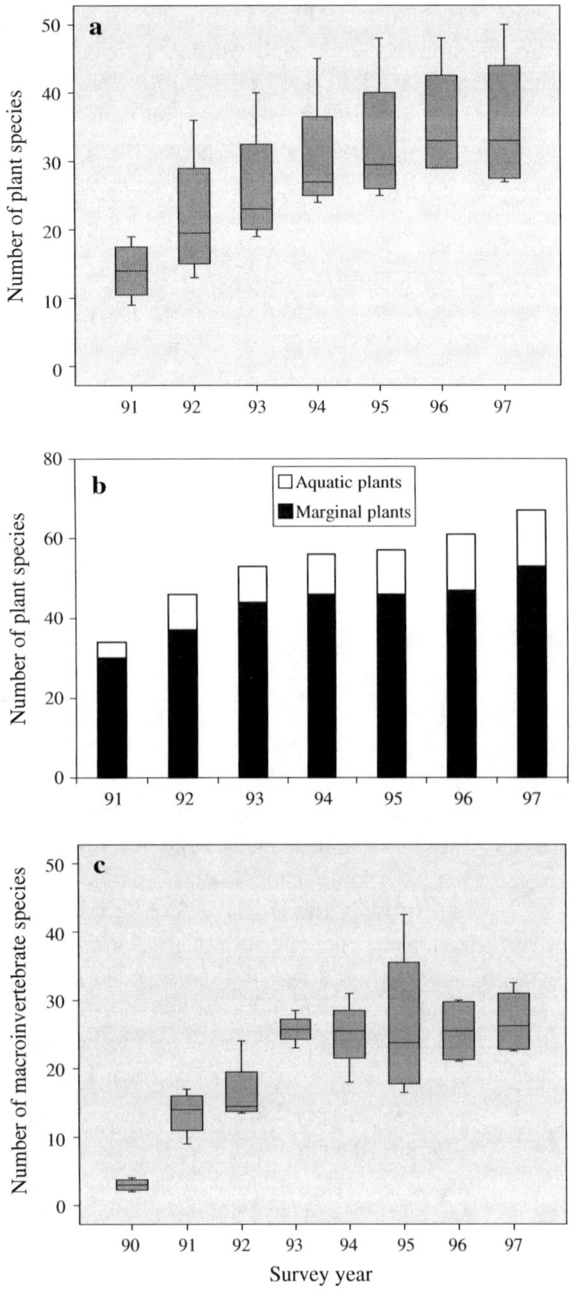

Fig. 1 Species richness from the Pinkhill Site (**a**) Plant richness from the four monitoring ponds (**b**) Plant richness from the whole site (**c**) Macroinvertebrate richness in the four monitoring ponds. Plots show median, inter-quartile and extreme values

(1992) and 110 (1997) macroinvertebrate species, and in total 165 macroinvertebrate species were recorded from all surveys over the 7 year period (Appendix 2).

Out of this total, eight species (all Coleoptera) are listed as Nationally Scarce in the UK. A further 13 could be considered locally uncommon. Most of these were Hemiptera, but local species in the orders Hirudinea, Coleoptera, Tricoptera and Hiudinea were also recorded.

Wetland birds

Between May 1990 and December 1997, 57 species of non-passerine wetland birds were recorded on the Pinkhill Meadow wetlands (excluding feral or escaped species). These included 24 species of wader and 33 other wetland species (Table 1). Most waterfowl showed marked seasonal patterns in their use of the site, the majority being present only in spring and summer.

Of the wading birds, the most commonly recorded species at Pinkhill Meadow were Lapwing *Vanellus vanellus* (L.), Snipe *Gallinago gallinago* (L.), Little Ringed Plover *Charadrius dubius* Redshank and Common Sandpiper *Actitis hypoleucos* (L.). All other waders were recorded much less frequently (Appendix 3). Three wading species (Lapwing, Little Ringed Plover, Redshank) and up to six species of waterfowl all bred on Pinkhill in one or more years (Table 1).

Comparison with other survey data

Wetland plants

Species-richness comparisons with ponds from other datasets suggest that the Pinkhill ponds supported unusually species-rich plant assemblages (Table 1). Compared to ponds in national UK datasets (which include ponds of all ages) the Pinkhill ponds supported, on average, two to three times more wetland plant species than the impaired countryside ponds of the IPD and LPS96 surveys (significance $P < 0.001$). Pinkhill typically supported ca. 30% more species than the high-quality reference sites of the National Pond Survey (NPS), although this difference was only marginally significant ($P = 0.03$, one-tailed test). The plant richness of the Main Pond also exceeded the maximum number of plants recorded in any pond survey.

Table 1 Comparison of the plant species richness of the four Pinkhill monitoring ponds with other survey data

	Number of plant species recorded: average (range)		
	Marginal species	Aquatic species	All wetland plant species
Pinkhill			
Pinkhill monitoring ponds (All 5 years old, $n = 4$)	28 (21–38)	5 (2–10)	33 (25–48)
National surveys			
National pond survey ($n = 102$)	18 (1–42)	5 (0–14)	23 (1–46)
Impacted pond database ($n = 148$)	11 (1–32)	3 (0–11)	14 (1–38)
LPS96 DETR ($n = 377$)[a]	8 (0–30)	2 (0–10)	10 (0–35)
New ponds only			
National Pond survey (<10 years old $n = 8$)	14 (3–28)	3 (1–5)	16 (4–30)
Impacted pond database (<10 years old $n = 12$)	13 (7–23)	3 (1–5)	16 (9–26)
LPS96 DETR ponds (<12 years old, $n = 26$)[a]	10 (0–24)	2 (0–4)	12 (0–27)
South Wales farm ponds (3–10 years old $n = 26$)[b]	7 (2–13)	3 (0–5)	10 (2–17)

[a] LPS96 = Lowland Pond Survey 1996, DETR = Department of the Environment, Transport and the Regions

[b] Modified from Gee et al. (1994)

Comparisons with the smaller datasets of new ponds (Table 1) shows that the Pinkhill ponds, again, supported at least two to three times more plant species than other new sites. This relationship was significant in all cases (all $P < 0.01$, except NPS where the relationship was week: $P = 0.046$, one-tailed test). The richness of new ponds in semi-natural areas (NPS survey) was lower than older ponds in these landscapes. For ponds in other countryside areas (IPD, LPS96), average plant richness was marginally greater in the new ponds.

Macroinvertebrates

The richness of Pinkhill's macroinvertebrate assemblages mirrored the richness of its plants (Table 2). Compared to national survey data, Pinkhill samples supported around double the number of invertebrate species typical of impaired wider countryside ponds ($P < 0.01$) and, on average, around a third more species than the high quality sites of the NPS (a marginally significant difference at $P = 0.046$, one-tailed test). The Pinkhill ponds also supported significantly more species ($P < 0.05$) than the subset of new ponds in these datasets. Comparison between the 1-min invertebrate sample of new Welsh farm ponds and the 1-min microhabitat samples from Pinkhill (see methods), showed a highly significant difference ($P < 0.001$). In high quality landscapes (NPS) new ponds typically supported fewer macroinvertebrate species than older ponds. In the more impaired ponds of the wider countryside (LPS96, IPD) new ponds were marginally richer in macroinvertebrates than older ponds.

Table 2 Comparison of the macroinvertebrate richness of the four Pinkhill monitoring ponds with other survey data

	No. invert. spp: average (range)
Pinkhill standard survey (3 min)	
Pinkhill ponds (5 years old, $n = 4$)	53 (33–85)
National surveys	
National pond survey ($n = 149$)	35 (6–78)
Impacted pond database ($n = 161$)	25 (2–64)
New ponds only	
National pond survey (<10 years old $n = 8$)	29 (17–39)
Impacted pond database (<10 years old, $n = 12$)	29 (15–52)
Pinkhill ponds, 1 min sample (5 years old, $n = 4$)	39 (23–62)
South Wales farm ponds 1 min. sample (3–10 years old, $n = 26$)[a]	14 (1–26)

[a] Modified from Gee & Smith (1995)

Wetland birds

Bird data from pond complexes comparable to Pinkhill are rarely published, with most studies of wetland birds being concerned with much larger sites. However, in terms of regional comparisons, the wading bird records represent just over half of the total number of wader species that had been seen in Oxfordshire since records were first kept. At its peak, the number of pairs of Little Ringed Plover and Redshank breeding at Pinkhill probably represented about 10% of the Oxfordshire's breeding population for these species. Overall wader densities were high for such a small area. For example, peak Redshank densities in Britain are around 100 pairs/km^2 (= 1 pair/hectare) in optimum habitat (Gibbons et al., 1993), and densities at Pinkhill were similar to this.

Environmental data comparisons

Compatible environmental data from new ponds in the three national datasets (NPS, IPD and LPS96) were combined and compared with data from the Pinkhill monitoring ponds. The results show that there were a few significant differences in terms of most environmental parameters including their area, water source and sediment characteristics. A major exception was a significant relationship with easting ($P < 0.001$), linked to Pinkhill's location in the southern lowlands. There were, however, no difference in pH or calcium concentrations suggesting that, the species-richness differences were unlikely to be related to the major SE (alkaline) to NW (acid) trends that broadly shape the UK's major bio-geographic zones. In terms of surrounding land use, the Pinkhill ponds were also relatively unusual within the dataset in being unshaded ($P < 0.05$) and located in an open (i.e., un-wooded) semi-natural landscape ($P < 0.01$). They had a significantly higher proportion of wetlands in their near vicinity than other ponds in the data-sets ($P = 0.01$). Unfortunately, one of the national datasets (LPS96) had only field meter water chemistry data (see methods), giving limited potential to compare water quality at the sites. However, conductivity data showed that Pinkhill ponds had significantly lower conductivity than other ponds in the datasets ($P < 0.01$). Other relationships between species richness and the physico-chemical variables were not significant.

Discussion

Pond colonisation

Data from the current survey show that new ponds can colonise quickly. Not only did plant and macroinvertebrate species accumulation in the Pinkhill monitoring ponds typically plateaux after only 3–4 years for macroinvertebrates and after 6 years for macrophytes, but new ponds in the Impacted Pond Database and Lowland Pond Survey 1996 were as species-rich as much older ponds in less than 10 and 12 years, respectively. Indeed, further analysis of LPS96 ponds (Williams et al., 1998) showed that 6–12-year-old ponds were significantly more species rich ($P < 0.01$) and supported more uncommon species ($P < 0.05$) than older ponds in this dataset.

The propensity for new ponds to colonise rapidly has long been recognised, in many cases colonisation may come from an existing seed, egg or spore bank in the soil or near surrounds, but authors from Darwin onwards have also noted the inherent mobility of freshwater taxa, which confers on individuals of many aquatic animal and plant species a strong potential for dispersal and colonisation (Darwin, 1859; Talling, 1951; Bilton et al., 2001).

Given that ponds are a natural habitat type, and that pond creation is likely to have been a common process through evolutionary history (Gray, 1988; Biggs et al., 1994), the rapid colonisation of new ponds may also be, in part, attributable to species adaptations to new pond conditions. In terms of their physico-chemical environment, new ponds are clearly different to older ponds: they are typically dominated by inorganic substrates, have a little vegetation cover, and may, at least in their early years, lack predation from higher predators, such as fish. A range of taxa appear to specifically thrive in such conditions. In the United Kingdom, new ball-clay pits, turf ponds and gravel pits have, for example, all been shown to support aquatic invertebrates or plants not found at later stages of succession. This includes uncommon plants, such as Lesser Water-plantain, *Baldellia ranunculoides*, damselflies, such as the Scarce Blue-tailed Damselfly, *Ischnura pumilio*, and rare water beetles such as *Helophorus longitarsus* (Barnes, 1983; Kennison, 1986; Foster, 1991; Fox & Cham, 1994). Other authors have found similar results in seasonal ponds: Fleury & Perrin (2004), for example, showed that pioneer plant assemblages of high conservation interest showed a rapid population increase in the first 2–3 years after ponds were created, followed by a progressive decline.

A third reason for the relative rapidity with which new ponds are colonised may be related to their nutrient status. Previous analysis of ponds in the LPS96 dataset (Williams et al., 1998), showed, for example, that new ponds had significantly lower Trophic Ranking Scores (*sensu* Palmer, 1992) than older ponds, suggesting that they were less enriched than more mature sites. It is possible that new ponds may lack the nutrient burden that accumulates in the sediments and water of many older countryside ponds and that the cleaner new ponds have, as a result, greater potential to support diverse plant and invertebrate assemblages. Such a suggestion is given some credence by the current analysis, which shows that new ponds supported similar numbers of plant and invertebrate species regardless of whether they were located in semi-natural or anthropogenically impaired landscapes. This contrasts with older ponds, where waterbodies located in semi-natural surrounds supported significantly more species than those in the wider lowland landscape (Williams et al., 1998).

Pond richness

It is clear from the current study that Pinkhill was unusually species rich, both at an individual pond scale and across the larger pond complex. In such new ponds where, it may be assumed, biological interactions are little developed, it seems likely that the particular richness of individual ponds should be explicable either in terms of (i) bottom-up effects produced by the physico-chemical environment created during or after pond construction, or (ii) stochastic processes influencing propagule arrival. Comparison of the Pinkhill ponds with other new ponds suggested that, in terms of their physical characteristics, the unusual richness of the Pinkhill ponds could not be explained in terms of difference in size, depth, substrate type or water source. However, differences in shading, land use and conductivity between the Pinkhill ponds and other new ponds indicated that some of the former's richness might be attributable to their open unshaded aspect, their semi-natural grassland surrounds and/or their low chemical conductivity. Of these, shade, is known to be associated with lower species richness, particularly for macrophyte assemblages (Gee et al., 1997; Craine & Orians, 2004; Biggs et al., 2005). Semi-natural grassland land use and conductivity (assuming the latter to be a surrogate for nutrient pollution) are also clearly plausible influences promoting the richness in the Pinkhill ponds. Both minimally impaired land use and water quality have been strongly positively correlated with species richness in ponds (Biggs et al., 2005; Menetrey et al., 2005; Williams et al., 1998).

Clearly, for the Pinkhill site to colonise so rapidly and richly, a wide range of plant and animal propagules must have been available to reach the site from other wetland areas. Comparisons with other new ponds in the current study showed that individual Pinkhill ponds were indeed unusual in having a significantly higher proportion of wetland in their surroundings than was typical of other new ponds. Thus in their near surrounds, the Pinkhill ponds were closely adjacent to (i) other ponds created in the new complex, (ii) the River Thames and its backwaters, and (iii) the large, but barren concrete-sided, Farmoor Reservoir. Further afield, Pinkhill may also have received plant and animal propagules from more distant ditches, streams, ponds and gravel pits which are widespread along the Thames Valley. Other studies, too, have emphasised the importance of propagule availability in ponds. In LPS96, ponds were shown to be significantly richer and to support more rare species when located on, or immediately adjacent to, floodplains and other traditionally wetland areas (Williams et al., 1998). The floristic assemblages of newly created turf ponds and species-richness of more mature ponds has also been shown to be positively related to the proximity of other ponds and wetlands in the neighbourhood (Moller & Rordam, 1985; Beltman et al., 1996; Linton & Goulder, 2000, 2003).

Site richness

In the 7 years following its creation the Pinkhill Meadow site as a whole supported at least 71 plant and 167 macroinvertebrate species: approximately 20% of all the UK's freshwater plants and macroinvertebrates in the groups assessed. The site was also valuable for birds; used both by a wide range of waders and waterfowl, and significant as a breeding area for species such as Redshank, Lapwing and Little Ringed Plover which have either declined markedly or are rare in the UK. For birds, the value of the site is likely to have been strongly influenced by its location. The Thames valley is a well-known flyway for migratory species, and the adjacent Farmoor Reservoir is one of the best areas for recording wetland birds in the county of Oxfordshire (Brucker et al., 1991). The value of Pinkhill as a whole for plants, macroinvertebrates and birds may, in addition, have been influenced by the new site's physical heterogenity and complexity (e.g., Froneman et al., 2001). The waterbody mosaic includes ponds that differ in size, substrate and water source but, more particularly, hydrological regime: it includes pools that are highly seasonal, semi-permanent ponds that dry up in drought years, as well as six large permanent ponds. Seasonality gradient has been clearly shown to drive community type in pond (as well as other freshwaters), in a number of studies at a range of landscape scales (Collinson et al., 1994; Schneider & Frost, 1996; Wellborn et al., 1996) and it seems probable that its range of waterbody hydroperiods contributed to the richness of the Pinkhill site.

Implications

We have argued elsewhere (Williams et al., 1997) that pond creation is a natural and ecologically valid method for maintaining pond biodiversity in the landscape. Man's creation of new ponds mimics age-old processes of natural pond formation, creating new sites that are of value in their own right and that eventually pass through a range of successional stages, each exploited by freshwater taxa.

The value of the current case study is that it suggests that it may be possible to use well-designed and targeted pond creation schemes to some considerable effect in the landscape. Wetland bird observations at Pinkhill suggest that through the development of a number of small-scale habitat-creation schemes it may be possible to significantly influence national breeding populations of wading birds such as Little Ringed Plover (national population ca. 1000 pairs), and perhaps, regional populations of Redshank (Oxfordshire's population ca. 30 pairs in 1997). However, for birds, habitat-creation work of this type seems most likely to succeed in areas where some wetland habitat already exists: for example, alongside rivers, in existing areas of damp grassland, or beside reservoirs and gravel-pits. In these areas, some feeding habitat may already be available, even when habitats are unsuitable for breeding. This appears to be the case at Pinkhill, at least for Little Ringed Plover and Redshank, which spend a certain amount of time feeding on the reservoir margin during the breeding season, and probably also visit sites further afield.

For wetland plant and aquatic macroinvertebrate species the Pinkhill data suggest that, by creating small pond complexes with semi-natural surrounds, good water quality and strong colonisation potential, it may be possible to create exceptionally biodiverse aquatic sites in very short periods of time. Ponds have recently been shown to be surprisingly significant as freshwater habitats, supporting a relatively high proportion of the total freshwater biodiversity present in a range of landscape types (Godreau et al., 1999; Williams et al., 2004). One implication from this finding is that pond creation might have the potential to be a more powerful ecological enhancement tool than is commonly credited. Ponds are relatively simple and cost effective habitats to create. The techniques for creating them are well developed and commonplace, and there are many areas of the landscape where the creation of high quality pond complexes is feasible. Thus, it may prove possible to use well-designed and located pond creation schemes not only to protect pond habitats, but also to enhance freshwater biodiversity across wider catchment areas.

Acknowledgements We would like to thank the many people and organisations that contributed to this work. Particularly, David Walker who undertook a considerable proportion of the invertebrate and chemical sampling, Thames Water Utilities who own the Pinkhill site, Mike Crafer (Thames Water Utilities) and Alistair Driver (Environment Agency) for instigating and funding the creation of Pinkhill Meadows. Richard Hellier (Environment Agency) for drawing up, and contributing to, the designs of the site, the many bird recorders who entered data into the Pinkhill and Farmoor Log books, Defra who gave permission to reanalyse unpublished data from the Lowland Pond Survey 1996 and Beat Oertli and two anonymous referees for helpful comments on an earlier draft of this article. In addition we are particularly indebted to two people: Alistair Driver, who achieved the near impossible by securing funding for the long-term monitoring of Pinkhill under NRA contract F01(91)2 383, and Bernard Johns (dec.

1992) from White Horse Contractors Ltd, who excavated most of the site. The complex is a memorial to his skill and our subsequent pond creation work has benefited greatly from the lessons he taught us.

References

Baker, J. M. R. & T. R. Halliday, 1999. Amphibian colonization of new ponds in an agricultural landscape. Herpetological Journal 9: 55–63.

Barnes, L. E., 1983. The colonisation of ball-clay ponds by macroinvertebrates and macrophytes. Freshwater Biology 13: 561–578.

Beltman B., T. van den Broek, K. van Maanen & K. Vaneveld, 1996. Measures to develop a rich-fen wetland landscape with a full range of successional stages. Ecological Engineering 7: 299–313.

Biggs, J., A. Corfield, D. Walker, M. Whitfield & P. Williams, 1994. New approaches to the management of ponds. British Wildlife 5: 273–287.

Biggs, J., P. Williams, M. Whitfield, P. Nicolet & A. Weatherby, 2005. 15 years of pond assessment in Britain: results and lessons learned from the work of Pond Conservation. Aquatic Conservation: Marine and Freshwater Ecosystems 15: 693–714.

Bilton D. T., J. R. Freeland & B. Okamura, 2001. Dispersal in freshwater invertebrates. Annual Review of Ecology and Systematics 32: 159–181.

Brucker J. W., A. G. Gosler & A. R. Heryet (eds) 1991. The Birds of Oxfordshire. Pisces Publications, Oxford.

Collinson, N. F., J. Biggs, A. Corfield, M. J. Hodson, D. Walker, M. Whitfield & P. J. Williams, 1994. Temporary and permanent ponds: an assessment of the effects of drying out on the conservation value of aquatic macroinvertebrate communities. Biological Conservation 74: 125–133.

Craine, S. I. & C. M. Orians, 2004. Pitch pine (*Pinus rigida* Mill.) invasion of Cape Cod pond shores alters abiotic environment and inhibits indigenous herbaceous species. Biological Conservation 116: 181–189.

Darwin, C., 1859. On the Origin of Species by Means of Natural Selection, or the Preservation of Favoured Races in the Struggle for Life, 6th edn. John Murray, London.

Davies, B. R., 2005. Developing a strategic approach to the protection of aquatic biodiversity. PhD thesis. Oxford Brookes University.

Fairchild, G. W., A. M. Faulds & J. F. Matta, 2000. Beetle assemblages in ponds: effects of habitat and site age. Freshwater Biology 44: 523–534.

Fleury, Z. & C. Strehler Perrin, 2004. Vegetation colonisation of temporary ponds newly dug in the marshes of the Grande Caricaie (lake of Neuchatel, Switzerland). Archives des Sciences 57: 105–112.

Foster, G. N., 1991. Conserving insects of aquatic and wetland habitats, with special reference to beetles. In Collins, N. M. & J. A. Thomas (eds), The Conservation of Insects and Their Habitats. Academic Press, London: 237–262.

Fox, A. D. & S. A. Cham, 1994. Status habitat use and conservation of the Scarce Blue-tailed damselfly *Ischnura pumilio* (Charpentier) (Odonata: Coenagrionidae) in Britain and Ireland. Biological Conservation 68: 115–122.

Froneman, A., M. J. Mangnall, R. M. Little & T. M. Crowe, 2001. Waterbird assemblages and associated habitat characteristics of farm ponds in the Western Cape, South Africa. Biodiversity and Conservation 10: 251–270.

Gee, J. H. R. & B. D. Smith, 1995. The conservation value of farm ponds: macroinvertebrates. Report to the Countryside Council for Wales.

Gee, J. H. R., K. M. Lee & S. Wyn Griffiths, 1994. The conservation and amenity value of farm ponds. Report to the Countryside Council for Wales.

Gee, J. H. R., B. D. Smith, K. M. Lee & S. Wyn Griffiths, 1997. The ecological basis of freshwater. Pond management for biodiversity. Aquatic conservation: Marine and Freshwater Ecosystems 7: 91–104.

Gibbons, D. W., J. Reid & R. A. Chapman, 1993. The New Atlas of Breeding Birds in Britain and Ireland: 1988–1991. T. & A. D. Poyser Ltd, London.

Godreau, V., G. Bornette, B. Frochot, C. Amoros, E. Castella, B. Oertli, F. Chambaud, D. Oberti & E. Craney, 1999. Biodiversity in the floodplain of Saone: a global approach. Biodiversity and Conservation 8: 839–864.

Gray, J., 1988. Evolution of the freshwater ecosystem: the fossil record. Palaeogeography, Palaeoclimatology, Palaeoecology 62: 1–214.

Hansson, L. A., C. Bronmark, P. A. Nilsson & K. Abjornsson, 2005. Conflicting demands on wetland ecosystem services: nutrient retention, biodiversity or both? Freshwater Biology 50: 705–714.

Kennison G. C. B., 1986. Preliminary observations on the plant colonisation of experimental turf ponds in a Broadland fen. Transactions of the Norfolk Naturalists Society 27: 193–198.

Linton, S. & R. Goulder, 2000. Botanical conservation value related to origin and management of ponds. Aquatic Conservation: Marine and Freshwater Ecosystems 10: 77–91.

Linton, S. & R. Goulder, 2003. Species richness of aquatic macrophytes in ponds related to number of species in neighbouring water bodies. Archiv für Hydrobiologie 157: 555–565.

Menetrey, N., L. Sager, B. Oertli & J. B. Lachavanne, 2005. Looking for metrics to assess the trophic state of ponds. Macroinvertebrates and amphibians. Aquatic Conservation: Marine And Freshwater Ecosystems 15: 653–664.

Moller, T. R. & C. P. Rordam, 1985. Species numbers of vascular plants in relation to area, isolation and age of ponds in Denmark. Oikos 45: 8–16.

Palmer, M., 1992. A Botanical Classification of Standing Waters in Great Britain and a Method for the Use of Macrophyte Flora in Assessing Changes in Water Quality. Joint Nature Conservation Committee, Peterborough, UK.

Pond Action, 1992. Experimental management of wetland habitats at Pinkhill Meadow. NRA National R&D project F01(91) 2 383.

Pond Action, 1998. A Guide to the Methods of the National Pond Survey. Pond Action, Oxford.

Preston C. D., D. A. Pearman & T. D. Dines (eds) 2002. New Atlas of the British and Irish Flora. Oxford University Press, Oxford.

Schneider, D. W. & T. M. Frost, 1996. Habitat duration and community structure in temporary ponds. Journal of the North American Benthological Society 15: 64–86.

Stumpel, A. H. P. & H. van der Voet, 1998. Characterizing the suitability of new ponds for amphibians. Amphibia-Reptilia 19: 125–142.

Talling, J. F., 1951. The element of chance in pond. Populations. The Naturalist 839: 157–170.

Wellborn G. A., D. K. Skelly & E. E. Werner, 1996. Mechanisms creating community structure across a freshwater habitat gradient. Annual Review of Ecology and Systematics 27: 337–363.

Williams, P., J. Biggs, A. Corfield, G. Fox, D. Walker & M. Whitfield, 1997. Designing new ponds for wildlife. British Wildlife 8: 137–150.

Williams, P., J. Biggs, C. J. Barr, C. P. Cummins, M. K. Gillespie, T. C. G. Rich, A. Baker, J. Baker, J. Beesley, A. Corfield, D. Dobson, A. S. Culling, G. Fox, D. C. Howard, K. Luursema, M. Rich, D. Samson, W. A. Scott, R. White & M. Whitfield, 1998. Lowland Pond Survey 1996. Department of the Environment, Transport and the Regions, London.

Williams, P., J. Biggs, M. Whitfield, A. Thorne, S. Bryant, G. Fox & P. Nicolet, 1999. The Pond Book: A Guide to the Management and Creation of Ponds. Ponds Conservation Trust, Oxford.

Williams P., M. Whitfield & J. Biggs, 2004. Comparative biodiversity of rivers, streams, ditches and ponds in an agricultural landscape in Southern England. Biological Conservation 115: 329–341.

Hydrobiologia (2008) 597:149–155
DOI 10.1007/s10750-007-9220-0

ECOLOGY OF EUROPEAN PONDS

Management of an ornamental pond as a conservation site for a threatened native fish species, crucian carp *Carassius carassius*

G. H. Copp · S. Warrington · K. J. Wesley

Abstract Ornamental ponds are important sites for conserving threatened native fish species (e.g. crucian carp *Carassius carassius* L.), but pond management plans rarely include considerations of native fishes. We developed and implemented a management plan for a small (0.8 h), ornamental estate pond in Hertfordshire (England) using historical information (aquatic plant and animal surveys) and a 9-year data set on climatic variables and crucian carp body condition. Crucian carp growth was not correlated with climatic variables, but body condition decreased with increasing temperature (in degree-days), which suggests that temperature influences on growth are counter-balanced by environmental factors. Management included the removal of one fish species (to eliminate hybridization with another species) and the introduction of two native species (to re-balance the fish assemblage), a reduction in floating aquatic plants (to reduce shading of the sediments) as well as the use of a chemical agent to compact the pond's fine sediments and barley straw to enhance invertebrate habitat and thus fish prey production.

Keywords Species diversity · Fish body condition · Hybridization · Goldfish · Barley straw

Guest editors: R. Céréghino, J. Biggs, B. Oertli and S. Declerck
The ecology of European ponds: defining the characteristics of a neglected freshwater habitat

Electronic supplementary material The online version of this article (doi:10.1007/s10750-007-9220-0) contains supplementary material, which is available to authorized users.

G. H. Copp (✉)
Salmon & Freshwater Fisheries Team, CEFAS, Pakefield Road, Lowestoft, Suffolk NR33 0HT, UK
e-mail: gordon.copp@cefas.co.uk

S. Warrington
National Trust, Westley Bottom, Bury St Edmunds, Suffolk IP33 3WD, UK

K. J. Wesley
Bedwell Fisheries Services, 22 Puttocks Lane, Welham Green, Hertfordshire AL9 7LP, UK

Introduction

Ponds, floodplain pools and small lakes sustain a disproportionately high number of aquatic plant and invertebrate species (e.g. Oertli et al., 2002; Williams et al., 2003). These small water bodies are also important for native fish species reproduction and thus their conservation (Copp, 1989, 1991; Wheeler, 2000), but their numbers are in decline throughout Europe (e.g. Steiner, 1988) and they are at risk from non-native species introductions (Adams, 2000; Copp et al., 2005). Thus, small water bodies are the subject of conservation and management initiatives at local (e.g. Conservators of Epping Forest, 2002), national (e.g. Everard et al., 1999; Schwevers et al., 1999) and international levels (e.g. European Commission's Council Directive 92/43/EEC [1] of 21 May 1992).

Ornamental and fishery ponds not accessible to the public are the most likely to contribute significantly to fish conservation initiatives (e.g. Copp et al., 2005). However, fish are usually ignored in pond surveys and conservation strategies (e.g. Everard et al., 1999; Oertli et al., 2002; Williams et al., 2003), though some exceptions exist (e.g. Schwevers et al., 1999; Conservators of Epping Forest, 2002).

A pond fish of particular conservation concern is crucian carp *Carassius carassius* L (Environment Agency, 2003), a cryptic species native to most of middle and northern Europe, including south-eastern England (Wheeler, 2000). Characteristic of small natural and artificial ponds and lakes (Holopainen et al., 1997; Wheeler, 2000), crucian carp is threatened in many parts of its range due to acidification (Holopainen & Ikari, 1992), loss of habitat (Copp, 1991; Schwevers et al., 1999), displacement by introduced gibel carp *C. gibelio* (Navodaru et al., 2002), and extirpation by introduced goldfish *C. auratus* L. through hybridization (Wheeler, 2000; Hänfling et al., 2005; Smartt 2007). Thus, the present article outlines a pond conservation management plan, with specific regard to native pond fishes (i.e. crucian carp), for Bayfordbury Lake (Hertfordshire, England: Lat:51:46:33 N, Lon: 0:05:48 W), and evaluates the pond's crucian carp growth response to pond management over a nine-year period (1992–2000).

Study site, material and methods

Bayfordbury Lake ($\approx$0.8 ha) is of particular interest for: (1) its location within 100 m of a small meteorological station, (2) being typical of ornamental estate ponds created in the 19th century; (3) the considerable historical information on the pond, and (4) its crucian carp population (the morphotype described by Wheeler, 2000). The pond has peaks in nitrate (up to 88 mg l^{-1}) in winter and in phosphate (up to 22 μg l^{-1}) in winter and mid-summer (Crispin & Izzo, 1995). Massive fish kills occurred in 1957–1958 (due to a pollution; Ward, 1978) and again in January 1992 (due to extended ice cover), when the fish assemblage consisted of crucian carp, roach *Rutilus rutilus* (L.), common bream *Abramis brama* (L.) and roach × bream hybrids.

Fish assemblage monitoring began in October 1992 (first and last weeks) and took place annually until 2000 using fyke nets, seine netting and electrofishing (see Copp et al., 2007). Fish were measured for standard length (SL) in mm and wet body weight (nearest g) and returned to the water. Crucian carp condition was assessed annually via three indices: (1) 'generalised body condition' (*sensu* Pitcher & Hart, 1992) using the slope parameter '*b*' (log-linear relationship of SL vs. wet weight in g); (2) Le Cren's (1951) mean body condition, $LK = w/w'$, where w is the observed body weight and w' is the expected weight as estimated from the log-linear model in '1' using the entire nine-year data set); and (3) Fulton's condition (plumpness) factor, $K = \mathrm{Wt}10^5 \cdot \mathrm{SL}^{-3}$, as per Mills and Eloranta (1985). Juvenile fish growth can be a good estimator of responses to local conditions (e.g. Copp et al., 2004), and this was taken as the mean SL at age 2 (see Copp et al., 2007). Cumulative degree-days (°Days) water temperatures were derived from local maximum air temperatures (water temperatures were assumed to be 3°C lower than air) for the vegetative period of each year (1 March–31 October), using the threshold of 12°C for cyprinids as per Mills & Mann (1985). Relationships between fish indices and °Days were tested using Spearman's rank correlation on the entire data set and separately on fish of age $\leq$2+ only.

In 1995, the growth (back-calculated from scales), sex ratio and external morphology of crucian carp were examined in detail (see Copp et al. 2007), the distributions of aquatic and marginal plants were plotted cartographically (Fig. 1), and the density (numbers per 3 min sample) and richness (S = number of species) of aquatic invertebrates were determined by routine 3-min kick-sampling and sweep-netting (in daytime) for open-water and vegetated areas during winter, spring and summer. Invertebrate density and S were averaged for seasons, with S adjusted (S') to animal density to permit among-season comparisons.

Results and discussion

Management of the pond's fish stocks began in October 1992, based on four principles: (1) fish stocks should not exceed the ecosystem's carrying capacity (i.e. available food resources, reduced oxygen levels during ice cover); (2) 'top-down' control of fish numbers should be maintained by

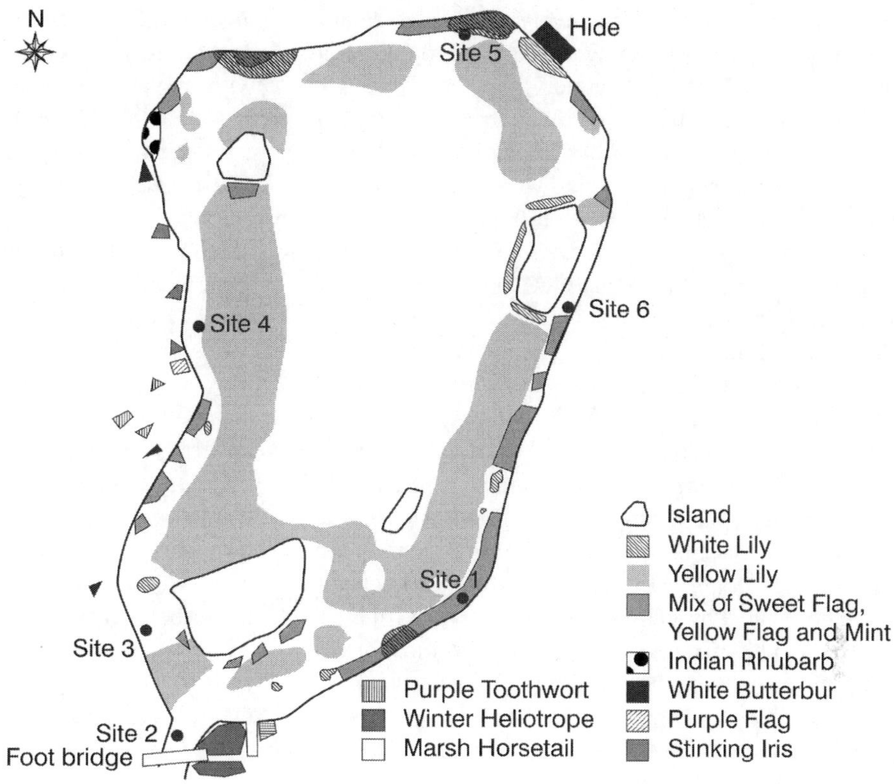

Fig. 1 Detailed map of Bayfordbury Lake (Hertfordshire, England), with vegetation zones indicated: white lily *Nymphaea alba* L., yellow lily *Nuphar lutea* (L.), sweet flag *Acorus calamus* L., white flag *Iris pseudacorus* L., water mint *Mentha aquatica* L., Indian rhubarb *Darmera peltata* (Torr.) Voss., purple flag *Iris versicolor* L., white butterbur *Petasites albus* (L.) Gaertner, winter heliotrope *Petasites fragrans* (Vill.) C. Presl, stinking iris *Iris foetidissima* L., purple toothwort *Lathraea clandestina* L. (a parasite on the roots of a poplar (*Populus* spp.)), marsh horsetail *Equisetum palustre* L

harvesting fish rather than by introducing a piscivorous fish species (the income helps sustain the pond's management, as does the harvest of excess yellow water lily *Nuphar lutea* (L.) tubers); (3) the existing prohibition on angling should be retained to reduce the risk of unauthorized fish introductions of angling amenity species (including piscivores), unwanted pet fish, especially non-native species; and (4) the conservation value of the pond's fish assemblage should be enhanced by favouring the crucian carp and roach populations through the removal of common bream, which were hybridizing with roach and thus undermining its genetic integrity (and commercial value), and the introduction of native fish species.

Development of the pond management plan began, as per Williams et al. (1999), with an assessment of historical information on the pond (e.g. Clinton-Baker, 1946; Beckett, 1961; Ward, 1978; Crispin & Izzo, 1995). The resulting plan had four main objectives: (1) Preserve and enhance species diversity of vegetation on the banks and islands, with protection and enhancement of notable species (e.g. *Lathraea clandestine*); (2) Maintain and enhance the pond through improvements to water quality and bottom habitat, the control of certain aquatic plant species (e.g. removal of yellow water lily tubers along pelagic periphery of beds to reduce siltation), and further enhancement of the conservation (and commercial) value of the fish assemblage; (3) Control and/or discourage pest bird species, e.g. Canada geese *Branta canadensis* (L.); and (4) Maintenance of existing conservation infrastructure (access routes, existing and new bird boxes, bird hide).

The fish surveys revealed crucian carp to represent about 19% of the catch in 1992 and 1993, rising to about 30% in 1995 and 1996, and then varied between 48 and 68% from 1997 to 2000 (see Electronic supplementary material). With the progressive removal of common bream in 1992–1994, the contribution of roach rose initially from 51% in 1992 to 77% in 1993, but then progressively dropped from 68–69% in 1995 and 1996 to about 21–32% in the following years. No common bream were recorded in catches after 1994. To compensate for the onset of cormorant predation in 1995, and to balance the fish assemblage, rudd *Scardinius erythrophthalmus* (L.) and tench *Tinca tinca* (L.) were introduced that year. These surface pelagic and mid-water benthic feeders, respectively, comprised minor proportions of the assemblage from 1998 to 2000 (on average, 15 and 4%, respectively).

The aquatic vegetation and invertebrates surveys in 1994–1995 revealed a total of 51 aquatic and marginal plant species of which only five were non-native (see Electronic supplementary material). Of the 17 purely aquatic species, the most notable was Canadian pondweed *Elodea canadensis* Michaux, which was introduced in *circa* 1847, spread across the pond during the 1860s to such an extent that eradication attempts by pond drain down were undertaken in 1863, 1864, 1867, 1868, after which the species was not recorded (Clinton-Baker, 1946) again until the 1990s, when it was observed in low abundance. Emergent aquatic species were dominated by sweet flag *Acorus calamus* L., water mint *Mentha aquatica* L. and reed mace *Typha latifolia* L., whereas submerged species were dominated by yellow water lily.

The aquatic macro-invertebrates were most abundant in spring (150.3 per 3 min. kick+sweep sample), followed by summer (121.3) and winter (99.8). Species number (S) was also greatest in spring (28 species), followed by winter (20) and summer (19), but S' was highest in winter (20.0), followed by spring (18.6) and summer (15.7). Total taxonomic richness was 36, with mean S'' per site being 8.7, 9.2 and 9.3 in winter, spring and summer, respectively (see Electronic supplementary material). Collated species records for the period 1996–2000 (see Electronic supplementary material) revealed that aquatic Coleoptera were the richest group, with 30 (37%) of the 81 taxa identified to species level, followed by aquatic Heteroptera. The 18 invertebrate taxonomic groups observed include taxa characteristically found in eutrophic and mesotrophic waters, reflecting the general nature of the pond.

Overall, aquatic taxonomic richness in Bayfordbury Lake (see Electronic supplementary material) was below average (mean = 30 species) for aquatic macrophytes but well above average for macro-invertebrates (mean = 25 species) relative to a widespread survey of lowland ponds ($n = 157$) in the UK (Ponds Conservation Trust, 2003). Contributions by aquatic Coleoptera (37%) and aquatic Heteroptera (16%) to invertebrate species richness in Bayfordbury Lake were similar to those (43% and 18%, respectively) from the national survey (Ponds Conservation Trust, 2003), but Odonata richness was greater at Bayfordbury (14% compared to 6%), which appears to possess many of the attributes associated with elevated invertebrate richness in ponds (Oertli et al., 2002): low altitude, shade (for Odonata), coverage of submerged (for Coleoptera) and floating (for Gastropoda) vegetation. Relative to other fish-bearing ponds, Odonata and Coleoptera taxonomic numbers in Bayfordbury Lake were low relative to some nearby ponds (Leeming & Warrington, 2004), being about half of the mean reported for lowland Swiss ponds (Oertli et al., 2002). Whereas, Bayfordbury Lake was relatively rich in Gastropoda and aquatic plant species (see Electronic supplementary material), with about a third more Gastropoda and twice the number of aquatic plant species predicted from species-area relationships for lowland Switzerland (Fig. 3 in Oertli et al., 2002). Therefore, the elevated nutrient levels in Bayfordbury Lake appear to be compensated in part by the habitat diversity, which is sustained by the taxonomic diversity of aquatic and marginal vegetation and the extensive area of terrestrial-aquatic interface.

Habitat enhancement measures and crucian carp growth patterns

Mitigation of the pond's nutrient enrichment and heavily-silted substratum was in two steps. Firstly, bales of barley straw were submerged around the water body (at 24–62 bales per ha) for 14 days in the summer of 1996, with a third of bales opened each 7 days to facilitate dispersal. During the decomposition of the barley straw (which is accentuated in summer), chemicals that inhibit algae growth are released. The exact mechanisms and chemicals involved remain unknown, but the straw bales favour aquatic invertebrates (CEH, 2004), acting as a substratum for both macro-invertebrates and zooplankton. Secondly, the pond's sediments were compacted in February 1998 using 25 kg per 100 m^2 treatments of Siltex®, which is a highly porous form of very small (<5 nm) calcium carbonate particles that reduce silt levels by decreasing organic matter and facilitating sedimentation (firmer bottom, reduced methane production) over the short-to-medium term (Broughton, 2000; Booker, 2003); this is favourable to epibenthos and should enhance food resources for fish.

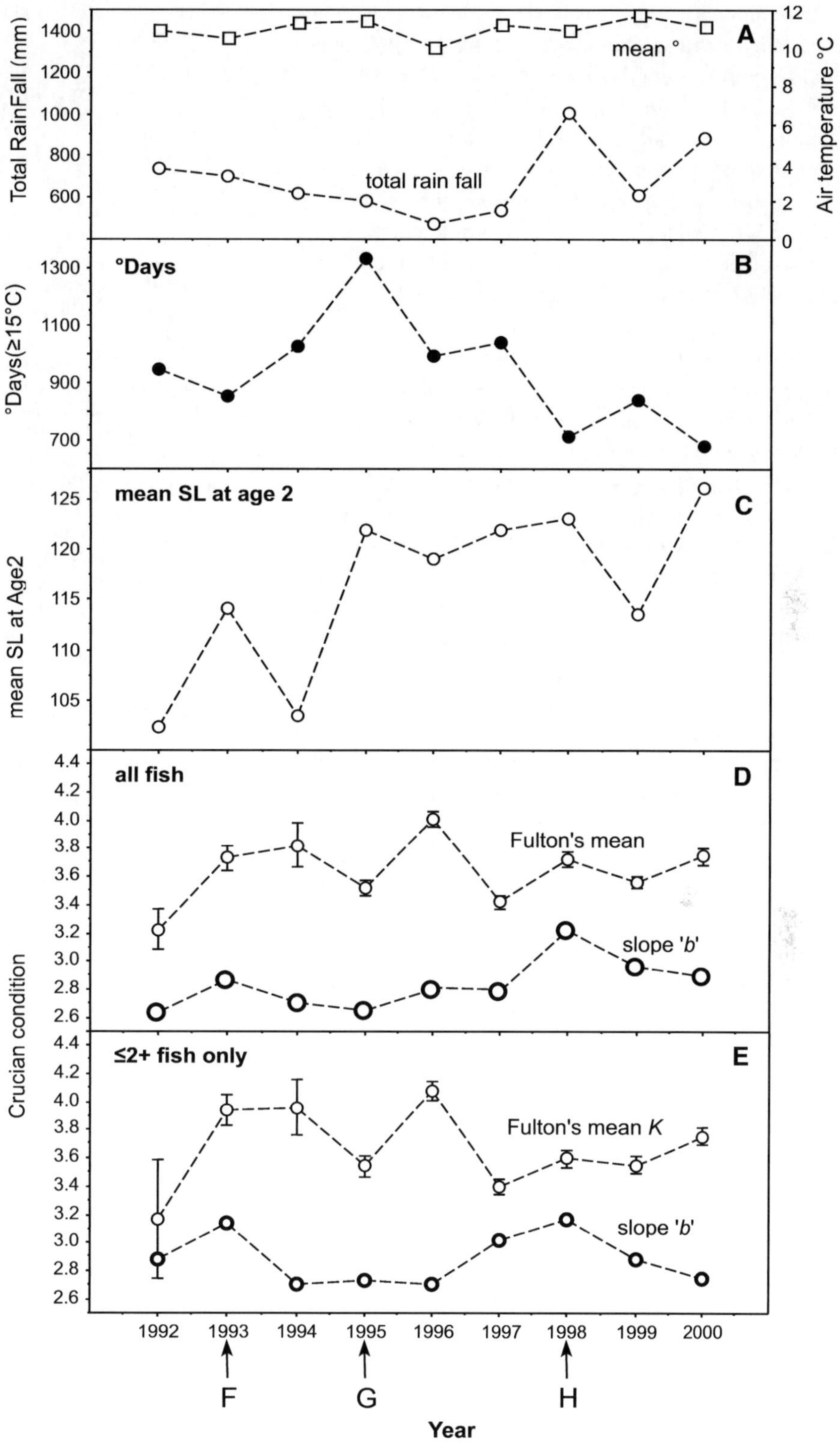

Fig. 2 Bayfordbury Lake (Hertfordshire, England) between 1992 and 2000: (**a**) Mean air temperature (°C) and total rainfall (mm) values (see Methods); (**b**) the number of degree-days (°D) ≥15°C of water temperature between 1 March and 31 October (see Methods); (**c**) back-calculated length at age 2 for all crucian carp; (**d**) slope regression '*b*' value and Fulton's condition index (*K*) for all crucian carp; (**e**) same as 'D' but for fish of age ≤2+ only; (**f**) Year that common bream removal began; (**g**) year when cormorants began to make annual feeding visits to the pond; and (**h**) year when Siltex® was applied in early March to compact the sediments

Crucian carp growth patterns in response to management practices were equivocal (Fig. 2). Crucian body length and length-to-weight conversions from the 9-year database were highly significant (see Electronic supplementary material). Body condition factors presented similar patterns for ages (Fig. 2d) and for age ≤2+ fish only (Fig. 2e). Juvenile growth, *K* and *FK* were not correlated with any environmental variable, but *b* (for all fish) decreased significantly with increasing °Days (Spearman's Rho $= 0.77$, $n = 9$, $P = 0.03$). These results may be purely artefact (e.g. from the use of air rather than water temperatures), however, the increase in *b* after 1996 and its decrease after 1998 suggest an inter-play between environmental factors and the management initiatives.

The onset of cormorant predation in 1995 had a considerable impact on the pond fishes (see Electronic supplementary material), with roach numbers decreasing thereafter, with losses observed between the fish samplings in early and late October, the month of cormorant visits. In late October, most fish were found deep in the marginal vegetation. The peak in *K* in 1997 and the rise in '*b*' (Fig. 2) in the three years after 1995 may be due to an overall reduced fish density in the pond, with crucian carp reaping the benefits of more abundant invertebrate food supplies. After the Siltex® application in 1998, crucian *K* increased slightly, whereas '*b*' decreased. Therefore, it remains unclear whether Siltex® enhanced the benthic environment sufficiently to be reflected in crucian condition, but these years were generally colder due to more rain, coinciding with lower °Days and a decreasing trend in '*b*' (Fig. 2). Pond management plans with specific regard to native fish species such as presented here would make a considerable contribution to the conservation of limnophilous species such as crucian carp, rudd and tench, which are the species most often impacted by habitat alternation, e.g. river regulation (Copp, 1991).

Acknowledgements We thank J. Harman and various groups of undergraduate students, but in particular M. Crispin, for assistance in the collection and processing of data and background information.

References

Adams, M. J., 2000. Pond permanence and the effects of exotic vertebrates on anurans. Ecological Applications 10: 559–568.

Beckett, K. A., 1961. A guide to the natural history of the Lake. Journal of the John Innes Society 8 (2nd series).

Booker, C., 2003. HSE outlaws an 800-year-old method of keeping water fresh. Christopher Bookers' Notebook, The Sunday Telegraph (London), 28 September 2003, p. 16.

Broughton, B., 2000. Chemical desilting of lakes (last accessed on 6 Sept. 2006: http://www.anglersnet.co.uk/authors/bruno09.htm)

CEH., 2004. Information sheet 1: Control of algae with barley straw. Centre for ecology & hydrology, Wallingford, 13 pp. Available at: http://www.nerc-wallingford.ac.uk/research/capm/pdf%20files/1%20Barley%20Straw.pdf (last accessed on 7 August 2007).

Clinton-Baker, L., 1946. Bayfordbury record book (a collection of written records of, and correspondence about, the estate from its foundation in 1759 up to 1869. Presented to C.D. Darlington by Lady Clinton Baker in 1946. Held at the John Innes Centre, Norfolk (Ref. SS1 A 2).

Conservators of Epping Forest, 2002. Epping Forest Annual Report of the Superintendent for 2001/2002. Corporation of London, Loughton, Essex, 45 pp.

Copp, G. H., 1989. The habitat diversity and fish reproductive function of floodplain ecosystems. Environmental Biology of Fishes 26: 1–26.

Copp, G. H., 1991. Typology of aquatic habitats in the Great Ouse, a small regulated lowland river. Regulated Rivers: Research & Management 6: 125–134.

Copp, G. H., J. Černý & V. Kováč, 2007. Growth and morphology of an endangered native freshwater fish, crucian carp *Carassius carassius*, in an English ornamental pond. Aquatic Conservation: Marine and Freshwater Ecosystems (in press).

Copp, G. H., M. G. Fox, M. Przybylski, F. Godinho & A. Vila-Gispert, 2004. Life-time growth patterns of pumpkinseed *Lepomis gibbosus* introduced to Europe relative to native North American populations. Folia Zoologica 53: 237–254.

Copp, G. H., K. Wesley & L. Vilizzi, 2005. Pathways of ornamental and aquarium fish introductions into urban ponds of Epping Forest (London, England): the human vector. Journal of Applied Ichthyology 21: 263–274.

Crispin, M. & M. Izzo, 1995. A management plan for Bayfordbury Lake. Unpublished report, Department of Environmental Sciences, University of Hertfordshire, Hatfield, 46 pp.

Environment Agency, 2003. Crucian Carp Field Guide. National Coarse Fish Centre, Environment Agency, Bristol, 2 pp.

Everard, M., B. Blackham, K. Rouen, W. Watson, A. Angell & A. Hull, 1999. How do we raise the profile of ponds? Freshwater Forum 12: 32–43.

Hänfling, B., P. Bolton, M. Harley & G. R. Carhalho, 2005. A molecular approach to detect hybridisation between crucian carp (*Carassius carassius*) and non indigenous carp species (*Carassius* spp. and *Cyprinus carpio*). Freshwater Biology 50: 403–417.

Holopainen, I. J. & A. Ikari, 1992. Ecophysiological effects of temporary acidification on crucian carp, *Carassius carassius* (L.): a case history of a forest pond in eastern Finland. Annales Zoologici Fennici 29: 29–38.

Holopainen, I. J., W. M. Tonn & T. C. Paszkowski, 1997. Tales of two fish: the dichotomous biology of crucian carp (*Carrassius carassius* (L.)) in Northern Europe. Annales Zoologica Fennici 34: 1–22.

Le Cren, E. D., 1951. The length-weight relationship and seasonal cycle in gonad weight and condition in the perch (*Perca fluviatilis*). Journal of Animal Ecology 20: 201–219.

Leeming, D. J. & S. Warrington, 2004. An aquatic invertebrate survey of Ickworth Park, Suffolk. Transactions of the Suffolk Natural History Society 40: 55–72.

Mills, C. A. & A. Eloranta, 1985. Reproductive strategies in the stone loach *Noemacheilus barbatulus*. Oikos 44: 341–349.

Mills, C. A. & R. H. K. Mann, 1985. Environmentally-induced fluctuations in year-class strength and their implications for management. Journal of Fish Biology 27 (Suppl. A): 209–226.

Navodaru, I., A. D. Buijse & M. Staras, 2002. Effects of hydrology and water quality on the fish community in Danube delta lakes. International Review of Hydrobiology 87: 329–348.

Oertli, B., D. A. Joye, E. Castella, R. Juge, D. Cambin & J.-B. Lachavanne, 2002. Does size matter? The relationship between pond area and biodiversity. Conservation Biology 104: 59–70.

Pitcher, T. J. & P. J. B. Hart, 1992. Fisheries Ecology. Chapman & Hall, London, 414 pp.

Ponds Conservation Trust, 2003. Aquatic ecosystems in the UK agricultural landscape. Report CSG 15. Defra, London (Available at: http://www.pondconservation.org.uk/; last accessed 7 August 2007).

Schwevers, U., B. Adam & C. Gumpinger, 1999. Zur Bedeutung von Auegewässern für die Fischfauna von Bundeswasserstraßen. Wasser und Boden 51(6): 35–39.

Smartt, J., 2007. A possible genetic basis for species replacement: preliminary results of interspecific hybridisation between native crucian carp *Carassius carassius* (L.) and introduced goldfish *Carassius auratus* (L.). Aquatic Invasions 2: 59–62.

Steiner, E., 1988. Teichwirtschaft und Naturschutz. Österreichs Fisherei 41: 142–149 (English summary).

Ward, G., 1978. Bayfordbury Lake. Occasional papers published by Hatfield Polytechnic, Department of BioSciences, Hatfield, Hertfordshire, 2 pp.

Wheeler, A. C., 2000. Status of the crucian carp, *Carassius carassius* (L.), in the UK. Fisheries Management and Ecology 7: 315–322.

Williams, P., J. Biggs, M. Whitfield, A. Thorne, S. Bryant, G. Fox & P. Nicolet, 1999. The pond book. A guide to the management and creation of ponds. The Ponds Conservation Trust, Oxford, UK, 105 pp.

Williams, P., M. Whitfield, J. Biggs, S. Bray, G. Fox, P. Nicolet & D. Sear, 2003. Comparative biodiversity of rivers, streams, ditches and ponds in an agricultural landscape in Southern England. Conservation Biology 115: 329–341.

Hydrobiologia (2009) 634:1–9
DOI 10.1007/s10750-009-9891-9

POND CONSERVATION

Pond conservation: from science to practice

Beat Oertli · Régis Céréghino · Andrew Hull · Rosa Miracle

Published online: 5 August 2009

Abstract In Europe, ponds are an exceptionally numerous and widely distributed landscape feature forming a major part of the continental freshwater resource and contributing significantly to freshwater biodiversity conservation. This has been reflected by a growing scientific concern over the first few years of the twenty-first century and is evidenced by an increasing number of academic publications on pond related topics, particularly those relating to biodiversity. It is essential, however, that this expanding scientific knowledge is widely disseminated to those involved with pond management and is then rapidly translated into action. Inevitably, the task of transferring science to practice remains a significant challenge. As a first step towards meeting this challenge the European Pond Conservation Network (EPCN), at its biennial meeting in 2008 in Valencia (Spain), made this the main theme of the conference together with two special workshops further encouraging exchanges between scientists, practitioners and policy makers. The papers selected for this special issue of *Hydrobiologia* (from over 120 communications presented) are all from the conference. They represent a diverse collection of themes from across the continent and North Africa and present new and original insights into topics as wide ranging as: pond biodiversity; human disturbance; landscape ecology; ecological assessment and monitoring; practical management measures; ecological restoration; hydrology and climate change; invasive species and threatened species. In all cases, the papers demonstrate an overriding need for the development of a tight link between scientific knowledge and management. Furthermore, scientific advances have to be beneficial for *on the ground* management and, vitally, have to be disseminated, communicated and implemented into local, national and international policy. As such, national and international networks (such as the EPCN) have a central role to play and have to develop a robust information and communication strategy which will enable the dissemination of best practice materials and advice across the continent and beyond. The work contained in this volume represents a step in the right direction and will help to ensure that ponds remain a characteristic and highly

Guest editors: B. Oertli, R. Céréghino, A. Hull & R. Miracle
Pond Conservation: From Science to Practice. 3rd Conference of the European Pond Conservation Network, Valencia, Spain, 14–16 May 2008

B. Oertli (✉)
University of Applied Sciences Western Switzerland, hepia Lullier, 150 route de Presinge, 1254 Jussy-Geneva, Switzerland
e-mail: beat.oertli@hesge.ch

R. Céréghino
University of Toulouse, Toulouse, France

A. Hull
Liverpool John Moores University, Liverpool, UK

R. Miracle
University of Valencia, Valencia, Spain

visible feature of the European landscape in the twenty-first century.

Keywords Biodiversity conservation · Management · Ponds · Small waterbodies · Temporary pools · Europe

A network of stakeholders for a neglected freshwater resource

The management and conservation of freshwater resources has traditionally focussed upon running water and larger waterbodies. In comparison small waterbodies, such as ponds, have long been overlooked. Recently, however, there has been a growing realization that these small wetland patches are important not only for biodiversity but also for a range of socio-economic activities linked to them. Ponds, of course, are very numerous and worldwide there are hundreds of millions (Downing et al., 2006). Their small size (1 m^2–5 ha) coupled with their great abundance has meant that these small waterbodies have a critical role to play in the global carbon cycle, as collectively they probably trap significantly more carbon than the worlds oceans (Downing et al., 2008). Furthermore, together, they host a high and unique biodiversity, particularly significant at the regional scale, when compared to other freshwater systems (Williams et al., 2003).

In Europe ponds form a significant part of the continental freshwater resource. In order to address this issue, a collaborative workshop was held in Geneva (Switzerland) in 2004 (Oertli et al., 2005a) resulting in the launch of an international forum—the "European Pond Conservation Network" (EPCN; www.europeanponds.org). The EPCN was established with the principal aim of coordinating research, policy and management through the promotion of awareness, understanding and conservation of ponds in a changing European landscape. Since the first workshop the network has grown steadily and successfully and has held an international conference every 2 years since the original meeting in Geneva. The second EPCN conference took place in Toulouse (France) in 2006 (Céréghino et al., 2008) and a third in Valencia (Spain) in 2008. This most recent EPCN conference was organised by the Government of the Valencian Region (Conselleria de Medi Ambient, Aigua Urbanisme i Habitatge of the Generalitat Valenciana). Its success can be measured by the 130 delegates attending from 25 countries (Austria, Belgium, Czech Republic, Denmark, Estonia, France, Germany, Greece, Hungary, Iceland, Ireland, Italy, Mexico, Morocco, Netherlands, Poland, Portugal, Slovakia, Slovenia, Spain, Sweden, Switzerland, Tunisia, UK, USA). The participants included scientists attached to universities and wetland research agencies together with a variety of other 'pond stakeholders' including field managers, local authority representatives, politicians and conservation NGO's. A total of 123 communications were presented, 38 as oral presentations and 85 as posters (which can be downloaded from http://campus.hesge.ch/epcn/posters_valencia08.asp). At the Valencia conference an important milestone was delivered with the publication of the "Pond Manifesto" (EPCN (European Pond Conservation Network), 2008). This vital document, drafted and appended during the first two conferences in 2004 and 2006 (EPCN (European Pond Conservation Network), 2007), presents the case for conserving European ponds and provides an outline strategy for much needed conservation action across and beyond the continent.

This special issue of *Hydrobiologia* represents a selection of the communications presented in Valencia and includes a broad range of topics covering research activity across Europe and North Africa.

A growing scientific interest for ponds and their biodiversity?

It has become increasingly clear that research activity focussing upon ponds and other small water bodies has risen significantly over the past few years. The number of publications in peer-reviewed journals is a good indicator of this progress and taking the number of publications addressing the topic "pond" into consideration (in the ISI Web of Knowledge database), the trend between 1991 and 2008 illustrates two defined periods before and after 2001 (Fig. 1). At the world level, the number of indexed publications is higher after 2001 than before (about 10% higher), but this is even more marked at the European level (about 40% higher).

Taking the two keywords "pond" and "biodiversity" into consideration, the trend since 1991 shows a different pattern, with a continuous and very sharp

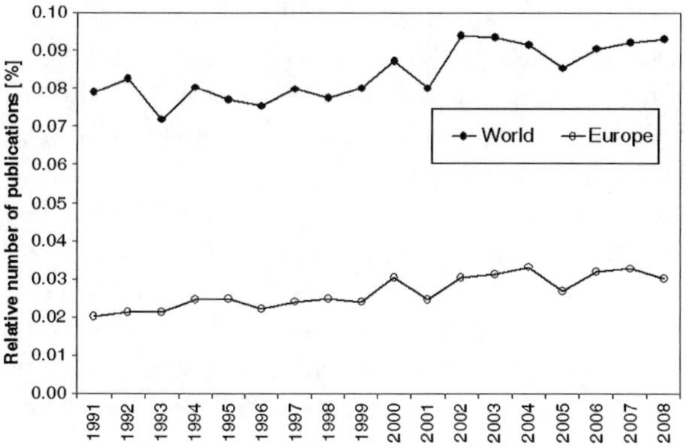

Fig. 1 Temporal dynamic between 1991 and 2008 of the number of "pond" publications indexed in ISI Web of Knowledge database (*upper line* world, *underline* Europe). Both trends underline higher proportions of "pond" publications subsequent to 2001. The number of "pond" publications is expressed in % (ratio on the total number of publications indexed), to avoid a possible bias linked to the continuous increase in the total number of publication indexed in the database. The search was realised on ISI Web of Knowledge—Science Citation Index Expanded (SCI-EXPANDED)—Topic "pond" or "ponds"

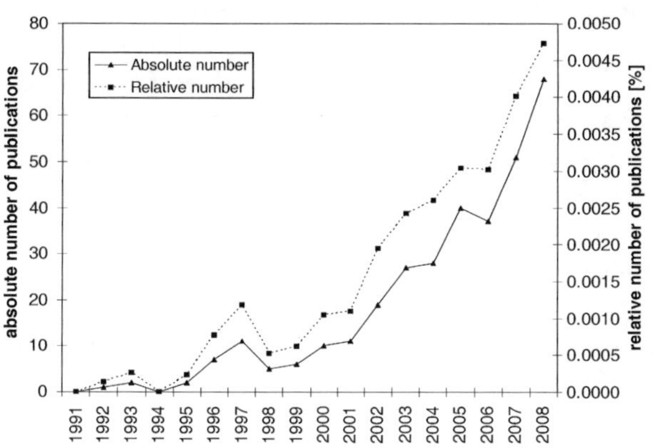

Fig. 2 Increase in the number of peer-reviewed publications addressing both topics "pond" and "biodiversity" from 1991 to 2008. The publications taken into consideration are those indexed in ISI Web of Knowledge database (keywords "pond" or "ponds" and "biodiversity"). Relative values expressed in % (ratio to the total number of publications indexed each year) are also presented; they evidence here a same increasing trend

increase (Fig. 2), even sharper since the EPCN was launched in 2004. In 2008, about 70 publications were indexed—seven times more than in 2000. The greatest part of this increase relates to a growing interest in biodiversity. Indeed, this trend can also be observed for other types of freshwater systems including rivers, streams and lakes (Fig. 3). Despite

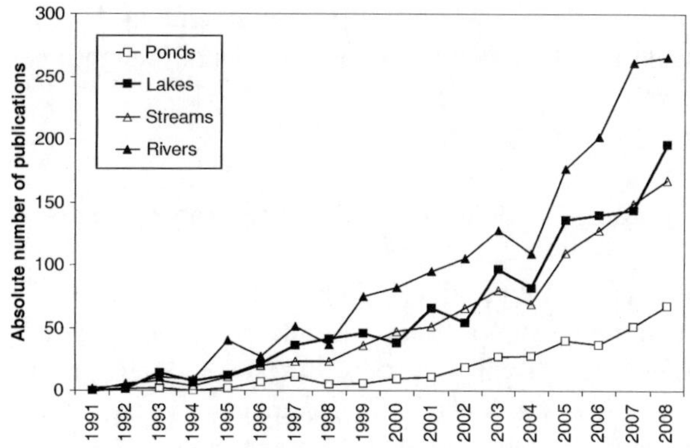

Fig. 3 Increase in the number of peer-reviewed publications addressing together the topic "Biodiversity" and one of the four types of freshwater systems ("ponds" or "lakes" or "rivers" or "streams") from 1991 to 2008. The publications taken into consideration are those indexed in ISI Web of Knowledge database. The relative values (ratio to total number of publications indexed) present the same trends and are therefore not presented here

these encouraging trends, there still remains a much greater interest for lakes and running waters and, of the four freshwater types, the proportion of pond publications still remains under 10% during the period, from 1991 to 2008. This stresses the necessity to undertake more pond research, especially if we accept the high value of ponds in the conservation of global freshwater biodiversity.

Inevitably, the increase in publications dealing with biodiversity is a consequence of the Rio Conference of 1992 and of the recognition that this topic is central to nature conservation. Nevertheless, it is depressing that scientific literature on freshwater biodiversity only increased sharply after 2002—10 years after Rio Conference—and underlines a poor relationship between policy and research.

Nevertheless by 2008, a large number of researchers are actively involved in investigating "pond biodiversity", as the Valencia conference illustrated. This underlines also an increasing size of the body of researchers. Accordingly, the EPCN has a critical role to play in facilitating and enhancing interaction and exchange between universities and research agencies and moving towards joint working in similar situations in different European countries. Another vital step to be undertaken by the EPCN is the communication of research outputs to all stakeholders and, principally, land owners, land managers and policy makers. The broad dissemination of information remains a key objective of the EPCN and needs to be undertaken at the local, national and international level.

Linking scientific knowledge and management

An encouraging development at the third EPCN conference in Valencia was the presence of representatives from a variety of stakeholder groups including researchers, field managers and local authorities. This diversity is also reflected by the communications presented which included many excellent case studies on pond management and biodiversity and this scientific exchange between scientist and managers provided a unique opportunity for serious networking. In order to enhance this process, a special workshop was dedicated to this topic—*Linking pond management to scientific knowledge*—with the aim of creating a better link between scientists and practitioners, and to improve the evidence base for pond conservation and management. Clearly, much knowledge on pond management is still held by practitioners working "on the ground" and in many cases this never reaches publication and therefore unavailable for others to learn from. Conversely, there is relatively little scientific research on pond management, or monitoring to assess the result of management activities. Clearly, both scientists and practitioners would both benefit from sharing best practice experiences and knowledge and one of the main themes of the workshop, was to discuss the key issues to develop pond management practices based on a sound scientific basis. One of the main questions considered was: *"How is it possible to improve the flow of information between management and research ?"*. Importantly, managers are also aware of good (and bad !) management practices and, all too often, these success stories are seldom highlighted. In order to tackle this issue a second workshop was held—*Pond management success stories*—to address positive and innovative approaches from across the continent. It was agreed that *success stories* need to be more widely disseminated to both managers, the scientific community and all other stakeholders. It was acknowledged that peer-reviewed journals play only a limited role here, only occasionally publishing such case studies (but see exceptions as Williams et al., 2008 or Rannap et al., 2009). Furthermore, peer-reviewed literature is often not read by end-users, partly because they don't have access to them, or because they do not read the English language. Inevitably, good practices and success stories are often published only in local journals, remaining for the most part in the grey literature. In future, it was felt that the web could play a central role in providing a platform for such case studies, centralising all type of literature and it was decided that this should be another key objective for the EPCN and the development of its website.

Special issue content

The papers selected for this special issue cover a variety of themes from different parts of Europe and North Africa, and in all cases demonstrate the link between scientific knowledge and management. The themes included in this volume are reflected by keywords including: biodiversity, human disturbance, landscape ecology, ecological assessment, practical management measures, ecological restoration, climate changes, invasive species, threatened species and monitoring.

Basic knowledge on pond biology and ecology is still full of gaps, and even areas regarding fauna and flora assemblages remain poorly understood. Pond **biodiversity** is particularly important at the regional scale (Oertli et al., 2002; Williams et al., 2003; Angélibert et al., 2006), but such regional studies are still scarce and there is an urgent need for studies from other parts of Europe. This is particularly important for regions where some of the most seriously threatened pond types can still be found (e.g. Mediterranean temporary pools and turloughs in the western extremities of the British Isles). Pinto-Cruz et al. (2009) demonstrate in this volume that knowledge on plant assemblages associated with Mediterranean temporary ponds (European community priority habitat Number 3170) is an important determinant for regional biodiversity conservation. In the light of these findings, similar investigation needs to be undertaken on other pond types, including newly created ponds designed for novel uses in the twenty-first century.

Over the past century, ponds have been faced with a great variety of **human disturbance** and no more so than during the past 50 years. Even today, and despite our greater knowledge, pond filling, water pollution and eutrophication continue to be unchecked in many parts of Europe and this presents perhaps one of the greatest challenges to managers who must learn to deal effectively with these deep rooted problems.

Disturbances also often occur at the local scale and here experimental approaches conducted directly on ponds can be very useful and can give accurate information to the manager for tackling the identified problem. For example Amami et al. (2009) and Sahib et al. (2009) considered the effect of small scale disturbances in a series of Mediterranean temporary pools (simulating that caused by herbivorous mammals). Results showed that the richness of plant communities was not affected by this disturbance (Sahib et al., 2009). Furthermore, the recovery from such disturbance was rapid and, after 2 years, the species composition of the vegetation in disturbed plots was similar to that of the control plots (Amami et al., 2009). These results do not necessarily mean that disturbance by herbivores, especially during the plants' growing season, will not have a significant impact on the vegetation and its richness, but rather that its impact can be modulated.

Landscape ecology is particularly relevant when managing small waterbodies. Ponds should never be considered in isolation and the growing realisation that pond landscapes—once common in many parts of Europe—are now seriously threatened, has serious implications for biodiversity. The acknowledgement of climate change and the northward movement of many species means that ponds have a vital role to play as stepping stones through the landscape. In the Landscape Protected Areas (LPA) of southern Estonia, the creation of several hundred ponds for the management of amphibian populations demonstrates how effectively conservation measures can be put in place (Rannap et al., 2009). The significance of conserving biodiversity at the regional scale was also demonstrated by Triest and Sierens (2009) who emphasize the role of regional pond habitat diversity for the preservation of *Ruppia* taxa and their unique haplotype diversity in extreme saline habitats. Ponds also provide a stepping stone for freshwater biodiversity linked to larger waterbodies providing a resting place for birds and mammals (e.g. beaver, otter) (e.g. Santoul et al., 2009).

Ecological assessment is a topic of applied research which is particularly useful for managers but exchanges between researchers and practitioner need to focus on the development of tools that respond both to international, national or regional policy (nature conservation and freshwater management) and also to the needs of practitioner. Such standardised tools have been developed for running waters or lakes, but are still scarce for ponds. Some tools have now been developed in selected regions of Europe (e.g. Biggs et al., 2000; Boix et al., 2005; Chovanec et al., 2005; Oertli et al., 2005b; Solimini et al., 2008; Indermuehle et al., 2009; Trigal et al., 2009). In the present issue, Della Bella & Mancini (2009) investigate how macroinvertebrates and diatoms communities in Italian permanent ponds can be useful bioindicators (taxa and metrics) to evaluate the effect of human impacts (water pollution or habitat alteration). In another study conducted in Switzerland, Sager & Lachavanne (2009) propose a method based on the aquatic plant communities to assess easily and to monitor the trophic state of ponds.

Pond managers also want scientific support for **practical management measures**. This is particularly important in relation to **threatened species**, especially those confined mainly to ponds. Experimental research brings an important contribution to the understanding of the ecology of species and to the response to management practices. Rannap et al. (2009) demonstrate how new ponds created within an existing network of several hundreds of ponds enhances the number of breeding sites of two threatened amphibian species—the crested newt (*Triturus cristatus* Laur.) and the common spadefoot toad (*Pelobates fuscus* Wagler). Research experiments conducted in the laboratory are also invaluable. Rhazi et al. (2009b) tested the competition between two plant species—a locally invasive clonal species (*Bolboschoenus maritimus*) and a rare threatened quillwort (*Isoetes setacea*)—typical in the Mediterranean temporary pools of southern France. They demonstrate that although the threatened species is eventually excluded by the invasive species, the threatened species can nevertheless survive in pioneer situations—practical advice that will be passed on to managers.

Management is often also directed to **ecological restoration.** Peretyatko et al. (2009) present an example conducted in ponds from Belgium: the effects of biomanipulation (fish removal) was assessed on different compartments of pond ecosystems (phytoplankton, zooplankton, submerged vegetation and nutrients). The results clearly indicated that such management intervention can be used, at least, for the short-term restoration of ecological water quality. The most drastic restoration measure is the creation of a new pond. This measure is often undertaken locally to compensate for the drastic loss in pond number at the regional scale. The creation of a new pond is also often much more appropriate (and less expensive) than the attempt of restoring a given degraded pond. Managers are aware of good practices and also of the benefits of such a management measure. Ruhí et al. (2009) present an assessment of the impact on macrofauna of the creation of new ponds in Catalonia (NE Iberian Peninsula). This work confirmed that the colonisation process is very rapid and 50% of the species found in the study were already captured 1–2 months after flooding. However, more time is needed for the establishment of specialized populations and in particular those linked to aquatic plants. Another particularly successful example of creating a new pond complex was detailed by Williams et al. (2008).

What is the **future for ponds in the European landscape of the twenty-first century ?** Local managers are at the forefront and are the first to be faced with these new challenges which have been brought about by an ever changing landscape, land use and local practice. Scientists have therefore to work closely not only with *natural* ponds, but also with the many hundreds of thousands of ponds created by human intervention across Europe during the past few centuries. Importantly, the era of pond digging is still prevalent and new demands for water storage, flood defence and biodiversity conservation are generating new pond types and, increasingly, managers want information about these new habitats and how they can be managed to maximise their utility. Even if these new ponds never replace vanishing pond types, they can provide some contribution for conserving regional freshwater biodiversity. Artificial ponds, a priori considered to have little or no interest, can reveal surprising results. Highway stormwater retention ponds have already been presented as biodiversity islands (Scher et al., 2004). Becerra-Jurado et al. (2009) in this volume highlighted the potential contribution of constructed ponds used for wastewater treatment to the landscape biodiversity in Ireland. Gravel pits constitute another example of an important artificial pond type. For example in an urban landscape in South-western France, a set of gravel pits captured more than half of the regional species pool of aquatic birds (Santoul et al., 2009). Urban or semi-urban ponds also play also a crucial role in enhancing the vital link between people and nature. Even urban ponds have to be managed, even if biodiversity remains often a secondary objective to esthetics or human health and welfare. Peretyatko et al. (2009) present a successful management intervention focussed on the prevention of noxious cyanobacterial bloom formation in ponds in the heart of Brussels.

Ponds in the twenty-first century will also have to face **global challenges, especially climate change**. Recently, completed models of temperature increases in Alpine ponds predict drastic changes in species assemblages and richness (Rosset et al., 2008). Changes in the hydrology will also be a key factor to investigate and Florencio et al. (2009) demonstrate in this volume how hydroperiod can have a decisive influence on the composition of macroinvertebrate assemblages in Mediterranean temporary ponds. This was also observed during first years of macrofauna colonisation of new excavated ponds (Ruhí et al., 2009). The same evidence came also from studies of aquatic plants from these Mediterranean temporary ponds: hydrology appears to be the primordial factor that structures and selects the species of a community (Sahib et al., 2009; Rhazi et al., 2009a); the hydrology modifies any disturbance effects and influences competition intensity through its impact on primary production. The expected reduction of humid years and of rainfall globally may lead to a decrease in the probability of survival of populations of characteristic pool species (Rhazi et al., 2009a). Such data will be particularly useful in the future, because we expect drastic changes in hydroperiod length. For example, a hydroperiod decrease of more than 52 days is predicted for Lake Kourna, a Mediterranean Temporary Pond from Crete (predictions with IPCC B2 and A2 climate scenarios) (Dimitriou et al., 2009). Investigations on the effect of the length of the hydroperiod have not only focussed upon Mediterranean waterbodies—in the future, we have also to consider the effect of changing hydrology on permanent ponds, which in turn can become temporary or can benefit (or suffer) from a large drawdown zone.

Globally, **invasive species** will be an increasing threat for biodiversity in the twenty-first century, due to a sharp rise in the geographical movement of species. This is particularly true for freshwater ecosystems (e.g. Ricciardi, 2006), including ponds. Aquatic plants, snails, turtles and fish species are moved around the world by aquarium hobbyists; many exotic, and potentially invasive species, often finish in a garden pond or in a natural pond close to human settlement. Spreading through other processes and of other taxa is also observed; for example Rodríguez-Pérez et al. (2009) have monitored the colonisation of the North American aquatic bug *Trichocorixa verticalis verticalis* (Hemiptera: Corixidae) in the wetlands of Doñana (southern Spain).

Long-term **monitoring** can provide particularly rich information, especially in the context of global change and many protected areas have now set up systems to monitor their biodiversity. In a few cases, this also includes the monitoring of pond networks, but to date this remains rare. With their small size and their relatively simple community structure, ponds provide an ideal sentry and early warning system (De Meester et al., 2005). To this effect in 2002, the Swiss National Park has begun a long-term monitoring programme of freshwater systems, including about 30 alpine ponds (Robinson & Oertli, 2009). Fahd et al. (2009) present in this volume a four-decade record (1964–2007) of the Copepods and Branchiopods associated with 50 temporary ponds in the Doñana National Park (SW Spain) and underline the usefulness of such long-term monitoring. Shorter time span can also already bring useful results, as demonstrated here by a 10-years continuous monitoring in a Mediterranean temporary pool facing fluctuating hydrology (Rhazi et al., 2009a).

Conclusion

In the early twenty-first century there is a growing interest for ponds, and this is clearly reflected by an increasing scientific activity, especially on the topic of biodiversity. It is vital to link this research with the needs expressed by managers, practitioners and policy makers. Furthermore, these scientific advances have to be beneficial for "on the ground" management and have to be disseminated, communicated and implemented into local, national and international policy. National and international networks (such as the EPCN) have a central role to play and have to develop a robust information and communication strategy which is disseminated across the continent and beyond. Conferences and workshops are unique occasions and help to disseminate knowledge and information and in this light the fourth EPCN conference is already scheduled for Berlin in June 2010. After the Valencia conference in 2008, which largely focussed on Mediterranean waterbodies, the gathering in Berlin will provide an opportunity to further promote pond conservation in northern Europe.

Acknowledgements We are very grateful to the "Valencia team" for the organisation of a very successful and useful meeting. In particular; special thanks goes to Ignacio Lacomba, Vicente Sancho, Benjamí Perez and all the other anonymous helpers who contributed to the excellent organisation and hospitality in their beautiful city. Thanks also for their supports to the Life-Nature project "Restoration of priority habitats for amphibians" (LIFE05/NAT/E/00060) and to the Conselleria de Medi Ambient, Aigua Urbanisme i Habitatge of the Generalitat Valenciana. We would also like to acknowledge the support from the MAVA Foundation for the organisation of the two workshops "*Pond management success stories*" and "*Linking pond management to scientific knowledge*". We would also like to thank Audrey Greenman for undertaking the data analysis for Figs. 1, 2 and 3 and, finally, thanks to the 60 reviewers who provided helpful comments on the manuscripts submitted to this special issue.

References

Amami, B., L. Rhazi, S. Bouahim, M. Rhazi & P. Grillas, 2009. Vegetation recolonization of a Mediterranean temporary pool in Morocco following small scale experimental disturbance. Hydrobiologia (this issue).

Angélibert, S., N. Indermuehle, D. Luchier, B. Oertli & J. Perfetta, 2006. Where hides the aquatic biodiversity in the Canton of Geneva (Switzerland)? Archives des Sciences 59: 225–234.

Becerra Jurado, G., M. Callanan, M. Gioria, J.-R. Baars, R. Harrington & M. Kelly-Quinn, 2009. Aquatic macroinvertebrate diversity of natural and wastewater treatment ponds: community structure and driving environmental factors. Hydrobiologia (this issue).

Biggs, J., P. Williams, M. Whitfield, G. Fox & P. Nicolet, 2000. Biological techniques of still water quality assessment: phase 3. Method development. Environment Agency, Bristol.

Boix, D., S. Gascon, J. Sala, M. Martinoy, J. Gifre & X. D. Quintana, 2005. A new index of water quality assessment in Mediterranean wetlands based on crustacean and insect assemblages: the case of Catalunya (NE Iberian peninsula). Aquatic conservation: marine and freshwater ecosystems 15: 635–651.

Céréghino, R., J. Biggs, S. Declerck & B. Oertli, 2008. The ecology of European ponds: defining the characteristics of a neglected freshwater habitat. Hydrobiologia 597: 1–6.

Chovanec, A., J. Waringer, M. Straif, W. Graf, W. Reckendorfer, A. Waringer-Löschenkohl, H. Waidbacher & H. Schultz, 2005. The floodplain index—a new approach for assessing the ecological status of river/floodplain-systems according to the EU Water Framework Directive. Large Rivers 15: 169–185.

De Meester, L., S. Declerck, R. Stoks, G. Louette, F. Van de Meutter, T. De Bie, E. Michels & L. Brendonck, 2005. Ponds and pools as model systems in conservation biology, ecology and evolutionary biology. Aquatic Conservation: Marine and Freshwater Ecosystems 15: 715–725.

Della Bella, V. & L. Mancini, 2009. Freshwater diatom and macroinvertebrate diversity of coastal permanent ponds along a gradient of human impact in a Mediterranean ecoregion. Hydrobiologia (this issue).

Dimitriou, E., E. Moussoulis, F. Stamati & N. Nikolaidis, 2009. Modeling hydrological characteristics of Mediterranean Temporary Ponds and potential impacts from climate change. Hydrobiologia (this issue).

Downing, J. A., J. J. Cole, J. J. Middelburg, R. G. Striegl, C. M. Duarte, P. Kortelainen, Y. T. Prairie & K. A. Laube, 2008. Sediment organic carbon burial in agriculturally eutrophic impoundments over the last century. Global Biogeochemical Cycles 22: GB1018.

Downing, J. A., Y. T. Prairie, J. J. Cole, C. M. Duarte, L. J. Tranvik, R. G. Striegl, W. H. McDowell, P. Kortelainen, N. F. Caraco, J. M. Melack & J. J. Middelburg, 2006. The global abundance and size distribution of lakes, ponds, and impoundments. Limnology and Oceanography 51: 2388–2397.

EPCN (European Pond Conservation Network), 2007. Developing the pond manifesto. Annales de Limnologie—International Journal of Limnology 43: 221–232.

EPCN (European Pond Conservation Network), 2008. The Pond Manifesto: 20 pp.

Fahd, K., A. Arechederra, M. Florencio, D. León & L. Serrano, 2009. Copepods and Branchiopods of temporary ponds in the Doñana Natural Area (SW Spain): a four-decade record (1964–2007). Hydrobiologia (this issue).

Florencio, M., L. Serrano, C. Gómez-Rodríguez, A. Millán & C. Díaz-Paniagua, 2009. Inter and intra-annual variations of macroinvertebrate assemblages are related to the hydroperiod in Mediterranean temporary ponds. Hydrobiologia (this issue).

Indermuehle, N., S. Angélibert, V. Rosset & B. Oertli, 2009. The pond biodiversity index "IBEM": a new tool for the rapid assessment of biodiversity in ponds from Switzerland. Part 2. Method description and examples of application. Limnetica (in press).

Oertli, B., D. Auderset Joye, E. Castella, R. Juge, D. Cambin & J.-B. Lachavanne, 2002. Does size matter? The relationship between pond area and biodiversity. Biological Conservation 104: 59–70.

Oertli, B., D. Auderset Joye, E. Castella, R. Juge, A. Lehmann & J.-B. Lachavanne, 2005a. PLOCH: a standardized method for sampling and assessing the biodiversity in ponds. Aquatic Conservation: Marine and Freshwater Ecosystems 15: 665–679.

Oertli, B., J. Biggs, R. Céréghino, P. Grillas, P. Joly & J.-B. Lachavanne, 2005b. Conservation and monitoring of pond biodiversity: introduction. Aquatic Conservation: Marine and Freshwater Ecosystems 15: 535–540.

Peretyatko, A., S. Teissier, S. De Backer & L. Triest, 2009. Restoration potential of biomanipulation for eutrophic peri-urban ponds: the role of zooplankton size and submerged macrophyte cover. Hydrobiologia (this issue).

Pinto-Cruz, C., J. A. Molina, M. Barbour, V. Silva & M. D. Espírito-Santo, 2009. Plant communities as a tool in temporary ponds conservation in SW Portugal. Hydrobiologia (this issue).

Rannap, R., A. Lõhmus & L. Briggs, 2009. Restoring ponds for amphibians: a success story. Hydrobiologia (this issue).

Rhazi, L., P. Grillas, M. Rhazi & J.-C. Aznar, 2009a. 10-years dynamics of vegetation in relation with fluctuating hydrology in a Mediterranean temporary pool (Western Morocco). Hydrobiologia (this issue).

Rhazi, M., P. Grillas, L. Rhazi, A. Charpentier & F. Médail, 2009b. Competition in microcosm between a clonal plant species (*Bolboschoenus maritimus*) and a rare quillwort (*Isoetes setacea*) from Mediterranean temporary pools of southern France. Hydrobiologia (this issue).

Ricciardi, A., 2006. Are modern biological invasions an unprecendent form of global change? Conservation Biology 21: 329–336.

Robinson, C. T. & B. Oertli, 2009. Long-term biomonitoring of alpine waters in the Swiss National Park. eco.mont 1: 23–34.

Rodríguez-Pérez, H., M. Florencio, C. Gómez-Rodríguez, A. J. Green, C. Díaz-Paniagua, L. Serrano, 2009. Monitoring the invasion of the aquatic bug *Trichocorixa verticalis verticalis* (Hemiptera: Corixidae) in the wetlands of Doñana National Park (SW Spain). Hydrobiologia (this issue).

Rosset, V., B. Oertli, S. Angélibert & N. Indermuehle, 2008. The local diversity of macroinvertebrates in alpine ponds as an indicator of global changes: a Gastropod case-study. Verhandlungen Internationale Vereinigung Limnologie 30: 482–484.

Ruhí, A., D. Boix, J. Sala, S. Gascón & X. D. Quintana, 2009. Spatial and temporal patterns of pioneer macrofauna in recently created ponds: taxonomic and functional approaches. Hydrobiologia (this issue).

Sager, L. & J.-B. Lachavanne, 2009. The M-TIP: a macrophyte based trophic index for ponds. Hydrobiologia (this issue).

Sahib, N., L. Rhazi, M. Rhazi & P. Grillas, 2009. Experimental study of the effect of hydrology and mechanical soil disturbance on plant communities in Mediterranean temporary pools in Western Morocco. Hydrobiologia (this issue).

Santoul, F., A. Gaujard, S. Angélibert, S. Mastrorillo & R. Céréghino, 2009. Gravel pits support waterbird diversity in an urban landscape. Hydrobiologia (this issue).

Scher, O., P. Chavaren, M. Despreaux & A. Thiéry, 2004. Highway stormwater detention ponds as biodiversity islands ? Archives des Sciences 57: 121–130.

Solimini, A., M. Bazzanti, A. Ruggiero & G. Carchini, 2008. Developing a multimetric index of ecological integrity based on macroinvertebrates of mountain ponds in central Italy. Hydrobiologia 597: 109–123.

Triest, L. & T. Sierens, 2009. High diversity of *Ruppia* meadows in saline ponds and lakes of the western Mediterranean. Hydrobiologia (this issue).

Trigal, C., F. García-Criado & C. Fernández-Aláez, 2009. Towards a multimetric index for ecological assessment of Mediterranean flatland ponds: the use of macroinvertebrates as bioindicators. Hydrobiologia 618: 109–123.

Williams, P., M. Whitfield & J. Biggs, 2008. How can we make new ponds biodiverse?—a case study monitored over 8 years. Hydrobiologia 597: 137–148.

Williams, P., M. Whitfield, J. Biggs, S. Bray, G. Fox, P. Nicolet & D. Sear, 2003. Comparative biodiversity of rivers, streams, ditches and ponds in an agricultural landscape in Southern England. Biological Conservation 115: 329–341.

Hydrobiologia (2009) 634:11–24
DOI 10.1007/s10750-009-9885-7

POND CONSERVATION

Plant communities as a tool in temporary ponds conservation in SW Portugal

C. Pinto-Cruz · J. A. Molina · M. Barbour · V. Silva · M. D. Espírito-Santo

Published online: 5 August 2009

Abstract Temporary ponds are seasonal wetlands annually subjected to extreme and unstable ecological conditions, neither truly aquatic nor truly terrestrial. This habitat and its flora have been poorly studied and documented because of the ephemeral character of the flora, the changeable annual weather that has a great effect on the small, herbaceous taxa and the declining abundance of temporary ponds. The objectives of this study are: (a) to define plant community diversity in terms of floristic composition of ephemeral wetlands in SW Portugal, (b) to identify temporary pond types according to their vegetation composition and (c) to identify those ponds that configure the European community priority habitat (3170* – Mediterranean temporary ponds).

Vegetation sampling was conducted in 29 ponds, identifying 168 species grouped among 15 plant communities. Soil texture, pH, organic C and N content were measured, but only N and percent of clay appear to be related with the distribution of each community type. The results showed that ephemeral wetlands could be classified into four type: vernal pools, marshlands, deep ponds and disturbed wetlands. Vernal pools correspond to the Mediterranean temporary ponds (3170*), protected as priority habitat under the EU Habitats Directive. Submersed *Isoetes* species (*Isoetes setaceum* and *Isoetes velatum*) represents, together with *Eryngium corniculatum*, the indicator species for vernal pools. We identify also indicator plant communities of this priority habitat, namely *I. setaceum* and *E. corniculatum–Baldellia ranunculoides* plant communities. In this region, the conservation of temporary ponds has so far been compatible with traditional agricultural activities, but today these ponds are endangered by the intensification of agriculture and the loss of traditional land use practices and by the development of tourism.

Keywords SW Portugal · Temporary ponds · Ephemeral vegetation · Pond typology

Guest editors: B. Oertli, R. Cereghino, A. Hull & R. Miracle
Pond Conservation: From Science to Practice. 3rd Conference of the European Pond Conservation Network, Valencia, Spain, 14–16 May 2008

C. Pinto-Cruz (✉)
Departamento de Biologia, ICAAM-Instituto de Ciências Agrárias e Ambientais Mediterrânicas, Universidade de Évora, Apartado 94, 7002-554 Évora, Portugal
e-mail: ccruz@uevora.pt

J. A. Molina
Departamento de Biología Vegetal II, Universidad Complutense de Madrid, E-28040 Madrid, Spain

M. Barbour
Department of Plant Sciences, University of California, Davis, CA 95616, USA

V. Silva · M. D. Espírito-Santo
Departamento de Protecção de Plantas e de Fitoecologia, CBAA-Centro de Botânica Aplicada à Agricultura, Instituto Superior de Agronomia, Tapada da Ajuda, 1349-017 Lisbon, Portugal

Introduction

The Mediterranean Basin has been recognised as a global biodiversity "hotspot" (Blondel & Arronson, 1999). Temporary ponds are classified among the most biologically and biogeographically interesting ecosystems in the Mediterranean region (Grillas et al., 2004). Temporary ponds (vernal pools) are unusual habitats, neither truly aquatic nor truly terrestrial. They are seasonal wetlands with annually alternating phases of flooding and drying in shallow depressions. The water-holding capacity is, in some cases, related with the underlining impervious substrate (Keeley & Zedler, 1998). Braun-Blanquet (1936) was among the first to point out the high biological value of temporary ponds, but new interest has grown among several ecologist in the Mediterranean region.

The ephemeral vegetation of temporary ponds is dominated mainly by annual and herbaceous perennials that appear during winter and spring months. The vegetation is diverse and is rich in annual hygrophytes, hemicryptophytes and geophytes (Brullo & Minissale, 1998; Barbour et al., 2003; Deil, 2005). Species composition and the dynamics of plant communities in Mediterranean temporary ponds are affected by inter- and intra-annual climatic conditions (Rhazi et al., 2001; Espírito-Santo & Arsénio, 2005). Studies on Portuguese vernal pools mainly remain at a descriptive level (Jansen & Menezes de Sequeira, 1999; Pinto-Gomes et al., 1999; Rosselló-Graell et al., 2000; Espírito-Santo & Arsénio, 2005; Rudner, 2005a, b).

Temporary ponds are extremely vulnerable habitats due to their small size, shallow depth of water, proximity to expanding urban areas and to intensive agriculture, industrialisation, development of tourism and their scattered and isolated distribution at a regional level. Temporary ponds have, for thousands of years, been compatible with and even favoured by traditional farming regimes, but modern agriculture obliterates them (Rhazi et al., 2001; Beja & Alcazar, 2003). They are recognized by the Ramsar Convention on Wetlands and some are priority habitats under the Habitats Directive of the European Community (Natura 2000 code: 3170* – Mediterranean temporary ponds).

Mediterranean temporary ponds are significant and highly sensitive ecosystems that should be studied more due to their vulnerability to human activities and climate changes (Stamati et al., 2008). We hypothesize that temporary pond types can be defined according to their plant community composition and that some communities or mixture of communities in a pond is related to their conservation status. The objectives of this study are: (a) to define plant community diversity in terms of floristic composition of ephemeral wetlands in SW Portugal, (b) to identify temporary pond types according to their vegetation composition and (c) to identify those ponds that configure the European community priority habitat (3170* – Mediterranean temporary ponds). It is important to define seasonal wetland typology to establish conservation priorities. The assessment of the conservation status can be used as a basis for management strategies.

Materials and methods

Study area

The study area is the coastal plain of southwest Portugal, which runs north–south for about 100 km long x 5–15 km wide, ranging 50–150 m above sea level (Fig. 1). This area hosts a large number of seasonal wetlands as a consequence of climatic, edaphic and topographic characteristics. The approximate pond density in the studied area is 0.28 per Km^2. Waters are soft to slightly hard, circum-neutral to slightly acidic and sometimes high in phosphates and nitrates (Beja & Alcazar, 2003).

This coastal platform is carved in Palaeozoic schist and covered by sandstone types (sands, sandstone and conglomerates as described by Neto et al., 2007). The lithology of the territory includes siliceous materials and base-poor soils. According to local weather data (INMG, 1991), the climate is Mediterranean with an oceanic influence. From north to south, the mean annual precipitation declines from 614 to 456 mm, falling mainly from October to March. Thus, aridity increases southwards. Winter and summer average temperatures are 11 and 20.5°C, respectively.

This area is administered as the Natural Park, of southwest Alentejo and Vicentina Coast, a large stretch of the Portuguese coastline subject to special protection. Nevertheless, a high percentage of the ephemeral wetlands are privately owned. The

Fig. 1 Location of the 29 sampled temporary ponds. Map layout created with Quantum GIS

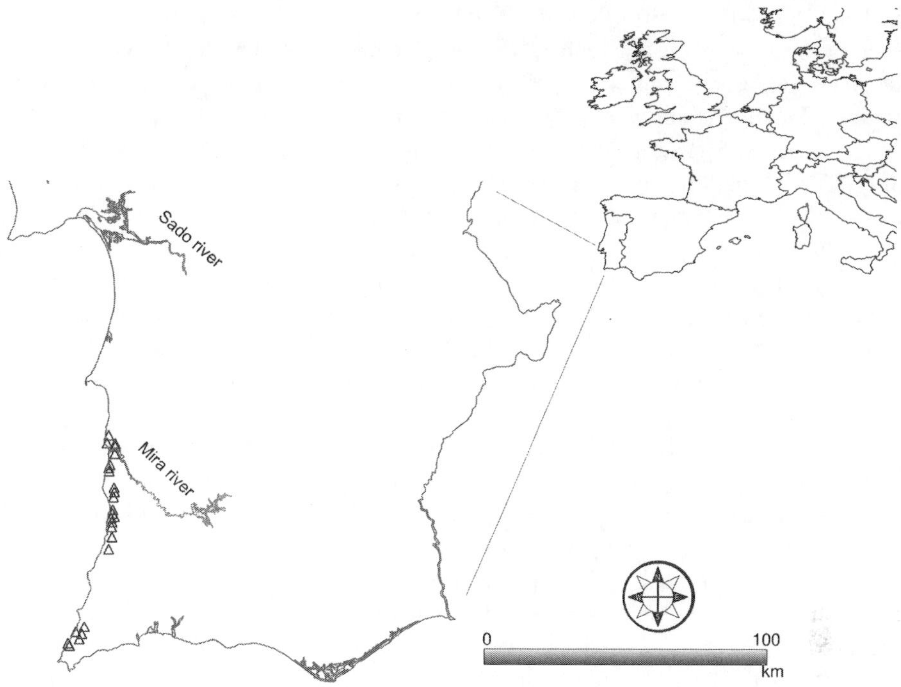

traditional land uses in the regions were extensive agriculture and livestock pastures in rotation. Presently, some 12,000 ha are administrated by a hydrological plan to further develop agricultural activities. In 2006 and 2007, fodder and maize occupied the largest areas, around 18–20% each (ABM, 2007, 2008).

Data collection

The vegetation was sampled in a stratified random manner to obtain broadly representative data (Kent & Coker, 1992). The stratification took into account lithology (sands and conglomerates) as well as the wetland's morphology (area 0.1–5 ha, depth 0.4–2 m). Field sampling was carried out in 29 seasonal wetlands. The study period extended from late winter (February) to early summer (June) in 2006 and 2007. Ponds were visited twice a year. Plant communities were surveyed in visually homogenous 4 m^2 quadrats, and each taxon's percent cover was recorded adapting Braun-Blanquet's (1964) method to allow conventional multivariate procedures (Podani, 2006). In each pond all physiognomically homogeneous patches of vegetation were sampled. Usually, three types, at different pond depths (margin, intermediate and deep parts), were present and each was sampled twice. Plant community types were named according to the syntaxonomical checklist of Rivas-Martínez et al. (2001, 2002a, b). Plant nomenclature follows Flora Iberica (Castroviejo et al., 1986–2008) and Nova Flora de Portugal (Franco, 1984; Franco & Rocha Afonso, 1994–2003).

In each pond, soil surface samples were collected with a hand probe and mixed for chemical and physical analyses. Soil samples were air-dried and sieved at 2 mm. Three fractions (sand, silt and clay) of soil texture were determined for each sample using the sedimentation method (Sedigraph 5100, Micrometrics Instrument Corporation). Standard soil analyses were carried out to determine the soil pH in a 1:2.5 soil–water suspension (glass electrode CRISON, Microph 2002), conductivity in a 1:5 suspension microprocessor conductivity meter LF 330 WTW and a standard conductivity cell Tetracon 325, and organic carbon by dry combustion analysed by an SC-144DR (LECO Instruments). Determination of nitrogen content was made after the ISO 14891:2002 standard (ISO/IDF, 2002).

Data analysis

The data set includes 302 *releve's* and 168 species. The raw matrix was analysed with the software package SYN-TAX 5.0 (Podani, 2001) and plant community types were recognized with Hierarchical Cluster Analysis using the weighted-pair group method (WPGMA) and Chord Distance as a dissimilarity measure. Similarity Analysis (ANOSIM) with Bray–Curtis similarity measure (PRIMER, version 6) was used to test for significant differences in plant community composition. Indicator species analysis (Dufrêne & Legendre, 1997) evaluated taxonomic consistency of plant communities by calculating species fidelity to each cluster, identifying species responsible for differences between plant groups (software PC-ORD 4).

Based on the relative abundance of plant communities in each pond, an incremental sum of squares, with standardized Euclidean distance, cluster analysis, was performed to group the ponds (SYN-TAX 5.0). Also, based on the relative abundance of plant communities, an initial detrended correspondence analysis (DCA) was carried out to determine the gradient lengths before deciding on the most appropriate method for analysis. Principal components analysis (PCA) was carried out to define pond groups using the program CANOCO 4.5. (ter Braak & Šmilauer, 2002).

Multiple discriminant analysis (MDA) (Legendre & Legendre, 1998) was used to determine the best set of variables that discriminate groups among the PCA clusters, based on several soil environmental variables: soil texture, pH, conductivity, organic carbon and N content. In order to eliminate those variables that provided insignificant information ($P = 0.01$), a forward stepwise procedure was applied to each variable. For this last analysis SPSS-15.0 program package was used. Prior to analysis, all data were either log 10 $(x + 1)$ (linear measurements) or arcsin [sqrt (x)] (percentages) transformed to improve normality.

Results

Plant community types

Ephemeral wetlands, taken as a group, are rich in species (168 taxa in our study) and in community types. Some 15 community types, putative associations or alliances, were identified by cluster analysis, using floristic information (Table 1). Significant differences ($P < 0.01$) from similarity analysis were observed for all communities. Indicator species analysis identified characteristics species for each community. The mean number of taxa per plant community (see Table 1) ranged from 5–6 in aquatic communities in wetland bottoms to 17–21 in ephemeral grasslands in wetland margins.

Aquatic vegetation is represented by crowfoot (L—*Ranunculus peltatus*) and water-milfoil (N—*Myriophyllum alterniflorum*) communities. Helophytic vegetation is constituted by low grasslands (*E—Eleocharis palustris–Glyceria declinata*), bulrushes (O—*Bolboschoenus maritimus*) and endemic rushes (*F—Juncus emmanuelis*) communities. Amphibious vernal pool vegetation includes Atlantic decumbent-floating vegetation (*D—Eleogiton fluitans–Juncus heterophyllus* communities), quillwort swards (*C—Juncus capitatus–Isoetes histrix* communities, *J—Isoetes setaceum* communities) and Mediterranean thistles (*I—Eryngium corniculatum–Baldellia ranunculoides* communities). Marshland communities are dominated by *H—Agrostis castellana*, together with *K—Eleocharis multicaulis* communities in the margins. Edge vegetation includes communities dominated by rushes (*B—Juncus rugosus*), perennial grasslands (*G—Phalaris coerulescens* and *H—A. castellana*) and annual grasslands (*M—Chaetopogon fasciculatus* communities) (Table 2).

Habitats and Communities

A previous numerical clustering allowed identifying three habitat groups and a subgroup namely: vernal pools (VP) within which deep ponds (L) can be identified as a subgroup, marshlands (M) and disturbed wetlands (D). The PCA ordination diagram (Fig. 2) also distinguished these units. Axis 1 separated the wetlands containing ephemeral temporary ponds' plant communities (e.g. *I. setaceum* communities, *E. corniculatum–B. ranunculoides* communities and *C. fasciculatus* communities) from the others. Deep ponds occurring in the far right side of the ordination diagram are defined by aquatic and bulrush communities such as *R. peltatus* communities, *M. alterniflorum* communities, and *B. maritimus* communities. In the left half part of the ordination diagram (axis 2), separated wetlands were characterised by typical marshland plant communities

Table 1 Synoptic table for 15 community types

Group	A	B	C	D	E	F	G	H	I	J	K	L	M	N	O	
No. of *relevés*	54	30	54	28	39	20	6	5	11	17	11	6	6	9	6	
Mean species n	12	15	17	9	7	10	15	10	11	10	12	6	21	5	9	**IV**
Ranunculus peltatus	–	0.1	–	0.2	2	0.3	–	–	0.2	–	–	**69*** **	–	3	–	83.4***
Callitriche stagnalis	–	0.1	0.1	2	2	0.1	0.2	–	–	–	–	**4***	–	0.2	–	20.4*
Callitriche brutia	–	–		1	1	–	–	–	–	–	–	–	–	–	–	
Isoetes histrix	0.3	1	**6*** **	–	–	–	–	–	–	–	–	–	1	–	–	43.4***
Lythrum borysthenicum	0.2	–	0.1	1	0.1	2	–	–	1	0.3	0.3	–	0.2	–	–	
Solenopsis laurentia	0.1	0.1	0.1	–	–	–	–	2	–	2	–	–	–	–	–	
Eryngium corniculatum	1	–	0.1	1	11	6	9	–	**35*** *	14	1	13	1	19	12	23**
Isoetes setaceum	–	–	0.1	0.4	2	3	–	–	3	**33*** **	3	–	1	1	1	54.7***
Isoetes velatum	0.2	1	0.1	–	2	0.3	–	–	3	**12*** **	1	–	–	–	–	34.1***
Illecebrum verticillatum	0.1	0.2	1	0.4	1	2	–	0.2	6	2	0.3	**6***	1	0.1	1	15.8*
Isolepis pseudosetacea	0.1	0.3	**3*** *	–	–	–	–	–	–	–	–	–	1	–	–	22.2**
Kickxia cirrhosa	0.2	1	–	–	–	–	–	–	–	–	–	–	1	–	–	
Cicendia filiformis	0.1	0.2	1	–	–	0.2	–	–	–	0.1	–	–	–	–	–	
Exaculum pusillum	–	0.2	0.1	0.1	–	–	–	–	0.4	**1***	0.1	–	0.1	–	–	17.5*
Chaetopogon fasciculatus	0.1	–	1	–	–	0.1	–	–	0.1	4	2	–	**35*** **	–	–	72.6***
Juncus bufonius	1	2	**14*** **	1	–	0.1	4	0.1	0.1	0.1	0.1	–	4	–	–	34.4***
Isolepis cernua	2	5	3	2	0.1	2	5	–	0.2	1	0.3	–	1	–	–	
Pulicaria paludosa	0.4	–	–	–	0.4	3	0.2	1	3	**9*** **	1	–	1	–	–	31.9***
Mentha pulegium	1	6	0.4	1	0.1	2	**3*** *	–	0.4	–	–	–	–	–	–	23.2**
Juncus pygmaeus	1	0.4	1	0.1	–	2	–	–	0.5	1	2	–	1	–	–	
Juncus capitatus	0.3	1	**7**	–	–	–	–	–	–	–	–	–	1	–	–	45***
Juncus tenageia	0.5	1	0	0.2	–	2	–	1	0.1	0.1	2	–	1	–	–	
Agrostis pourretii	0.2	1	1	0.1	0.1	0.3	3	0.4	0.3	–	–	–	0.2	–	–	
Lythrum hyssopifolia	0.4	1	1	0.2	0.2	0.2	1	–	1	0.1	1	–	0.2	–	1	
Carlina racemosa	0.1	0.1	0.1	—	–	–	3	–	–	–	–	–	0.2	–	–	
Hypericum humifusum	0.3	0.3	1	–	–	–	–	–	–	–	–	–	3	–	–	
Antinoria agrostidea	–	–	–	–	–	–	–	–	1	–	2	–	0.2	–	–	
Baldellia ranunculoides	3	1	0.3	5	6	**17*** *	4	1	4	5	3	–	0.3	5	4	21**
Myriophyllum alterniflorum	–	–	–	2	1	–	–	–	–	–	–	–	–	**61*** **	–	92.1***

Table 1 continued

Group	A	B	C	D	E	F	G	H	I	J	K	L	M	N	O	
Eleocharis multicaulis	3	2	0.2	0.1	–	0.1	–	–	1	–	**46********	3	1	–	1	66.9***
Juncus heterophyllus	1	1	0.3	**28********	3	3	–	–	2	–	–	–	–	0.3	–	40.9***
Hydrocotyle vulgaris	2	3	0.2	2	–	–	–	–	–	–	2	–	–	–	–	
Eleogiton fluitans	1	0.2	–	**7****	0.4	–	–	–	–	–	–	–	–	–	–	28.8**
Juncus bulbosus	3	2	0.5	1	–	–	–	–	–	–	2	–	–	–	–	
Littorella uniflora	–	–	–	–	–	–	–	–	2	**5***	–	–	–	–	–	
Hypericum elodes	1	1	0.3	1	–	–	–	–	–	–	2	–	–	–	–	
Myosotis lusitanica	0.1	–	0.1	–	–	–	–	–	–	–	–	–	–	–	–	
Eleocharis palustris	1	0.1	0.3	2	**43********	8	10	–	8	0.4	–	2	–	5	6	33.5***
Glyceria declinata	0.1	1	0.1	6	**22********	4	1	–	0.5	–	0.1	8	–	3	–	30.1***
Bolboschoenus maritimus	–	–	–	0.1	0.2	0.3	–	–	0.1	–	–	–	–	0.3	**38********	91.2***
Ranunculus ophioglossifolius	0.4	1	0.1	1	0.2	0.4	**5****	–	0.3	–	–	–	–	–	–	25.4**
Alisma lanceolatum	–	–	–	0.1	–	2	3	–	–	–	–	–	–	–	–	
Agrostis stolonifera	**49********	11	3	7	2	1	0.3	–	8	5	11	1	0.3	1	2	35.8***
Phalaris coerulescens	1	0.1	1	–	0.3	1	**35********	5	1	1	0.2	–	0.3	–	0.3	56.2***
Carum verticillatum	1	0.1	0.1	–	1	4	–	4	2	1	4	13	2	–	**11***	18.4*
Holcus lanatus	5	**5****	2	0.2	–	0.1	–	8	–	–	1	2	–	–	–	21.1**
Lythrum junceum	1	1	1	1	0.2	0.4	**5****	–	0.2	–	0.4	–	–	–	–	23.7**
Juncus acutiflorus subsp. *rugosus*	0.1	**11****	0.1	–	–	–	–	–	–	–	–	–	–	–	–	29**
Chamaemelum nobile	0.1	–	–	–	–	0.2	0.2	1	0.3	2	1	–	4	–	–	
Silene laeta	1	1	2	–	–	–	0.1	0.4	–	–	–	–	3	–	–	
Trifolium dubium	0.3	–	1	–	–	–	0.3	0.3	–	–	–	–	**3****	–	–	33.1**
Juncus effusus	2	0.2	–	0.1	–	–	–	–	–	–	0.1	–	–	–	–	
Lotus uliginosus	1	1	0.4	0.3	0.1	–	–	–	–	–	–	–	–	–	–	
Juncus acutiflorus subsp. *acutiflorus*	0.1	0.2	0.2	–	–	0.3	–	**7****	–	1	0.2	–	–	–	–	27.6**
Juncus emmanuelis	3	1	1	1	1	**26****	8	1	0.3	0.2	8	–	3	–	2	29.6**
Cynodon dactylon	2	8	2	2	0.2	2	–	3	4	2	2	–	2	1	5	
Leontodon taraxacoides subsp. *taraxacoides*	1	4	5	0.4	0.1	1	–	–	3	1	2	–	**10********	–	2	30.4***

Table 1 continued

Group	A	B	C	D	E	F	G	H	I	J	K	L	M	N	O	
Paspalum paspalodes	2	0.3	1	11	1	1	–	–	1	0.2	1	–	–	–	1	
Juncus maritimus	–	–	–	–	–	–	–	4	–	–	–	–	–	–	**19****	25.2**
Plantago coronopus	1	0	6	0.1	–	–	5	1	–	1	–	–	3	–	–	
Lotus hispidus	1	2	2	1	0.1	–	0.3	4	0.3	0.1	0	–	**10****	—	–	22.6**
Myosotis debilis	0.4	1	0	0.3	1	2	1	–	2	1	1	–	1	–	1	
Dittrichia viscosa subsp. *revoluta*	1	6	1	0.4	0.3	0.1	0.2	1	0.3	0.1	1	–	**3***	–	–	14.8*
Cotula coronopifolia	0.2	0.2	0.2	1	1	1	**2****	—	3	–	–	–	0.2	–	–	21.4**
Chamaemelum mixtum	0.2	2	3	0.1	–	0.1	0.3	–	0.1	1	–	–	**4****	–	–	24.8**
Anagallis tenella	1	**5****	0.3	0.4	–	–	–	–	–	–	2	–	–	–	–	25.5**
Polypogon maritimus	0.1	0.1	0.1	1	0.3	1	0.3	0.4	2	2	–	–	–	–	2	
Briza minor	1	0.2	2	–	–	0.3	1	0.2	–	–	–	–	**3*****	–	1	38.6***
Briza maxima	1	1	2	–	–	–	–	0.2	–	–	0.1	–	1	–	3	
Anagallis arvensis	1	2	**3*****	0.1	–	0.4	–	–	–	0.1	0.2	–	1	–	–	28.6***
Gaudinia fragilis	0.3	0.2	0.4	–	–	–	–	–	–	–	–	–	**6*****	–	0.2	65.6***
Ornithopus pinnatus	0	2	**4*****	–	–	–	–	–	–	–	–	–	1	–	–	43.4***
Vulpia muralis	0.3	0.2	2	–	–	–	0.3	2	–	–	–	–	**3****	–	–	27.4**
Ranunculus trilobus	0.3	0.3	1	0.1	0.1	0.2	1	–	1	0.1	–	–	–	–	–	
Panicum repens	1	1	1	2	0.3	–	–	–	–	–	0.1	–	–	–	–	
Lolium multiflorum	0.1	0.1	1	0.1	0.1	0.3	**2*****	–	–	0.1	–	–	–	–	–	29.8***
Hyacinthoides vicentina subsp. *transtagana*	0.2	–	0.3	–	–	1	–	0.4	1	2	0.1	–	–	–	–	
Potentilla erecta	0.1	3	0.1	–	–	–	–	–	–	–	1	–	–	–	–	
Lolium rigidum	0.1	–	0.1	–	0.1	–	**2***	–	0.3	0.2	–	–	0.2	–	–	20.2*
Serapias lingua	1	–	1	–	–	–	–	–	–	–	–	–	**2****	–	–	29.1**
Nitella sp.	–	–	–	1	1	–	–	–	–	–	–	–	–	1	–	
Agrostis castellana	–	0.2	0.1	–	–	0.1	–	**37*****	–	–	1	–	1	–	–	88.8***
Lobelia urens	1	1	0.1	–	–	–	–	–	–	–	0.1	–	–	–	–	
Pinus pinaster	0.1	2	0.1	–	–	–	–	–	–	–	0.2	–	–	–	–	
Juncus hybridus	0.1	1	1	–	–	–	–	–	–	–	–	–	–	–	–	
Centaurium maritimum	–	0.1	0.1	–	–	–	0.2	–	–	–	–	–	**1***	–	–	20.8*
Trifolium campestre	–	–	0.1	–	–	–	0.3	**1***	–	–	–	–	1	–	–	22*

Table 1 continued

Group	A	B	C	D	E	F	G	H	I	J	K	L	M	N	O	
Elatine macropoda	–	–	–	–	–	0.4	–	–	–	0.2	0.1	–	–	–	–	
Ranunculus saniculifolius	–	0.1	–	0.2	1	–	–	–	–	–	–	–	–	–	–	
Pinguicula lusitanica	–	–	1	–	–	–	–	–	–	–	–	–	–	–	–	
Chara sp.	–	–	–	–	–	1	–	–	–	–	–	**1****	–	–	–	26.6**

Letters A to O identify community types as named in Table 2. Indicator values (IV) of those taxa identified by indicator species analysis as being significant indicator species are shown in the last column. *Bold-faced* values in other columns indicate the group (community type) having maximum value for that species. Less abundant frequent species were excluded from table layout

* $P < 0.05$, ** $P < 0.01$, *** $P < 0.001$

such as *E. multicaulis* communities and *J. rugosus* communities (lower left quadrant) from other wetlands poorly characterised from a vegetational viewpoint (e.g. *E. palustris–G. declinata* communities and *E. fluitans–J. heterophyllus* communities). In fact, disturbed wetlands present the lower values in plant-community richness in contrast to vernal pools with the higher community richness (Table 3). In terms of soil features, temporary ponds vary from sandy clay loam to loamy sand soils. The pH values range from moderately (6.3) to strongly (4.4) acidic and N levels ranged between 0.11 and 0.78%. Mean soil values per habitat showed that marshlands were high in fine particles and that disturbed wetlands were high in C and N (Table 3).

Relation between habitat type and environmental variables

The MDA showed that % clay (Wilks' $\lambda = 0.759$) and N content (Wilks' $\lambda = 0.740$) were the only entered variables (Table 4). A total Wilks' λ value of 0.457 ($P < 0.001$) shows the reasonable discriminant power of the model. The discriminant function 1 (DF1) ($\lambda_1 = 0.9$) was mainly correlated (0.688) with soil clay, and it explained 85.5% of total variance, whereas the second discriminant function (DF2) ($\lambda_2 = 0.15$) was mainly correlated (0.364) with soil N content and explained 14.5% of total variance.

A plot of two canonical discriminant functions (Fig. 3) showed relative good, although unequal, separation of the centroids for the four habitat types. The high discriminant weight of soil clay % (DF1) was responsible for the segregation between marshlands (negative values) and vernal pool (VP) habitats (positive values). Whereas, DF2 revealed a weak separation of vernal pool and deep pond habitats, DF2 also differentiated disturbed wetlands by soil N content (positive values).

Discriminant functions appear to have a good classification with 70.4% of the original cases correctly classified. However, the robustness of this classification resulted from the fitness of only two groups (VP and M). Classification data for each habitat are presented in Table 5. Analysing the cross-validated data, deep ponds' vegetation was the worst-classified group, with 100% of misclassified cases.

Table 2 Physiognomy and preferential habitat traits of the 15 community (Comm.) types

Community type	Physiognomy	Habitat
A—*Agrostis stolonifera* Comm.	Stoloniferous perennial grasslands	Intermediate part in marshlands
B—*Juncus rugosus* Comm.	Rhizomatous rushes	Margins of vernal pools and marshlands
C—*Juncus capitatus–Isoetes hystrix* Comm.	Ephemeral quillwort swards	Margins of vernal pools
D—*Eleogiton fluitans–Juncus heterophyllus* Comm.	Decumbent floating vegetation	Deep part of marshlands and vernal pools
E—*Eleocharis palustris–Glyceria declinata* Comm.	Low helophytic grasslands	Deep part of vernal pools and marshlands
F—*Juncus emmanuelis* Comm.	Rhizomatous rushes	Intermediate part of vernal pools
G—*Phalaris coerulescens* Comm.	Perennial grassland	Disturbed wetlands
H—*Agrostis castellana* Comm.	Perennial grassland	Margins of vernal pools
I—*Eryngium corniculatum–Baldellia ranunculoides* Comm.	Swampland thistle forb	Deep part of vernal pools
J—*Isoetes setaceum* Comm.	Ephemeral quillwort swards	Deep part of vernal pools
K—*Eleocharis multicaulis* Comm.	Turf grasslands	Margins of marshlands
L—*Ranunculus peltatus* Comm.	Ephemeral aquatic crowfeets	Deep part of the vernal pools and deep ponds
M—*Chaetopogon fasciculatus* Comm.	Annual grasslands	Margins of vernal pools
N—*Myriophyllum alterniflorum* Comm.	Rooted submerged pondweeds	Deep part of the deep ponds
O—*Bolboschoenus maritimus* Comm.	Bulrushes	Intermediate part of the deep ponds

Discussion

Seasonal Mediterranean wetlands in SW Europe encompass a wide vegetation and community type richness that include both annual and perennial vegetations mainly dominated by grasses and herbs (Deil, 2005). Our study revealed 15 community types floristically distinct for a territory of about 300 km^2 in the coastal plain in SW Portugal with a typical Mediterranean climate. These results are similar to those carried out in other Atlantic–Mediterranean areas (Pinto-Gomes et al., 1999; Deil, 2005; Espírito-Santo & Arsénio, 2005; Bagella et al., 2007). Most of the vegetation corresponded to perennial plant communities ($n = 9$), some annual ($n = 4$), and few are perennating ($n = 2$) (Table 2). The dominance of perennial communities in studied temporary ponds seems to be in contradiction with the established concept that Mediterranean temporary ponds are dominated by annuals (Rhazi et al., 2006; Arguimbau, 2008; Della Bella et al., 2008). In our case study, we found annual plant communities in both ephemeral aquatic habitats mainly at the upper margin. The same pattern of annual vegetation distribution has been documented in other Mediterranean–Atlantic areas (Molina, 2005, Rhazi et al., 2006).

Numerical analysis revealed three main habitats: ponds, marshlands and disturbed wetlands. According to Oertli et al. (2005) in the definition of ponds, the area size of some studied ponds (>2 ha) is oversized, stretching out their segregation from vernal pools (VP) into a 4th group, named deep ponds (L). It is noteworthy that disturbed wetlands (D) result both from the degradation of vernal pools and marshlands (M), by human influence intensification. We infer from these considerations that Mediterranean seasonal wetlands include a variety of habitats not always well classified (Deil, 2005). However, our study shows that plant communities can help to correctly classify habitat types for the case of Mediterranean seasonal wetland ecosystems. Vernal pools showed the greatest diversity in number of community types, and they are definined by hydroseres containing annual grasslands of *C. fasciculatus*, ephemeral quillwort swards of *Isoetes setaceum*, and Mediterranean palustrian thistle forbs of *E. corniculatum*. Deep ponds share with the latter habitats VP vegetation which occurs in their intermediate and marginal belts, but deep ponds also support communities with a longer flooding period such as *Myriophyllum alterniflorum* and *B. maritimus*. This fact supports the differentiation of VP and L habitats and vegetation. Marshland habitats have been

Fig. 2 Principal component analysis plot of habitats in relation to plant communities. *VP* vernal pool, *M* marshlands, *L* deep ponds, *D* disturbed wetlands. The first principal component explains 46.2% of total variance and the first two components combined explain 62.2% of total variance ($n = 29$)

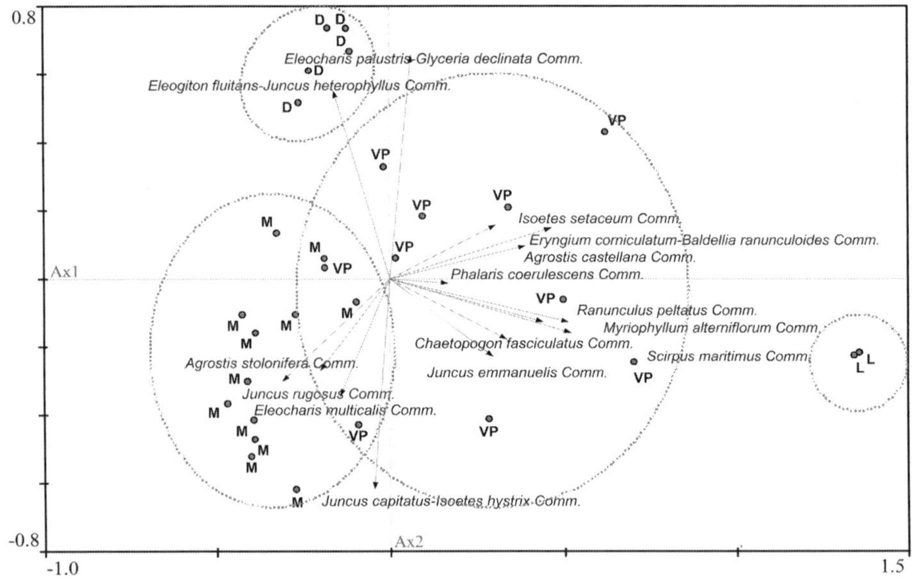

Table 3 Summary of habitats variables

	Species	Communities	Soil particle size (%)		pH	Conductivity	Organic carbon	Nitrogen
	n	*n*	Sand	Clay				
VP	129	13	61.6 (4.62)	20.9 (2.62)	5.3 (0.06)	156.8 (35.43)	1.966 (0.299)	0.295 (0.030)
M	109	8	83.9 (1.99)	8.8 (1.18)	5.4 (0.06)	160.1 (25.61)	1.757 (0.222)	0.358 (0.017)
L	102	11	64.1 (7.65)	20 (4.77)	5.6 (0.08)	199.2 (54.67)	2.269 (0.406)	0.265 (0.029)
D	63	6	66 (6.33)	21.5 (4.04)	5.1 (0.14)	357.9 (77.37)	2.808 (0.4)	0.537 (0.053)

VP vernal pool ($n = 34$), *M* marshlands ($n = 39$), *L* deep ponds ($n = 8$), *D* disturbed wetlands ($n = 11$)

Plant species and community diversity, mean values and ranges of physical and chemical variables. The standard error of the mean (SE) is given in brackets

Table 4 Standardized canonical discriminant function coefficients and correlations between discriminating variables and standardized canonical discriminant functions

	Canonical coefficients		Correlations	
	Axis 1	Axis 2	Axis 1	Axis 2
N content	−1.000	0.600	−0.471	0.882*
Soil clay (%)	1.029	0.549	0.514	0.858*

*$P < 0.05$

found more closely related to perennial grasslands (*Agrostis stolonifera* Comm.) and turf vegetation (*E. multicaulis* Comm.). Disturbed wetlands were related to low helophytic grasslands with the *E. palustris–G. declinata* community and to decumbent floating vegetation with the *E. fluitans–J. heterophyllus* community. Both communities play the role of pioneer in the succession dynamics of studied pools.

Although the mean number of taxa per plant community is not very high (Table 1), the total species diversity in studied ecosystems overall is high because of high species turnover from community to community. This fact is supported by the results of similarity analysis, in which all communities are significantly different. This fact can be related to high β-diversity (Williams et al., 2003; Magurran, 2004). In terms of community richness, the pattern of reduced richness with increasing length of flooding has also been described in different wetland ecosystems (Bauder, 2000; Barbour et al., 2005; Cherry & Gough, 2006; Edvardsen & Økland, 2006; Lumbreras et al., 2009). Only a small number of species are adapted to extensive flooding. A consistent explanation to this was provided by Bliss & Zedler (1998) that studied the effects of different inundation conditions in seed bank germination. According to

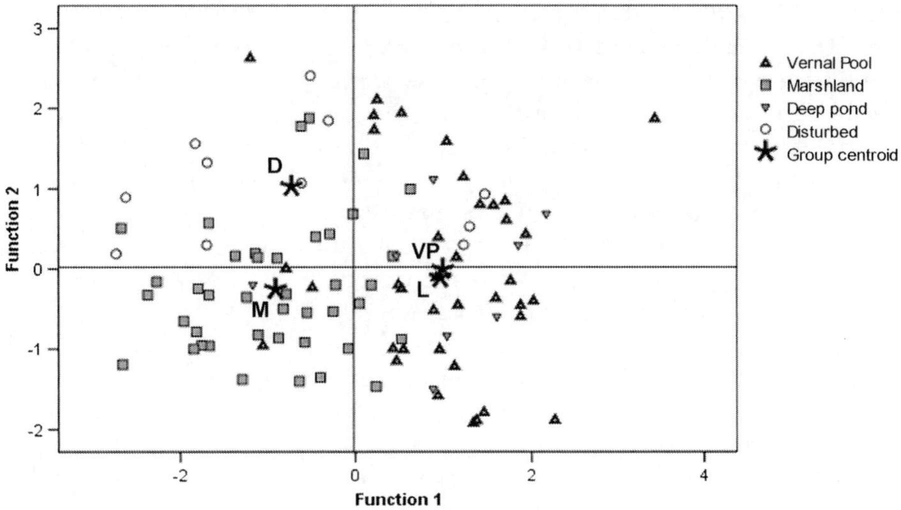

Fig. 3 Plot of the two canonical discriminant functions. *VP* vernal pool, *M* marshlands, *L* deep ponds, *D* disturbed wetlands

Table 5 MDA classification results

Habitat	Predicted group membership			
	VP	M	L	D
Original (%)				
VP	**84.2**	7.9	0	7.9
M	12.2	**82.9**	0	4.9
L	87.5	12.5	**0**	0
D	27.3	45.5	0	**27.3**
Cross validated (%)				
VP	**81.6**	7.9	0	10.5
M	17.1	**78**	0	4.9
L	87.5	12.5	**0**	0
D	27.3	54.5	0	**18.2**

VP vernal pools, *M* marshlands, *L* deep ponds, *D* disturbed wetlands

the same authors the inundation plays a large role in keeping non-pool competitors out of the ponds.

Our results revealed that soil texture and N content are two soil parameters that are significantly correlated to habitat type. Discriminant analysis provides good classification for vernal pools and marshland, but not for deep pond habitat. These divergent results may be due to insufficient data for deep ponds and disturbed wetlands. Vernal pool and marshland were much more frequently sampled as they appear to be the more abundant habitats. The higher clay % in vernal pools soil increases their water-holding capacity (Bonner et al., 1997). Nevertheless, in the summer, vernal pools in comparison with marshlands present a more desiccated aspect.

Untested variables not quantified in our study, such as hydroperiod have been shown by others (Rhazi et al., 2001; Grillas et al., 2004; Pyke, 2004; Serrano & Zunzunegui, 2008; Stamati et al., 2008; Waterkeyn et al., 2008) to be another driving factor. Clearly, more studies on soils profiles and water tables in these ecosystems are required. The intensive land uses like overgrazing or high technology agriculture in the territory can transform vernal pools and marshlands into disturbed ponds edaphically characterised, as shown our results, by a high content in N as a consequence of cumulative effects in the catchments area (Sileika et al., 2005; Szajdak et al., 2006). The regression of vernal pool habitats, due to anthropic disturbance, affects the abundance of rare communities and species in a specific area; in particular, we can refer to *I. setaceum*. The change from natural to disturbed wetlands leads to habitat eutrophication as well as a loss in vegetation richness. Moreover, the plant community becomes composed of more wide spread species, diminishing the individualization between described plant communities.

In terms of conservation, it is important to have a good classification of the natural values. Our study demonstrates that both species and plant communities can be used as tools to define wetland habitat typology, as the results clearly show, vernal pool and marshland habitats are strongly separated by floristic attributes. The ponds designated as vernal pools correspond to the Mediterranean temporary ponds (3170*), protected as a priority habitat under the EU Habitats Directive. In our case study, we identified *I. setaceum* communities and *E. corniculatum–B. ranunculoides* communities as their characteristic vegetation. Furthermore, submersed *Isoetes* species (*I. setaceum* and *I. velatum*) represents, together with *E. corniculatum*, the indicator species for vernal pools. *I. setaceum* is clearly a bioindicator species for Mediterranean temporary ponds. Nevertheless, this species is not included in the Mediterranean temporary ponds definition of the Interpretation Manual of European Union Habitats (EC, 2007). The importance of *Isoetes* species in this habitat was emphasized earlier by Quézel (1998) and Espírito-Santo & Arsénio (2005). Conversely, deep ponds have no priority conservation policy, even though they include almost all the communities of vernal pools plus vegetation tolerant of a longer flooded period. Thus, in terms of conservation policy, it is important to take deep ponds into account for the sake of the highly important plant species and communities within them.

The present trend in Mediterranean temporary ponds is clearly regressive, lacking recognition of their value and function which causes them to be readily destroyed or transformed (EC, 2008). In terms of threats, several authors (Zunzunegui et al., 1998; Brinson & Malvárez, 2002; De Meester et al., 2005; Zacharias et al., 2007) are unanimous in listing changes in drainage and the amount of silting as two of the most important factors that can lead to the loss of these seasonal wetlands. In our study area, the transition from traditional to intensive agriculture, with abandonment of traditional land use practices, represents the major threat. Nevertheless, recently we can observe an increased awareness of the value of temporary ponds, and our results, by identifying habitat indicator plant species and communities, present a helpful tool to establish habitat conservation priorities and strategies.

Acknowledgements The authors are grateful for all the logistical support from the Associação de Beneficiários do Mira. We also like to acknowledge Paula Matono for data analysis suggestions and comments. Special thanks to Rute Caraça for the collaboration in the fieldwork. We would also like to thank the two anonymous referees for the useful and valuable comments.

References

ABM (Associação de Beneficiários do Mira), 2007, 2008. Relatório e contas do exercício de 2006 e 2007. Odemira.

Arguimbau, P. F., 2008. Vascular flora associated to Mediterranean temporary ponds on the island of Minorca. Anales del Jardín Botánico de Madrid 65: 393–414.

Bagella S., M. C. Caria, E. Farris & R. Filigheddu, 2007. Issues related to the classification of mediterranean temporary wet habitats according with the European Union Habitats Directive. Fitosociologia, 44(1).

Barbour, M., A. Solomesch, C. Witham, R. Holland, R. McDonald, S. Cilliers, J. A. Molina, J. Buck & J. Hillman, 2003. Vernal pool vegetation of California, variation within pools. Madroño 50: 129–146.

Barbour, M. G., A. Solomeshch, R. Holland, C. Witham, R. Macdonald, S. Cilliers, J. A. Molina, J. Buck & J. Hillman, 2005. Vernal pool vegetation of California: communities of long-inundated deep habitats. Phytocoenologia 35: 177–200.

Bauder, E. T., 2000. Inundation effects on small-scale distributions in San Diego, California vernal pools. Aquatic Ecology 34: 43–61.

Beja, P. & R. Alcazar, 2003. Conservation of Mediterranean temporary ponds under agricultural intensification: an evaluation using amphibians. Biological Conservation 114: 317–326.

Bliss, S. A. & P. H. Zedler, 1998. The germination process in vernal pools: sensitivity to environmental conditions and effects on community structure. Oecologia 113: 67–73.

Blondel, J. & J. Arronson, 1999. Biology and wildlife of the Mediterranean region. Oxford University Press, London: 328.

Bonner, L., W. Diehl & R. Altig, 1997. Physical, chemical and biological dynamics of five temporary dystrophic forest pools in central Mississippi. Hydrobiologia 353: 77–89.

Braun-Blanquet, J., 1936. Un joyau floristique et phytosociologique «L'Isoetion» méditerranéen. SIGMA Communication 42: 1–23.

Braun-Blanquet, J., 1964. Pflanzensoziologie, 3rd ed. Grundzüge der Vegetationskunde. Springer, Vienna, New York.

Brinson, M. & A. I. Malvárez, 2002. Temperate freshwater wetlands: types, status, and threats. Environmental Conservation 29: 115–133.

Brullo, S. & P. Minissale, 1998. Considerazione sintassonomiche sula classe *Isoeto-Nanojuncetea*. Itinera Geobotánica 11: 263–290.

Castroviejo, S. et al. (eds), 1986–2008. Flora Ibérica, Vols. I, II, III, IV, V, VI, VII (I/II), VIII, X, XIV, XV, XVIII, XXI. Real Jardín Botánico, CSIC, Madrid.

Cherry, J. A. & L. Gough, 2006. Temporary floating island formation maintains wetland plant species richness: the role of the seed bank. Aquatic Botany 85: 29–36.

De Meester, L., S. Declerck, R. Stoks, G. Louette, F. De Meutter, T. De Bie, E. Michels & L. Brendonck, 2005. Ponds and pools as model system in conservation biology, ecology and evolutionary biology. Aquatic Conservation: Marine and Freshwater Ecosystems 15: 715–725.

Deil, U., 2005. A review on habitats, plant traits and vegetation of ephemeral wetlands – a global perspective. Phytocoenologia 35: 533–705.

Della Bella, V., M. Bazzanti, M. G. Dowgiallo & M. Iberite, 2008. Macrophyte diversity and physico-chemical characteristics of Tyrrhenian coast ponds in central Italy: implications for conservation. Hydrobiologia 597: 85–95.

Dufrêne, M. & P. Legendre, 1997. Species assemblages and indicator species: the need for a flexible asymmetrical approach. Ecological Monographs 67: 345–366.

EC (European Commission), 2007. The Interpretation Manual of European Union Habitats – EUR27 [online]. European Commission DG Environment, Brussels. Available from: http://ec.europa.eu/environment/nature/legislation/habitatsdirective/docs/2007_07_im.pdf.

EC (European Commission), 2008. Management of Natura 2000 habitats. *Mediterranean temporary ponds—3170. European Commission Technical Report [online]. Available from: http://ec.europa.eu/environment/nature/natura2000/management/habitats/pdf/3170_Mediterranean_temporary_ponds.pdf.

Edvardsen, A. & R. H. Økland, 2006. Variation in plant species richness in and adjacent to 64 ponds in SE Norwegian agricultural landscapes. Aquatic Botany 85: 79–91.

Espírito-Santo, D. & P. Arsénio, 2005. Influence of land use on the composition of plant communities from seasonal pond ecosystems in the Guadiana Valley Natural Park (Portugal). Phytocoenologia 35(2–3): 267–281.

Franco, J. A., 1984. Nova Flora de Portugal, Vol. II. Edição de Autor, Lisboa.

Franco, J. A. & M. L. Rocha Afonso, 1994–2003. Nova Flora de Portugal, Vol. III (I–III). Escolar Editora, Lisboa.

Grillas P., P. Gauthier, N. Yavercoski & C. Perennou, 2004. Mediterranean temporary pools I – issues relating to conservation, functioning and management. Station Biologique de la Tour du Valat, Arles: 120 pp.

INMG (Instituto Nacional de Meteorologia e Geofísica), 1991. O Clima de Portugal. Normais climatológicas da região de "Alentejo e Algarve", correspondentes a 1951–1980. Fasc XLIV, Vol. 4. Lisboa.

ISO/IDF (International Organization for Standardization), 2002. Milk and milk products. Determination of nitrogen content. Routine method using combustion according to the Dumas principle ISO/IDF International Standard (ISO/IDF), No. 14891:2002(E), 185:2002(E), Geneva (Switzerland). International Dairy Federation, Brussels, Belgium.

Jansen, J. & M. Menezes de Sequeira, 1999. The vegetation of shallow waters and seasonally inundated habitats (Littorelletea and Isoeta-Nanojuncetea) in higher parts of the Serra da Estrela, Portugal. Mitteilungen des Badischen Landesvereins für Naturkunde und Naturschutz 17(2): 449–462.

Keeley J. E. & P. H. Zedler, 1998. Characterization and global distribution of vernal pools. In C. W. Witham et al. (eds), Ecology, Conservation, and Management of Vernal Pool Ecosystems. Proceedings from a 1996 Conference. California Native Plant Society, Sacramento, CA: 1–14.

Kent, M. & P. Coker, 1992. Vegetation Description and Analysis: A Practical Approach. Wiley, England: 363 pp.

Legendre, P. & L. Legendre, 1998. Numerical Ecology, 2nd ed. Elsevier, Amsterdam.

Lumbreras, A., A. Olives, J. R. Quintana, C. Pardo & J. A. Molina, 2009. Ecology of aquatic *Ranunculus* communities under the Mediterranean climate. Aquatic Botany 90: 59–66.

Magurran, A. E. 2004. Measuring Biological Diversity. Blackwell Publishing, 256 pp.

Molina, J. A., 2005. The vegetation of temporary ponds with Isoetes in the Iberian Peninsula. Phytocoenologia 35: 219–230.

Neto, C., J. Capelo, C. Sérgio & J. C. Costa, 2007. The Adiantetea class on the cliffs of SW Portugal and of the Azores. Phytocoenologia 37(2): 221–237.

Oertli, B., J. Biggs, R. Céréghino, P. Grillas, P. Joly & J.-B. Lachavanne, 2005. Conservation and monitoring of pond biodiversity: introduction. Aquatic Conservation: Marine and Freshwater Ecosystems 15: 535–540.

Pinto-Gomes, C., A. García Fuentes, A. de Almeida Leite & P. Cardoso Gonçalves, 1999. Charcos temporários mediterrânicos do Barrocal Algarvio: diversidade e conservação. Quercetea 1: 53–64.

Podani, J., 2001. SYN-TAX 2000. Computer programs for data analysis in ecology and systematics. User's manual. Scientia, Budapest, UH.

Podani, J., 2006. Braun-Blanquet's legacy and data analysis in vegetation science. Journal of Vegetation Science 17: 113–117.

Pyke, C. R., 2004. Simulating vernal pool hydrologic regimes for two locations in California, USA. Ecological Modelling 173: 109–127.

Quézel, P., 1998. La végétation des mares transitoires à Isoetes en région méditerranéenne, intérêt patrimonial et conservation. Ecologia Mediterranea 24(2): 111–117.

Rhazi, L., P. Grillas, A. M. Toure & L. Tan Ham, 2001. Impact of land use in catchment and human activities on water, sediment and vegetation of Mediterranean temporary pools. C.R. Acad. Sci. Paris Sciences de la vie 324: 165–177.

Rhazi, L., M. Rhazi, P. Grillas & D. El Khyari, 2006. Richness and structure of plant communities in temporary pools from western Morocco: influence of human activities. Hydrobiologia 570: 197–203.

Rivas-Martínez, S., F. Fernández-González, J. Loidi, M. Lousã & A. Penas, 2001. Syntaxonomical checklist of vascular plant communities of Spain and Portugal to association level. Itinera Geobotanica 14: 5–341.

Rivas-Martínez, S., T. E. Díaz, F. Fernández-González, J. Izco, J. Loidi, M. Lousã & A. Penas, 2002a. Vascular plant communities of Spain and Portugal. Addenda to the syntaxonomical checklist of 2001. Itinera Geobotanica 15(1): 5–432.

Rivas-Martínez, S., T. E. Díaz, F. Fernández-González, J. Izco, J. Loidi, M. Lousã & A. Penas, 2002b. Vascular plant communities of Spain and Portugal. Addenda to the syntaxonomical checklist of 2001. Itinera Geobotanica 15(2): 433–922.

Rosselló-Graell, A., D. Draper & C. Tauleigne Gomes, 2000. Conservation status of mediterranean temporary ponds in Campo Militar de Santa Margarida (Ribatejo, Portugal). Portugaliae Acta Biologica 19: 191–199.

Rudner, M., 2005a. Environmental patterns and plant communities of the ephemeral wetland vegetation in two areas of the Southwestern Iberian Peninsula. Phytocoenologia 35(2–3): 231–266.

Rudner, M., 2005b. Seasonal and interannual dynamics in dwarf rush vegetation in the Southwestern Iberian Peninsula. Phytocoenologia 35(2–3): 403–420.

Serrano, L. & M. Zunzunegui, 2008. The relevance of preserving temporary ponds during drought: hydrological and vegetation changes over a 16-year period in the Doñana National Park (south-west Spain). Aquatic Conservation: Marine and Freshwater Ecosystems 18: 261–279.

Sileika, A. S., K. Gaigalis, G. Kutra & A. Smitiene, 2005. Factors affecting N and P losses from small catchments (Lithuania). Environmental Monitoring and Assessment 102: 359–374.

Stamati, F., N. Nikolaidis, E. Dimitriou & T. Koussouris, 2008. Hydro-geochemical aspects of Mediterranean Temporary Ponds in Western Crete. Journal of Environmental Quality 37: 164–173.

Szajdak, L., I. Yczyñska-Ba£oniak & M. Szczepañski, 2006. Efficiency of the small pond as a biogeochemical barrier to decrease different kinds of nitrogen in the agricultural landscape. Polish Journal of Soil Science 39: 11–20.

ter Braak, C. J. F & P. Smilauer, 2002. CANOCO reference manual and user's guide to Canoco for Windows: Software for Canonical Community Ordination (version 4.5). Microcomputer Power, Ithaca, NY, USA.

Waterkeyn, A., P. Grillas, B. Vanschoenwinkel & L. Brendonck, 2008. Invertebrate community patterns in Mediterranean temporary wetlands along hydroperiod and salinity gradients. Freshwater Biology 53: 1808–1822.

Williams, P., M. Whitfield, J. Biggs, S. Bray, G. Fox, P. Nicolet & D. A. Sear, 2003. Comparative biodiversity of rivers, streams, ditches and ponds in an agricultural landscape in Southern England. Biological Conservation 115: 329–341.

Zacharias, I., E. Dimitriou, A. Dekker & E. Dorsman, 2007. Overview of temporary ponds in the Mediterranean region: threats, management and conservation issues. Journal of Environmental Biology 28: 1–9.

Zunzunegui, M., M. C. Diaz Barradas & F. Garcia Novo, 1998. Vegetation fluctuation in Mediterranean Dune Ponds in relation to rainfall variation and water extraction. Applied Vegetation Science 1: 151–160.

Hydrobiologia (2009) 634:25–41
DOI 10.1007/s10750-009-9890-x

POND CONSERVATION

Freshwater diatom and macroinvertebrate diversity of coastal permanent ponds along a gradient of human impact in a Mediterranean eco-region

Valentina Della Bella · Laura Mancini

Published online: 5 August 2009

Abstract Mediterranean coastal areas are characterised by heavily transformed landscapes and an ever-increasing number of ponds are subjected to strong alterations. Although benthic diatoms and macroinvertebrates are widely used as indicators in freshwater ecosystems, little is still known about the diatom communities of lowland freshwater ponds in the Mediterranean region, and, furthermore, there are few macroinvertebrate-based methods to assess their ecological quality, especially in Italy. This article undertakes an analysis of benthic diatom and macroinvertebrate communities of permanent freshwater ponds, selected along a gradient of anthropogenic pressures, to identify community indicators (taxa and/or metrics) useful to evaluate the effect of human impacts. A series of 21 ponds were sampled along Tyrrhenian coast in central Italy. Five of these ponds, in a good conservations status and surrounded by woodland were selected as 'reference sites' for macroinvertebrates and epipelic diatoms. The remaining sixteen ponds were located in an agricultural landscape subject to different levels of human impact. The total number of macroinvertebrate taxa found in each pond was significantly higher in reference sites than in both the intermediate and heavily degraded ones, whereas the diatom species richness did not result in a good community variable to evaluate the pond ecological quality. The analysis revealed a substantial difference among the compositions of diatom communities between reference ponds and degraded ponds. The former were characterised by the presence of several species belonging to genera, such as *Pinnularia* sp., *Eunotia* sp., *Stauroneis* sp., *Neidium* sp., all of which were mostly absent from degraded ponds. Furthermore, the taxonomic richnesses of some macroinvetebrate groups (Odonata, Ephemeroptera, Trichoptera, Coleoptera), and taxa composition attributes of macroinvertebrate communities (total abundance, percentages of top three dominant taxa, percentages of Pleidae, Ancylidae, Hirudinea, Hydracarina) significantly correlated with variables linked with anthropogenic pressures. The results of the investigation suggested that diatoms tended more to reflect water chemistry through changes in community structure, whereas invertebrates responded to physical habitat changes primarily through changes in taxonomic richness. The methodologies developed for the analysis of freshwater benthic diatom and macroinvertebrate communities may have a considerable

Electronic supplementary material The online version of this article (doi:10.1007/s10750-009-9890-x) contains supplementary material, which is available to authorized users.

Guest editors: B. Oertli, R. Cereghino, A. Hull & R. Miracle
Pond Conservation: From Science to Practice. 3rd Conference of the European Pond Conservation Network, Valencia, Spain, 14–16 May 2008

V. Della Bella (✉) · L. Mancini
Department of Environment and Primary Prevention, National Institute of Health, Viale Regina Elena 299, 00161 Rome, Italy
e-mail: valentina.dellabella@iss.it

potential as a tool for assessing the ecological status of this type of water body, complying with the European Union Water Framework Directive 2000/60/EC.

Keywords Algae · Bacillariophyceae · Macrofauna · Lowland ponds · Environmental modifications and land cover conversions · Pond conservation status

Introduction

Ponds are now widely recognised as an important biodiversity resource (Nicolet et al., 2004; Della Bella et al., 2005; Oertli et al., 2005), especially at landscape scale (Williams et al., 2004). Moreover, recent developments have contributed to a better understanding of their economic value, in particular, the role of ponds in delivering ecosystem services (EPCN, 2008). Although broadly distributed across the regions of Europe, ponds and others small water bodies are seriously threatened by anthropogenic pressures such as increasing urbanisation and agricultural development which has resulted in both a sharp decline in numbers and ecological quality (Wood et al., 2003; Biggs et al., 2005; EPCN, 2008). This situation can be witnessed along the Tyrrhenian coast of central Italy where many ponds and other small water bodies have been subjected to severe pressure as a result of environmental modifications and land cover conversion. This highly vulnerable coastal system is faced with increasing urban development and intensive agriculture practices which has a negative impact upon river basins and small water bodies (Mancini & Arcà, 2000).

In view of the above, it is vital that pristine ponds are maintained and that degraded ponds are restored wherever possible. The protection strategies of inland surface water bodies in Europe is outlined in the current European regulation on water, the Water Framework Directive 2000/60/EC (WFD) which establishes a Framework for Community Action in the Field of Water Policy (CEC, 2000). The EU WFD aims, among others, to achieve good ecological status for inland waters. Small still waters and wetlands, ecologically and functionally important elements of aquatic ecosystems, are acknowledged as playing a strategic role in the achievement of WFD objectives (CEC, 2000, 2005) and the restoration and creation of wetlands and ponds are included in the listed actions to achieve its objectives. In its assessment of water bodies, the Directive defines different biological quality elements, including macroinvertebrates and benthic algae, and these are the biological indicators taken into account in this research.

Diatoms are often the main element of phytobenthos biodiversity in inland waters, representing an important component in freshwater ecosystems and are one of the most important groups of algae for monitoring purposes (Kelly et al., 1998; Mancini, 2005; King et al., 2006). Although they are widely used as indicators in rivers (for a review see Prygiel, 1999), lakes (Blanco et al., 2004; DeNicola and Eyto, 2004) and large wetlands (Gell et al., 2002; Gaiser et al., 2005; Wang et al., 2006), little is known about the benthic diatom communities of Mediterranean coastal ponds (Della Bella et al., 2007).

Macroinvertebrates play key roles in the freshwater ecosystem functioning and are often the main components of animal biodiversity in small water bodies. Together with algae, they are the most often recommended group of organisms for water monitoring activities (Hellawell, 1986). While consolidated standard methods based on macroinvertebrates exist for flowing (Armitage et al., 1983; De Pauw & Vanhooren, 1983; Alba-Tercedor & Sánchez-Ortega, 1988; Ghetti, 1997) and standing waters (Wiederholm, 1980; Verneaux et al., 2004; Rossaro et al., 2007), and for broad wetland areas in Australia and North America (Hicks & Nedeau, 2000; Apfelbeck, 2001; Helgen & Gernes, 2001), there are still few macroinvertebrate-based methods proposed for assessing ecological quality of small lentic bodies in Europe (Biggs et al., 2000; Menetrey et al., 2005; Solimini et al., 2008), and, in particular, of lowland ponds in the Mediterranean eco-region (Trigal et al., 2009).

Although some studies on both macroinvertebrates and diatoms in other aquatic ecosystems have shown that they could provide consistent and complementary information on environmental quality (Apfelbeck, 2001; Triest et al., 2001; Chessman et al., 2006; Chipps et al., 2006; Feio et al., 2007), to date these biological components are not usually studied together in ponds, as recommended by the integrated approach of the WFD. Among the different typologies of wetlands, this study has focussed on lowland freshwater ponds. These ecosystems represent one of the biotopes most severely threatened by human impact and, therefore, worthy of urgent restoration actions.

The main purpose of this study is to carry out an analysis of both benthic diatom and macroinvertebrate communities of permanent ponds, selected along a gradient of anthropogenic pressures in the Mediterranean coastal area of central Italy to analyse the relationships between diatom and macroinvertebrate diversity and environmental variables (surrounding land use, pond conservation status, habitat condition, human disturbances and water chemistry), and to identify the community indicators (taxa and/or metrics) useful to evaluate the effect of human impact. The methodologies developed in this study may also provide valuable tools for assessing the ecological quality of these aquatic ecosystems.

Methods

Study area and site selection

The study area is located in central Italy, a region containing several lakes (volcanic, coastal, man-made and Apennine lakes), small still waters and wetlands, and a number of designated protected areas have been created specifically to protect these lentic freshwater ecosystems.

Among the permanent freshwater lowland ponds of the Tyrrhenian coast of central Italy having a surface area less than one hectare, a homogeneous group of ponds were selected, based on water permanence and conductivity, low altitude and a surface area between 500 and 8,000 m^2. Temporary ponds have been excluded from the sample because previous studies indicated differences in the structure and composition of the biological communities, due to the effects of wet phase duration (Della Bella et al., 2005, 2008). A preliminary survey of the study sites was undertaken in spring 2007 following pond identification through cartographic search, previous studies made on potential suitable ponds, information gathered from staff attached to the Protected Areas together with information from local landowners. For each pond, surrounding land use up to 500 m radius was evaluated including the pond habitat condition, the presence/absence of human disturbances, and the application of ECELS index. This is a rapid methodology, developed and applied in Spain to assess the conservation status of Mediterranean wetlands (Sala et al., 2004), and is based upon five main components: (1) basin littoral morphology, (2) human activity, (3) water characteristics, (4) emergent vegetation and (5) hydrophytic vegetation. Although this index was developed as a tool to measure only the conservation status of wetlands, through the evaluation of hydromorphological, biological and physico-chemical characteristics, it can be used to establish the ecological status of wetlands, as considered under the WFD, together with assessments of other quality elements. From this evaluation, the percentage of land use under different categories (woodland, other natural vegetation such as shrub and grass, pasture, agricultural and artificial surface, water basins) surrounding the basin within a radius of 500 m was measured. In addition, the riparian vegetation width (as minimum distance between pond and land use conversion from wood or natural vegetation) using field observations, aerial photo interpretation and Corine Land Cover 2000 (CLC2000) was also measured. The percentage of pond surface covered by aquatic macrophyte (emergent and submerged) was estimated, and the percentage of perimeter covered by emergent vegetation, shore slope, and recorded the number of macrophyte species for each pond were noted. The maximum depth and area of ponds was assessed with a graduated pole and a measuring tape and, finally, the number of visible human disturbances was assessed and included, for example, the presence of exotic species, irrigation use of pond, pond shore cleaning, livestock grazing, presence of rubbish, fire, transport infrastructures within 100 m, pond digging, salinisation, farm activities, pesticides and herbicides use (see Table 1).

Out of this, 21 ponds were selected along the Tyrrhenian coast of the Tuscany and Latium Regions, a heavily transformed landscape which has been subject to strong alterations, due to environmental changes and land use conversions (Fig. 1). All study ponds are located inside Protected Areas: Maremma Regional Park (Tuscany); Litorale Romano Natural State Reserve (Latium), and in particular Macchiagrande di Ponte Galeria (SIC IT 6030025) and LIPU Oasis Castel di Guido; Decima-Malafede Natural Reserve; Presidential Estate of Castelporziano (SIC IT6030027-8, IT6030084); Wood of Foglino (SIC IT6030047). Out of the 21 sites, five were undegraded/unimpaired ponds and were identified as 'reference sites', surrounded by woodlands, in a high/good conservation status and without human disturbance. The other 16 ponds were located in the

Table 1 Pond variables taken into account in the study and their codes

Environmental variables	Code
Land use surrounding (Buffer radius = 500 m)	
Agriculture, artificial surface, pasture	AG_ART_P
Woodland and natural vegetation	BOS_NAT
Water bodies	ACQ
Conservation status	
ECELS index	ECELS
Habitat condition	
Macrophyte cover	CMAC
N° of macrophyte species	NVEG
Shore slope	RIVE
Riparian zone width	FRIP
Pond morphometry	
Depth	PROF
Surface	SUP
Water characteristics	
pH	PH
Dissolved oxygen	DO
Nitrate	NO_3^-
Phosphate	PO_4^{3-}
Conductivity	COND
BOD_5	
Number of human disturbances	DIST
Exotic species	
Irrigation use	
Shore cleaning	
Livestock grazing	
Rubbish	
Fire	
Roads	
Farm activities	
Salinisation	
Use of pesticides, herbicides	
Pond digging	

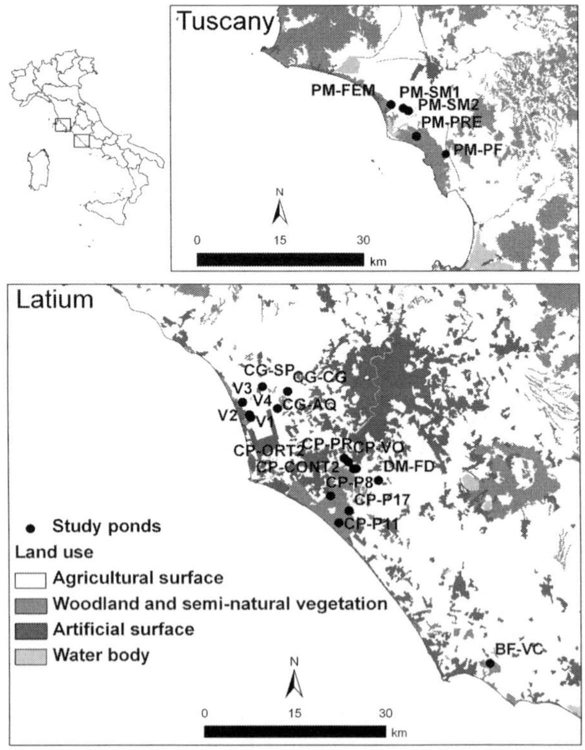

Fig. 1 Study area, studied ponds location and land use surrounding. Reference ponds are PM-PRE, CP-P11; CP-P8, CP-17, BF-VC

agricultural landscapes and were subject to varying levels of human alteration.

Sampling and laboratory methods

Floristic and faunistic materials and water samples were taken from all the ponds between late spring and early summer 2007, covering the period of the year generally characterised by the highest richness of diatom and macroinvertebrate species (Bazzanti et al., 1996; Dell'Uomo, 2004; Della Bella et al., 2005, 2007; Trigal et al., 2006). Conductivity, pH and dissolved oxygen were recorded by field meters (Table 1). Water samples were collected and analysed in laboratory for nitrate (NO_3), phosphate (PO_4^{3-}) contents and BOD_5, as per standard methods reported in IRSA (1994) and APHA (1998).

Diatoms

Diatom sampling, sample treatment, and laboratory analysis were carried out according to the European recommendations (Kelly et al., 1998; EN 1394, 2003; EN 14407, 2004, King et al., 2006) and national guidelines (APAT, 2008a). Epipelic forms present on the upper surface of the littoral sediment were sampled with a pipette, and five samples were taken for each pond. This sampling substratum was chosen because cobbles and emergent macrophytes were not present in all the water bodies visited, whereas it is very important that comparisons between water bodies are based on samples from the same substratum. The epipelon was the only substratum available in all the study ponds. In standing waters where the substrata types recommended for sampling diatoms are not an option, the use of the collection of soft sediment is justified (King et al., 2006).

Diatom samples were immediately examined in the laboratory or fixed and stored in 4% formalin. In order to identify the diatom frustules, the diatom valves were cleaned using hydrogen peroxide to eliminate organic matter and with hydrochloric acid to dissolve calcium carbonate. Clean diatom frustules were mounted in a synthetic resin with high refraction index (Naphrax©) and up to 400 valves were counted and identified to species or variety level in each sample using a light microscope with 1,000× magnification. Measurements were made with the aid of image analysis software (Leica IM1000). Images of diatoms were digitised using a video camera (Leica DC 300) connected to a microscope (Optiphot-2) and to a computer. The main references for diatom taxonomy were Krammer & Lange-Bertalot (1986, 1988, 1991a, b, 2000), Krammer (2000), Lange-Bertalot (2001) and Prygiel & Coste (2000). Their nomenclature is used here, but is updated to include changes. Qualitative and structural attributes of algae communities evaluated as candidate metrics for assessing biotic responses to pond alteration were (1) species richness, (2) species composition: relative abundance of some genera (Hill et al., 2000; Chipps et al., 2006; Watchorn et al., 2008).

Macroinvertebrates

Macroinvertebrates were qualitatively and quantitatively sampled with a hand net (dimension: 20 × 27 cm; mesh size: 0.5 mm) which was worked over replicates of a known surface of 0.135 m^2. The area of each replicate sample was obtained by multiplying the width of the hand net by the sweep length (0.5 m). The number of replicates was calculated in proportion to the surface area of the each pond according to some surface area classes: a minimum of five replicates for surface lower than 1,000 m^2 (eight ponds); seven replicates for surface between 1,000 and 3,000 m^2 (five ponds), 10 replicates for surface between 3,000 and 4,000 m^2 (five ponds), 12 replicates for surface between 4,000 and 5,000 m^2 (one pond) and up to a maximum of 17 replicates for surface of 8,000 m^2 (two ponds).

The pond was netted in all the mesohabitats according to their relative extent within each pond following a 'multihabitat' sampling methodology previously applied in streams (Hering et al., 2004; APAT, 2008b). The concept of mesohabitat has been successfully adopted in recent studies on ponds (Biggs et al., 2005; Della Bella et al., 2005; Oertli et al., 2005). Mesohabitats identified in this study, defined as visually distinct and easily identifiable habitats within the freshwater body, were: filamentous algae; aquatic emergent and submersed vegetation; fine and coarse organic matter; roots of living riparian vegetation; dead wood; sediment dominated by silt and clay (diameter lower than 6 μm) or sand (between 6 μm and 2 mm) or gravel (between 2 mm and 2 cm). Material was preserved in alcohol (75%) until sorting. Invertebrate macrofauna was identified at the same level in each pond. Macroinvertebrates were usually identified to genus and species level, sometimes to family, and rarely to a higher taxonomic level (i.e. Oligochaeta, Turbellaria, Hirudinea and Hydracarina). Samples from different substrates were then pooled obtaining one combined sample for each site for next data analysis. Invertebrate 'density' was calculated by the number of sampled individuals divided by the total surface area of all replicate samples taken in each pond, and was expressed as individuals per square metre (ind/m^2). Some macroinvertebrate responses to pond alteration were evaluated including a variety of qualitative and structural attributes of communities known to respond to environmental degradation (Hicks & Nedeau, 2000; Biggs et al., 2000; Apfelbeck, 2001; Helgen & Gernes, 2001; Tangen et al., 2003; Menetrey et al., 2005; Gerecke and Lehmann, 2005; Solimini et al., 2008). In detail, we evaluated total taxa and family richness, some groups richness at genus and family level (Odonata, Ephemeroptera, Trichoptera, Coleoptera), total abundance (ind/m^2), percentage of top three dominant taxa, and percentages of some family and higher groups (Corixidae, Pleidae, Ancylidae, Hirudinea, Hydracrina). Community richness was calculated at family and a lower taxonomical level in order to evaluate the possibility in reduction of taxon identification effort.

Data analysis

Both multivariate and univariate approaches in statistical analyses were used to analyse the data (Reynoldson et al., 1997). Principal Component Analysis (PCA) was employed, based on environmental characteristics of the 21 study ponds, to summarise variations among sites and to highlight environmental gradients. Before the analysis, all variables were standardised following $X_{st} = (X - \overline{X})/SD$. The counts of each diatom taxon were expressed into relative abundance as a percentage of the total valves counted and identified in each sample. The macroivertabrate density data were expressed as individuals/m^2. In order to relate the diatoms and macroinvertebrate data to environmental data, a Canonical Correspondence Analysis (CCA), with Monte Carlo permutation tests, was applied. All the 18 environmental variables were included in the analysis. The ordination analyses were performed at the species level for diatoms and at the taxa and family level for macrofauna. The removal of all the taxa with less than two occurrences reduced the size of the original data set from 196 to 128 species and varieties of diatoms, and from 74 to 55 taxa of macroinvertabrate. Before the analyses, the relative abundance of diatom species were arcsin $\sqrt{p}$ transformed and macroinvertabrate densities were log $(x + 1)$ transformed to stabilise the variance and environmental variables were standardised as for PCA above. Besides observed total macroinvertebrate taxa of study ponds, the real richness estimator Chao1 (Chao, 1984) was used to take into account a possible underestimation of the real value due to sampling. The nonparametric Kruskal–Wallis analysis was used to test significant differences in observed and estimated richness, and in surface area, among pond groups, and Wilcoxon Matched Pairs Test to check significant differences between diatom and invertebrate richness. Spearman rank correlation (r_s) with Bonferroni's correction was used to explore relationships between two taxonomic richness (diatom and invertebrate), between both richness and sampling date and number of replicate samples, and between environmental and community variables or metrics. Data analyses were conducted with STATISTICA (version 5), PRIMER 5 (version 5.2.0), EstimateS (version 8.0.0) and PC-ORD (version 3.09) for Windows softwares.

Results

Environmental characteristics of ponds

The first two components extracted in the PCA performed on land use, pond habitat condition, number of human disturbances, ECELS index, and morphology and water variables, accounted for 45.8% of variance in the original data (Fig. 2). The ordination analysis PCA pointed out a clear separation among reference ponds, the most degraded ponds and the ponds with intermediate level of human alteration along the first axis (PC1). The PC1 axis represented the environmental gradient positively correlated with the number of human disturbances, percentages of agriculture and artificial surface in the land use surrounding, shore slope, conductivity, and negatively with ECELS index, percentage of land use occupied by woodland and natural vegetation, number of macrophyte species and percentage of macrophyte cover. Three pond groups (reference ponds, intermediately impaired ponds and grossly impaired ponds) did not significantly differ in area. In detail the results of 18 environmental characteristics of ponds taken into account in the study are reported in Table 2 and in Appendices 1 and 2—Supplementary Material.

Diatom species richness and assemblages

A total of 196 species and varieties of diatoms belonging to 53 genera were identified (Table 3). The ponds with intermediate level of human alteration had significantly higher number of species than reference and the most degraded ponds (Kruskal–Wallis test: $H_{2,N=21} = 8.12$; $P < 0.05$) (Fig. 3). The total number of species found in samples of each study pond was significantly correlated with pond surface area ($r_s = 0.64$; $P < 0.001$) and macrophyte cover ($r_s = 0.63$; $P < 0.01$). No significant relationship was found between diatom number of species and sampling date.

The CCA revealed a substantial difference among diatom communities of reference ponds, degraded ponds, and ponds with intermediate level of human alteration (Fig. 4). The first two axes were significant (Monte Carlo test $P < 0.05$) and explained 25.5% of cumulative variance in diatom data. Indeed, communities of undegraded reference ponds were characterised

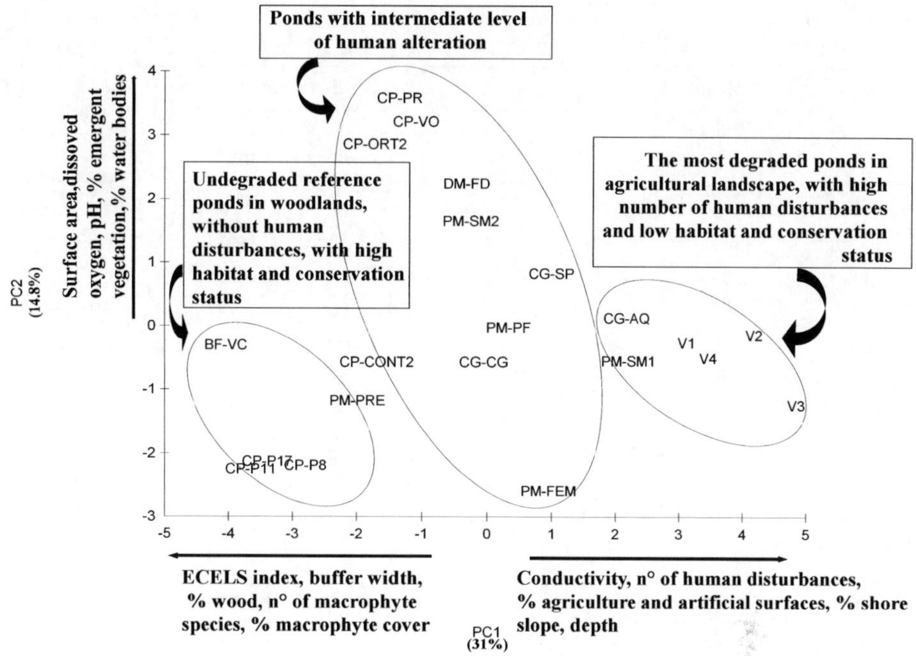

Fig. 2 PCA performed on environmental variables of studied ponds. *Arrows* indicate the correlation (r_s) with a significance at least $P < 0.05$ between axis pond scores and environmental variables

by a presence of several species belonging to genera, such as *Pinnularia* sp., *Eunotia* sp., *Neidium* sp., *Stauroneis* sp., which were almost totally absent from disturbed ponds. In contrast, the most impaired ponds were characterised by species of *Fragilaria* and *Pseudostaurosira* genera, such as *Pseudostaurosira brevistriata*, *Fragilaria ulna* var. *arcus* and *Fragilaria* cf. *nanana*.

The relative abundances of species belonging to genera *Fragilaria* were significantly correlated with surrounding land use, in particular, positively correlated with agriculture, pasture and artificial surfaces, and negatively correlated with surfaces occupied by woodland and natural vegetation ($r_s = 0.81$, $P < 0.0001$). Species abundances of this genera were negatively correlated also with the ECELS index, used to evaluate pond conservation status ($r_s = -0.78$, $P < 0.0001$). A positive relationship was found between this genera abundances and number of human disturbances present ($r_s = 0.72$, $P < 0.001$), pond water conductivity ($r_s = 0.79$, $P < 0.0001$) and shore slope ($r_s = 0.71$, $P < 0.001$). On the contrary, the total of abundances of species belonging to genera *Eunotia*, *Stauroneis*, *Neidium* and *Pinnularia*, were positively correlated with the ECELS index ($r_s = 0.62$, $P < 0.005$), riparian zone width ($r_s = 0.58$, $P < 0.01$), woodland and natural vegetation surface ($r_s = 0.50$, $P < 0.05$), and negatively correlated with pond water conductivity ($r_s = 0.57$, $P < 0.05$), pastureland, agricultural and artificial surfaces ($r_s = 0.52$, $P < 0.05$).

Macroinvertebrates

A total of 74 taxa, belonging to 15 main zoological groups were collected from the 21 study ponds (Table 4). Macrofauna was qualitatively dominated by insects with 60 taxa, most of which belonged to Coleoptera (a total of 27 taxa) and Diptera (11 taxa), and, secondly, to the Hemiptera (10 taxa) and Odonata (nine taxa). The observed and estimated total number of macroinvertebrate taxa are strongly correlated to each other ($r_s = 0.96$, $P < 0.0001$), and richness was not significantly correlated with diatom richness. There were significant differences between diatom and two invertebrate richness (Wilcoxon test $P < 0.001$). Unlike diatoms, the reference ponds had significantly higher number of observed (Kruskal–Wallis test: $H_{2,N=21} = 13.71$; $P < 0.001$) and estimated ($H_{2,N=21} = 12.83$; $P < 0.005$) number of taxa of macroinvertebrates than both the most degraded ponds and ponds with intermediate level of human alteration (Fig. 5). The total number of taxa of macroinvertebrates found in each pond as positively correlated with the ECELS index ($r_s = 0.83$, $P < 0.0001$), number and cover of macrophytes ($r_s = 0.84$, $P < 0.0001$ and $r_s = 0.64$, $P < 0.001$,

Table 2 Median, mean, minimum and maximum of the distribution of pond environmental variables in reference, intermediate and impaired groups of ponds

	Morpholgy			Water chemistry					
Variable:	PROF	SUP	RIVE	NO_3^-	PO_4^{3-}	PH	COND	DO	BOD_5
Units:	cm	m^2	%	mg/l	mg/l		µS	mg/l	
Reference ponds									
Median	55	800	5	1.22	0.45	7.83	173.8	8.60	7
Mean	99	2226	23	1.39	0.52	8.10	156.3	8.68	8
Min.	45	690	5	0.86	0.14	7.14	73.0	6.97	3
Max.	200	8000	80	2.00	0.84	9.95	222.0	10.70	14
Ponds with intermediate level of human alteration									
Median	150	3195	30	0.77	0.47	8.49	611.0	8.95	8
Mean	144	3349	25.5	1.32	0.52	8.96	867.3	9.93	8.1
Min.	60	700	5	0.50	0.11	8.19	337.0	5.25	3
Max.	300	8000	50	5.44	1.08	10.25	2960.0	15.87	16
Very degraded pond									
Median	200	1100	99	12.88	0.51	8.71	1361.0	9.82	7.5
Mean	210	1042	97.5	13.64	0.61	8.67	1329.0	9.89	8.7
Min.	110	500	90	0.63	0.12	8.27	1114.0	7.85	5
Max.	300	1600	99	28.62	1.29	8.90	1460.0	12.36	14

	Habitat condition				Conservation status	Human Disturbances	Land use		
Variable:	NVEG	CMAC	VEM	FRIP	ECELS	DIST	ACQ	ARG_ART_P	BOS_NAT
Units:		%	%	m	Value	N°	%	%	%
Reference ponds									
Median	7	80	5	700	80	0	0	0	100
Mean	7.6	55	25	900	83	0	0	0	100
Min.	5	5	0	500	75	0	0	0	100
Max.	14	90	90	2000	95	1	0	0	100
Ponds with intermediate level of human alteration									
Median	4	80	80	20	61.5	4	1	78	22
Mean	7.8	64	56	18.4	60.5	4	3	71	27
Min.	3	20	0	1	40	2	0	10	0
Max.	19	95	99	50	74	6	10	95	90
Very degraded pond									
Median	1	0	10	0	17	7	0	97	2
Mean	1.5	8	27	0.83	13.3	6	1	96	4
Min.	0	0	0	0	2	2	0	87	0
Max.	4	30	99	3	21	9	5	99	13

For code variables see Table 1

respectively), percentages of woodland and natural vegetation surfaces ($r_s = 0.79$, $P < 0.001$), riparian zone width ($r_s = 0.71$, $P < 0.001$), and negatively with pastureland, agricultural and artificial surfaces ($r_s = 0.77$, $P < 0.001$), conductivity ($r_s = 0.67$, $P < 0.001$) and shore slope ($r_s = 0.60$, $P < 0.005$). No significant relationship was found between the total number of macroinvertebrate taxa and sampling date, nor between total number of macroinvertebrate taxa and number of replicate samples taken per pond.

As for diatoms, the CCA analyses performed on densities (ind/m^2) of macroinvertebrates at family

Table 3 Number of diatom species and varieties identified for each genera

Genera	Number of species
Nitzschia	30
Navicula	28
Gomphonema	13
Amphora	8
Pinnularia, Stauroneis	7
Eunotia, Cyclotella, Cymbella, Gyrosigma, Fragilaria, Neidium, Planothidium	5
Achnanthidium, Craticula, Caloneis, Surirella	4
Sellaphora, Tryblionella	3
Diadesmis, Encyonema, Encyonopsis, Epithemia, Mayamaea, Cymatopleura, Rhopalodia, Anomoeoneis, Luticola, Diploneis, Cocconeis	2
Brachysira, Achnanthes, Cymbopleura, Denticula, Adlafia, Bacillaria, Cyclostephanos, Hippodonta, Synedra, Staurosira, Rhoicosphenia, Rhizosolenia, Pseudostaurosira, Fallacia, Mastogloia, Discostella, Hanztschia, Parlibellus, Ulnaria, Fistulifera, Eucocconeis, Eolimna, Frustulia	1
Total No. of species	196

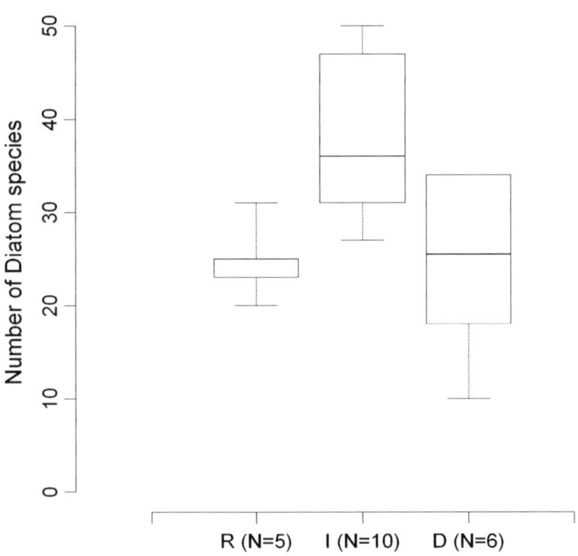

Fig. 3 *Box plot* of number of diatom species found in reference ponds (R), the most degraded ponds (D) and ponds with intermediate level of human alteration (I). *Box* is interquartile range (25–75%), *black line* is median value, and *whiskers* are minimum and maximum values

and at lower taxonomic level revealed a substantial difference between communities of reference ponds and those of ponds with intermediate level of human alteration and the most degraded ones. Alternatively, the analysis did not highlight a clear separation between the latter two groups of ponds. Figure 6 shows the plot of CCA performed on family densities. The first two axes explained 32.3% of cumulative variance but were not significant after Monte Carlo permutation test. The evaluated invertebrate metrics were significantly correlated with pond conservation status (ECELS index), land use (percent woodland and natural vegetation, and percent agriculture, pasture and artificial surfaces), shore slope, conductivity, nitrate and number of human disturbances (Table 5). Among the tested richness metrics, the total number of taxa and family had the strongest correlation with pond conservation status (ECELS Index). Total richness metrics were also strongly correlated with percentage of woodland and natural vegetation in land use surrounding. Then, the number of taxa and family of Odonata and EOT (Ephemeroptera, Odonata and Trichoptera) highly responded to pond conservation status level and the surrounding land use variables. The number of Coleoptera was weakly correlated, and only at genera level. In general, taxa composition metrics showed to be slightly less significant than most of the richness metrics. Among the tested composition metrics, the top three dominant taxa and percentage of Pleidae were strongly correlated with the ECELS index. The percentage of Pleidae and Hydracarina had the highest positive correlation with the percentage of woodland and natural vegetation in the surrounding land use. Moreover, a very high correlation was found also between the total density of all the sampled macroinverterbrate individuals (ind/m^2) and the ECELS Index and the percentage of surrounding land use variables (Table 5).

Discussion

This study contributed to a better knowledge of benthic diatom of ponds. Indeed, this pond biological component in the Mediterranean eco-region is still little studied at both national (Della Bella et al., 2007) and international levels, especially associated with the macroinvertebrate component analysis. We

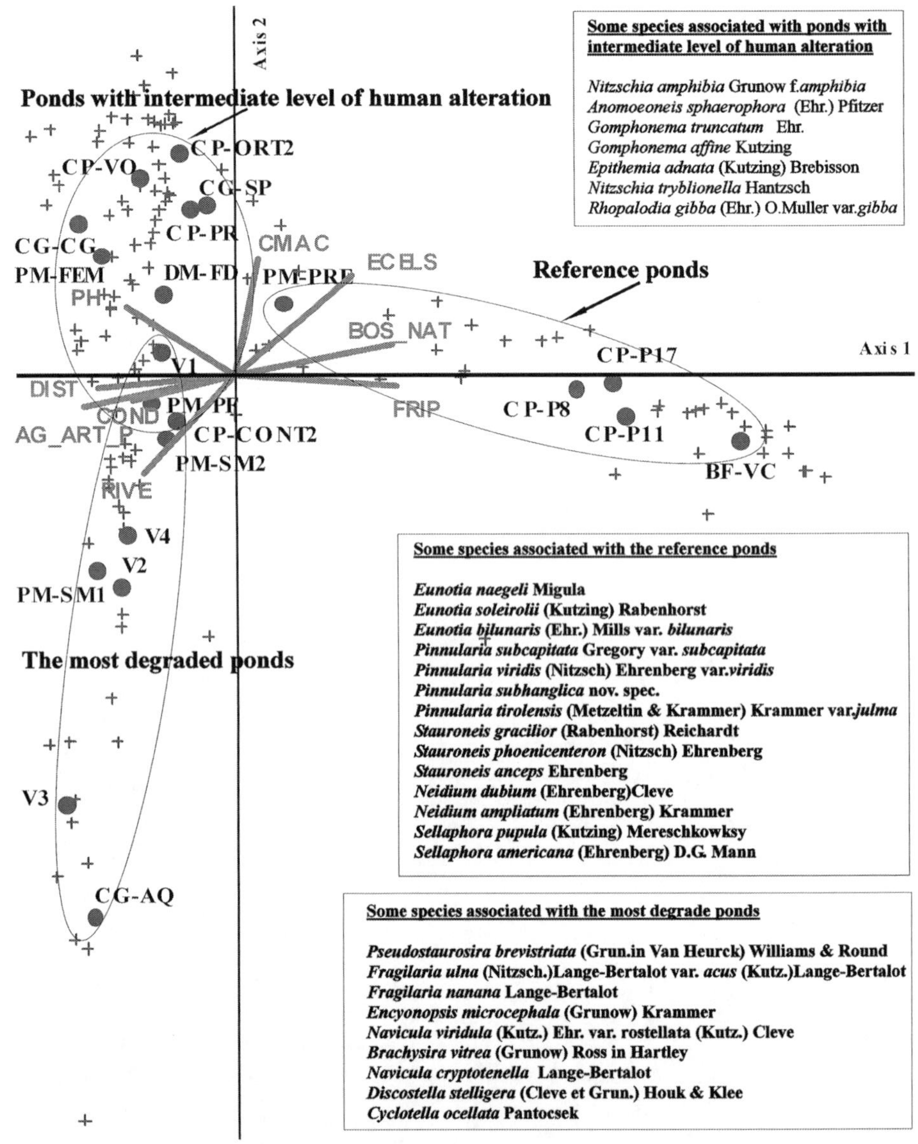

Fig. 4 CCA performed on relative abundances of diatom species and varieties found in each study ponds. For variable and pond codes see Table 1 and Fig. 1, respectively

analysed for the first time the epipelic diatom assemblages and associated aquatic macrofauna of some of the few last remnant pristine ponds along the Thyrrenian coast of central Italy. They are still not or minimally impaired by human activities, and we used them as reference condition to compare with those of impacted ponds.

Principal Component Analysis highlighted that the five reference ponds with a good/high conservation status and surrounded by woodland and shrubs showed the highest number of macrophytes and greatest macrophyte cover. In addition, they had a lower conductivity than degraded ponds with high number of human disturbances, high shore slope, and high percentage of agriculture in its surroundings. Although the highest nitrate contents were found in some of most impaired ponds, the present analysis did not indicate the water column nutrients as an important environmental gradient from low impact site to disturbed sites. Some studies on wetlands also found that conductivity, or specific conductance, were positively correlated with the percentage of agriculture land or negatively with the percentage of natural areas (Stewart et al., 2000; Carrino-Kyker & Swanson, 2007). Wetlands, when influenced by agricultural activity, often have significantly higher conductivity

Table 4 List of main zoological groups collected in studied ponds, with their total number of identified systematic units and their identification level

Zoological groups	Identification taxonomic level	Number of taxa
TURBELLARIA	Class	1
OLIGOCHAETA	Class	1
HIRUDINEA	Class	1
ISOPODA	Genera	1
DECAPODA	Species	2
HYDRACARINA	Order	1
EPHEMEROPTERA	Genera (2)/Species (1)	2
ODONATA	Family (1)/Genera (7)/Species (1)	9
HEMIPTERA	Genera (5)/Species (5)	10
TRICHOPTERA	Genera	1
COLEOPTERA	Family (8)/Genera (17)/Species (2)	27
DIPTERA	Family (7)/subFamily (2)/Species (2)	11
GASTROPODA	Family (1)/Genera (3)/Species (1)	5
BIVALVIA	Species	1
BRIOZOA	Class	1
Total taxa		75

The number of taxa for each identification level is indicated in brackets

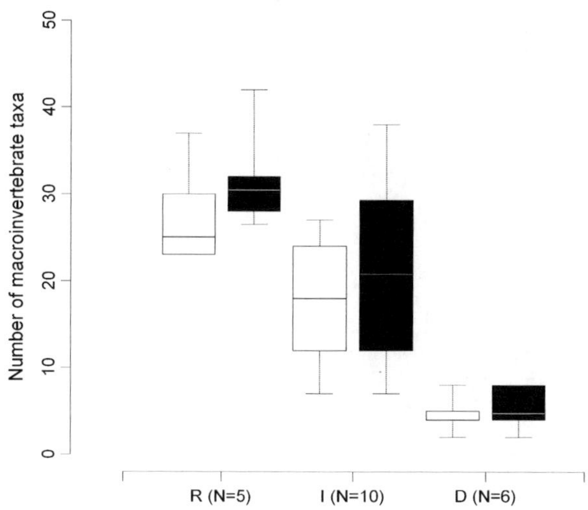

Fig. 5 *Box plot* of number of macroinvertebrate taxa in reference ponds (R), the most degraded ponds (D) and ponds with intermediate level of human alteration (I). *White boxes* indicate observed number of taxa, *black boxes* indicate estimated number of taxa with Chao1. *Box* is interquartile range (25–75%), *black* and *white line* their median values, and *whiskers* are minimum and maximum values

than reference wetlands (Helgen & Gernes, 2001; Chipps et al., 2006). Our reference ponds were characterised by conductivity lower than 225 μS/cm^{-1}, confirming these findings.

The total number of diatom species found in each pond did not provide a good community variable to evaluate pond ecological quality. Other studies have shown only minor and statistically insignificant reductions in diatom species richness between low and high disturbance sites of wetlands (Chipps et al., 2006) and lakes (Cohen et al., 1993). Indeed, reference ponds did not show a higher number of species than heavily degraded ponds. On the contrary, the diatom species number was significantly higher in ponds with an intermediate human impact. This result supports the intermediate disturbance hypothesis that predicted that environments under intermediate disturbances create more variable living conditions (e.g. available niche space, food resources, competition), which allow high levels of species diversity (Connell, 1978; Watchorn et al., 2008). Diatom species probably had a competitive advantage as a result of the habitat disturbance filling the available niche. Although Mackey & Currie (2001) suggest that the intermediate disturbance hypothesis has generally not been demonstrated, especially when few disturbances levels were examined, some researchers have reported increases in diatom richness under moderate stress (van Dam, 1982; Stevenson, 1984; Hill et al., 2000).

The CCA analysis indicated that diatom communities of minimally impaired ponds are characterised by the presence and abundance of some species belonging to genera *Pinnularia* sp. *Eunotia* sp.,

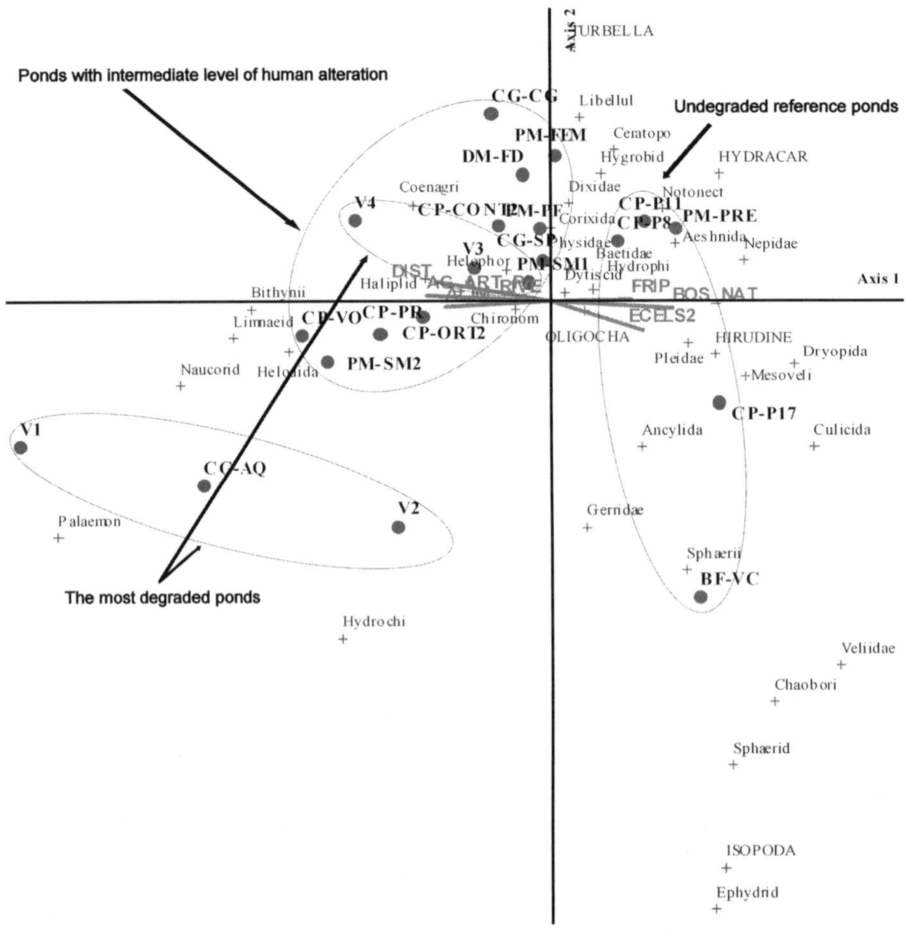

Fig. 6 CCA performed on relative abundances of macroinvertebrate family found in each study ponds. For variable and pond codes see Table 1 and Fig. 1, respectively. TURBELLA = Turbellaria; Libellul = Libellulidae; Ceratopo = Ceratopogonidae; Hygrodib = Hygrobidae; HYDRACAR = HYDRACAR INA; Coenagri = Coenagrionidae; Notonect = Notonectidae; Corixida = Corixidae; Aeshniida = Aeshnidae; Helophor = ; Helophoridae; Hydrophi = Hydrophilidae; Haliplid = Haliplidae; Dytiscid = Dytiscid; Bithiynii = Bithyniidae; Chironom = Chironomidae; Limnaeid = Limnaeidae; OLIGOCHA = OLIGOCHAETA; HIRUDINE; HIRUDINEA; Naucorid = Naucoridae; Helodida = Helodidae; Dryopida = Dryopidae; Ancylida = Ancylidae; Mesoveli = Mesovelia;Cuicida = Culicidae; Palaemon = Palaemonidae; Hydrochi = Hydrochidae; Sphaerii = Sphaeridiidae; Chaobori = Chaoboridae; Sphaerid = Sphaeridae; Ephydrid = Ephydridae

Neidium sp., *Stauroneis* sp., almost totally absent from degraded ponds. *Pinnularia* and *Eunotia* genera were inhabiting typically pristine water ecosystems such as Italian Alpine springs with siliceous substrate, and are important taxonomic groups in these habitats (Cantonati et al., 2005, 2006). On the other hand, diatom communities of the most disturbed ponds are characterised by the presence and high relative abundance of species belonging to genera *Fragilaria* and *Pseudostaurosira*. In their paleolimnological reconstruction using diatoms from sediments of the Swan Lake (Southern Ontario, Canada), Watchorn et al. (2006) found very similar changes in diatom community composition through time rather than in space (among sites). The diatom flora changed from an assemblage dominated by larger, benthic, acid-tolerant species including *Neidium* spp., *Eunotia* sp., *Pinnularia* spp., *Stauroneis* spp. and *Sellaphora pupula* to an assemblage characterised by smaller, and facultative planktonic taxa-like Fragilariaceae-type. The authors directly linked these changes in diatom assemblages to regional deforestation and agricultural activities associated with European settlements. In this study, these groups of species were

Table 5 Spearman rank correlation coefficients (r_s) for the relationships between invertebrate metrics and environmental variables in the studied ponds

Metrics	Variables									
	Conservation status	Land use		Morphology	Habitat condition			Water chemistry		Disturbances
	ECELS index	ARG_ART_P	BOS_NAT	RIVE	NVEG	CMAC	FRIP	NO3	COND	DIST
Taxa richness										
Total number of taxa	0.83 ****	−0.77***	0.79***	−0.60*	0.84****	0.64*	0.71**		−0.63**	
Total number of family	0.85****	−0.75***	0.76***	−0.62*	0.90****	0.71**	0.68**		−0.68**	
Number of Odonata taxa	0.70**	−0.72**	0.77***		0.55*	0.68**	0.57*		−0.55*	
Number of Odonata family	0.73**	−0.73**	0.78***		0.67**	0.70**	0.55*			
Number of EOT taxa	0.68**	−0.66**	0.71**			0.74***	0.56*	−0.65*	−0.58*	
Number of EOT family	0.74***	−0.66**	0.71**		0.71**	0.78***	0.56*	−0.58*	−0.58*	
Number of Coleoptera taxa	0.62*	−0.60*	0.56*	−0.55*	0.69**		0.56*			
Number of Coleoptera family					0.58*					
Taxa composition										
Percent top 3 Dominant taxa	−0.76***				−0.71**	−0.64*				0.58*
Percent Corixidae				−0.58*						
Percent Pleidae	0.73**	−0.83****	0.84****	−0.60*	0.63*		0.75***			−0.55*
Percent Ancylidae	0.67**	−0.58*	0.59*		0.70**					
Percent Hirudinea + Bivalvia	0.62*	−0.60*	0.61*	−0.67**	0.68**		0.66*			
Percent Hydracarina	0.61*	−0.80***	0.81***				0.57*			−0.69**
Taxa abundance										
Total density (ind/m^2)	0.81****	−0.93****	0.94****	−0.74**	0.67**		0.83****		−0.70**	−0.66*

Only $|r_s| \geq 0.55$ are reported. *Total density* number of all the sampled individuals divided to total surface area of all the replicate samples taken in each pond

**** $P < 0.00001$; *** $P < 0.0001$; ** $P < 0.001$; * $P < 0.01$

correlated with pond water conductivity, percentage land use, pond conservation status, human disturbances, confirming they could be candidate metrics to assess the ecological quality of these types of habitat.

Concerning the multivariate approach, diatom communities were more clearly associated with each of the three pond types of reference, intermediate and highly altered ponds than macroinvertebrate assemblages. In fact, invertebrate communities did not point out a clear separation between ponds with intermediate level of human alteration and the most degraded ones. As this algae group is represented by several species and might provide a better ecological resolution, this analytical method could be promising for detecting human impact using diatoms.

On the contrary, the univariate approach indicated that among the considered attributes of macroinvertebrate community, several metrics significantly correlated with variables associated with human impact. Contrary to diatoms, the number of macroinvertebrate taxa was significantly higher in reference ponds than both intermediate and heavily degraded ponds. This finding does not support the intermediate disturbance hypothesis (Connell, 1978). The observed difference between macroinvertebrate and diatom richness response could depend on the different spatial scales involved. Because of their larger size compared to microalgae, macroinvertebrate are more sensitive to physical habitat characteristics such as the morphological alterations of the habitat conditions (presence of steep shore, destruction of riparian belt, loss of mesohabitats). Differences between diatom and macroinvertebrate responses were also found in studies on river ecosystems (Triest et al., 2001; Chessman et al., 2006; Feio et al., 2007). Our findings confirmed several studies that show decreasing macroinvertebrate diversity with an increasing anthropogenic alteration of wetlands and ponds, and that identified the taxonomic richness as a good community variable to evaluate their ecological integrity (Biggs et al., 2000; Apfelbeck, 2001; Hicks & Nedeau, 2000).

In this study, almost all the evaluated taxonomic richness attributes of macroinvertebrate communities significantly correlated with variables reflecting human disturbance. According to previous studies (Biggs et al., 2000; Menetrey et al., 2005; Solimini et al., 2008) , the total number of taxa and family, number of Odonata, number of Ephemeroptera, Odonata, Trichoptera, and number of Coleoptera significantly showed correlation with pond impairment. Pond conservation status (ECELS Index) had the strongest correlation with total macroinvertebrate richness metric which, in turn, highly responded also to variables associated with the surrounding land use. Among the other tested metrics, the number of Odonata and number of Ephemeroptera, Odonata and Trichoptera were also strongly correlated with pond conservation status level and the percentage of surrounding woodland and natural vegetation. The response of evaluated qualitative metrics obtained at family and a lower taxonomical level were not very different, except for the case of Coleoptera. In our study their richness correlated with variables linked with anthropogenic pressures only when they were identified to lower level than family.

Taxa composition metrics tended to show only slightly less significance than most of taxa richness-based metrics. The proportion of the three most abundant taxa (percent top three dominant taxa) in the studied communities was negatively correlated with pond conservation status and habitat condition variables, in agreement with previous studies which found an increase of dominance by few groups along the anthropogenic disturbance gradient (Hicks & Nedeau, 2000; Apfelbeck, 2001). Surprisingly, in our study we did not find, as expected, a correlation between relative abundance of water boatmen (Corixidae) and site degradation variables. This hemipteran group of herbivores and detritivores are tolerant to low dissolved oxygen in water and tend to increase with eutrophication impact (Hicks & Nedeau, 2000; Solimini et al., 2008), whereas the degraded ponds in our study were not clearly characterised by deoxygenated and eutrophic waters. On the other hand, percentage of Pleidae correlated with land use categories, riparian zone width and conservation status of ponds, where this group was represented exclusively by the pigmy back swimmer *Plea minutissima*. This predator and carnivorous species feed on small crustaceans such as *Cyclops*, Ostracoda, small *Gammarus*, and also larvae of aquatic insects such as Ephemeroptera, Chironomidae, and mosquitoes (Vafaei, 2004). Usually, higher proportion of predator bugs occurred in higher-quality wetlands (Helgen & Gernes, 2001). Also the percentage of another group of predators, the adults of Hydracarina, showed a positive correlation with the percentage of surface occupied by woodland and natural vegetation around ponds. This group is particularly sensitive to pollution and contaminant and was not usually found in ecosystems altered for water abstraction (Gerecke & Lehmann, 2005; Cantonati et al., 2006). Finally, we found an expected decrease of total macroinvertebrate density along with study site degradation. This community attribute showed a highly positive correlation with conservation status, percentage woodland and natural vegetation and riparian buffer width. Most often, organism density provided variable response to impact and is rarely used as considerable a metric for data analysis, but could be decreased with loss of habitat, siltation, and toxic substance presence (Hicks & Nedeau, 2000; Apfelbeck, 2001).

Our results suggested that the methodologies we developed for the analysis of freshwater benthic Diatom and macroinvertebrate communities may have a considerable potential as a tool for assessing the ecological status of this type of water body, complying with the WFD 2000/60/EC. Diatoms tended more to reflect water chemistry through changes in community structure, whereas invertebrates responded to physical habitat changes primarily through changes in taxonomic richness. Moreover, further applied studies, in particular on diatom component in the Mediterranean eco-region are necessary, especially for lowland freshwater ponds.

Acknowledgements The study was supported by a fellowship from Accademia Nazionale dei Lincei to V.D.B. We wish to thank E. Alleva for suggestions and continuing support. We thank J. Sala, S. Gascón and D. Boix for their help in the application of ECELS index. We are grateful to Luc Ector and to the instructors of the 1st Workshop 'Ecology & taxonomy of benthic diatoms in oligotrophic freshwater habitats' (Lake Tovel Limnological Station of the Trentino Nature Science Museum, 16–20 June 2008), in particular to H. Lange-Bertalot, R. Pienitz, B. Van de Vijver and K. Buczkò for their precious help in the taxonomic determination of some diatom species. We also thank, for their help in the field work, F. Russo, S. Ciadamidaro and C. Marinilli. We are also grateful to C. Rondinini for critical reading of the earlier version of the manuscript. Finally, we wish to thank the Presidential Estate of Castelporziano, Maremma Regional Park, Litorale Romano Natural State Reserve, LIPU Oasi Castel di Guido, Wood of Foglino, and landowners for the help in the identification of ponds and for granting us permission to conduct our research within their natural reserves and private properties. First version of this article was improved based on the suggestions of B. Oertli and two anonymous reviewers.

References

Alba-Tercedor, J. & A. Sánchez-Ortega, 1988. Un metodo rapido y simple para evaluar le calidad biologicas de las aguas corrientes basado en el de Hellawell (1978). Limnetica 4: 51–56.

APAT, 2008a. Protocollo di campionamento e analisi delle diatomee bentoniche dei corsi d'acqua. In APAT (ed.). Metodi Biologici per le acque dolci. Parte I [available on line http://www.apat.gov.it/site/it-IT/APAT/Pubblicazioni/Altre_Pubblicazioni.html].

APAT, 2008b. Protocollo di campionamento dei macroinvertebrati bentonici dei corsi d'acqua guadabili. In APAT (ed.). Metodi Biologici per le acque dolci. Parte I [available on line http://www.apat.gov.it/site/it-IT/APAT/Pubblicazioni/Altre_Pubblicazioni.html].

Apfelbeck, R. S., 2001. Development of biocriteria for wetland in Montana. In Rader, R. B., D. P. Batzer & S. A. Wissinger (eds), Bioassessment and Management of North American Freshwater Wetlands. Wiley, New York.

APHA (American Public Health Association), 1998. Standard methods for Examination of water and waste water. American Public Health Association, Washington, DC.

Armitage, P. D., D. Moss, J. F. Wright & M. T. Furse, 1983. The performance of a new biological water quality score system based on macroinvertebrates over a wide range of unpolluted running-water sites. Water Research 17: 333–347.

Bazzanti, M., S. Baldoni & M. Seminara, 1996. Invertebrate macrofauna of a temporary pond in Central Italy: composition, community parameters and temporal succession. Archiv für Hydrobiologie 137: 77–94.

Biggs J, Williams P, Whitfield M, Fox G, Nicolet P. 2000. Biological techniques of still water quality assessment. Phase 3. Method development. R&D Technical Report E110, Environment Agency, Bristol.

Biggs, J., P. Williams, M. Whitfield, P. Nicolet & A. Weatherby, 2005. 15 years of pond assessment in Britain: results and lessons learned from the work of Pond Conservation. Aquatic Conservation: Marine and Freshwater Ecosystems 15: 693–714.

Blanco, S., L. Ector & E. Bécares, 2004. Epiphytic diatoms as water quality indicators in Spanish shallow lakes. Vie Milieu 54: 71–79.

Cantonati, M., E. Bertuzzi, R. Gerecke, K. Ortler & D. Spitale, 2005. Long-term ecological research in springs of the Italian Alps: six years of standardised sampling. Verhangen International Verein Limnology 29: 907–911.

Cantonati, M., R. Gerecke & E. Bertuzzi, 2006. Spring of the Alps—sensitive ecosystems to environmental change: from biodiversity assessments to long-term studies. Hydrobiologia 562: 59–96.

Carrino-Kyker, S. R. & A. K. Swanson, 2007. Seasonal physicochemical characteristics of thirty northern Ohio temporary pools along gradients of GIS-delineated human land-use. Wetlands 27: 749–760.

CEC, 2000. Council of European Communities Directive 2000/60/EEC of 23 October 2000 establishing a framework for community action in the field of water policy. Official Journal of European Communities, L327/1.

CEC, 2005. Common Implementation Strategy for the Water Framework Directive (2000/60/EC). Wetlans Horizontal Guidance. Guidance Document N12. The role of wetlands in the Water Framework Directive. Official Journal of European Communities, Luxembourg.

Chao, A., 1984. Nonparametric estimation of the number of classes in a population. Scandinavian Journal of Statistics 11: 265–270.

Chessman, B. C., K. A. Fryirsb & G. J. Brierley, 2006. Linking geomorphic character, behaviour and condition to fluvial biodiversity: implications for river management. Aquatic Conservation: Marine and Freshwater Ecosystems 16: 267–288.

Chipps, S. T., D. E. Hubbard, K. B. Werlin, N. J. Haugerud, K. A. Powell & J. Thompson, 2006. Association between wetland disturbance and biological attributes in floodplain wetlands. Wetlands 26: 497–508.

Cohen, A. S., R. Bills, C. Z. Cocquyt & A. G. Caljon, 1993. The impact of sediment pollution on biodiversity in Lake Tanganyika. Conservation Biology 7: 667–677.

Connell, J. H., 1978. Diversity in tropical rain forests and coral reefs. Science 199: 1302–1310.

Della Bella, V., M. Bazzanti & F. Chiarotti, 2005. Macroinvertebrate diversity and conservation status of Mediterranean ponds in Italy: water permanence and mesohabitat influence. Aquatic Conservation: Marine and Freshwater Ecosystems 15: 583–600.

Della Bella, V., C. Puccinelli, S. Marcheggiani & L. Mancini, 2007. Benthic diatom communities and their relationship to water variables in wetlands of central Italy. Annales de Limnologie/International Journal of Limnology 43(2): 89–99.

Della Bella, V., M. Bazzanti, M. G. Dowgiallo & M. Iberite, 2008. Macrophyte diversity and physico-chemical characteristics of Tyrrhenian coast ponds in central Italy: implications for conservation. Hydrobiologia 597: 85–95.

Dell'Uomo, A., 2004. L'Indice Diatomico di Eutrofizzazione/Polluzione (EPI-D) nel monitoraggio delle acque correnti Linee Guida. APAT, CTN AIM, Roma.

DeNicola, D. M. & E. Eyto, 2004. Using epilithic algal communities to assess trophic status in Irish lakes. Journal of Phycology 40: 481–495.

De Pauw, N. & G. Vanhooren, 1983. Method for biological quality assessment of watercourses in Belgium. Hydrobiologia 100: 153–168.

EN 13946, 2003. Water quality. Guidance standard for the routine sampling and pre-treatment of benthic diatom samples from rivers. European Committee for Standardization, Brussels.

EN 14407, 2004. Water quality. Guidance standard for the identification, enumeration and interpretation of benthic diatom samples from running waters. European Committee for Standardization, Brussels.

EPCN, 2008. The Pond Manifesto. European Pond Conservation Network [available at www.europeanponds.org].

Feio, M. J., S. F. P. Almeida, S. C. Craveiro & A. J. Calado, 2007. Diatoms and macroinvertebrates provide consistent and complementary information on environmental quality. Fundamental and Applied Limnology/Archiv für Hydrobiologie 169: 247–258.

Gaiser, E., A. Wachnicka, P. Ruiz, F. Tobias & M. Ross, 2005. Diatom indicators of ecosystem change in subtropical coastal wetlands. In Bortone, S. A. (ed.), Estuarine Indicators. CRC Press, Boca Raton, FL: 127–144.

Gell, P., I. R. Sluiter & J. Fluin, 2002. Seasonal and interannual variations in diatom assemblages in Murray River connected wetlands in north-west Victoria, Australia. Marine & Freshwater Research 53: 981–992.

Gerecke, R. & E. O. Lehmann, 2005. Towards a long term monitoring of Central European water mite faunas (Acari: Hydracnidia and Halacaridae)—considerations on the background of data from 1900 to 2000. Limnologica 35: 45–51.

Ghetti, P.F., 1997. Indice Biotico Esteso (IBE). I macroinvertebrati nel controllo della qualità degli ambienti di acque correnti. Provincia Autonoma di Trento.

Helgen, J. C. & M. Gernes, 2001. Monitoring the condition of wetlands: indexes of biological integrity using invertebrates and vegetation. In Rader, R. B., D. P. Batzer & S. A. Wissinger (eds), Bioassessment and Management of North American Freshwater Wetlands. Wiley, New York: 167–185.

Hellawell, J. M., 1986. Biological Indicators of Freshwater Pollution and environmental management. Elsevier, New York.

Hering, D., O. Moog, L. Sandin & P. F. M. Verdonschot, 2004. Overview and application of the AQEM assessment system. Hydrobiologia 516: 1–20.

Hicks, A. L. & E. J. Nedeau, 2000. New England freshwater wetlands invertebrate biomonitoring protocol (NEFWIBP). A manual for volunteers. University of Massachusetts, Executive Office of Environmental Affairs, Massachusetts.

Hill, B. H., A. T. Herlihy, P. R. Kaufmann, R. J. Stevenson, F. H. McCornick & C. B. Johnson, 2000. Use of periphyton assemblage data as an index of biotic integrity. Journal of the North American Benthological Society 19: 50–67.

IRSA, 1994. Metodi Analitici per le Acque. Istituto Poligrafico e Zecca dello Stato, CNR, Rome.

Kelly, M. G., A. Cazaubon, E. Coring, A. Dell'Uomo, L. Ector, B. Goldsmith, H. Guasch, J. Hurlimann, A. Jarlman, B. Kawecka, J. Kwandrans, R. Laugaste, E. A. Lindstrom, M. Leitao, P. Marvan, J. Padisàk, E. Pipp, J. Prygiel, E. Rott, S. Sabater, H. van Dam & J. Vizinet, 1998. Recommendations for the routine sampling of diatoms for water quality assessments in Europe. Journal of Applied Phycology 10: 215–224.

King, L., G. Clarke, H. Bennion, M. Kelly & M. Yallop, 2006. Recommendations for sampling littoral diatoms in lakes for ecological status assessments. Journal of Applied Phycology 18: 15–25.

Krammer, K., 2000. The genus Pinnularia. In Lange-Bertalot, H. (ed.), Diatoms of European Inland Waters and Comparable Habitats, Vol. 1. A.R.G. Gantner Verlag, Germany.

Krammer, K. & H. Lange-Bertalot, 1986. Bacillariophyceae. Teil: Naviculaceae; Bacillariaceae, Epithemiaceae, Surirellaceae; Centrales, Fragilariaceae, Eunotiaceae; Achnathaceae. Kritische Erganzungen zu Navicula und Gomphonema; and french translation of the keys. Süβwasserflora von Mitteleuropa, 2/1, Fischer, Stuttgart.

Krammer, K. & H. Lange-Bertalot, 1988. Bacillariophyceae. Teil: Naviculaceae; Bacillariaceae, Epithemiaceae, Surirellaceae; Centrales, Fragilariaceae, Eunotiaceae; Achnathaceae. Kritische Erganzungen zu Navicula und Gomphonema; and french translation of the keys. Süβwasserflora von Mitteleuropa, 2/2, Fischer, Stuttgart.

Krammer, K. & H. Lange-Bertalot, 1991a. Bacillariophyceae. Teil: Naviculaceae; Bacillariaceae, Epithemiaceae, Surirellaceae; Centrales, Fragilariaceae, Eunotiaceae; Achnathaceae. Kritische Erganzungen zu Navicula und Gomphonema; and french translation of the keys. Süβwasserflora von Mitteleuropa, 2/3, Fischer, Stuttgart.

Krammer, K. & H. Lange-Bertalot, 1991b. Bacillariophyceae. Teil: Naviculaceae; Bacillariaceae, Epithemiaceae, Surirellaceae; Centrales, Fragilariaceae, Eunotiaceae; Achnathaceae. Kritische Erganzungen zu Navicula und Gomphonema; and french translation of the keys. Süβwasserflora von Mitteleuropa, 2/4, Fischer, Stuttgart.

Krammer, K. & H. Lange-Bertalot, 2000. Bacillariophyceae. Teil: Naviculaceae; Bacillariaceae, Epithemiaceae, Surirellaceae; Centrales, Fragilariaceae, Eunotiaceae; Achnathaceae. Kritische Erganzungen zu Navicula und Gomphonema; and french translation of the keys. Süβwasserflora von Mitteleuropa, 2/5, Fischer, Stuttgart.

Lange-Bertalot, H., 2001. Navicula sensu stricto: 10 genera separated from Navicula sensu stricto, Frustulia. In Lange-Bertalot, H. (ed.), Diatoms of Europe: Diatoms of the European Inland Waters and Comparable Habitats. A.R.G. Gantner Verlag, Germany.

Mackey, R. L. & D. J. Currie, 2001. The diversity-disturbance relationship: is it generally strong and peaked? Ecology 82: 3479–3492.

Mancini, L., 2005. Organization of biological monitoring in the European Union. In Ziglio, G., M. Siligardi & G. Flaim (eds), Biological Monitoring of Rivers. Wiley, London: 171–201.

Mancini, L. & G. Arcà (eds), 2000. Carta della qualità biologica dei corsi d'acqua della Regione Lazio. Regione Lazio, Istituto Superiore di Sanità, Roma.

Menetrey, N., L. Sager, B. Oertli & J. B. Lachavanne, 2005. Looking for metrics to assess the trophic state of ponds. Macroinvertebrates and amphibians. Aquatic Conservation: Marine and Freshwater Ecosystems 15: 653–664.

Nicolet, P., J. Biggs, G. Fox, M. J. Hodson, C. Reynolds, M. Withfield & P. Williams, 2004. The wetland plant and macroinvertebrate assemblages of temporary ponds in England and Wales. Biological Conservation 120: 265–282.

Oertli, B., D. Auderset Joye, E. Castella, R. Juge, A. Lehmann & J. B. Lachavanne, 2005. POLCH: a standirdized method for sampling and assessing the biodiversity ini ponds. Aquatic Conservation: Marine and Freshwater Ecosystems 15: 665–679.

Prygiel, J. & M. Coste, 2000. Guide metodologique pour la mise en oeuvre de l'Indice Biologique Diatomées, NFT 90-354.

Prygiel, J., M. Coste & J. Bukowska, 1999. Review of the major diatom-based techniques for the quality assessment of rivers. State of art in Europe. In Prygiel, J., B. A. Whitton & J. Bukowska (eds), Use of Algae for Monitoring Rivers, III. Proceeding of International Symposium, Douai, France 29 September-1 October 1997. Agence de l'Eau Artois-Picardie: 122–127.

Reynoldson, T. B., R. H. Norris, V. H. Resh, K. E. Day & D. M. Rosenberg, 1997. The reference condition: a comparison of multimetric and multivariate approaches to assess water quality impairment using benthic macroinvertebrates. Journal of the North American Benthological Society 16: 833–852.

Rossaro, B., L. Marziali, A. C. Cardoso, A. Solimini, G. Free & R. Giacchini, 2007. A biotic index using benthic macroinvertebrates for Italian lakes. Ecological Indicators 7: 412–429.

Sala, J., S. Gascòn, D. Boix, J. Gesti & X. D. Quintana, 2004. Proposal of a rapid methodology to assess the conservation status of Mediterranean wetlands and its application in Catalunya. Archive des Sciences 57: 141–152.

Solimini, A. G., M. Bazzanti, A. Ruggiero & G. Carchini, 2008. Developing a multimetric index of ecological integrity based on macroinvertebrates of mountain ponds in central Italy. Hydrobiologia 597: 109–123.

Stevenson, R. J., 1984. Epilithic and epipelic diatoms in the Sandusky River, with emphasis on species diversity and water pollution. Hydrobiologia 114: 161–175.

Stewart, P. M., J. T. Butcher & T. O. Swinford, 2000. Land use, habitat, and water quality effects on macroinvertebrate communities in three watersheds of a Lake Michigan associated marsh system. Aquatic Ecosystem Health and Management 3: 179–189.

Tangen, B. A., M. G. Butlerv & M. J. Ell, 2003. Weak correspondence between macroinvertebrate assemblages and land use in prairie pothole region wetlands, USA. Wetlands 23: 104–115.

Triest, L., P. Kaur, S. Heylen & N. De Pauw, 2001. Comparative monitoring of diatoms, macroinvertebrate and macrophytes in the Woluwe River (Brussels, Belgium). Aquatic Ecology 35: 183–194.

Trigal, C., F. Garcıa-Criado & C. Fernández-Aláez, 2006. Among-habitat and temporal variability of selected macroinvertebrate based metrics in a Mediterranean shallow lake (NW Spain). Hydrobiologia 563: 371–384.

Trigal, C., F. Garcıa-Criado & C. Fernández-Aláez, 2009. Towards a multimetric index for ecological assessment of Mediterranean flatland ponds: the use of macroinvertebrates as bioindicators. Hydrobiologia 618: 109–123.

Vafaei, R., 2004. Biology of Plea minutissima found in the central provinces of Iran. NABS Annual meeting, Vancouver, British Columbia, June 2004.

van Dam, H., 1982. On the use of measures of structure and diversity in applied diatom ecology. Nowa Hedwigia 73: 97–115.

Verneaux, V., J. Verneaux, A. Schmitt, C. Lovy & J. C. Lambert, 2004. The Lake Biotic Index (LBI): an applied method for assessing the biological quality of lakes using macrobenthos; the Lake Châlain (French Jura) as an example. Annales de Limnologie/International Journal of Limnology 40: 1–9.

Wang, Y.-K., R. J. Stevenson, P. R. Sweet & J. DiFranco, 2006. Developing and testing diatom indicators for wetlands in the Casco Bay watershed, Maine, USA. Hydrobiologia 561: 191–206.

Watchorn, M. A., P. B. Hamilton, T. W. Anderson, H. M. Roe & R. T. Patterson, 2008. Diatoms and pollen as indicators of water quality and land-use change: a case study from the Oak Ridges Moraine, Southern Ontario, Canada. Journal of Paleolimnology 39: 491–509.

Wiederholm, T., 1980. Use of benthos in lake monitoring. Journal of the Water Pollution Control Federation 52: 537–547.

Williams, P., M. Whitfield, J. Biggs, S. Bray, G. Fox, P. Nicolet & D. Sear, 2004. Comparative biodiversity of rivers, streams, ditches and ponds in an agricultural landscape in Southern England. Biological Conservation 115: 329–341.

Wood, R. J., M. T. Greenwood & M. D. Agnew, 2003. Pond biodiversity and habitat loss in the UK. Area 35: 206–216.

Hydrobiologia (2009) 634:43–63
DOI 10.1007/s10750-009-9899-1

POND CONSERVATION

The M-NIP: a macrophyte-based Nutrient Index for Ponds

Lionel Sager · Jean-Bernard Lachavanne

Published online: 13 August 2009

Abstract In Swiss ponds, eutrophication represents one of the major threats to biodiversity. A biological method to assess the trophic state would, therefore, be particularly useful for monitoring purposes. Macrophytes have already been successfully used to evaluate the trophic state of rivers and lakes. Considering their colonizing abilities and their roles in pond ecosystem structure and function, macrophytes should be included in any assessment methods as required by the European Water Framework Directive. Vegetation survey and water quality data for 114 permanent ponds throughout Switzerland were analysed to define indicator values for 113 species including 47 with well-defined ecological response to total water phosphorus (TP). Using indicator values and species cover, a Macrophyte Nutrient Index for Ponds (M-NIP) was calculated for each site and assessed with both the original pond data set and a limited validation data set. The resulting index performed better when considering only species with narrow responses to TP gradient and was more applicable, but less accurate when including all species. Despite these limitations, the M-NIP is a valuable and easy tool to assess and monitor the nutrient status of Swiss ponds and was shown to be robust and relatively sensitive to slight changes in phosphorus loading with a validation subset.

Keywords Aquatic vegetation · Bioassessment · Eutrophication · Water quality · EU Water Framework Directive

Guest editors: B. Oertli, R. Cereghino, A. Hull & R. Miracle
Pond Conservation: From Science to Practice. 3rd Conference of the European Pond Conservation Network, Valencia, Spain, 14–16 May 2008

L. Sager (✉) · J.-B. Lachavanne
Laboratory of Ecology and Aquatic Biology, University of Geneva, Ch. des Clochettes 18, 1206 Geneva, Switzerland
e-mail: lionel.sager@gmail.com

Introduction

The ecological assessment of surface water quality is one of the main environmental concerns in many countries. In the European Union (EU), the Member States have set a common standard with ambitious objectives—the Water Framework Directive (WFD)—which aims to achieve at least a "good" ecological and physico-chemical status for all surface water and ground water bodies by 2015 (Bundi et al., 2000; Communities, 2000; Irmer, 2000). Although the WFD aims to protect all inland surface waters, ponds are not specifically mentioned in the Directive and for most Member States a lower size of 50 ha has been applied for standing waters to be included in monitoring programs (Davies et al., 2008). However, ponds are now increasingly recognized as very significant components of ecological quality, notably in term of their contribution to local and regional biodiversity (Murphy, 2002, Oertli et al., 2002, 2005;

Williams et al., 2003) and, as such, require adapted tools and assessment methods supported by robust scientific knowledge (EPCN, 2007).

Among the threats to surface water, eutrophication, notably through diffuse pollution linked to the intensification of the agriculture (Havens et al., 2001), is still an important and even growing problem for freshwaters and coastal oceans (Carpenter et al., 1999; Smith et al., 1999; Bronmark & Hansson, 2002; Vadeboncoeur et al., 2003; Craft et al. 2007). With habitat destruction, eutrophication represents one of the major threats to the sustainability of biodiversity of most freshwater ecosystems; therefore, an assessment method of the nutrient status based on bioindicators and specifically designed for ponds could be a valuable tool.

The macrophyte community is one of the target groups required by the WFD in the assessment methods for lakes. In shallow systems like ponds, this group should also be included in any assessment method as it has an important potential of colonization and plays an important role in the structure and function of the freshwater ecosystem (Adams & Sand-Jensen, 1991). Moreover, macrophytes have already been widely used and are effective in the assessment and monitoring of various kinds of freshwaters ecosystems (see for e.g., Seddon, 1972; Kohler, 1975; Lehmann & Lachavanne, 1999; Melzer, 1999; Schneider & Melzer, 2003; Meilinger et al., 2005; Stelzer et al., 2005; Clayton & Edwards, 2006; Haury et al., 2006). Additionally, several trophic indexes based on macrophytes and the trophic profile of species already exists for lakes and rivers (e.g., Landolt, 1977; Melzer, 1988; Bornette et al., 1994; Robach et al., 1996; Holmes et al., 1998; Willby et al., 2000) and can serve as a basis for comparison with a pond index. Other advantages of macrophytes as bioindicator groups are the large number of taxa occurring in ponds as well as the relatively low-time investment in data acquisition, which would be particularly valuable for large scale programs (Palmer et al., 1992).

In order to build a Macrophytes-based Nutrient Index for Ponds (M-NIP), it is necessary to characterize the trophic state of the sites used to define the ecological profile of species. Eutrophication is primarily described as a regular increase of the primary productivity following larger inputs of inorganic nutrients (Naumann, 1927, 1932). Dodds (2006) gives a more general definition of the eutrophication process as an increase in nutritive factors leading to higher rates of whole system metabolism considering both the heterotrophic and the autotrophic metabolism. Independent of these definitions, the increase in nutrient concentrations enhances algal productivity and reduces light penetration in the water column, and hence the depth of colonization by submerged macrophytes, which can completely disappear with over enrichment (Phillips et al., 1978; Balls et al., 1989). The continuity of the eutrophication process complicates the establishment of well-defined limits between distinct trophic states, as well as the assignment of biological indicators values to a particular trophic state (Sondergaard et al., 2005). As a continuous measure of the whole system, metabolism is time and resource consuming and is hardly possible to perform on a large scale. For this reason, the water concentration of the main nutrients has often been used as a surrogate to define the trophic state of freshwater ecosystems (Vollenweider & Kerekes, 1982). This surrogate approach was shown to be conclusive, at least in low altitude ponds, as the water nutrient concentration significantly predicted the net periphytic primary productivity measured in nine ponds included in the present study (Sager, 2009).

By using a data set including water physicochemistry and standardized macrophytes data for 114 ponds located throughout Switzerland [including 80 from the previous PLOCH study of Oertli et al. (2005)], the present study aims to:

- Characterize a nutrient profile of each macrophyte species using water chemistry data for ponds in which the species occurs. This profile represents the range of nutrient concentration expressed in trophic categories, where the species was recorded even if nutrient concentration is only one of the factors likely to contribute to occurrence and abundance of particular plant species. Therefore, this nutrient profile is only designed to be used for an assessment at the scale of the whole site and not to define the micro-conditions at the level of a single plant stem or macrophyte bed.
- Develop and calculate different versions of the nutrient index for a site (M-NIP) based on the nutrient profile, tolerance, and abundance of the species.

- Assess the ability of the different versions of M-NIP to correctly classify ponds in the corresponding category of nutrient status expressed in trophic state. The time investment and prerequisite knowledge of the sampler were also considered in the assessment of the different versions of the index.

In addition, an evaluation of the applicability and reproducibility of the index on newly sampled sites is presented on a subset of ponds that were not used to characterize the nutrient profile of the species.

Methods

Study area and field survey

The study area was located in Switzerland, a country of 41,244 km^2 located in central Europe, a large proportion of which incorporates the Alpine mountain chain. Despite its small size, Switzerland harbors an important variety of environmental conditions and a strong altitudinal gradient. We built a database containing the vegetation survey and environmental parameters of a set of 114 permanent ponds and small lakes located in four altitudinal belts of vegetation (see Fig. 1 and Table 1). All ponds were sampled in the summer between 1996 and 2005 (one sampling date per pond) with the standardized method developed by Oertli et al. (2005). The ponds in this data set varied in size from 6 to 96,200 m^2 (mean: 7,959 m^2, median: 2,328 m^2) and covered an altitudinal range from 210 to 2,757 m.a.s.l. (mean: 957 m, median: 642 m).

Macrophytes sampling

Vegetation sampling was carried out in square quadrats of 0.25 m^2 disposed equidistantly along transects perpendicular to the longest axis of the waterbody, and located at regular intervals according to its surface area. The total number of quadrats sampled per pond (n) was related to the water surface area (m^2) by a relationship determined by Oertli et al. (2005; $n = 1.96 - 2.8 * \log10(\text{area}) + 2.6 * (\log10(\text{area}))$ and ranged from 5 to 460 (mean = 65, median = 38). Such a strategy allows at least 70% of the real species richness to be recorded. A species list as well as water depth was drawn up for each quadrat, and this standardized list of species was completed by the observation of species located outside the quadrats. Only aquatic species were taken into account, especially the 254 species of vascular plants (Spermatophyta and Pteridophyta) listed in the highest humidity class ($F = 5$) of the Landolt (1977) index of ecological

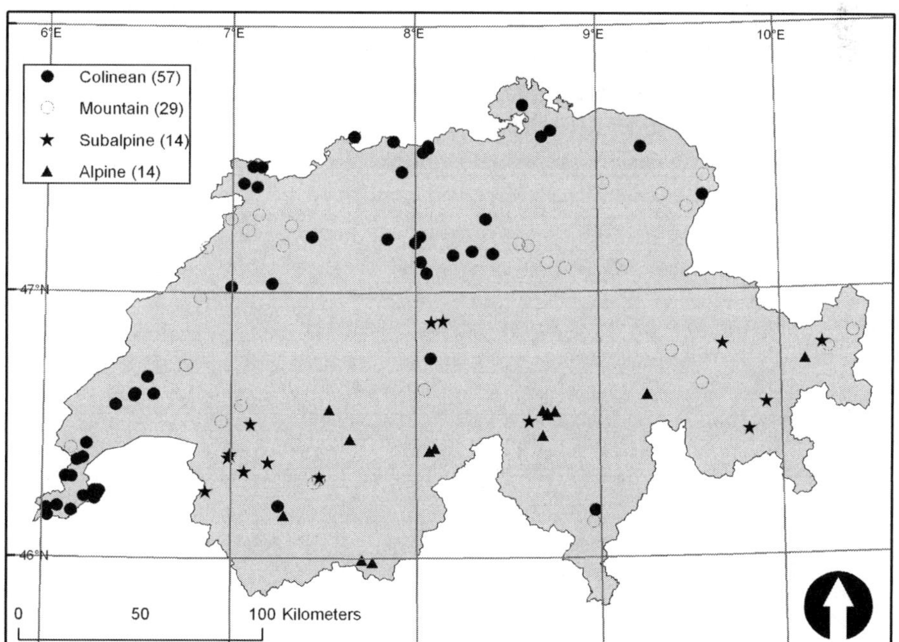

Fig. 1 Study area and locations of the 114 ponds throughout Switzerland. *Symbols* represent the four altitudinal vegetation belts with the number of sites in *brackets*

Table 1 Main morphometric and physico-chemical characteristics of the 114 ponds used to define the nutrient profiles of species and to calibrate the trophic index by site

	N	Mean	Mean std. error	Median	Minimum	Maximum
Mean depth (cm)	114	154	15	106	16	904
Sinuosity	114	1.52	0.04	1.37	0.99	3.27
Area (m^2)	114	7,959	1,320	2,328	6	96,200
Altitude (m)	114	957	59	642	210	2,757
TP (μg/l)	114	73.224	10.117	32.750	0	611
N_{min} (mg l^{-1})	114	1.060	0.135	0.551	0.036	8.790
Total hardness (méq $CaCO_3$)	112	180.2	13.1	182.5	0.8	884
Transparency *S* (cm)	113	41	2	47	4	60
Conductivity *W* (μS cm^{-1})	108	360.3	22.1	360.0	6.2	1,367

value. This standardized list was completed with 22 additional vascular species listed as $F = 4$ in the Landolt index: *Agrostis stolonifera* L., *Carex canescens* L., *Carex flava* L., *Carex lepidocarpa* Tausch, *Carex nigra* (L.) Reichard, *Eleocharis acicularis* (L.) Roem. & Schult., *Eleocharis quinqueflora* (Hartmann) O. Schwarz, *Equisetum palustre* L., *Galium palustre* L., *Juncus articulatus* L., *Juncus conglomeratus* L., *Juncus effusus* L., *Juncus filiformis* L., *Juncus inflexus* L., *Lysimachia nummularia* L., *Lythrum salicaria* L., *Lysimachia vulgaris* L., *Mentha longifolia* (L.) Huds., *Myosotis scorpioides* L., *Ranunculus repens* L., *Rorippa palustris* (L.) Besser, and *Scirpus sylvaticus* L. A plant quantity index *Q* was recorded for each species, and is the cube of the class of cover of the species, as explained in Table 2. This *Q* value is considered to be a good descriptor of the extent of a species actually present at a site (Schneider & Melzer, 2003). The size difference between sites does not seem to influence the distribution of species. The study of Oertli et al. (2002) conducted on 80 ponds included in our data set showed that among the 19 species of plants observed in >16% of the ponds, only two show a significant preference for large sites while the others are indifferent to the size.

Table 2 Correspondence between the percentage of the quadrats occupied by a species and the percentage cover classes also express as plant quantity index (*Q*) by cubing the class value

% Quadrats occupied	Covering classes	PQI (*Q*)
0–1	1	1
1–5	2	8
5–25	3	27
25–50	4	64
>50	5	125

Water physico-chemistry

Water physico-chemistry was measured in winter, when biological activity is at its minimum intensity and the concentration of nutrients in their inorganic form tends to be at its highest level (Linton & Goulder, 2000). Water sampling was carried out at the deeper central point of each pond by drilling a hole in the ice cover. Water samples were taken with a sampling bottle at 20 cm below the surface, and then immediately stocked in acid pre-rinse PE plastic bottles before being stored in the dark in a refrigerated box. Unfiltered samples were kept for total phosphorus (TP) analyses. TP was determined after potassium persulphate digestion at 121°C and under pressure for half an hour; soluble reactive phosphorus (SRP) was further measured by the ascorbate acid/molybdenum blue method (APHA et al., 1998). For this study, we only used the concentration of TP even if other physico-chemical parameters have been measured.

Classification of the ponds in trophic categories

For measuring the trophic state, water nutrient concentration was used as a surrogate for primary production, as this approach has been demonstrated to be effective by measuring the primary productivity in situ on a subset of nine of the ponds considered here (Sager, 2009). However, numerous studies have shown that the water column is not the single nutrient source for aquatic vegetation and rooted species can use the content of sediments and interstitial water for nutrient supply (Barko et al., 1991; Moore et al., 1994; Wigand et al., 1997; Vermeer et al., 2003; Engelhardt, 2006, review in Lacoul & Freedman, 2006), but the nutrient concentration of sediments or interstitial water can be highly variable within a waterbody (Wigand et al., 1997). As, we want to obtain a general value by site, the water column concentration of nutrients is, practically, more suitable than multiple measurements of the sediments. In effect, water column concentrations are generally more homogeneous at the pond scale through facilitated diffusion. For these reasons, even if the water concentration is not a measure of all available nutrients for plant growth, we consider it as representative of the actual conditions prevailing in the water body, and reliable for setting the nutrient profiles of macrophytes species usable at the site scale.

From the water concentrations of TP measured in winter, each pond was classified into a trophic state (oligotrophic, mesotrophic, eutrophic, and hypertrophic). As the most of the ponds were only surveyed once for winter physico-chemistry of the water, we choose to smooth the little variations between sites by using classes of trophic states defined by TP concentrations rather than the raw values of TP. Two different chemical scales for defining the trophic status were evaluated separately: the OECD criteria for defining the trophic state of lakes (Vollenweider & Kerekes, 1982) and a scale of ecological quality specifically defined for shallow lakes (Sondergaard et al., 2005). The latter sets limits between the first three trophic categories at a higher TP concentration than the OECD scale, and could be potentially more adapted to ponds. In addition, the shallow lake scale distinguishes a fifth category for bad status, when the TP concentration goes beyond 200 μg/l. The ranges of nutrient concentrations for these two scales are given in Table 3 along with the number of sites in each category.

Attribution of bioindication to species

In order to identify an indicator value (IV) for each species, we followed a procedure similar to that used by Schneider & Melzer (2003) in rivers. We set up a histogram for every species that showed its distribution by nutrient categories. Similarly to Schneider & Melzer (2003) and Friedrich (1990), a 20 points distribution was used to allow direct comparisons between species. Species present in less than three ponds were systematically excluded and those which occurred only in 3–6 ponds were carefully examined. An IV was calculated for the species with enough occurrences using weighted averaging (Eq. 1).

$$\mathrm{IV}_a = \frac{\sum_{i=1}^{n} O_{ai} \cdot T_i}{\sum_{i=1}^{n} O_{ai}} \qquad (1)$$

where IV_a is the indicator value of species a, O_{ai} is the number of occurrences of species a in the trophic category i, and T_i is the value of the trophic category i (from 1 = oligotrophic to 4 = hypertrophic).

One indicator value (IV) by species was calculated for each considered chemical scale of trophy, namely, IV-P and IV-Ps for the OECD scale and the shallow lakes. These IV by species are neither to design for a single use nor to define micro-conditions at the scale of a single stem. Instead, they were used on survey results fulfilling all the requirements enumerated thereafter to compute a reliable M-NIP at the pond scale.

Table 3 Ranges of water concentration for TP by trophic state for the scales used in this study

Trophic scale Classification	Oligotrophic High	Mesotrophic Good	Eutrophic Moderate	Hypertrophic Poor	Bad	*n*
OECD TP (μg/l)	0–10	10–35	35–100	>100		
n	24	35	32	23		114
Shallow lakes TP (μg/l)	0–25	25–50	50–100	100–200	>200	
n	49	22	17	15	11	114

n indicates the number of sites in the data set for each trophic category. The scales used are those proposed by OECD (Vollenweider & Kerekes, 1982) and Sondergaard et al. (2005) for shallow lakes

Table 4 Correspondence between the range of ecological amplitude and the weighting factors attributed to indicator species

a_a	W
0–0.2	16
0.2–0.4	8
0.4–0.6	4
0.6–0.8	2
>0.8	1

In order to express the ecological tolerance or amplitude of a species to a given factor, we calculated the root-mean-square-deviation weighted by the number of occurrences in each nutrient category (Eq. 2). This amplitude of tolerance permitted to weight the contributions of the species to the index by giving higher influence to the species with a narrow spectrum. The correspondence between ranges of a_a and weighting factors (W) are given in Table 4.

$$a_a = \sqrt{\frac{\sum_{i=1}^{n} (T_i - IV_a)^2 \cdot O_{ai}}{\sum_{i=1}^{n} O_{ai}}} \tag{2}$$

where $a_a$is the amplitude of species a, T_i is the value of the nutrient category i (from 1 = oligotrophic to 4 = hypertrophic), IV_a is the indicator value of species a, and O_{ai} is the number of occurrences of species a in the trophic category i.

Determination of the M-NIP and assessment of its accuracy

The M-NIP equation (Eq. 3) corresponded to one of the macrophyte trophic index (TIM) of Schneider & Melzer (2003), which was also the term used in the saprobic index of Zelinka & Marvan (1961).

$$\text{M-NIP} = \frac{\sum_{i=1}^{n} \text{IV}_a \cdot W_a \cdot Q_a}{\sum_{i=1}^{n} W_a \cdot Q_a} \tag{3}$$

where M-NIP is Macrophytes Nutrient Index for Ponds, IV_a is the indicator value of species a, W_a is the weighting factor of species a, and Q_a is the plant quantity of species a in the pond.

Depending on the amplitude of tolerance of a species expressed the weighting factors (W), two distinct types of M-NIP were calculated: one considering all species (M-NIP) and the other considering only species with a weight above one (M-NIP $-$ $W > 1$).

This calculation gave an M-NIP value for a given pond that could theoretically range from one to four. The subdivision in classes of trophic state was made further.

In order to assess the accuracy of the M-NIP for a given site, the weighted standard deviation of the indicator values of species present in the pond was calculated. If this rate of scatter (SC, Eq. 4) exceeded a fixed threshold, the computed M-NIP was not valid and could be used for the determination of the nutrient status. The thresholds were fixed as about half of the extent of the M-NIP values for the category with the narrower range. For this reason, the threshold for SC differed between the different indexes tested.

$$\text{sc} = \sqrt{\frac{\sum_{a=1}^{n} (\text{IV}_a - \text{M} - \text{NIP})^2 \cdot W_a \cdot Q_a}{(n-1)\sum_{a=1}^{n} W_a \cdot Q_a}} \tag{4}$$

Prerequisite for a consistent calculation of the M-NIP

In order to ensure that the M-NIP can be use as a reliable indicator of trophic state, the requirements are necessary as follows:

- Sampling of vegetation must have been made with the standardized method described above and during the main vegetation period (June–September), including the standardization of sampling intensity and the extrapolation of plant quantity (Q) using only species observed within quadrats.
- At least two indicator species must occur in the pond.
- The sum of the plant quantities Q must be at least 35 for the M-NIP's (one common and one infrequent species) and nine for the M-NIP $-$ $W > 1$'s (one rare and one infrequent species), since this index includes only species with narrower tolerance.
- The rate of SC must be inferior to the fixed threshold of confidence.

Several versions of the M-NIP, based on different subgroups of species, were assessed, these were:

(1) M-XNIP (with X for the nutrient used to classify the sites by trophic states) consider all the observed species with a valid IV.
(2) M-XNIP-WC consider only the species classified as aquatic or helophyte in the red list of fern and flowering plants of Switzerland (Moser

et al., 2002). This corresponds to the pool used for the M-XNIP without Characea (WC) and few others species not classified as aquatic or helophytes.

(3) M-XNIP-AQ consider only aquatic species. This version corresponds to the group used for the M-XNIP-WC but without the helophytes species.

(4) M-XNIP-SUB consider all the submerged species including the Characea.

This choice of testing multiple indexes was dictated by the need to find the most reliable group of species by considering the effectiveness of the index to correctly reclass ponds in the corresponding nutrient status. We also considered the ecological meaning of the subgroup of macrophytes and the easiness of applicability by end users in term of time investment and skills for species identification.

Defining the ranges of M-NIP values by trophic category

The valid M-NIP values for the ponds used to define the nutrient profile of the species and fulfilling all the pre-requisite conditions were box-plotted by trophic categories to define the maximum rate of SC for considering the index result as reliable. Thresholds of SC, corresponding approximately to half of the range of M-NIP values between trophic classes, were defined separately for each tested trophic scale and macrophytes groups.

After removing the sites with a too high SC to be reliable, M-NIP values by site were box-plotted versus trophic categories. Significance of the differences between classes was assessed by a non-parametric Mann–Whitney (MW) test between the groups of M-NIP values from adjacent trophic categories. When the M-NIP values were mostly overlapping between two contiguous categories, the index was considered unable to distinguish between these two nutrient status and the MW test was performed between groups of values from non-adjacent categories (e.g., oligotrophic and eutrophic).

Validation and test on the M-NIP index

In order to test the usability of the M-NIP, the indexes were calculated with data from surveys not included in the data set used to calculate the IV by species. As the calibration data set was just large enough to define ecological values for species, only five sites were set aside for this task. Among these, we included two sites previously incorporated in the building process but with new data obtained from other surveys performed between 1 and 10 years after the initial study. This validation step permitted us to assess if the calculated index corresponded to the nutrient status determined by water physico-chemistry. It also allowed us to estimate the stability of the index for two sites that were sampled in two subsequent years and to assess the monitoring ability of the index value over a longer time period for one pond. For this purpose, the M-PNIP values of resample sites were compared along with the species lists and physico-chemical data.

Results

Bioindication of the species

A total of 168 macrophytes species and 1,702 observations were recorded in 114 ponds used to define the indicator value (IV) by species. Among this set, 113 species representing 96.4% of the observations, had sufficient occurrences to compute an indicator value. These include 45 species that were at the lower limit of inclusion with only 3–6 occurrences in the present data set. The ranges of the IV by trophic scale were 1.2–3.67 for the IV-P and 1–4.33 for the IV-Ps.

Depending on their amplitude on the histogram of occurrences by trophic state, the IV's of the species were weighted to further calculate an index by site. A large part of the species fulfilling the conditions to compute a reliable IV can be classified as eurytrophe as they are able to grow on a wide range of nutrient concentrations. Depending on the chemical parameter and scale considered to define the trophic state, 66 (IV-P)–92 (IV-Ps) species showed a wide tolerance and an inherent minimum weighting factor in the index ($W = 1$). Consequently, IV of at least 21 species and at most 47 could be further used to compute the M-NIP $- W > 1$. The full list of species present in at least three sites is given in Table 5 with their corresponding IV and W for the two chemical scales used to define the nutrient status.

Table 5 Indicator values (IV) of the species and amplitude of tolerance expressed by the weighting factors (*W*) for the two chemical scales used to define the trophic state (see Table 3)

Species names	*n*	IV-P	*W*–P	IV-Ps	*W*-Ps	GF	EG
Acorus calamus **L.**	3	**3.67**	**4**	4.33	1	e	Aq
Agrostis stolonifera L.	9	1.89	1	1.89	1	s	Mar
Alisma lanceolatum With.	7	2.86	1	2.43	1	e	Aq
Alisma plantago-aquatica L.	34	2.82	1	2.38	1	e	Aq
Alnus glutinosa **(L.) Gaertn.**	8	**2.88**	**2**	2.63	1	–	For
Alopecurus aequalis Sobol.	3	2.33	1	1.67	4	e	Mar
Berula erecta **(Huds.) Coville**	6	**2**	**4**	1.17	8	e	Aq
Callitriche cophocarpa **Sendtn.**	3	**3.67**	**4**	4	1	fl	Aq
Callitriche palustris **L.**	4	**1.5**	**4**	1	16	fl	Aq
Callitriche stagnalis Scop.	3	3	1	2.67	1	fl	Aq
Caltha palustris L.	32	2.22	1	1.91	1	e	Mar
Cardamine amara L.	6	2	1	2.2	1	e	Mar
Carex acutiformis Ehrh.	34	2.65	1	2.24	1	e	Mar
Carex canescens L.	9	2.78	1	2.44	1	e	Mar
Carex diandra Schrank	3	3	1	2.33	1	e	Mar
Carex elata All.	45	2.78	1	2.36	1	e	Mar
Carex flava aggr.	3	2	1	1.67	1	e	Mar
Carex flava **L.**	18	**2.06**	**2**	1.39	1	e	Mar
Carex lepidocarpa Tausch	6	2	1	0.83	4	e	Mar
Carex limosa **L.**	3	**1.33**	**4**	1	16	e	Mar
Carex nigra (L.) Reichard	29	1.9	1	1.63	1	e	Mar
Carex paniculata L.	12	2.17	1	2	1	e	Mar
Carex pseudocyperus **L.**	6	**2.83**	**8**	2.17	1	s	Mar
Carex riparia **Curtis**	5	**3**	**16**	2.2	2	e	Mar
Carex rostrata Stokes	34	2.26	1	1.79	1	e	Aq
Carex vesicaria L.	25	2.88	1	2.68	1	e	Mar
Ceratophyllum demersum L.	4	2.5	1	2.5	1	s	Aq
Chara contraria **A. Braun**	3	**2.33**	**4**	1.33	4	s	Aq
Chara globularis **Thuillier**	18	**2.33**	**2**	1.83	1	s	Aq
Chara major **Vaillant**	3	**2**	**16**	0.67	8	s	Aq
Chara vulgaris **L.**	19	**2.05**	**2**	1.42	1	s	Aq
Eleocharis austriaca **Hayek**	8	**2.75**	**4**	2.13	4	e	Mar
Eleocharis palustris **(L.) Roem. & Schult.**	17	**2.71**	**2**	2.12	1	e	Mar
Eleocharis palustris **aggr.**	4	**3.5**	**4**	3.5	1	e	Mar
Eleocharis quinqueflora (Hartmann) O. Schwarz	3	1.67	1	1.33	4	e	Mar
Eleocharis uniglumis **(Link) Schult.**	7	**1.71**	**4**	0.86	16	e	Mar
Elodea canadensis Michx.	13	3.15	1	3.15	1	s	Aq
Epilobium palustre L.	10	2.3	1	2.2	1	e	Mar
Equisetum fluviatile L.	23	2.35	1	1.96	1	e	Aq
Equisetum palustre **L.**	26	**2.23**	**2**	1.58	1	e	Mar
Eriophorum angustifolium Honck.	13	1.69	1	1.27	1	e	Mar
Eriophorum scheuchzeri **Hoppe**	5	**1.2**	**8**	1	16	e	Mt
Galium palustre L.	34	2.74	1	2.38	1	e	Mar
Glyceria fluitans (L.) R. Br.	15	3.07	1	3.2	1	e	Aq

Table 5 continued

Species names	*n*	IV-P	*W*–P	IV-Ps	*W*-Ps	GF	EG
Glyceria maxima **(Hartm.) Holmb.**	5	**2.4**	**4**	2	1	e	Aq
Glyceria notata Chevall.	11	3.09	1	3.09	1	e	Aq
Groenlandia densa **(L.) Fourr.**	3	**1.33**	**4**	1	16	s	Aq
Hippuris vulgaris **L.**	5	**2.6**	**2**	2.4	2	e	Aq
Hydrocharis morsus-ranae **L.**	6	**2.67**	**2**	2.17	1	fl	Aq
Hydrocotyle vulgaris L.	3	2.33	1	1.33	1	e	Mar
Iris pseudacorus L.	43	2.84	1	2.56	1	e	Mar
Juncus articulatus L.	38	2.61	1	2.13	1	e	Mar
Juncus bulbosus **L.**	4	**2.25**	**4**	1.75	1	e	Mar
Juncus conglomeratus **L.**	21	**2.9**	**2**	2.57	1	e	Mar
Juncus effusus L.	40	2.9	1	2.58	1	e	Mar
Juncus filiformis L.	14	2.36	1	1.69	1	e	Mt
Juncus inflexus L.	26	2.46	1	1.96	1	e	Mar
Lemna minor L.	33	2.79	1	2.61	1	ff	Aq
Lemna trisulca L.	9	2.67	1	2.33	1	s	Aq
Lycopus europaeus **L. s.str.**	39	**3**	**2**	2.64	1	e	Mar
Lysimachia nummularia **L.**	18	**3.11**	**2**	3	1	e	For
Lysimachia vulgaris L.	41	2.85	1	2.56	1	e	Mar
Lythrum salicaria **L.**	43	**2.84**	**2**	2.49	1	e	Mar
Mentha aquatica L.	53	2.72	1	2.25	1	e	Mar
Mentha longifolia (L.) Huds.	10	2.5	1	2.2	1	e	Mar
Menyanthes trifoliata L.	14	2.64	1	2.43	1	e	Aq
Myosotis scorpioides **L.**	16	**2.69**	**2**	2.31	1	e	Mar
Myriophyllum spicatum L.	11	2.64	1	2.27	1	s	Aq
Myriophyllum verticillatum L.	4	2.75	1	2.5	1	s	Aq
Nasturtium officinale R. Br.	3	2.33	1	2.67	1	e	Aq
Nuphar lutea (L.) **Sm.**	9	**3.22**	**2**	3.11	1	fl	Aq
Nymphaea alba L.	22	2.91	1	2.64	1	fl	Aq
Nymphoides peltata (S. G. Gmel.) Kuntze	3	3.33	1	3.67	1	fl	Aq
Pedicularis palustris L.	5	2.4	1	1.8	1	e	Mar
Phalaris arundinacea **L.**	25	**2.72**	**2**	2.32	1	e	Mar
Phragmites australis (Cav.) Steud.	55	2.58	1	2.13	1	e	Aq
Poa palustris L.	9	2.44	1	2.44	1	e	Mar
Polygonum amphibium L.	14	3.07	1	2.93	1	ff	Mar
Potamogeton alpinus Balb.	12	2.33	1	2.25	1	fl	Aq
Potamogeton crispus L.	5	2.6	1	2.4	1	s	Aq
Potamogeton filiformis **Pers.**	3	**1.33**	**4**	1	16	s	Aq
Potamogeton gr pusillus	31	2.68	1	2.19	1	s	Aq
Potamogeton lucens **L.**	9	**2.56**	**4**	2	1	s	Aq
Potamogeton natans L.	30	2.2	1	1.83	1	fl	Aq
Potamogeton pectinatus **L.**	12	**2.17**	**2**	1.42	2	s	Aq
Potamogeton perfoliatus **L.**	3	**2**	**16**	0.67	8	s	Aq
Potamogeton pusillus L.	5	2.6	1	3	1	s	Aq
Potentilla palustris (L.) Scop.	9	2.56	1	1.89	1	e	Mar
Ranunculus flammula L.	9	2.67	1	2.11	1	e	Mar

Table 5 continued

Species names	*n*	IV-P	*W*–P	IV-Ps	*W*-Ps	GF	EG
***Ranunculus lingua* L.**	5	**2.6**	**2**	2.6	1	e	Aq
***Ranunculus repens* L.**	6	**2.83**	**2**	2.33	1	e	Rd
***Ranunculus trichophyllus* Chaix s.str.**	16	**2.19**	**2**	1.38	2	s	Aq
***Ranunculus trichophyllus* subsp. *eradicatus* (Laest.) C. D. K. Cook**	3	**1.33**	**4**	1	16	s	Aq
Rorippa palustris (L.) Besser	5	2.2	1	2	1	e	Mar
Salix cinerea L.	22	2.91	1	2.59	1	-	Mar
***Saxifraga stellaris* L.**	4	**1.25**	**4**	1	16	e	Mt
Schoenoplectus lacustris (L.) Palla	22	2.55	1	2.32	1	e	Aq
Schoenoplectus tabernaemontani (C. C. Gmel.) Palla	11	2.36	1	2.36	1	e	Aq
Scirpus sylvaticus L.	15	2.87	1	2.6	1	e	Mar
***Scutellaria galericulata* L.**	12	**3.17**	**2**	3	1	e	Mar
Sparganium angustifolium Michx.	6	1.67	1	1.33	1	fl	Aq
Sparganium erectum L. s.str.	14	2.86	1	2.93	1	e	Aq
***Sparganium erectum* subsp. *microcarpum* (Neuman) Domin**	5	**3**	**2**	2.6	1	e	Aq
***Spirodela polyrhiza* (L.) Schleid.**	5	**3.4**	**2**	4	1	ff	Aq
***Thelypteris palustris* Schott**	3	**3.33**	**4**	3.33	4	e	Mar
***Typha angustifolia* L.**	19	**2.74**	**2**	2.21	1	e	Aq
Typha latifolia L.	52	2.87	1	2.62	1	e	Aq
Utricularia australis R. Br.	16	2.88	1	2.81	1	s	Aq
***Utricularia minor* L.**	3	**2.67**	**4**	2	1	s	Aq
***Utricularia ochroleuca* R. W. Hartm.**	3	**2.67**	**4**	2	1	s	Aq
Veronica anagallis-aquatica L.	8	2.25	1	2	1	e	Aq
Veronica beccabunga L.	22	2.64	1	2.41	1	e	Aq
Veronica scutellata L.	6	3	1	3	1	e	Mar

In bold, species with an IV-P with $W > 1$. *n* number of occurrences within the data set. *GF* growth forms following Landolt (1977) completed for stoneworts and emerged species with *s* submerged, *fl* floating plants, *ff* free-floating and *e* emerged. *EG* ecological groups according to Moser et al. (2002) and completed for stonewort with *Aq* aquatic, *Mar* marsh, *Mt* mountain, *Rd* ruderal and *For* forest

M-NIP by site and SC thresholds

The M-NIP variants were computed for all the 114 sites used to define the nutrient profile of the species. None of the versions could be calculated for all the sites but overall the M-NIP version that incorporated the species with a low weight ($W = 1$) obviously fulfilled more often the conditions for a reliable index. On the other hand, the indexes considering subgroups of macrophytes could less often be calculated and particularly when only taking species with a weight above one into account.

According to the distribution of the M-NIP value by chemically defined trophic categories, the banding of index values by trophic states were defined separately for each index variant. For the index based on IV-Ps, the index values were overlapping between the trophic categories and mostly distributed over a small range of values leading to low SC thresholds. Therefore, this scale was considered as non-conclusive for a correct classification of the sites in trophic categories and not evaluated further.

The M-NIP based on the TP scale of OECD (M-PNIP) performed better than the one for shallow lakes and a clear pattern of correct classification appeared. The ranges of index values attributed to each nutrient status and the subsequent SC threshold are given in Table 6. After removing the sites with SC values

Table 6 Banding of the M-NIP values into trophic categories

M-NIP type	Oligotrophic 1	Mesotrophic 2	Eutrophic 3	Hypertrophic 4	SC thresholds	Ranges/differences
M-PNIP	1–2	2–2.5	2.5–2.9	2.9–4	0.2	M-NIP values
	1	0.5	0.4	1.1		Differences
M-PNIP − $W > 1$	1–1.8	1.8–2.5	2.5–3.1	3.1- 4	0.3	M-NIP values
	0.8	0.7	0.6	0.9		Differences
M-PNIP-WC	1–1.9	1.9-2.6	2.6-2.95	2.95-4	0.2	M-NIP values
	0.9	0.7	0.35	1.05		Differences
M-PNIP-WC − $W > 1$	1–1.8	1.8–2.55	2.55–3.05	3.05–4	0.25	M-NIP values
	0.8	0.75	0.5	0.95		Differences
M-PNIP-AQ − $W > 1$	1–1.8	1.8–2.3	2.3–2.95	2.95–4	0.25	M-NIP values
	0.8	0.5	0.65	1.05		Differences
M-PNIP-SUB	1–1.8	1.8–2.45	2.45–2.9	2.9–4	0.25	M-NIP values
	0.8	0.65	0.45	1.1		Differences

superior to the thresholds, the box plots of the M-NIP values by site versus nutrient status expressed by trophic categories (Fig. 2) showed distinctly the pattern of classification in trophic categories, especially for the index versions considering only species with a weight above one. The number of sites by trophic categories fulfilling the conditions for a reliable index value is given in Table 7.

In order to illustrate the different type of IV-P obtained, examples for few species are given below.

Potamogeton lucens L.

This species was recorded from nine ponds. Four were classified as mesotrophic and five as eutrophic (Fig. 3a) according to the OECD criteria and with TP concentrations ranging between 12 and 76 µg P/l. Based on these observations, an IV-P of 2.56 with a weighting factor (*W*) of 4 was calculated. In lakes, Lachavanne et al. (1988) observed similar optima in meso-eutrophic sites for *P. lucens*, likewise Melzer (1999) classified this species in the indicator group 3.5 on a scale of 5 points. In rivers, Schneider & Melzer (2003) obtained an IV but also an amplitude very close to our observation (IV 2.65/*W* 4). In the IBMR of Haury et al. (2006), the species score (Csi) for this species is slightly worse and tally at the site scale with a score of poor to bad status. Similarly, the general index of ecological values of Landolt (1977) classified *P. lucens* as four on a five point scale of affinity or tolerance to nutritive substance ("Nährstoffzahl", N). Our observations corroborate the prior classification of *P. lucens* as nutrient tolerant species linked to meso to eutrophic conditions. In effect, both the IV of Schneider & Melzer (2003) and our calculated IV-P are in the lower part of the range of values for eutrophic sites and no occurrences were observed in oligotrophic or hypertrophic ponds.

Ranunculus trichophyllus Chaix s.str.

Ranunculus trichophyllus Chaix s.str. occured at 16 ponds with TP concentrations ranging from 3 to 63 µg P/l. Two were classified as oligotrophic, nine as mesotrophic, and five as eutrophic (Fig. 3b) according to the OECD criteria. Based on these observations, an IV-P of 2.19 with a weighting factor (*W*) of 2 was calculated. Similarly to this IV-P, Haury et al. (2006) calculated a Csi of 11 out of 20 points, in the range for a good status of IBMR at the site level. Other authors classify *R. trichophyllus* with a higher affinity for nutrients, with an IV of, respectively, 2.7 and a wide tolerance for Schneider & Melzer (2003), four on five for Landolt (1977), and 4.5 corresponding to the eighth category on a scale of nine in the macrophyte index (MI) of Melzer (1999). Our data clearly support a shift downward of one trophic category with the calculated IV-P corresponding to an optimum in mesotrophic ponds. However, almost one-third of the sites, where *R. trichophyllus* was observed, were classified as eutrophic. This, along with the large amplitude of tolerance, indicates that the species can also growth in eutrophic conditions as shown by the above indexes, but seems to have its optimum in mesotrophic ponds.

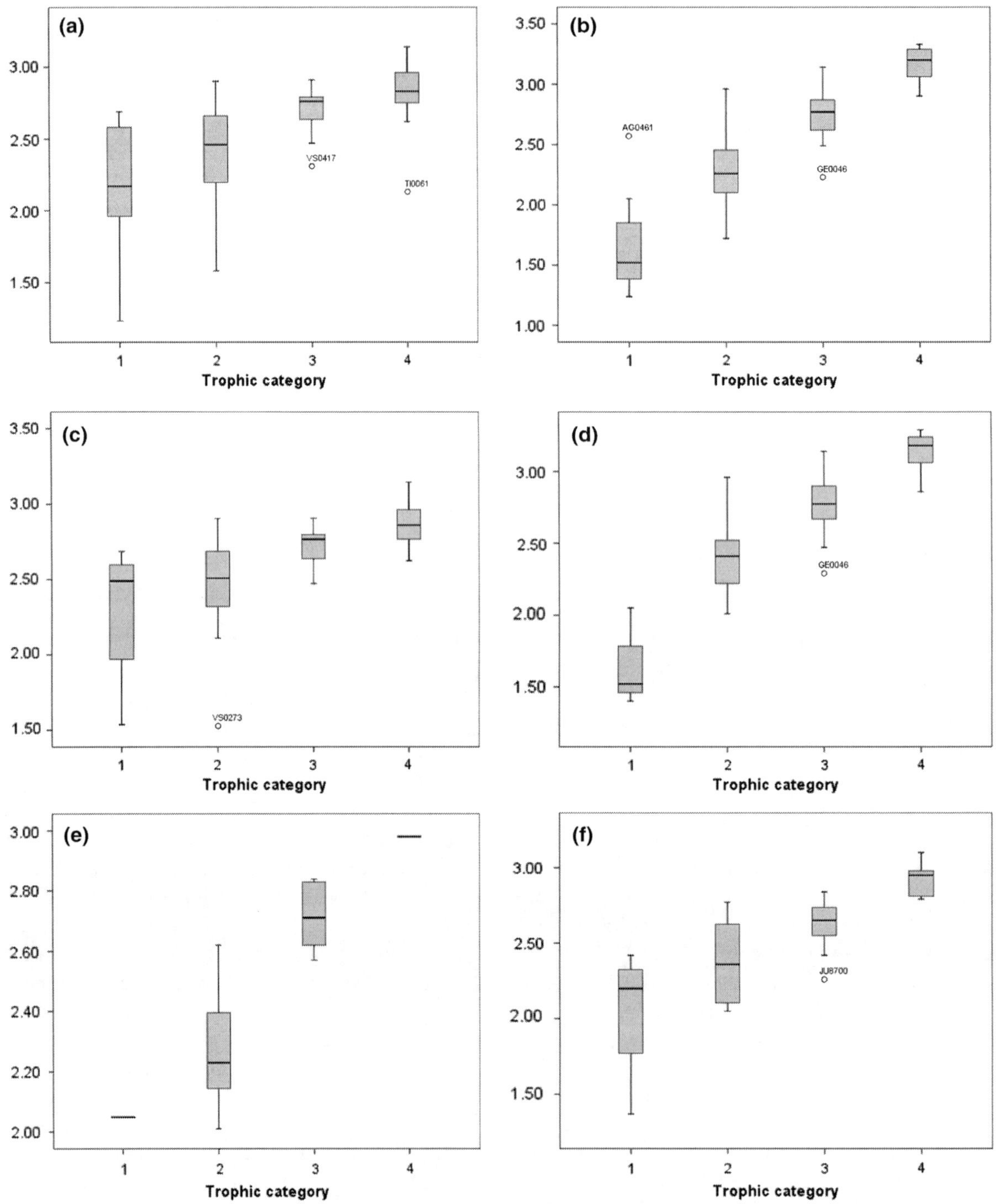

Fig. 2 Box plot of the M-PNIP index values (with an SC inferior to the thresholds defined in Table 6) by trophic categories based on TP (TP OECD trophic scale) with **a** M-PNIP, **b** M-PNIP – $W > 1$, **c** M-PNIP-WC, **d** M-PNIP-WC – $W > 1$, and **e** M-PNIP-SUB and 1: oligotrophic, 2: mesotrophic, 3: eutrophic, and 4: hypertrophic

Table 7 Number of sites fulfilling the conditions for a valid M-NIP computation in each trophic category defined by the banding of the index values (Table 6)

M-NIP		Oligotrophic	Mesotrophic	Eutrophic	Hypertrophic	*n*
M-PNIP	M-PNIP	8	25	63	8	104
	P-OECD	*21*	*31*	*31*	*21*	
M-PNIP − $W > 1$	M-PNIP	6	17	29	5	57
	P-OECD	*7*	*19*	*25*	*6*	
M-PNIP-WC	M-PNIP	4	34	53	5	96
	P-OECD	*18*	*28*	*30*	*20*	
M-PNIP-WC − $W > 1$	M-PNIP	3	15	23	5	46
	P-OECD	*4*	*13*	*24*	*5*	
M-PNIP-AQ − $W > 1$	M-PNIP	–	5	8	1	14
	P-OECD	*1*	*7*	*5*	*1*	
M-PNIP-SUB	M-PNIP	1	14	17	3	35
	P-OECD	*4*	*15*	*11*	*5*	

In italic, the number of ponds by trophic state defined by the chemical TP scale

Juncus bulbosus L.

Juncus bulbosus L. was observed in four ponds only, three were mesotrophic and one eutrophic, the resulting IV-P is 2.25 with a weighting of 4 (Fig. 3c). This value, even if calculated with few observations, clearly link this species to mesotrophic ponds, which is also in accordance with the mean *N* value (three out of five) given by Landolt (1977). The Csi of 16 out of 20 points obtained from river data by Haury et al. (2006) was a step above and fully in the range for the status "very good" of IBMR at the site level.

Berula erecta (Huds.) Coville

This species has been found in six ponds, four were mesotrophic, one oligotrophic, and the last stand in the lower part of TP concentration, indicating a eutrophic state (Fig. 3d). TP concentrations of the six sites ranged between 3 and 42 μg P/l. The IV-TP for *Berula erecta* is exactly two with a moderate amplitude ($W = 4$), making this species a good indicators of mesotrophic ponds. Both Landolt (1977) and Haury et al. (2006) assign also ecological values corresponding to meso-eutrophic and good conditions with an *N* value of 3 and a Csi species score of 14, respectively. However, the IV of 2.65 obtained in rivers by Schneider & Melzer (2003) classifies this species in the lower part of values indicating eutrophic conditions at the site level.

Carex riparia Curtis

The five occurrences of this sedge in the set of ponds were all in eutrophic sites with TP concentrations ranging between 39 and 76 μg/l (Fig. 3e). This narrow spectrum assigns a maximum weight *W* of 16 to that species and a strong link with eutrophic ponds (IV = 3) even if only five observations were available. Landolt (1977) gives also a median value of affinity to nutrients for *C. riparia* ($N = 3$) which support our calculated IV-P value.

Groenlandia densa (L.) Fourr

With only three observations, this species was poorly represented in the data set. However, TP concentrations of the colonized ponds were narrow, between 10 and 20 μg/l, two sites were at the upper limit of TP for oligotrophy, while the latter was clearly mesotrophic (Fig. 3f). The resulting IV-P of 1.33 and a weighting factor (*W*) of four make *G. densa* a good indicator of oligotrophic to oligo-mesotrophic ponds. The index proposed in the literature for this species supports this result, both Landolt (1977) and Schneider & Melzer (2003) obtain indicator values corresponding to the oligo-mesotrophic category with an *N* value of 2 and an IV of 1.83, respectively. The Csi score of Haury et al. (2006) is slightly worse, and with a value of 11 it corresponds to a moderate status for the IBMR at the site level.

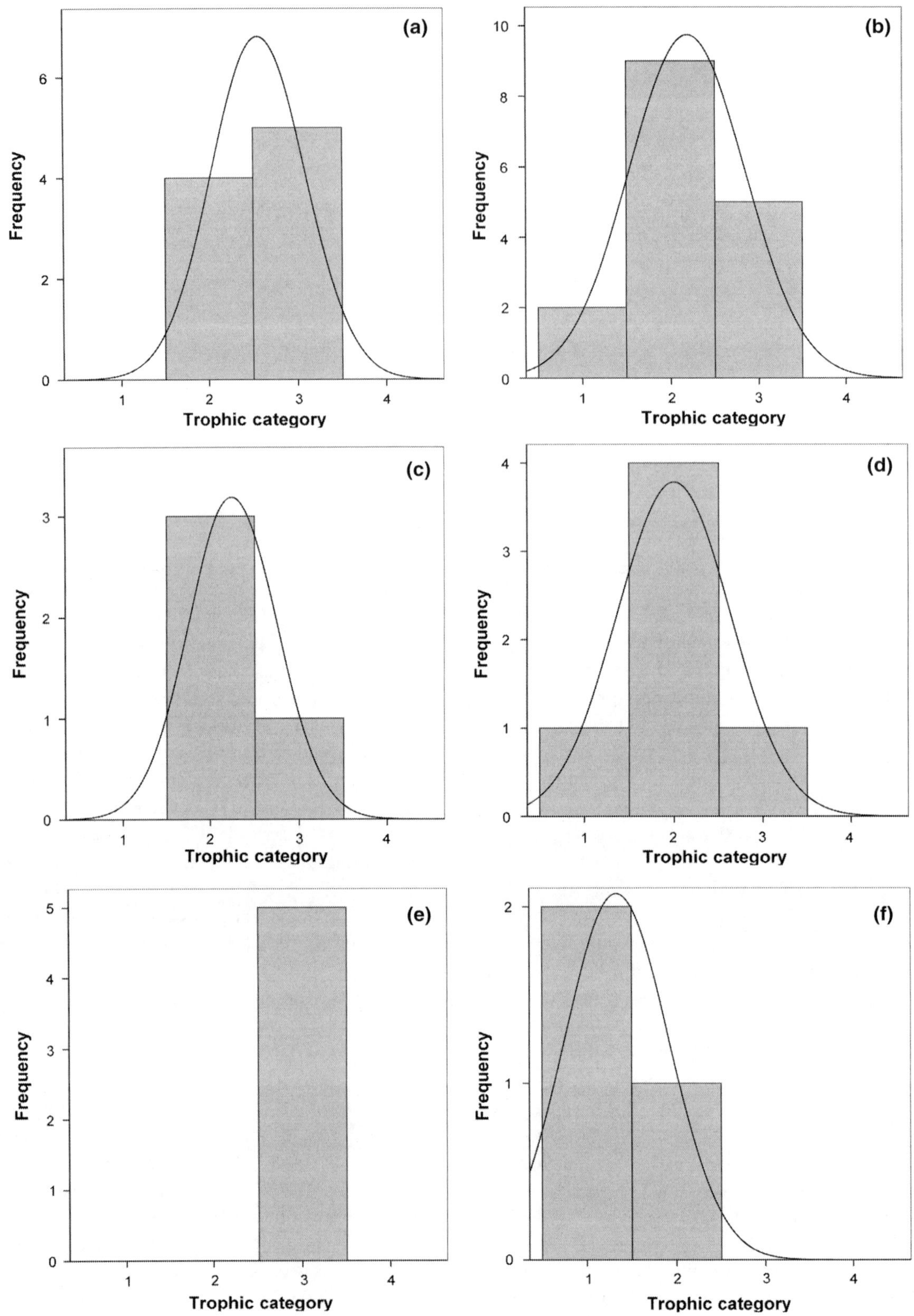
(a)
(b)
(c)
(d)
(e)
(f)
Frequency
Trophic category
1
2
3
4
0
2
4
6
8
10

◀ **Fig. 3** Histogram of occurrences by trophic category for **a** *Potamogeton lucens* L. (IV = 2.56/W = 4), **b** *Ranunculus trichophyllus* Chaix s.str. (IV = 2.19/W = 2), **c** *Juncus bulbosus* L. (IV = 2.25/W = 4), **d** *Berula erecta* (Huds.) Coville (IV = 2/W = 4), **e** *Carex riparia* Curtis (IV = 3/W = 16), and **f** *Groenlandia densa* (L.) Fourr. (IV = 1.33/W = 4)

Performance assessment and selection of the M-NIP index

The M-NIP derived from the trophic scale defined with TP (M-PNIP) give the most reliable index as it classifies with significant differences ponds of different chemically defined trophic categories (Table 8). However, when taking into account species with a wide amplitude ($W = 1$), the M-NIP values overlapped between categories of trophy leading to a high rate of misclassification. Table 9 summarizes the proportion of matching classification between the trophic categories defined with the M-PNIP's and the trophic state chemically defined with TP concentrations.

When the IV of all the macrophytes species present were taken into account, the sets of M-PNIP by chemically defined trophic categories were significantly different, except between oligotrophic and mesotrophic categories, where the M-PNIP values were largely overlapping. However, even if the M-PNIP values were significantly different between trophic categories, the rate of correct reclassification was overall low (53.8%), particularly for the ponds chemically defined as oligotrophic or hypertrophic which were upgraded up to eutrophic category (71.4%) or downgraded by one category (65.9%), respectively, leading to an over-representation of the eutrophic state.

Table 8 Significance of the Mann–Whitney tests performed between sets of M-NIP values grouped by chemically defined trophic categories

M-PNIP variant	o–m	m–e	e–h	o–e
M-PNIP	0.091	**0.000**	**0.011**	**0.000**
M-PNIP − $W > 1$	**0.004**	**0.000**	**0.001**	
M-PNIP-WC	0.132	**0.000**	**0.003**	**0.000**
M-PNIP-WC − $W > 1$	**0.005**	**0.000**	**0.004**	
M-PNIP-AQ − $W > 1$	ne	**0.009**	ne	
M-PNIP-SUB	0.293	**0.027**	**0.007**	**0.003**

The values in bold indicate significant differences between set at the 0.05 threshold. *o* oligotrophic, *m* mesotrophic, *e* eutrophic, *h* hypertrophic. *ne* not evaluated as there were not enough cases to perform the statistical test

The performance of the index was much better when considering only the species with a weighting factor above one (M-PNIP − $W > 1$). In that case, the sets of M-NIP values by trophic categories were always significantly different and the rates of correct reclassification overall reached 80.7% and 66.7–92% for single trophic categories. However, with this reduced pool of indicator species, only 57 out of 114 ponds fulfilled the conditions to calculate a reliable index.

When the Characea species were not included in the computation (M-PNIP-WC), the index values remained significantly different between trophic states (MW test, Table 8). Again, when counting the species with $W = 1$, the M-PNIP-WC values were not significantly different between sites chemically defined as oligotrophic and mesotrophic. In addition, the range of M-PNIP values became narrower for the mesotrophic and eutrophic categories (Table 6; Fig. 3), while the ratios of correct reclassification were low for the oligotrophic and hypertrophic ponds (Table 9). The M-PNIP-WC − $W > 1$ performed better and each set of index values by trophic categories was significantly different from the adjacent sets. With this index, the ratio of correct reclassification reached 82.6% overall and never fell under 75% for a single trophic category, while the lag between the TP scale and the M-PNIP index never exceeded one trophic category. Nonetheless, the conditions to calculate a reliable index were fulfilled for a smaller subset of ponds with only 46 sites out of 114 with valid values.

Two additional M-PNIP indexes based on subgroups of macrophytes gave values significantly different between trophic categories: the M-PNIP-AQ − $W > 1$ [species with a weighting factor superior to one and classified as aquatic in the red list of fern and flowering plants of Switzerland of Moser et al. (2002)], and the M-PNIP-SUB based on submerged species only. Even so, with these smaller numbers of species, more sites did not meet the conditions to calculate a reliable index, mainly due to an insufficient number of species with valid IV. For instance, with the M-PNIP-AQ − $W > 1$, 100 sites had an unreliable index and the valid values allocated mainly the 14 remaining ponds to mesotrophic (7) and eutrophic (5) categories (Table 7). This very low

Table 9 Number of sites and rates of matching classification between the trophic categories defined with the M-PNIP's and the trophic state chemically defined with TP concentrations

Index variants		Total	G1	G2	O	G1	G2	M	G1	G2	E	G1	G2	H	G1	G2	Total
M-PNIP	*n*	56	39	9	6	7	8	15	16	0	28	3	0	7	13	1	104
	%	53.8	37.5	8.7	28.6	33.3	38.1	48.4	51.6	0.0	90.3	9.7	0.0	33.3	61.9	4.8	
M-PNIP-W > 1	*n*	46	10	1	5	1	1	14	5	0	23	2	0	4	2	0	57
	%	80.7	17.5	1.8	71.4	14.3	14.3	73.7	26.3	0.0	92.0	8.0	0.0	66.7	33.3	0.0	
M-PNIP-WC	*n*	52	41	3	3	12	3	18	10	0	26	4	0	5	15	0	96
	%	54.2	42.7	3.1	16.7	66.7	16.7	64.3	35.7	0.0	86.7	13.3	0.0	25.0	75.0	0.0	
M-PNIP-WC − *W* > 1	*n*	38	8	0	3	1	0	11	2	0	20	4	0	4	1	0	46
	%	82.6	17.4	0.0	75.0	25.0	0.0	84.6	15.4	0.0	83.3	16.7	0.0	80.0	20.0	0.0	
M-PNIP-AQ − *W* > 1	*n*	10	4	0	0	1	0	4	3	0	5	0	0	1	0	0	14
	%	71.4	28.6	0.0	0.0	100.0	0.0	57.1	42.9	0.0	100.0	0.0	0.0	100.0	0.0	0.0	
M-PNIP-SUB	*n*	22	13	0	1	3	0	9	6	0	9	2	0	3	2	0	35
	%	62.9	37.1	0.0	25.0	75.0	0.0	60.0	40.0	0.0	81.8	18.2	0.0	60.0	40.0	0.0	

O oligotrophic, *M* mesotrophic, *E* eutrophic, *H* hypertrophic. G1 and G2 count the number of case with a gap of respectively one or two trophic category between the M-PNIP and the chemical scale

Table 10 Calculated M-NIP values of the validation data set and corresponding trophic categories

Site code	*GE0010_95*	GE0010_05	GE0010_06	*GE0048_05*	GE0048_06	ZH0002	GE0044	GE4408
TP-CATEG	*M*	E	M	*M*	E	E	E	H
M-PNIP	*2.73*	2.81	2.74	*2.71*	2.79	2.76	2.75	2.98
	E	E	E	*E*	E	E	E	H
M-PNIP − *W* > 1	*2.26*	3.00	2.72	*2.71*	2.81	2.85		3.33*
	M	E	E	*E*	E	E		H*
M-PNIP-WC	*2.79*	2.81	2.74	*2.71*	2.78	2.76	2.75	2.98
	E	E	E	*E*	E	E	E	E
M-PNIP-WC − *W* > 1	*2.55*	3.00	2.72	*2.71*	2.80	2.85		3.33
	M	E	E	*E*	E	E		H
M-PNIP-AQ − *W* > 1	*2.62*	3.09*		_2.61*_	2.59	2.74*		
	E	H*		_E*_	E	E*		
M-PNIP-SUB	*2.77*	2.73	2.70	*2.77*	2.88	2.72		2.52*
	E	E	E	*E*	E	E		E*

In italic the two surveys that were incorporated in the calibration process on the M-NIP. Asterisk (*) indicates unreliable index with an SC superior to the thresholds defined in the Table 6. *O* oligotrophic, *M* mesotrophic, *E* eutrophic, *H* hypertrophic

applicability makes this index inappropriate for assessment purpose even if the rate of matching classification between the chemical TP scale and the index was quiet good (71.4%). With the M-PNIP-SUB variant, the full range of trophic categories was represented but both the rate of matching classifications (62.9%) and the number of valid values were too low (35 ponds out of 114) to retain this index further.

Validation and application of the M-NIP index

Among the five sites used for the validation process, three were not in the set of ponds used to set the indicator values and the remaining two were new surveys of ponds already incorporated in the building process of the index. When possible, the M-NIP values were computed for the six indexes based on TP concentration (M-PNIP) selected for significance during the previous steps (Table 6). The calculated M-PNIP by site and the trophic categories either determined by the indicator values or with the TP concentration are given in Table 10.

For all three newly surveyed pond, the M-PNIP and M-PNIP $-$ $W > 1$ correctly reclassified the sites in the trophic categories determined by TP. However, for one site, the SC of the M-PNIP $-$ $W > 1$ was above the threshold of 0.3 and, therefore, the index cannot be considered as reliable. As there were no Characea species at these three sites, the versions of the index without stoneworts (M-PNIP-WC) gave the same index value. For that reason, only the index ranges defined for the general index were applied to define the trophic status. The index taking only the aquatic species into account (M-NIP-AQ $-$ $W > 1$) could be calculated for one site but with an SC above the threshold even if it fitted in the range of the corresponding trophic category. Finally, the version of the index taking into account submerged species only (M-NIP-SUB) correctly reclassified one pond, while the index and SC values were out of the range for another and could not be calculated on the last site.

The trophic category of the pair of sites surveyed two subsequent years varied by one class between the two sampling occasions according to TP concentrations. The M-NIP values followed the same trend but remained, however, in the range of the same trophic category. This seems to indicate that the M-NIP value was able to detect slight variations in the chemical status of waterbody but with lower amplitude. This reduced response of the index to phosphorus load could possibly be explained by the resilience properties of the macrophytes community. In effect, the small decrease of the measured TP concentration between the consecutive survey of 2005 and 2006 lead to re-assign the pond GE0010 to the mesotrophic category that it had in 1995, but the M-PNIP index continues to indicate eutrophic conditions in 2006. Moreover, with a longer time period between assessments, the trophic classification based on M-PNIP follows the variation in TP concentration. In effect, between the 1995 and 2005 surveys, the physico-chemical data showed a shift from mesotrophic to eutrophic condition that was also indicated by the two more accurate versions of the MI, the M-PNIP $-$ $W > 1$ and the M-PNIP-WC $-$ $W > 1$.

Discussion

Significance and limitation of the indicator values (IV) by species

The nutrient profile of a large part of the species could be derived from the pond data set. Among the species dismissed, due to an insufficient number of observations, some are known for their narrow trophic profile in rivers or lakes and their inclusion could have potentially improved the performance and applicability of the index. This is notably the case of *Potamogeton plantagineus* Roem. & Schult that occurred in only one oligotrophic pond within the data set but is known for a high affinity to oligotrophic conditions from other studies and coded with an IV of 1.05 corresponding to the oligotrophic category in the river index of Schneider & Melzer (2003). Similarly, *Ranunculus circinatus* Sibth. occurred in two ponds across the whole data set, both classified as mesotrophic according to the TP concentration, and IV of Schneider & Melzer (2003) as well as *N* value of Landolt (1977) indicate a similar nutrient profile that could validate this IV. On the other hand, *Zannichellia palustris* L. was also observed in two mesotrophic ponds, but the trophic profile found by other authors are one or two degrees higher, with an IV of 2.93 in rivers (Schneider & Melzer 2003) corresponding to eutrophic conditions and a maximum *N* value for Landolt (1977) indicating nutrient rich conditions, respectively. In addition, some of the species with enough observations to contribute to a reliable index also indicated nutrient conditions differing from the profile established from lakes and rivers data, in fact, what was expected from an index based on pond data. For all these reasons, we have decided to strictly exclude any of the species with less than three observations within the present data set, even when the nutrient conditions in colonized ponds were concordant with the trophic category assigned to species from rivers or lakes or even from an expert judgment. For the index calibration step, these exclusions have only a slight influence on the results as discarded species never occupy more than two sites. By contrast, for an assessment of newly sampled ponds, the greater the number of coded species, and especially of stenotrophic species, the more chance to compute a valid M-NIP value. In order to improve the M-NIP, it is, therefore, highly recommended to include data from more sites in the calibration set. In addition, this would also allow refining of the trophic profiles for species already coded.

An important aspect of the nutrient profile of species is the amplitude of tolerance (a) transcribed in weighting factors for the M-NIP and SC calculations. This amplitude was wide for many species, indicating either eurytrophe species with a low bioindication potential or a too small number of observations to bring out a distinct optima. Other factors, among which the type of chemical data used to define the profile, contribute to the relatively wide amplitude observed for most species. In effect, the mean water concentration of nutrient is expressed at the site scale and does not take into account the variability of the water chemistry within the pond. For large sites with an important sinuosity and an irregular morphometry, this spatial variability of the physico-chemicals conditions can be quite important. Moreover, the content of the sediment was not measured for this study and this important source of nutrient for rooted species can show important variations even for a similar concentration of nutrients in the water. The lack of sediment data made it difficult to disentangle the effects of the variability of water chemistry from the influence of the sediment content, which both probably contribute to widen the amplitude of nutrient profiles based on mean values of water chemistry. The amplitude of tolerance expresses, however, the range of mean water nutrient concentration where the species was observed. The fact that free-floating species also had wide amplitudes, even wider that some emerged species, indicates that the observed tolerance is linked to parameters measured in the water and not only to variations or parameters not taken into account by the chemical data. This increase in the amplitude of tolerance to nutrient conditions leads to less accurate profiles of bioindication by individual species; nonetheless, the accuracy of the IV remains sufficient for an assessment at the pond level.

M-NIP as an assessment tool

The wide tolerance of most of the species also has implications on the M-NIP accuracy and particularly when the species with the larger amplitude are taken into account ($W = 1$). Despite the limitations of indicator values and considering the goal to obtain a tool for assessing the trophic status of the whole pond, the indexes adopted are able to classify most of the ponds by trophic categories albeit with an error rate relatively important. In effect, the M-NIP variants always integrate the nutrient profiles of several species with indicator values that should not diverge beyond a threshold of confidence expressed by the rate of SC. When the species with the widest amplitude were discarded ($W = 1$), the indexes classified the ponds in trophic categories matching relatively well the classifications based on TP concentrations. However, these variants of M-NIP were not applicable to a large number of sites, mainly due to the lack of valid IV for many species. By contrast, when keeping all species, the increase in the overlap between ranges of M-NIP values belonging to adjacent trophic categories led to higher rates of misclassification and the index accuracy dropped considerably. The rate of misclassification increased particularly for the oligotrophic and hypertrophic ponds pushed, respectively, up or down by the indicator values of tolerant species, despite their lower weight in the index. The performance assessment of the M-NIP variants presented in the results and summarized in Tables 8 and 9 permits to establish a preferential order for the use of the indexes taking the rate of correct reclassification and the time investment as evaluation criteria. Knowing that the identification of Characea at the species level requires time and expertise often less available to site managers, we propose to apply a first division depending on the presence of this group of macrophytes in an assessed site:

- When Characea are observed, the surveyor records it and collects samples by quadrat. However, the M-PNIP-WC $- W > 1$ is first computed and Characea species are identified to calculate the M-PNIP $- W > 1$ only if the first index cannot be calculated or does not fulfill all requirements. If the lack of species with $W > 1$ means that both versions cannot be calculated, the less accurate but more often applicable M-PNIP-WC and M-PNIP indexes can be used instead. The results of the two latter variants must be interpreted with the limitations linked to their lower precision.
- If no Characea species are observed, the M-PNIP $- W > 1$ is used first. If it cannot be

calculated or is unreliable, the M-PNIP is used instead and interpreted with the limitations linked to the lower accuracy of this nutrient index.

Conclusions

Ecological indicators must have an ecological meaning, be easy to use and reproducible, and sensitive to moderates changes in environmental conditions. The proposed M-NIP index based on TP concentration fulfill the first three conditions: first, macrophytes are sensitive to trophic conditions, second, the index is easy to obtain with the nutrient profiles and cover of the species observed during a survey, and third, the sampling design used to define the nutrient profiles and calculate the index values is standardized and fully reproducible. The fourth condition is, however, only partly fulfilled. In effect, both the rate and the amplitude of misclassification between the physico-chemical scale and macrophytes index are relatively high for the two more applicable variants of the M-PNIP taking all species into account. Moreover, over short time intervals, the macrophyte index response to variation in nutrients concentration is measurable but limited in amplitude. This reduced response of the biotic index was observed on two ponds sampled over two subsequent years that vary by one trophic category between the two sampling occasions according to TP, but remained in the same category with the M-PNIP. By contrast, with a longer time period between two surveys, the macrophytes index seems to respond to an increase in nutrient concentration with the same amplitude as the TP scale. This latter observation was made on the single pond where such data were available, therefore, it still needs to be confirmed with other sites but, for monitoring purposes, seems promising.

Despite these limitations regarding the accuracy and delay in the response of the index, the M-NIP based on concentration of TP (M-PNIP) makes up a good indicator of the pond nutrient status that can be easily used for site assessment or monitoring. This index is, thus, a reliable metric of eutrophication to be integrated in a multimetric index to assess the water quality of Swiss ponds. Nonetheless, the index does not fulfill the requirements of the WFD to perform the assessment by a measure of the deviation from a set of reference sites considered as unimpacted. Indeed, as a pond can be naturally eutrophic the M-NIP index does not necessarily express degradations or human influences. Nevertheless, water quality is one of the aspects to be taken into account by the WFD. A biotic index assessing specifically the trophic state is a valuable complementary descriptor to an approach conducted in parallel, which is based on the comparison of composition and abundance of the macrophytes communities between references conditions and assessed site.

As stated earlier, a greater data set needs to be available in order to improve the accuracy and applicability of the index by incorporating more species and refining their IV. Moreover, a greater number of additional records would also enable the ranges of values by categories of nutrient status to be more precisely refined. Specifically, it would allow ranges of index values by biogeographic and altitudinal regions to be defined, which was unfortunately not possible with our data set as there were not enough ponds for each type of trophic categories.

Acknowledgments We thank our colleagues Nathalie Menetrey, Dominique Auderset Joye, Christiane Ilg, Gilles Carron, and Amael Paillex for their help with the fieldwork. Thanks also to Pascale Nicolet for improving the english style and valuable comments on the manuscript. We are also grateful to the "*Bourse Augustin Lombard*" of the "*Société de physique et d'histoire naturelle de Genève (SPHN)*" and the "*Office of Environment and Energy*" of Kanton Luzern for their financial support. Finally, thanks to the SECOE of the Canton de Genève and SESA of the Canton de Vaud for providing us some of the chemical analyses of the water samples. This paper was realized with data in part collected in a study financially supported by the "*Swiss Agency for the Environment, Forests and Landscape*".

References

Adams, M. S. & K. Sand-Jensen, 1991. Introduction: ecology of submersed aquatic macrophytes. Aquatic Botany 41: 1–4.

APHA, American Public Health Association, American Water Works Association, and W. P. C. Federation, 1998. Standard Methods for the Examination of Water and Wastewater, 20th edn. APHA, AWWA and WPCF, Washington, DC 20005-2605.

Balls, H., B. Moss & K. Irvine, 1989. The loss of submerged plants with eutrophication.1. Experimental-design, water chemistry, aquatic plant and phytoplankton biomass in experiments carried out in ponds in the Norfolk Broadland. Freshwater Biology 22: 71–87.

Barko, J. W., D. Gunnison & S. R. Carpenter, 1991. Sediment interactions with submersed macrophyte growth and community Dynamics. Aquatic Botany 41: 41–65.

Bornette, G., C. Henry, M. H. Barrat & C. Amoros, 1994. Theoretical habitat templets, species traits, and species richness – aquatic macrophytes in the Upper Rhone River and its floodplain. Freshwater Biology 31: 487–505.

Bronmark, C. & L. A. Hansson, 2002. Environmental issues in lakes and ponds: current state and perspectives. Environmental Conservation 29: 290–307.

Bundi, U., A. Peter, A. Frutiger, M. Hutte, P. Liechti & U. Sieber, 2000. Scientific base and modular concept for comprehensive assessment of streams in Switzerland. Hydrobiologia 422: 477–487.

Carpenter, S. R., D. Ludwig & W. A. Brock, 1999. Management of eutrophication for lakes subject to potentially irreversible change. Ecological Applications 9: 751–771.

Clayton, J. & T. Edwards, 2006. Aquatic plants as environmental indicators of ecological condition in New Zealand lakes. Hydrobiologia 570: 147–151.

Communities, C. o. E., 2000. WFD, Directive 2000/60/EC of the European Parliament and of the Council establishing a framework for the Community action in the field of water policy. CEC L 327:0001-0073.

Craft, C., K. Krull & S. Graham, 2007. Ecological indicators of nutrient enrichment, freshwater wetlands, Midwestern United States (U.S.). Ecological Indicators 7: 733–750.

Davies, B., J. Biggs, P. Williams, M. Whitfield, P. Nicolet, D. Sear, S. Bray & S. Maund, 2008. Comparative biodiversity of aquatic habitats in the European agricultural landscape. Agriculture Ecosystems & Environment 125(1–4): 1–8.

Dodds, W. K., 2006. Eutrophication and trophic state in rivers and streams. Limnology and Oceanography 51: 671–680.

Engelhardt, K. A. M., 2006. Relating effect and response traits in submersed aquatic macrophytes. Ecological Applications 16: 1808–1820.

E.P.C.N. European Pond Conservation Network, 2007. Developing the Pond Manifesto. Annales de Limnologie – International Journal of Limnology 43: 221–232.

Friedrich, G., 1990. A revision of the saprobic system. Journal for Water and Wastewater Research [Zeitschrift Fur Wasser Und Abwasser Forschung] 23: 141–152.

Haury, J., M. C. Peltre, M. Tremolieres, J. Barbe, G. Thiebaut, I. Bernez, H. Daniel, P. Chatenet, G. Haan-Archipof, S. Muller, A. Dutartreg, C. Laplace-Treyture, A. Cazaubon & E. Lambert-Servien, 2006. A new method to assess water trophy and organic pollution – the Macrophyte Biological Index for Rivers (IBMR): its application to different types of river and pollution. Hydrobiologia 570: 153–158.

Havens, K. E., J. Hauxwell, A. C. Tyler, S. Thomas, K. J. McGlathery, J. Cebrian, I. Valiela, A. D. Steinman & S.-J. Hwang, 2001. Complex interactions between autotrophs in shallow marine and freshwater ecosystems: implications for community responses to nutrient stress. Environmental Pollution 113: 95–107.

Holmes, N. T. H., P. J. Boon & T. A. Rowell, 1998. A revised classification system for British rivers based on their aquatic plant communities. Aquatic Conservation – Marine and Freshwater Ecosystems 8: 555–578.

Irmer, U., 2000. The new EC Framework Water Directive: assessment of the chemical and ecological status of surface waters. Acta Hydrochimica Et Hydrobiologica 28: 7–14.

Kohler, A., 1975. Submerse Makrophyten und ihre Gesellschaften als Indikatoren der Gewässerbelastung. Beiträge zur Naturkundlichen Forschung in Südwestdeutschland 34: 149–159.

Lachavanne, J.-B., J. Perfetta, A. Noetzlin, R. Juge & B. Lods-Crozet, 1988. Etude chorologique et écologique des macrophytes des lacs suisses en fonction de leur altitude et de leur niveau trophique. 1976–1985, Deuxième édition 1988, Rapport final. FNRS et Université de Genève, Geneva.

Lacoul, P. & B. Freedman, 2006. Environmental influences on aquatic plants in freshwater ecosystems. Environmental Reviews 14: 89–136.

Landolt, E., 1977. Ökologische Zeigerwerte zur Schweizer Flora. Veröffentl. Geobot. Inst. ETH, Stiftung Rübel Zürich 64: 208 pp.

Lehmann, A. & J. B. Lachavanne, 1999. Changes in the water quality of Lake Geneva indicated by submerged macrophytes. Freshwater Biology 42: 457–466.

Linton, S. & R. Goulder, 2000. Botanical conservation value related to origin and management of ponds. Aquatic Conservation 10: 77–91.

Meilinger, P., S. Schneider & A. Melzer, 2005. The reference index method for the macrophyte-based assessment of rivers – a contribution to the implementation of the European Water Framework Directive in Germany. International Review of Hydrobiology 90: 322–342.

Melzer, A., 1988. Der Makrophytenindex-Eine biologische Methode zur Ermittlung der Nährstoffbelastung von Seen. Habilitationsschrift der TU München. TU München, München.

Melzer, A., 1999. Aquatic macrophytes as tools for lake management. Hydrobiologia 396: 181–190.

Moore, B. C., J. E. Lafer & W. H. Funk, 1994. Influence of aquatic macrophytes on phosphorus and sediment porewater chemistry in a fresh-water wetland. Aquatic Botany 49: 137–148.

Moser, D. M., A. Gygax, B. Bäumler, N. Wyler & R. Palese (eds), 2002. Liste rouge des espèces menacées de Suisse. Fougères et plantes à fleurs. Office fédéral de l'environnement, des forêts et du paysage, Berne., Bern.

Murphy, K. J., 2002. Plant communities and plant diversity in softwater lakes of northern Europe. Aquatic Botany 73: 287–324.

Naumann, E., 1927. Ziel und Hauptprobleme der regionalen Limnologie. Botaniska Notiser 8: 1–103.

Naumann, E., 1932. Grundzüge der regionalen Limnologie. Schweizerbartsche Verlagsbuchhandlung, Stuttgart.

Oertli, B., D. Auderset Joye, E. Castella, R. Juge, D. Cambin & J.-B. Lachavanne, 2002. Does size matter? The relationship between pond area and biodiversity. Biological Conservation 104: 59–70.

Oertli, B., D. Auderset Joye, E. Castella, R. Juge, A. Lehmann & J. B. Lachavanne, 2005. PLOCH: a standardised method for sampling and assessing the biodiversity in ponds. Aquatic Conservation: Marine and Freshwater Ecosystems (submitted).

Palmer, M. A., S. L. Bell & I. Butterfield, 1992. A botanical classification of standing waters in Britain – applications for conservation and monitoring. Aquatic Conservation – Marine and Freshwater Ecosystems 2: 125–143.

Phillips, G. L., D. Eminson & B. Moss, 1978. A mechanism to account for macrophyte decline in progressively eutrophicated freshwaters. Aquatic Botany 4: 103–126.

Robach, F., G. Thiebaut, M. Tremolieres & S. Muller, 1996. A reference system for continental running waters: plant communities as bioindicators of increasing eutrophication in alkaline and acidic waters in north-east France. Hydrobiologia 340: 67–76.

Sager, L., 2009. Measuring the trophic status of ponds: relationships between summer rate of periphytic net primary productivity and water physico-chemistry. Water Research 43: 1667–1679.

Schneider, S. & A. Melzer, 2003. The Trophic Index of Macrophytes (TIM) – a new tool for indicating the trophic state of running waters. International Review of Hydrobiology 88: 49–67.

Seddon, B., 1972. Aquatic macrophytes as limnological indicators. Freshwater Biology 2(2): 107–130.

Smith, V. H., G. D. Tilman & J. C. Nekola, 1999. Eutrophication: impacts of excess nutrient inputs on freshwater, marine, and terrestrial ecosystems. Environmental Pollution 100: 179–196.

Sondergaard, M., E. Jeppesen, J. P. Jensen & S. L. Amsinck, 2005. Water framework directive: ecological classification of Danish lakes. Journal of Applied Ecology 42: 616–629.

Stelzer, D., S. Schneider & A. Melzer, 2005. Macrophyte-based assessment of lakes – a contribution to the implementation of the European Water Framework Directive in Germany. International Review of Hydrobiology 90: 223–237.

Vadeboncoeur, Y., E. Jeppesen, M. J. Vander Zanden, H. H. Schierup, K. Christoffersen & D. M. Lodge, 2003. From Greenland to green lakes: cultural eutrophication and the loss of benthic pathways in lakes. Limnology and Oceanography 48: 1408–1418.

Vermeer, C. P., M. Escher, R. Portielje & J. J. M. de Klein, 2003. Nitrogen uptake and translocation by *Chara*. Aquatic Botany 76: 245–258.

Vollenweider, R. A. & J. Kerekes, 1982. Eutrophication of waters: monitoring, assessment and control. OECD Cooperative Programme on Monitoring of Inland Water (Eutrophication Control). OECD, Paris, France.

Wigand, C., J. C. Stevenson & J. C. Cornwell, 1997. Effects of different submersed macrophytes on sediment biogeochemistry. Aquatic Botany 56: 233–244.

Willby, N. J., V. J. Abernethy & B. O. L. Demars, 2000. Attribute-based classification of European hydrophytes and its relationship to habitat utilization. Freshwater Biology 43: 43–74.

Williams, P., M. Whitfield, J. Biggs, S. Bray, G. Fox, P. Nicolet & D. Sear, 2003. Comparative biodiversity of rivers, streams, ditches and ponds in an agricultural landscape in Southern England. Biological Conservation 115: 329–341.

Zelinka, M. & P. Marvan, 1961. Zur Praezisierung der biologischen Klassifikation der Reinheit fliessender Gewaesser. Archives for Hydrobiology 57: 389–407.

Hydrobiologia (2009) 634:65–76
DOI 10.1007/s10750-009-9895-5

POND CONSERVATION

Vegetation recolonisation of a Mediterranean temporary pool in Morocco following small-scale experimental disturbance

Btissam Amami · Laïla Rhazi · Siham Bouahim · Mouhssine Rhazi · Patrick Grillas

Published online: 5 August 2009

Abstract Disturbances are key factors in the dynamics and species richness of plant communities. They create regeneration niches allowing the growth of new individuals in patches submitted to lower intensity of competition. In Mediterranean temporary pools, the intense summer drought constitutes for communities a large-scale disturbance whose intensity varies along the topographical and hydrological gradient between the centre and the edges. In this context, the importance of small-scale disturbance, such as those created by trampling and rooting herbivores in temporary pools, is poorly known. The recolonisation of small bare patches of a woodland temporary pool in western Morocco was studied experimentally in the field. The experiment was carried out using nine small control plots and nine experimental plots (sterilisation of the soil) distributed along the topographical gradient (centre, intermediate and edge zones). The area covered by plant species, and the water levels, were recorded for the plots over two successive hydrological cycles (2006/2007 and 2007/2008). The effects of natural history traits (size of seeds, presence or absence of dispersal mechanisms and annual/perennial) on the success of recolonisation of individual species were analysed. The results show that the experimental plots were rapidly recolonised. The community composition apparently was affected by the very dry conditions during the first year of the experiment, when annual species were largely absent and the clonal perennial species (*Bolboschoenus maritimus* and *Eleocharis palustris*) were dominant in the centre and intermediate zones, whilst not a single species colonised the edge zone. In the second year, less dry hydrological conditions allowed annual plants to appear in all three zones. After 2 years, the species composition of the vegetation in the experimental plots was similar to that of the unsterilised (control) plots. The abundance of plants in the centre zone was identical for experimental and control plots; in the intermediate

Guest editors: B. Oertli, R. Cereghino, A. Hull & R. Miracle
Pond Conservation: From Science to Practice. 3rd Conference of the European Pond Conservation Network, Valencia, Spain, 14–16 May 2008

B. Amami · L. Rhazi · S. Bouahim
Laboratory of Aquatic Ecology and Environment, Hassan II Aïn Chock University, BP 5366, Maarif, Casablanca, Morocco

P. Grillas (✉)
Tour du Valat, Research Centre for the Conservation of Mediterranean Wetlands, Le Sambuc, 13200 Arles, France
e-mail: grillas@tourduvalat.org

M. Rhazi
Department of Biology, Faculty of Sciences and Techniques of Errachidia, Moulay Ismail University, BP 509, Boutalamine, Errachidia, Morocco

B. Amami · S. Bouahim
Institute of Evolution Sciences, University of Montpellier II – CNRS, Case 061, 34095 Montpellier Cedex 05, France

and edge zones, the species' abundance was lower in the experimental plots than in the control plots, suggesting an incomplete return to the reference condition (control state). Differences in abundance of species were uncorrelated with the size of seeds or to the annual/perennial nature of the plants, but were particularly dependent on the hydrological conditions, which favoured lateral colonisation by perennials (runners, rhizomes). These results show that recovery from the minor disturbances can be rapid in Mediterranean temporary pools.

Keywords Temporary pool · Hydrological conditions · Disturbance · Richness · Vegetation cover · Recolonisation · Seed size

Introduction

Natural ecosystems are subjected to a wide range of natural and anthropogenic pressures, which affect the structure and dynamics of their communities (White & Pickett, 1985). A number of studies have investigated the effects of disturbances on the ecological processes involved, particularly plant succession (Connell & Slatyer, 1977; Crain et al., 2008) and competition (Bertness & Shumway, 1993), as well as the mechanisms involved in the resilience of habitats following disturbances, such as the presence of permanent seedbanks (Zedler, 2000), dispersal and colonisation (Zedler, 2000; Crain et al., 2008). Disturbances create regeneration niches within communities (Johnstone, 1986) that are occupied according to the seeds present (Morzaria-Luna & Zedler, 2007) and hence on dispersal capacity (Fraterrigo & Rusak, 2008), and the longevity of the seeds (Bliss & Zedler, 1998; Wetzel, 2001) as well as the conditions influencing their recruitment into the vegetation (Noe & Zedler, 2000; Chase & Leibold, 2003).

In contrast to large-scale disturbances (fire, extreme drought, floods, etc.), whose effects on the vegetation are relatively well-known (Barrat-Segretain et al., 1998; Zavala et al., 2000), the importance of minor disturbances in the regeneration of plants in Mediterranean ecosystems has been little studied. Studies carried out into the effects of disturbed microsites on the establishment of vegetation have mostly focussed on old fields and mountain forests (Lavorel et al., 1994; Herrera, 1997; Manzaneda et al., 2005).

The response of vegetation to disturbances not only depends on the nature and intensity of the disturbances (Grime, 1985; Airoldi, 1998), but also depends on the intensity of the stress exerted by the environment (Airoldi, 1998; Bisigato et al., 2008) and the availability of resources (Airoldi, 1998). The process of recolonisation by communities is under the control of, first, stochastic factors associated with dispersal (Bisigato et al., 2008) and, second, the species' life history traits, such as seed size (Pearson et al., 2002) and whether reproduction is sexual or asexual (Barrat-Segretain et al., 1998; Riis, 2008). A number of studies have attempted to combine species' life history traits with habitat variables in order to predict the dispersal, colonisation and survival of species within communities (Fröborg & Eriksson, 1997; Fenner, 2000; Morzaria-Luna & Zedler, 2007).

The size of seeds and the distance to seed-producing plants have often been used to estimate dispersal and hence the capacity for colonising disturbed habitats (Eriksson, 2000; Wolters & Bakker, 2002). Seed size affects the process of colonisation and the composition of communities in natural habitats (Fröborg & Eriksson, 1997; Fenner, 2000). A comparison of germination in relation to seed size (Leishman, 2001; Fenner, 2000) shows that large seeds are more likely to germinate successfully in conditions that are critical for survival (Westoby et al., 1996; Fenner, 2000), such as drought, with a significant role of chance also involved (Coomes & Grubb, 2003). Plants with small seeds are usually considered to be the best colonisers and to be most dependent on disturbances (Fenner, 2000; Coomes & Grubb, 2003). They disperse over a wide spatial area, allowing them to occupy vacant patches due to their high degree of persistence in the soil seedbank and their prolific production of seeds (Fenner, 2000; Coomes & Grubb, 2003). Distance from seed-producing plants affects the colonisation of disturbed habitats (Eriksson, 2000; Wolters & Bakker, 2002). Generally, it is the nearest species which colonise rapidly (Barrat-Segretain & Bornette, 2000). The dispersal mechanisms involved are varied and include wind (Neff & Baldwin, 2005), animals (Vanschoenwinkel et al., 2008; Soons et al., 2008) and water (Barrat-Segretain & Bornette, 2000). Specialised structures (hooks, wings, pappus, etc.) adapted to specific dispersal vectors are fairly often present on the seeds (Van den Broek et al., 2005; Cousens et al., 2008; Soons et al., 2008), facilitating their dispersal, sometimes over long distances. In Mediterranean wetlands, temporary pools constitute rich ecosystems with a high degree of biodiversity and many rare species (Grillas et al., 2004). Temporary pools provide a good model for studying plant community dynamics due to their species richness, the varied disturbance regime (duration and frequency of inundation) along steep environmental gradients and state of isolation within contrasting types of dry landscape (forest or farmland). Disturbances play a major role in the dynamics of the vegetation, where species with short life-cycles are dominant (Medail et al., 1998; Rhazi et al., 2006). First, the alternation of wet and dry phases over the course of the annual cycle is equivalent, for the vegetation, to large-scale disturbances leading to the destruction of many individuals. Second, both wild (especially wild boar) and domestic herbivores also create disturbances on a smaller scale by grazing, trampling and rooting. These small-scale disturbances are frequent in temporary pools which are intensively used by cattle and wild boar. Their frequency and intensity are higher in spring and their effects on the species richness of the communities are poorly known. Grazing is often mentioned as a key factor for the conservation of rare plant species in Mediterranean temporary pools (Quézel, 1998). The study of the regeneration of the vegetation after small-scale disturbance is, therefore, of interest for a better understanding of the role of disturbance in vegetation and for assessing the resilience of these communities and the implications for conservation.

The objective of this study was to test the following hypotheses:

(1) Following disturbance, the recolonisation of a patch is possible and rapid via immigration from neighbouring undisturbed patches.
(2) The order of arrival of species in a disturbed patch depends on:
    - their abundance in the main populations at undisturbed patches
    - annual or perennial nature of the species
    - the size of the seeds and the presence of mechanisms facilitating dispersal; small-seed species with or without dispersal mechanisms will colonise more rapidly than species with bigger seeds and without dispersal mechanisms.

Materials and methods

Study area

The Benslimane region (western Morocco) is situated on the Atlantic coast between Rabat and Casablanca. The bioclimate is semi-arid Mediterranean with mild winters, mean precipitation 450 mm/year, mean minimum temperature 7.5°C and mean maximum temperature 29.5°C (Zidane, 1990). This region is characterised by its great abundance of temporary pools (2% of the total surface area of the region) with a wide range of size, shape, depth and location (Rhazi et al., 2006). Despite this variety, the temporary pools have features in common, associated with the hydrological regime (alternating dry and wet phases) and the composition of the vegetation. Within this system of temporary pools, a site of surface area 2,900 m^2 (33°38.497′ N; 07°05.242′ W) situated in the Benslimane cork oak *Quercus suber* woodland, was chosen for this study. This pool is grazed by cattle, sheep and goats and used as a feeding place by wild boar. Three fairly distinct zones of vegetation can be distinguished along the topographical gradient (Rhazi et al., 2001, 2006): the centre dominated by hydrophytes, an intermediate zone where semi-aquatic species predominate, and an edge zone dominated by terrestrial species.

The 2 years of the study were hydrologically different. The year 2006–2007 was very dry, with a rainfall total of 115 mm (between September 2006 and August 2007), a maximum depth of water in the pool of 5 cm and an inundation period of 3 weeks (in January). The year 2007–2008 was less dry in comparison, with 271 mm of rain (60% of the annual mean), a maximum depth of water in the pool of 10 cm (January) and an inundation period of about 2 months (15 December to 15 February).

Field experiments

An experiment was carried out at the pool in 2006–2007 in order to understand the process of recolonisation by vegetation following disturbance. For this, 18

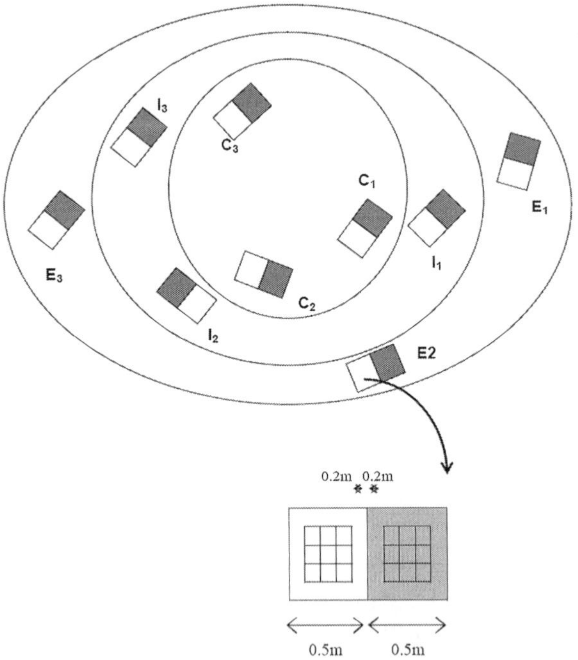

Fig. 1 Location of the experimental plots in the different belts of vernal pool (*E* edge, *I* intermediate and *C* centre; E_1 first replicate, E_2 second replicate and E_3 third replicate). Each square represents a single 0.5 m × 0.5 m plot. The *white* and *grey squares* represent the control and experimental plots, respectively. On the right, plot of 0.5 m × 0.5 m containing the quadrats (0.3 × 0.3 m) divided into nine square of 0.1 × 0.1 m

plots (nine controls, nine experimental) each 0.5 m × 0.5 m, were set up in pairs (Fig. 1) along the topographical gradient, with three replicates per zone (edge, intermediate and centre). In the nine experimental plots, the top 16 cm of soil was removed, heated to 200°C in an autoclave for 3 days to destroy the seed bank (Hanley et al., 2001; Rhazi et al., 2004), and then replaced in the plots in the field. The nine 'control' plots were kept intact throughout the duration of the experiment (2 years: 2006–2007 and 2007–2008). In order to detect any viable seeds remaining in the soil after the heat treatment, samples of soil (1 kg/sample) from the experimental plots were placed in eight pots (18 cm × 18 cm × 13 cm deep) in the laboratory, and kept in conditions favourable for germination, with daily watering, from February to June. Germinations were counted each week and any seedlings were removed after identification.

At the centre of each plot (0.5 × 0.5 m), the vegetation was measured using 0.3 × 0.3 m quadrats on four dates in 2007 (March, April, May and June) and five dates in 2008 (February, March, April, May and June). Water levels were measured on the same dates and using the same quadrats as for vegetation measurements. The ground cover of each species was estimated in each of nine squares (0.1 × 0.1 m) marked out within the 0.3 × 0.3 m quadrats (Fig. 1). This protocol left a 20 cm buffer zone between the experimental and control plot quadrats. For each species, abundance per plot was calculated as its frequency (between 0 and 9) in the quadrat (0.3 × 0.3 m); mean abundance was calculated for each zone separately for the 2 years.

For each of the 18 plots, the total species richness was calculated in each year as the cumulative number of species recorded on all dates when the vegetation was measured. For these same plots, the mean vegetation cover was also calculated separately for the 2 years.

For each species recorded in the vegetation, its annual/perennial nature was determined from the Flora of North Africa (Maire 1952–1987) and the Flora of Morocco (Fennane et al., 1999, 2007), the size of the seeds (length and width) and the presence or otherwise of dispersal structures were noted with reference to on-line databases relating to seeds: http://www2.dijon.inra.fr/hyppa/hyppa-f/hyppa_f.htm (free access), http://www.seedimages.com/ (limited access), http://www.seedatlas.nl (limited access) and to an atlas of seeds (Beijerinck, 1976).

The differences, between the 'control' and 'experimental' plots and between years, in total richness, richness in annuals and perennials, and vegetation cover, were examined using non-parametric Kruskal–Wallis tests. The relationship between the abundance of species in the experimental plots and the control plots in the second year was tested using linear regressions carried out separately for each of the three zones of the pool (centre, intermediate and edge). Relationships between the residuals of the regressions and the size of seeds were examined using linear regressions. Differences between residuals were tested between annual and perennial species and between species whose seeds do and do not have dispersal structures (Kruskal–Wallis).

Results

Over the 2 years during which the field experiment was carried out, a total of 35 species (21 annuals and 14 perennials) were recorded in all the plots, with 27 species present in the control plots of which 15 (56%) were annuals and 12 (44%) were perennials, and 30 species in the experimental plots of which 18 (60%) were annuals and 12 (40%) were perennials. During the first year, 17 species (seven annuals and 10 perennials) were recorded in total for all the plots; all the species were found in the control plots and three (perennials) in the experimental plots. In the second year, 33 species were found in total for whole the plots (20 annuals and 13 perennials), with 25 species in the control plots (14 annuals and 11 perennials) and 30 species in the experimental plots (18 annuals and 12 perennials).

In all the eight pots containing the soil that had been subjected to high temperature (200°C for 3 days), only a single germination (of *Lotus hispidus*) was observed during the whole period of the experiment (February–June).

Post-disturbance recolonisation

During the first year (2007), there was significantly less vegetation cover in the experimental plots than in the control plots (Fig. 2, $\chi^2 = 6.02$; df = 1; $P = 0.01$). In the second year (2008), there was no significant difference in vegetation cover between the two treatments ($\chi^2 = 0.32$; df = 1; $P = 0.56$) (Fig. 2).

Extent of cover by annuals was significantly greater in the second year than in the first, in the control plots as well as the experimental plots (Table 1). Extent of cover by perennials showed a significant increase between the 2 years only in the experimental plots and not in the control plots (Table 1). Total species richness (Fig. 3) was significantly less in the experimental plots than in the

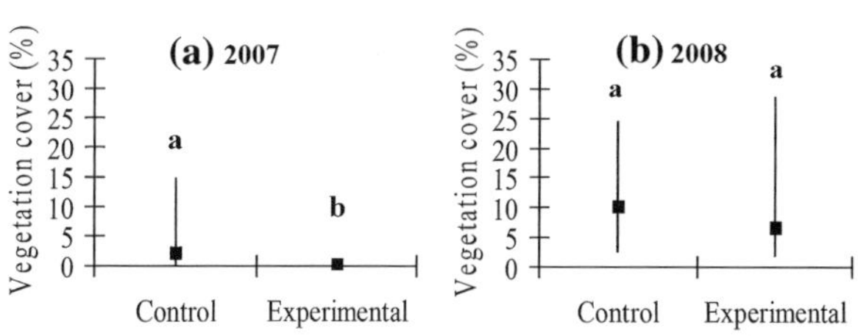

Fig. 2 Variation of vegetation cover (%) in the control and experimental treatments in **a** 2007 and **b** 2008. The median, the min and the max for each treatment are shown on the graph; the different letters on the graph mean significant difference between the treatments ($P < 0.05$)

control plots in the first year ($\chi^2 = 5.65$; df = 1; $P = 0.02$), but there was no significant difference in the second year ($\chi^2 = 0.23$; df = 1; $P = 0.62$). Richness in annuals was significantly greater in 2008 than in 2007, in the control plots as well as in the experimental plots (Table 1). However, richness in perennials showed a significant increase only in the experimental plots and not in the control plots (Table 1).

Secondary succession

First year

In 2007, only three species (all perennials) appeared, in low numbers, in the experimental plots, and 17 species in the control plots. The species appearing in the centre zone were *Bolboschoenus maritimus* and *Eleocharis palustris* in the experimental plots (Table 2) and *Heliotropium supinum* and *E. palustris* in the controls. In the intermediate zone (Table 3), *B. maritimus* and *Narcissus viridiflorus* were present in the experimental plots, whilst *B. maritimus*, *Leontodon saxatile* and *E. palustris* were present in the controls. No species appeared in the experimental plots in the outer zone, whereas *L. saxatile* and *Scilla autumnalis* were the most abundant amongst 13 species (including eight perennials) that were present in the control plots (Table 4).

The establishment of *B. maritimus* and *E. palustris* in the experimental plots clearly took place by means of vegetative spread from individuals established around the edge.

Second year

In 2008, 13 species appeared in the experimental plots in the centre zone (compared with seven in the control plots), seven species in the intermediate zone (compared with nine in controls) and 20 species in the edge zone (compared with 18 in controls; Tables 2, 3, 4).

In the centre zone, seven species were present in both experimental and control plots and six were found only in the experimental plots (Table 2). The most abundant species were the same in both treatments: *Ranunculus baudotii*, *Heliotropium supinum* and *B. maritimus*. In the intermediate zone, six species were common to both experimental and

Table 1 Comparison of the vegetation cover and the species richness of annual and perennial plants in control and experimental treatments with three quartiles: the median, and the lower (25%) and upper quartiles (75%) (Kruskal–Wallis test)

Plots	Test			2007			2008		
	χ^2	df	*P*	25%	50%	75%	25%	50%	75%
Control									
Annual species cover	11.67	1	***	0	0.1	2.1	3.7	11	18.3
Perennial species cover	1.87	1	ns	0.2	2.3	11.4	4.3	7.1	18.8
Annual species richness	8.55	1	**	0	1	2.5	2.5	3	5
Perennial species richness	1.47	1	ns	0.5	2	4	2	3	5
Experimental									
Annual species cover	14.6	1	***	0	0	0	4.6	10.4	24.9
Perennial species cover	11.09	1	***	0	0.1	0.4	1.6	4.5	6.6
Annual species richness	14.68	1	***	0	0	0	2	5	5.5
Perennial species richness	12	1	***	0	0.2	1	0	3	5

ns not significant

*** $P < 0.001$, ** $P < 0.01$

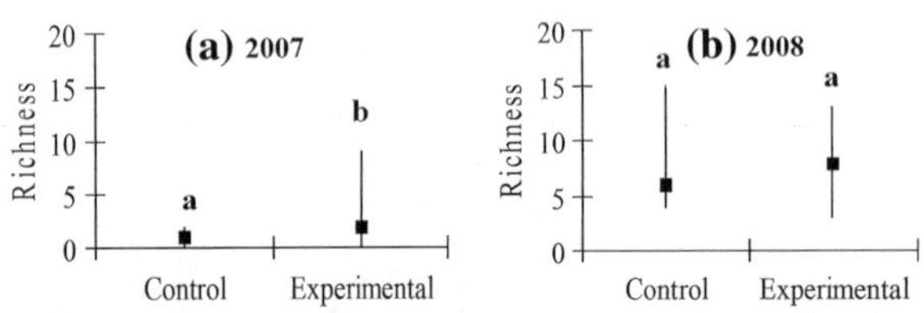

Fig. 3 Variation of the species richness in the control and the experimental treatment in **a** 2007 and **b** 2008. The median, the min and the max for each treatment are shown on the graph; the different letters on the graph mean significant difference between the treatments ($P < 0.05$)

control plots, of which *R. baudotii* and *B. maritimus* were the most abundant. Three species were present only in the control and a single species only in the experimental plots (Table 3). In the edge zone, 15 species were common to experimental and control plots. Eight species were found in only one of the two plot types (three in controls and five in experimental, Table 4) mostly at low levels of abundance.

In each zone, the abundance of species in the experimental plots was significantly correlated with their abundance in the control plots (Fig. 4). The slope and r^2 of these correlations decreased from the centre (slope = 1.09, $r^2 = 0.89$; $P < 0.0001$) to the edge (slope = 0.60, $r^2 = 0.42$, $P < 0.0001$; Fig. 4). Some species did not fit the regression lines very closely, such as *Glyceria fluitans* in the centre zone (Fig. 4a), where it was more abundant in the experimental plots than in the controls. This is also the case in the intermediate zone for *Pulicaria arabica* and *Ranunculus baudotii*, and in the edge zone for *Lolium rigidum*, *Pulicaria arabica* and *Plantago coronopus*, which were more abundant in the experimental plots than in the control plots. Only *Scilla autumnalis* in the edge zone was more abundant in the controls (Fig. 4c).

The residuals of the correlations were not significantly correlated with seed size ($P > 0.05$) and were not significantly different between species with or without seed dispersal structures ($P > 0.05$) or between annual and perennial species ($P > 0.05$).

Discussion

Post-disturbance recolonisation

The field experiment showed that the pool vegetation quickly recolonised the disturbed patches, with considerable differences between years and zones. The two successive years (2007 and 2008) were very dry (25% of mean rainfall) and dry (60% of the mean), respectively, resulting in poor vegetation growth in the pool. Annuals, which are generally predominant in the vegetation of temporary pools (Medail et al., 1998; Grillas et al., 2004) occurred at very low levels of abundance in the pool in 2007, with 41% of the total species richness compared with 81% recorded at the pool over the 10-year period

Table 2 Total richness and the mean abundance of species found in the control and experimental treatments located at the centre of the vernal pool-during 2007 and 2008

	Life span	Abundance of species in the centre			
		2007		2008	
		Control	Experimental	Control	Experimental
Ranunculus baudotii	A			9	9
Heliotropium supinum	A	0.3		5.67	7.67
Bolboschoenus maritimus	P		0.3	2.33	4
Pulicaria arabica	P			2	2.67
Eleocharis palustris	P	1	0.7	1.67	0.67
Damasonium stellatum	A			0.67	0.67
Glyceria fluitans	A			0.33	3.67
Agrostis salmantica	A				0.33
Isoetes velata	P				0.33
Leontodon saxatilis	A				0.33
Myriophyllum alterniflorum	A				0.33
Polygonum aviculare	A				0.33
Rumex crispus	P				0.67
Total richness		2	2	7	13

Each species had a specific life cycle: perennial (P) and annual (A)

Table 3 Total richness and the mean abundance species within control and experimental plots located at the intermediate belt of a vernal pool-during 2007 and 2008

	Life span	Abundance of species in the intermediate belt			
		2007		2008	
		Control	Experimental	Control	Experimental
Ranunculus baudotii	A			9	9
Bolboschoenus maritimus	P	8.7	2	8.67	5
Leontodon saxatilis	A	1.3		6	2.67
Eleocharis palustris	P	5.7		5.33	2.33
Isoetes velata	P			4	1.67
Glyceria fluitans	A			1	
Pulicaria arabica	P			0.33	4.67
Corrigiola litoralis	A			0.33	
Scilla autumnalis	P			0.33	
Baldelia ranunculoides	P				0.33
Narcissus viridiflorus	P		0.3		
Total richness		3	2	9	7

Each species had a specific life cycle: perennial (P) and annual (A)

1997–2006 (Rhazi, unpublished data). During the first year, the experimental patches were characterised by a significantly lower species richness and significantly less extensive vegetation cover than in the controls (Figs. 2, 3).

The first established species at the experimental plots in the first year were the clonal perennials, *Bolboschoenus maritimus* and *Eleocharis palustris* (in the intermediate and centre zones), which colonised vegetatively by means of rhizomes and runners. These species originated in the neighbouring vegetation and colonised the experimental patches via a border effect (Peripherical colonisation; Barrat-Segretain & Bornette, 2000; Crain et al., 2008). Annual plants were completely absent from the experimental patches in the first year.

The absence of any recruitment of annual plants in the experimental patches in the first year is explained by the severe lack of rainfall (110 mm = 25% of the mean). The drought was comparatively less severe in the

Table 4 Total richness and the mean abundance species within control and experimental plots located at the edge belt of a vernal pool-during 2007 and 2008

	Life span	Abundance of species in the edge belt			
		2007		2008	
		Control	Experimental	Control	Experimental
Scilla autumnalis	P	7		7.67	1
Leontodon saxatilis	A	7.7		6.33	6
Lolium rigidum	A	0.3		5	7.33
Filago gallica	A	0.7		4	3.67
Narcissus viridiflorus	P	2		3.33	1.67
Lythrum hyssopifolia	A			3.33	0.67
Lolium perenne	P	2.3		2.67	3
Carlina racemosa	P	4		2.67	1.67
Pulicaria arabica	P	3		2.33	3.67
Ranunculus baudotii	A			2.33	1
Polypogon monspeliensis	A			2.33	0.33
Carex divisa	P	2.7		2.33	
Plantago coronopus	A	1.7		1.33	3.67
Cistus monspeliensis	P			1.33	0.33
Cynodon dactylon	P			1	2.33
Tolpis barbata	A			1	
Trifolium campestre	A			0.67	
Lathyrus angulatus	A			0.33	0.33
Illecebrum verticillatum	A	0.3			1
Juncus bufonius	A				1
Corrigiola litoralis	A				0.33
Crassula tillaea	A				0.33
Polygonum aviculare	A				0.33
Isoetes histrix	P	1.3			
Baldelia ranunculoides	P	1			
Rumex bucephalophorus	A	0.7			
Total richness		13	0	18	20

Each species had a specific life cycle: perennial (P) and annual (A)

second year (271 mm = 60% of the mean). Climatic constraints have been recognised as a decisive factor in the selection of species following disturbances (Lavorel et al., 1994). Also, in temporary pools or deserts (Clauss & Venable, 2000; Angert et al., 2007), annual species adapted to unpredictable conditions have developed life history strategies that allow them not to appear every year and to remain dormant (Bonis, 1993).

The rate of colonisation observed during the first year, therefore, represents a minimum, since it is possible that some species had already dispersed onto the experimental patches but had not been able to develop there.

In the second year, the species richness and vegetation cover in the experimental patches greatly increased, reaching values similar to those for the controls (Figs. 2, 3). Species richness and vegetation cover also increased between the 2 years on the control plots, especially for annuals (Table 1).

The arrival of these annual species in the experimental patches could be associated with different dispersal mechanisms of variable importance, such as transport by water after the first rainfall of the autumn, by the wind, by the movements of mammals (wild and domestic herbivores) and by invertebrates (for example ants, which are abundant on the site). The arrival rate of plant species at the experimental patches probably varies depending on the dispersal mechanisms. For some species, climatic and hydrological conditions may play a part. For such species,

Fig. 4 Correlation (linear regression) between the abundance of species during 2008 in control and experimental treatment; the species were remote from the regression line are identified: *G.f*: *Glyceria fluitans*; *P.a*: *Pulicaria arabica*; *P.c*: *Plantago coronopus*; *L.r*: *Lolium rigidum*; *S.a*: *Scilla autumnalis*; **a** centre belt, **b** intermediate belt, **c** edge belt

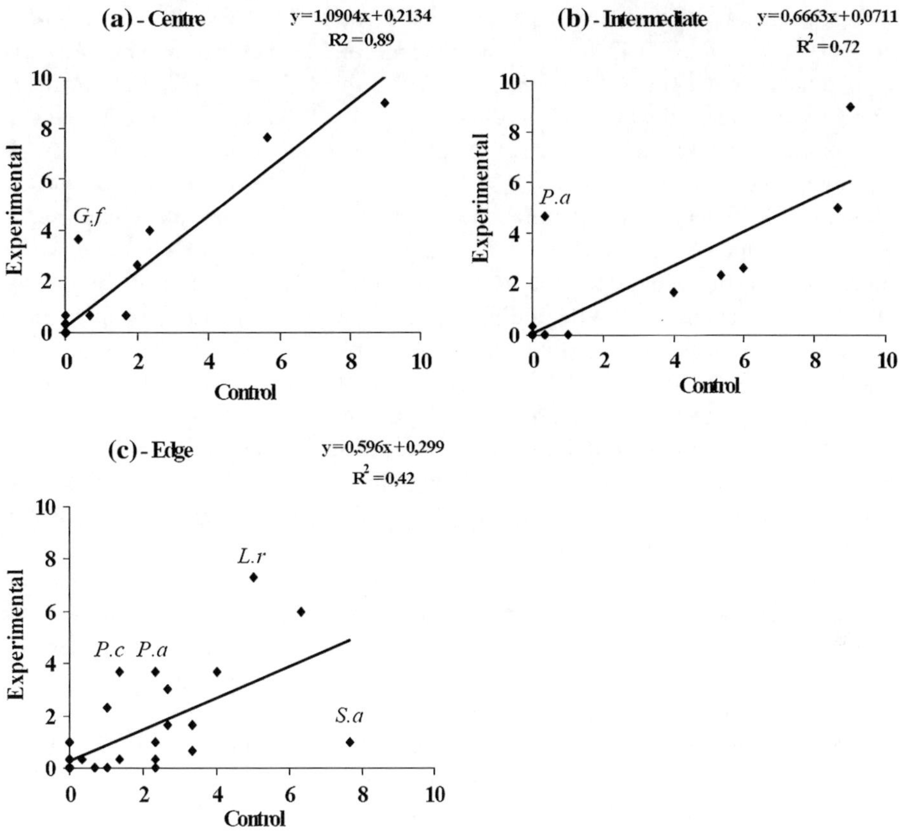

hydrochory was impossible in the first year and probably remained insignificant in the second year. Ectozoochory was probably facilitated in the second year when the dampness of the soil favoured the adherence of the sediment and the seeds contained in it to animals (Vanschoenwinkel et al., 2008) which would have transported them from one patch to another within the pool. Rhazi et al. (2001) found high densities of seeds at this site (91,600 ± 44,450 seeds/m^2 in the centre of the pool, 109,355 ± 44,448 seeds/m^2 in the intermediate zone and 136,066 ± 70,861 seeds/m^2 in the edge zone of the pool).

The development of the vegetation in the experimental patches is interpreted as post-disturbance recolonisation via the arrival of propagules. Sterilisation of the soil at 200°C destroyed the seed bank, as confirmed by the laboratory test in which only a single germination (*Lotus hispidus*) was obtained over 5 months. It is possible that this single germination resulted from contamination in the greenhouse by unsterilised (untreated) sediment. Exposure of seeds to high temperatures for several days' duration affects the pre-germination and growth of seeds in the soil (Hanley & Fenner, 1998; Hanley et al., 2001). The temperature used in this experiment was greater than or equal to that recommended for sterilising soils for the purposes of seed bank studies (Rhazi et al., 2004).

The degree of similarity between the experimental and control plots was measured for each zone using the linear correlation between the abundance of each species in the two treatments. The slope (a) of the regression line gives a measure of the similarity in abundance of the species (equal in the case where $a = 1$), and R^2 measures the scatter (variance) of the individual species around this regression line. In the centre of the pool, the slope ($a = 1.09$) and ($R^2 = 0.89$) of the regression line show that the experimental plots had almost returned to their original (control) condition. For the intermediate and edge zones, the abundance of species in the experimental patches was always lower than in the controls (slopes 0.5 and 0.6, respectively, Fig. 4) and the variance around the overall pattern was greater ($R^2 = 0.72$ and $R^2 = 0.42$ for the intermediate and edge zones, respectively). These results indicate that after 2 years, the original condition of the community returned more quickly in the centre of the pool than at the edge.

A possible explanation is that the species richness of the vegetation increases from the centre to the edge, with a concomitant increase in the diversity of life history traits and thus of the individual responses of species to disturbances (Lenssen et al., 1999; Rhazi et al., 2001; Collinge, 2003). Another hypothesis is that hydrological conditions (depth and duration of inundation) will influence the speed of recovery of the vegetation along the topographical gradient. The mechanisms involved could be linked with the less intense interspecific competition resulting from low species richness, the proportions of perennials and annuals in the vegetation, and differences in primary production along the topographical and hydromorphic gradient (which would be particularly noticeable in dry years). Competition from clonal perennials (*Bolboschoenus maritimus*, *Eleocharis palustris*) is probably greatest in the intermediate zone, where they had become established in the first year, and it could have restricted the germination and establishment of species in the patches (Grime, 1973; Rhazi et al., 2001). The drought could have restricted the appearance of species especially in the edge zone, and conversely favoured primary production and the production of seeds in the centre thanks to the wetter conditions. The hypothesis that there is a greater degree of local dispersal (at a ½ m scale) of seeds in the centre compared with the edge cannot be rejected, in particular, in relation to the flooded or saturated phase (which was not observed at the edge over the 2 years of study). The transport of seeds by animals (ectozoochory) was probably facilitated in the centre where the sediments remain damp and sticky for longer periods.

Some species diverge from the general pattern shown by the correlations between the abundance of species in experimental and control plots (Fig. 4). This is the case for *Glyceria fluitans*, which was four times as abundant in the experimental patches as in the controls (Fig. 4a). This species, which is very water-demanding and abundant in the seedbank, was low in abundance in the centre of the pool in 2008 compared with average or wet years (Rhazi et al., 2001). Similarly, *Pulicaria arabica* in the intermediate zone (Fig. 4b) and *Lolium rigidum*, *Plantago coronopus* and *P. arabica* in the edge zone (Fig. 4c) were more abundant in experimental than in control patches. The greater abundance of these species in experimental patches may results from their efficient dispersal, which could be associated with the small size of their seeds (less than 1.8/1.2 mm = length/width), and also with the presence of a dispersal structure (pappus) in the case of *P. arabica* (Asteraceae), which becomes abundant during the dry phase. However, the analysis of the plant traits did not show any significant effect on the rate of colonisation after disturbance.

The abundance of *Scilla autumnalis* was low in the experimental patches, whilst it was more abundant in the controls (edge zone, Fig. 4c). This is a bulbous perennial plant which produces few large seeds (3 mm/2.1 mm = length/width) and hence has a poor capacity for dispersal by seeds and almost none by vegetative spread.

Implications for the recovery of temporary pools from disturbances

Over a fairly short period of time (2 years), the vegetation in the experimental patches was able to redevelop quickly and was similar to the vegetation of the nearby control patches. This demonstrates the effects of dispersal from close proximity in the process of recovery from small-scale local disturbances (Manzaneda et al., 2005) which are generally frequent in temporary pools. This result, linked with the scale of the disturbances, reflects the resilience of these habitats following disturbances (Angeler & Moreno, 2007). It is the nearest and the relatively most abundant species which quickly become established. However, their development is subject to hydrological stress, which acts as an environmental filter (Middleton, 1999), and also depends on the species' life history traits (Lavorel & Garnier, 2002; Lake, 2003; Angeler & Moreno, 2007). Superimposed on these local, small-scale disturbances are the large-scale disturbances that are exerted by the climate. In a Mediterranean climate, the frequency of droughts or the occurrence of dry periods during the phase of plant growth constitute major abiotic constraints (Rey & Alcantara, 2000), which determine the selection of species and hence the composition of post-disturbance communities.

Acknowledgments We thank Deirdre Flanagan for help in English, Dr S. D. Muller (University of Montpellier 2) for supporting the project, Florence Daubigney for her logistical and technical support and two anonymous referees for constructive comments which helped in a significant improvement of manuscript. This project has been achieved with the financial support of the EGIDE-CMIFM program (PHC Volubilis AI-N° MA/07/172) and was partly funded by the Fondation Tour du Valat and Fondation MAVA.

References

Airoldi, L., 1998. Roles of disturbance, sediment stress, and substratum retention on spatial dominance in algal turf. Ecology 79: 2759–2770.

Angeler, D. G. & J. M. Moreno, 2007. Zooplancton community resilience after press-type anthropogenic stress in temporary ponds. Ecological Applications 17: 1105–1115.

Angert, A. L., T. E. Huxman, G. A. Barron-Gafford, K. L. Gerst & D. L. Venable, 2007. Linking growth strategies to long-term population dynamics in desert annuals. Journal of Ecology 95: 321–331.

Barrat-Segretain, M. H. & G. Bornette, 2000. Regeneration and colonization abilities of aquatic plant fragments: effects of disturbance seasonality. Hydrobiologia 421: 31–39.

Barrat-Segretain, M. H., G. Bornette & A. Hering-Vilas-Bôas, 1998. Comparative abilities of vegetative regeneration among aquatic plants growing in disturbed habitats. Aquatic Botany 60: 201–211.

Beijerinck, W., 1976. Zadenatlas der Nederlandsche flora ten behoeve van de botanie, palaeontologie, bodemcultuur en warenkennis. Backhuys & Meesters, Amsterdam.

Bertness, M. D. & S. W. Shumway, 1993. Competition and facilitation in marsh plants. The American Naturalist 142: 718–724.

Bisigato, A. J., P. E. Villagra, J. O. Ares & B. E. Rossi, 2008. Vegetation heterogeneity in Monte Desert ecosystems: a multi-scale approach linking patterns and processes. Journal of Arid Environments 73: 182–191.

Bliss, S. A. & P. H. Zedler, 1998. The germination process in vernal pools: sensitivity to environmental conditions and effects on community structure. Oecologia 113: 67–73.

Bonis, A., 1993. Dynamique des populations et mécanismes de coexistence des populations de macrophytes immergées en marais temporaires. Thèse Doctorat de l'Université de Montpellier II.

Chase, J. M. & M. A. Leibold, 2003. Ecological Niches: Linking Classical and Contemporary Approaches. University of Chicago Press, Chicago, IL.

Clauss, M. J. & D. L. Venable, 2000. Seed germination in desert annuals: an empirical test of adaptive bet hedging. The American Naturalist 155: 168–186.

Collinge, S. K., 2003. Germination, early growth, and flowering of a vernal pool annual in response to soil moisture and salinity. Madaroño 50(2): 83–93.

Connell, J. H. & R. O. Slatyer, 1977. Mechanisms of succession in natural communities and their role in community stability and organization. The American Naturalist 111: 1119–1144.

Coomes, D. A. & P. J. Grubb, 2003. Colonization, tolerance, competition and seed-size variation within functional groups. Trends in Ecology & Evolution 18: 283–291.

Cousens, R., C. Dytham & R. Law, 2008. Dispersal in Plants: A Population Perspective. Oxford University Press, Oxford.

Crain, C. M., L. K. Albertson & M. D. Bertness, 2008. Secondary succession dynamics in estuarine marshes across landscape-scale salinity gradients. Ecology 89: 2889–2899.

Eriksson, O., 2000. Seed dispersal and colonization ability of plants – assessment and implications for conservation. Folia Geobotanica 35: 115–123.

Fennane, M., M. Ibn Tattou, J. Mathez, A. Ouyahya & J. El Oualidi (eds), 1999. Flore pratique du Maroc. Manuel de détermination des plantes vasculaires, Vol. 1. Travaux de l'Institut Scientifique, Série Botanique N° 36. Rabat.

Fennane, M., M. Ibn Tattou, A. Ouyahya & J. El Oualidi (eds), 2007. Flore Pratique du Maroc, Manuel de détermination des plantes vasculaires, Vol. 2. Travaux de l'Institut Scientifique, Série Botanique N° 38. Rabat.

Fenner, M. (ed.), 2000. Seeds: The Ecology of Regeneration in Plant Communities. CABI Publishing, Wallingford, UK.

Fraterrigo, J. M. & A. J. Rusak, 2008. Disturbance-driven changes in the variability of patterns and processes. Ecology Letters 11: 756–770.

Fröborg, H. & O. Eriksson, 1997. Local colonization and extinction of field layer plants in a deciduous forest and their dependence upon life history features. Journal of Vegetation Science 8: 395–400.

Grillas, P., P. Gauthier, N. Yavercovski & C. Perennou (eds), 2004. Mediterranean Temporary Pools. Vol. 1. Issues Relating to Conservation, Functioning and Management. Tour du Valat, Arles.

Grime, J. P., 1973. Competitive exclusion in herbaceous vegetation. Nature 242: 344–347.

Grime, J. P., 1985. Towards a functional description of vegetation. In White, J. (ed.), The Population Structure of Vegetation. Dr Junk Publishers, Dordrecht: 503–514.

Hanley, M. E. & M. Fenner, 1998. Pre-germination temperature and the survivorship and onward growth of Mediterranean fire-following plant species. Acta Oecologica 19: 181–187.

Hanley, M. E., M. Fenner & G. Ne'eman, 2001. Pregermination heat shock and seedling growth of fire-following Fabaceae from four Mediterranean-climate regions. Acta Oecologica 22: 315–320.

Herrera, J., 1997. Effects of disturbances on the reproductive potential of *Lavandula stoechas*, a Mediterranean sclerophyllous shurb. Ecography 20: 88–95.

Johnstone, I. M., 1986. Plant invasion windows: a time-based classification of invasion potential. Biological Reviews 61: 369–394.

Lake, P. S., 2003. Ecological effects by drought in flowing waters. Freshwater Biology 48: 1161–1172.

Lavorel, S. & E. Garnier, 2002. Predicting the effects of environmental changes on plant community composition and ecosystem functioning: revisiting the Holy Grail. Functional Ecology 16: 545–556.

Lavorel, S., J. Lepart, M. Debussche, J. Lebreton & J. Beffy, 1994. Small scale disturbances and the maintenance of species diversity in Mediterranean old fields. Oikos 70: 455–473.

Leishman, M. R., 2001. Does the seed size/number tradeoff model determine plant community structure? An assessment of the model mechanisms and their generality. Oikos 93: 294–302.

Lenssen, J., F. Menten, W. V. D. Putten & K. Bloom, 1999. Control of plant species richness and zonation of functional groups along a freshwater flooding gradient. Oikos 86: 523–534.

Maire, R., 1952–1987. Flore de L'Afrique du Nord, Lechevalier, Vol. 16. Paris.

Manzaneda, A. J., U. Sperens & M. B. Garcia, 2005. Effects of microsite disturbances and herbivory on seedling performance in the perennial herb *Helleborus foetidus* (Ranunculaceae). Plant Ecology 179: 73–82.

Medail, F., H. Michaud, J. Molina, G. Paradis & R. Loisel, 1998. Conservation de la flore et de la végétation des mares temporaires dulçaquicole et oligotrophes de France Méditerranéenne. Ecologia Mediterranea 24: 119–134.

Middleton, B. A., 1999. Wetland Restoration, Flood Pulsing, and Disturbance Dynamics. John Wiley and Sons, New York.

Morzaria-Luna, H. N. & J. B. Zedler, 2007. Does seed availability limit plant establishment during salt marsh restoration? Estuaries 30: 12–25.

Neff, K. P. & A. H. Baldwin, 2005. Seed dispersal into wetlands: techniques and results for a restored tidal freshwater marsh. Wetlands 25: 392–404.

Noe, G. B. & J. B. Zedler, 2000. Differential effects of four abiotic factors on the germination of salt marsh annuals. American Journal of Botany 87: 1679–1692.

Pearson, T. R. H., D. F. R. P. Burslem, C. E. Mulins & J. W. Dalling, 2002. Germination ecology of neotropical pioneers: interacting effects of environmental conditions and seed size. Ecology 83: 2798–2807.

Quézel, P., 1998. La vegetation des mares transitoires à Isoetes en region méditerranéenne, intérêt patrimonial et conservation. Ecologia Mediterranea 24(2): 111–117.

Rey, P. & J. M. Alcantara, 2000. Recruitment dynamics of a fleshy-fruited plant (*Olea europaea*), connecting patterns of seed dispersal to seedling establishment. Journal of Ecology 88: 622–633.

Rhazi, L., P. Grillas, L. Tan Ham & D. El Khyari, 2001. The seed bank and the between years dynamics of the vegetation of a Mediterranean temporary pool (NW Morocco). Ecologia Mediterranea 27: 69–88.

Rhazi, M., P. Grillas, A. Charpentier & F. Medail, 2004. Experimental management of Mediterranean temporary pools for conservation of the rare quillwort *Isoetes setacea*. Biological Conservation 118: 675–684.

Rhazi, L., M. Rhazi, P. Grillas & D. El Khyari, 2006. Richness and structure of plant communities in temporary pools from western Morocco: influence of human activities. Hydrobiologia 570: 197–203.

Riis, T., 2008. Dispersal and colonisation of plants in lowland streams: success rates and bottlenecks. Hydrobiologia 596: 341–351.

Soons, M. B., C. Van der Vlugt, B. Van Lith, G. W. Heil & M. Klaassen, 2008. Small seed size increases the potential for dispersal of wetland plants by ducks. Journal of Ecology 96: 619–627.

Van den Broek, T., R. Van Diggelen & R. Bobbink, 2005. Variation in seed buoyancy of species in wetland ecosystems with different flooding dynamics. Journal of Vegetation Science 16: 579–586.

Vanschoenwinkel, B., A. Waterkeyn, T. Vandecaetsbeek, O. Pineau, P. Grillas & L. Brendonck, 2008. Dispersal of freshwater invertebrates by large terrestrial mammals: a case study with wild boar (*Sus scrofa*) in Mediterranean wetlands. Freshwater Biology 53: 2264–2273.

Westoby, M., M. R. Leishman & J. Lord, 1996. Comparative ecology of seed size and dispersal. Philosophical Transactions of the Royal Society London. Series B 351: 1309–1318.

Wetzel, R. G., 2001. Limnology: Lake and River Ecosystems, 3rd ed. Academic Press, San Diego.

White, P. S. & S. T. A. Pickett, 1985. Natural disturbance and patch dynamics: an introduction. In Pickett, S. T. A. & P. S. White (eds), The Ecology of Natural Disturbance and Patch Dynamics. Academic Press, London: 3–13.

Wolters, M. & J. P. Bakker, 2002. Soil seed bank and driftline composition along a successional gradient on a temperate salt marsh. Applied Vegetation Science 5: 55–62.

Zavala, M. A., J. M. Espelta & J. Retana, 2000. Constraints and trade-offs in the Mediterranean plant communities: the case of holm-oak-aleppo pine forest. The Botanical Review 66(1): 119–149.

Zedler, J. B., 2000. Progress in wetland restoration ecology. Trends in Ecology & Evolution 15: 402–407.

Zidane, L., 1990. Etude bioclimatique et étude phyto-écologique des forêts de la province de Benslimane "Ouest Marocain" (Bioclimatic Study and Phyto-ecological Study of the Woodland of the West Moroccan (Province of Benslimane). These 3ème cycle. Univ. Mohammed V, Rabat: 187 pp.

Hydrobiologia (2009) 634:77–86
DOI 10.1007/s10750-009-9894-6

POND CONSERVATION

Experimental study of the effect of hydrology and mechanical soil disturbance on plant communities in Mediterranean temporary pools in Western Morocco

Nargis Sahib · Laïla Rhazi · Mouhssine Rhazi · Patrick Grillas

Published online: 4 August 2009

Abstract Physical soil disturbance and the hydrology of temporary pools affect the biomass, species composition and richness of plant communities. Disturbance liberates sites for the random recruitment of new individuals. The addition of seeds modifies the structure of the communities. In order to verify these hypotheses concerning the vegetation of temporary pools, an experiment was carried out using 72 soil samples collected from a pool in Western Morocco and placed in containers. Three types of laboratory treatments were applied, each combined with control treatments: soil disturbance (control/disturbed), hydrology (flooded, saturated and dry) and seed addition (sowing/no sowing). The total biomass, the annual and perennial species richness were calculated for each sample to test the effects of disturbance, hydrology and seed addition on the biomass and species richness of the various plant communities. The results show that disturbance reduces the total biomass, especially of perennials, but without significantly increasing the richness of annuals. Seed addition does not affect the total biomass and reduces total richness only in saturated soil, where biomass production is high. The most extreme stress conditions (drought and flooding) limit the abundance of species and therefore competition.

Keywords Temporary pools · Biomass · Richness · Disturbance · Competition · Morocco

Guest editors: B. Oertli, R. Cereghino, A. Hull & R. Miracle
Pond Conservation: From Science to Practice. 3rd Conference of the European Pond Conservation Network, Valencia, Spain, 14–16 May 2008

N. Sahib · L. Rhazi
Laboratory of Aquatic Ecology and Environment, Hassan II Aïn Chock University, BP 5366, Maarif, Casablanca, Morocco

M. Rhazi
Department of Biology, Moulay Ismail University, Faculty of Sciences and Techniques of Errachidia, Boutalamine, BP 509, Errachidia, Morocco

P. Grillas (✉)
Tour du Valat, Research Centre for the Conservation of Mediterranean Wetlands, Le Sambuc, 13200 Arles, France
e-mail: grillas@tourduvalat.org

Introduction

In aquatic ecosystems, the establishment, development and species diversity of plant communities are often controlled by a combination of internal and external forces (Mitsch & Gosselink, 2000). The most important external forces include periodic disturbances (fire, herbivores, etc.) and hydrological stress (long-term drought-flooding cycles) which differentially favour particular species (Keddy & Fraser, 2000; Trémolières, 2004; Bornette et al., 2008). Stress is defined as external constraints that limit the rate of biomass production of some or all plant species and disturbance as a mechanism that limits biomass through varying levels of destruction (Grime, 2001). The effects of stress and disturbance are accentuated in habitats that are deficient in resources and, consequently, have a low competition capacity (Gaudet & Keddy, 1995; Van Eck et al. 2005).

The importance of these two factors (stress and disturbance) in the regulation of species distribution and community dynamics is widely known (Koning, 2005; Devictor et al., 2007). Disturbance leads to a considerable change to the structure of communities which can in turn affect population dynamics (Elderd & Doak, 2006). Vegetation responds to disturbance and stress in function of their intensity (Bisigato et al., 2008). Disturbance can act as a force for maintaining biodiversity in aquatic systems (Lepori & Hjerdt, 2006). However, if the disturbance increases in intensity and frequency, it leads to local extinction and a reduction in species richness, whereas an increase in the frequency of stress leads to an increase in richness due to the limitation of competition (Tabacchi & Planty-Tabacchi, 2005).

Trampling is a form of disturbance that affects vegetation both directly and indirectly (Liddle, 1975; Kobayashi et al., 1997); it has a selective impact on the abundance of species (Gallet & Rozé, 2001) and modifies their composition and structure (Liddle, 1997) by creating regeneration patches where seeds can germinate and become established without any significant competition effect (Chambers, 1995; Siemann & Rogers, 2003). The regeneration of plant communities within patches can also occur by means of the storage organs (stolons and rhizomes) from individuals in the area of the disturbance (Barrat-Segretain & Bornette, 2000). The mechanisms enabling species occurrence in disturbed systems involve a trade-off between species' competition capabilities and their tolerance of disturbance (Kneitel & Chase, 2004).

The availability of seeds is a determining factor in the regeneration of disturbed habitats (Tilman, 1997; Zobel et al., 2000; Foster & Tilman, 2003; Foster & Dickson, 2004). In aquatic habitats, the presence of durable seed stocks makes easy regeneration (Van der Valk & Davis, 1978), nevertheless the density of seed banks fluctuates from year to year in function of reproduction phenology and environmental constraints (Leck & Simpson, 1995, Capon & Brock, 2005). After a favourable year, the seed banks of certain species increase in the sediment (Bonis et al., 1995). It remains debatable how the over-abundance of a particular species within the seed bank affects the vegetation composition. Variable results have been obtained in experiments involving the addition of seeds to terrestrial habitats. The addition of seeds to grassland communities had a positive effect on vegetation cover (Tilman, 1997) and biomass production (productivity), whereas no effect was shown in other studies (Wilsey & Polley, 2003; Mouquet et al., 2004).

In the Mediterranean region, temporary pools are wetland habitats with high biodiversity and numerous rare species, particularly in Morocco (Grillas et al., 2004). They are submitted to multiple pressures, both natural and from anthropogenic origin, influencing the structure and composition of their plant communities (Deil, 2005). The interannual rainfall variability of the Mediterranean climate results in considerable variations in the depth and duration of the flooding of the pools (Thiéry, 1991; Grillas et al., 2004). The disturbances caused by the alternating flooded and dry phases that occur during the course of the annual cycle lead to the destruction of many individual plants and result in a high level of species diversity (Médail et al., 1998). In addition, the low productivity resulting principally from summer drought enables the coexistence of plant species that are typically annual, relatively uncompetitive and small in size (Médail et al., 1998). Disturbances generally lead to a reduction in the biomass of perennials, thus, favouring higher species richness, especially annuals, in the plant communities (Rhazi et al., 2005).

In temporary pools, the interannual hydrological variations determine community composition and seed production (Bliss & Zedler, 1998, Grillas & Battedou, 1998). Seed banks are continually modified by the cumulative effects of environmental pressures that have differential impacts on the germination, development and reproduction of species (Bonis et al., 1993, 1995; Brock & Rogers, 1998).

The aim of this study was to test experimentally the following hypotheses:

1. Physical disturbances of the soil affect the structure of communities by reducing the biomass of perennials and increasing the richness of annual species;
2. The abundance of seeds in the seed bank influences the composition of the vegetation;

3. Hydrology (water level variation) plays an essential role in the selection of species.

Materials and methods

Experimental design

The study was conducted on the vegetation of a temporary pool located at Benslimane, Western Morocco (W33°38′697″; N007°05′215″; elevation: 257 m; area: 3760 m^2) in a Cork-oak forest. A set of 72 soil samples (18 cm × 18 cm; 5 cm depth) were randomly collected in December 2006, using a metallic frame (18 cm × 18 cm) within two distinct belts (Rhazi, 2001), the edge and the centre (Fig. 1). Each soil sample was carefully transferred into a plastic container (18 cm × 18 cm × 10 cm depth), to maintain the whole clod soil intact (avoiding mixing). Each container was considered as containing a representative sample of the plant community (Grillas et al., 1992) where the dormant forms dominated (seed bank and bulbs). The samples (72) were placed in the laboratory, subdivided into two sets of 36 samples which were, respectively, submitted to the following treatments (Fig. 1):

Disturbed In the top soil, 5 cm of soil was mixed manually to disturb the vertical structure of the seed bank and to homogenise the soil. This treatment is a simulation of the disturbance made by herbivores by trampling on wet soils and burrowing (wild boar).

Control Sediment core remained intact (no mixing). Each set of 36 samples was secondly separated into three subsets of 12 samples, which were, respectively, submitted to one of the following treatments from early January 2006 to late June 2007:

Flooded (F) During the whole experiment, the soil was kept flooded with 3 cm of water above soil surface.

Wet (W) Samples were watered daily to maintain high humidity of soil (minimum of 95% saturation).

Dry (D) Soil was watered twice a week (15% of saturation).

Finally, each subset of 12 samples was separated into two groups of six samples that were submitted, respectively, to one of the following treatments:

Sowing (seed addition) Under a binocular microscope, 100 seeds of either *Illecebrum verticillatum* or *Spergularia salina* were carefully selected and added into each sample. *I. verticillatum* (an amphibious annual species) seeds were added to the flooded and wet treatments and *S. salina* seeds (a terrestrial annual species) to the dry treatment. The seeds of both species were collected in the previous autumn in the same pool (September 2005) at the end of plant life cycle.

No sowing (control) The samples were randomly placed in the laboratory and randomized weekly during the experiment.

The vegetation of each sample was harvested on three successive dates (March, April and June), respectively, on one randomly selected third of the surface area (3 squares 6 cm × 6 cm, selected from a grid of nine squares). The vegetation was cut at soil surface, separated by species and dried at 60°C for 48 h and the dry weight of each species was measured. The biomass considered for each species was the maximum biomass of the three dates. The total biomass and the contribution of annuals and perennials to the total biomass were calculated. The annual and perennial life span was determined according to the flora of North Africa (Maire, 1952–1987) and flora of Morocco (Fennane et al., 1999, 2007). For each sample and each sampling date (March, April and June), the species richness, as well as the annual and perennial species richness, were calculated.

Data analysis

The differences between the biomass of the 'control' and 'disturbed' treatments and between the species richness of the different hydrological treatments were tested using analysis of variance (ANOVA) ($n = 72$). The effects of hydrology, disturbance, sowing and there interaction on the biomass (total, annuals and perennials) and the species richness (total, annuals and perennials) were tested using multiple regressions excluding the biomass and richness of the sowed species (*I. verticillatum* and *S. salina*) ($n = 72$).

The effects of 'hydrology' and 'sowing' and their interaction on species richness were tested by multiple regressions. The effects of sowing on the species richness (total, annuals and perennials) and on the biomass production (total, annuals and perennials)

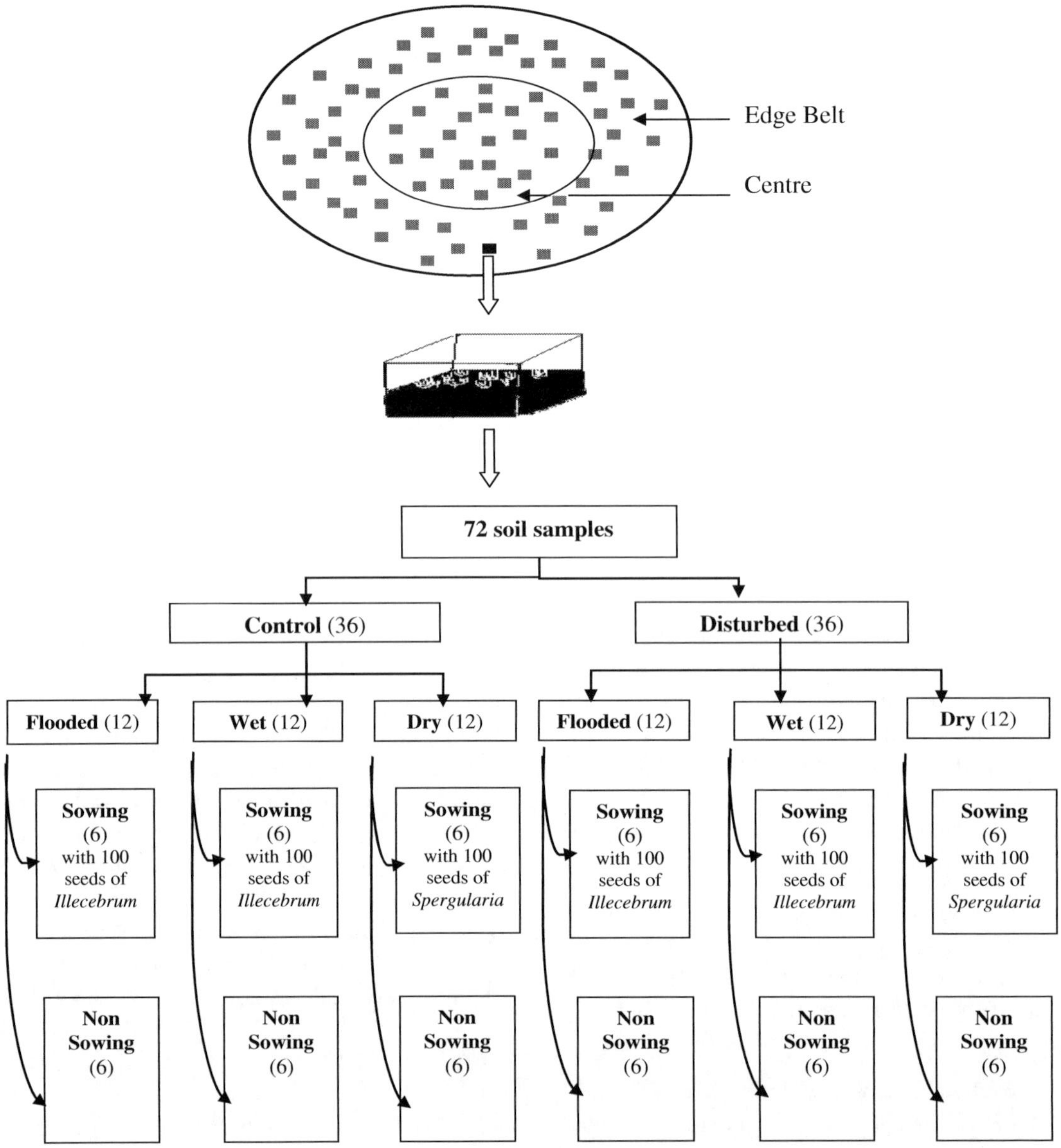

Fig. 1 Schematic representation of the protocol; the figures given in brackets correspond to the number of replicates per treatment in the analyses

were tested using ANOVA tests for the three hydrological treatments together ($n = 72$) and also separately (flooded, wet, dry) ($n = 24$). The biomass and richness of the sowed species (*I. verticillatum* and *S. salina*) were not included in the analysis.

The effect of the 'disturbed' treatment on the abundance of each occurring species was tested using Kruskal–Wallis non-parametric tests for each hydrological treatment. The abundance per sample for each species was calculated as the ratio of the species biomass over the total biomass.

Results

Biomass production

The total biomass produced (324 cm^2 area) was significantly affected (multiple regression $F = 14.43$; df = 6; $P < 0.01$) by hydrology ($F = 32.56$; df = 2; $P < 0.01$) and by disturbance ($F = 19.5$; df = 1; $P < 0.01$). No significant effect, however, was found for sowing and the interaction between the three factors ($P > 0.05$). The hydrology factor explains 65% of the total variance and 23% of the disturbance. The perennial species biomass was significantly affected (multiple regression $F = 18.3$; df = 6; $P < 0.01$) by hydrology ($F = 48.3$; df = 2; $P < 0.01$) and by disturbance ($F = 9.5$; df = 1; $P < 0.01$); but no significant effect of sowing and of the interaction between these factors ($P > 0.05$). Concerning the biomass of annuals (multiple regression $F = 5.5$; df = 6; $P < 0.01$), a significant effect was found only for hydrology ($F = 4.17$; df = 2; $P < 0.01$).

The total species biomass per sample was significantly higher (25%) in the control than in the disturbed treatment (Table 1). There was no significant difference between the biomass of annual species in the two treatments ($P > 0.05$); however, the biomass of perennials was significantly lower in the 'disturbed' treatment (Table 1) than in the control. No significant correlation was found between the biomass of annual and perennial species ($P > 0.05$).

No significant difference was found between the 'sowing' treatment and the control (no sowing) on the biomass of annuals, of perennials and total, even when considering each hydrological treatment separately (Table 2).

Community richness

During the experiment, 28 species germinated, most of them being annuals (71%) (Table 3). The characteristic of aquatic and amphibious species represented 82% and the terrestrial species, 18%. The total number of species recorded in each treatment was 20 (14 annuals and 6 perennials) in the 'control' treatment and 25 (18 annuals and 7 perennials) in the 'disturbed' treatment. Seventeen species occurred in both treatments, three were only found in the 'control' samples (*Agrostis salmantica*, *Sagina apetala* and *Scilla autumnalis*) and eight were solely present in the 'disturbed' samples (*Lotus hispidus*, *Myosotis sicula*, *Apium inundatum*, *Baldellia ranunculoides*, *Pilularia minuta*, *Eleocharis palustris*, *Daucus carota* and *Chara* sp.).

The hydrology factor had a significant effect on total species richness, as well as on the perennial and annual species richness. No significant effect on species richness was found for the 'disturbed' and 'sowing' treatment or their interaction with the hydrology factor (Table 4). Hydrology explains 78% of the total variance. Total, annual and perennial species richness were significantly higher in the 'wet' treatment when compared to the 'flooded' and 'dry' treatments (Fig. 2).

Table 1 Results of the comparison (ANOVA) between the 'disturbed' and 'control' treatments of the biomass per sample of the annuals, the perennial species and of total biomass

	ANOVA			Means (g.) ± standard error	
	F	df	*P*	Control	Disturbed
Total biomass	10.04	1	<0.01	0.87 ± 0.05	0.64 ± 0.04
Perennial biomass	4.43	1	<0.05	0.76 ± 0.05	0.6 ± 0.04
Annual biomass	1.97	1	ns	0. 4 ± 0.27	0. 2 ± 0.08

ns Not significant

Species response to disturbance

Within the 'wet' treatment, the disturbance had a significant effect only on the abundance of *Pulicaria arabica* ($\chi^2 = 4.02$; df = 1; $P = 0.04$) and *Glyceria fluitans* ($\chi^2 = 4.66$; df = 1; $P < 0.05$), both species being less abundant in the 'disturbed' treatment than in the 'control' (Table 5). Within the 'flooded' treatment, *Bolboschoenus maritimus* was the only one species with a significantly lower abundance in the disturbed treatment than in the control ($\chi^2 = 6.43$; df = 1; $P = 0.01$). The disturbance factor had no significant effect on the abundance of any species of the 'dry' treatment (Table 5).

Effect of sowing on community richness

The total richness was not significantly affected by sowing ($F = 1.23$; df = 1; $P > 0.05$), whereas hydrology significantly affected total richness ($F = 4.87$; df = 2; $P = 0.01$) as did the interaction of hydrology and sowing ($F = 4.30$; df = 2; $P < 0.05$). Within the 'wet' treatment, the species richness of the annuals, perennials and total was significantly lower in the 'sowing' treatment (*I. verticillatum*) (Table 2) than in the control (no sowing). The addition of *I. verticillatum* seeds in the 'flooded' treatment did not have a significant effect on the total, perennial and annual species richness (Table 2). There was no effect on the species richness with the addition of *S. salina* in the 'dry' treatment also (Table 2).

The abundance of *I. verticillatum* was significantly higher in the 'wet' treatment ($\chi^2 = 7.97$; df = 1; $P < 0.01$) when compared to the 'flooded' treatment.

Table 2 Results of the comparison (variance analysis) of the differences in the biomass (gr) and species richness of annuals, perennials and total between sowing (S) and no sowing (Ns) treatments in each of the hydrological treatments ('wet', 'flooded' and 'dry'); seeds added were *Illecebrum* for the 'wet' and 'flooded' treatments and *Spergularia* for the 'dry' treatment

	Wet			Flooded			Dry		
	F	*P*	Comparison	*F*	*P*	Comparison	*F*	*P*	Comparison
Total biomass	0.3	>0.05	ns	0.33	>0.05	ns	0.12	>0.05	ns
Annual biomass	2.9	>0.05	ns	0.33	>0.05	ns	2.12	>0.05	ns
Perennial biomass	1.12	>0.05	ns	0.09	>0.05	ns	0.53	>0.05	ns
Total richness	5.69	0.01	$(Ns)^a(S)^b$	0.88	>0.05	ns	1.43	>0.05	ns
Annual richness	4.47	<0.05	$(Ns)^a(S)^b$	2.58	>0.05	ns	1.9	>0.05	ns
Perennial richness	4.59	<0.05	$(Ns)^a(S)^b$	0.22	>0.05	ns	1.41	>0.05	ns

Different letters in the comparison column indicate a significant difference between the treatments; for each analyse df = 1

ns Not significant

No significant difference was found in the abundance of *Illecebrum* between the disturbed and control treatments ($\chi^2 = 1.59$; df = 1; $P > 0.05$). *I. verticillatum* abundance was significantly higher within the sowed samples ($\chi^2 = 17.12$; df = 1; $P < 0.01$). The abundance of *I. verticillatum* in the disturbed samples was not significantly different between the 'wet' and 'flooded' treatments ($\chi^2 = 0.94$; df = 1; $P = 0.33$). In the control samples, the abundance of *I. verticillatum* was significantly higher in the 'wet' treatment than in the 'flooded' treatment ($\chi^2 = 8.93$; df = 1; $P<0.05$).

Discussion

Effect of disturbance and hydrology on the community

The experiments carried out in the laboratory resulted in the expression of 28 species, mostly annuals (71%). The predominance of annual species in the plant community is characteristic of temporary pools (Zedler, 1987; Grillas et al., 2004; Rhazi et al., 2006; Fraga i Arguimbau, 2008). Annual species frequently predominate in case of unpredictable periods and durations of flooding and drought (Mitchell & Rogers, 1985; Brock, 1986; Médail et al., 1998). The total number of species expressed (28) in the experiment was close to the number of species (31) found in the (30 × 30 cm) quadrats used to monitor the vegetation of the pool (Rhazi, 2001). This slight difference can be explained by the low surface area sampled for the experiment (18 × 18 cm) (Grillas et al., 1992), but it can still be assumed that the number of microcosms used in this study give an accurate picture of the field.

The species richness of the communities in the microcosms can be explained by hydrology alone (78% of variance explained); this factor is often critical for the diversity of plant communities in wetland zones (Keddy & Reznicek, 1986). The richness of communities was higher in saturated soil, with respect to both annuals and perennials, and low in dry or flooded soils (Fig. 2). The alternate conditions of drought and flooding that characterise the annual cycle of temporary pools limit potentially the growth of plants (Bonis et al., 1993; Brewer et al., 1997; Bornette et al., 1998; Trémolières, 2004; Koning, 2005; Bornette et al., 2008). Conversely, saturated conditions are temporarily highly favourable to the development of a large number of species within the communities (Koning, 2005); hence, the high species richness was found in Mediterranean temporary pools.

Biomass production in the microcosms was explained first by hydrology (65% of variance explained) and second by mechanical disturbance of soil (23% of variance explained). The biomass levels produced remained low, in particular, with regard to annual plants, which were predominant in the community in terms of the number of species. The low biomass of annuals could either be due to their small size or to light conditions in the laboratory being lower than natural conditions.

Soil disturbance led to a significant reduction in biomass, particularly of perennials. Physical disturbances are known to alter the storage organs generally responsible for high biomass production (Fahrig

Table 3 List of the species that occurred in each of the six treatments combining hydrological (wet/flooded/dry) and mechanical disturbance (disturbed/control) treatments (72 samples in total); the life cycle of each species (annual or perennial is given from literature, for more details see Materials and methods section)

	Life cycle	Disturbed			Control		
		Wet	Flooded	Dry	Wet	Flooded	Dry
Agrostis salmantica (Lag.) Kunth	A						x
Apium inundatum (L) Reichenb.Fil.	A		x				
Baldellia ranunculoides Lam.	A		x				
Bolboschoenus maritimus L.	P	x			x	x	
Callitriche brutia Petagna.	A	x	x	x	x	x	x
Chara vulgaris Linnaeus.	A		x				
Daucus carotta L.	A	x					
Elatine brochonii Clavaud.	A		x			x	
Eleocharis palustris (L) Roem. & Schult.	P	x					
Exaculum pusillum (Lam.) Caruel.	A			x	x		
Glyceria fluitans (L) R.Br.	A	x	x	x	x	x	x
Illecebrum verticillatum L.	A	x	x	x	x	x	x
Isoetes velata A.Braun.	P	x	x	x	x	x	x
Juncus pygmaeus L.C.M.Richard.	A	x	x	x	x	x	x
Kichxia commutata (Reichenb) Fritch	A	x			x		x
Lotus hispidus DC.	A	x		x			
Lythrum hyssopifolia L.	A		x				x
Mentha pulegium L.	P		x	x	x	x	x
Myosotis sicula Guss.	A		x				
Narcisus viridiflorus Schousb.	P	x		x	x		
Pilularia minuta Durieu	P		x				
Polypogon monspeliensis L.	A	x		x	x		x
Pulicaria arabica (L) Cass.	P	x	x	x	x	x	x
Ranunculus ophioglossifolius Vill.	A	x	x	x	x		x
Ranunculus peltatus Schrank	A	x	x	x			x
Sagina apelata Ard.	A				x		
Scilla autumnalis L.	P				x		
Spergularia salina J. Presl & C. Presl	A	x		x			x

Table 4 Multiple regressions testing the effect of hydrology, disturbance, sowing and their interactions on the species richness in annuals, perennials and total

	Total richness			Annual richness			Perennial richness		
	F	df	*P*	*F*	df	*P*	*F*	df	*P*
Hydrology	4.30	2	<0.01	3.65	2	<0.05	3.77	2	<0.05
Disturbance	0.47	1	ns	0.61	1	ns	0.19	1	ns
Sowing	0.06	1	ns	0.14	1	ns	2.34	1	ns
Hydro*Disturb*Sow	0.93	2	ns	0.50	2	ns	2.43	2	ns

ns Not significant

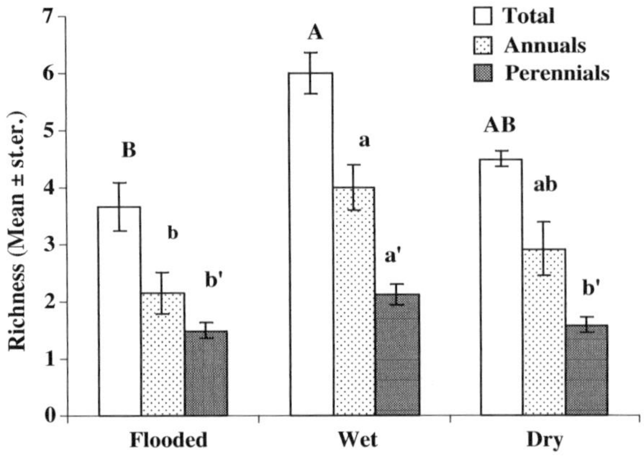

Fig. 2 Effects of the hydrological treatments on the total species richness and the numbers of annuals and perennials; the differences are tested by ANOVAs and different letters indicate significant differences between treatments

et al., 1994; Winkler & Fischer, 2001). The species negatively affected by disturbance were the perennials *B. maritimus* and *P. arabica*, together with the annual *G. fluitans*. The expected result was an increase in the species richness of annuals following the reduction in the biomass of perennials. However, no significant effect of soil disturbance on the species richness of annuals was revealed (Table 4). This could be due to:

1. The burying of seeds after mixing of the sediment. Burial considerably reduces the germination of seeds not only because of the lack of light but also by raising to the surface older seeds with lower germinating power (Bonis et al., 1993; Devictor et al., 2007).
2. The biomass of perennials was too low to induce competition between species (production too low and experiment duration too short).

Effect of seed addition on community structure

Adding seeds of *Illecebrum* and *Spergularia* did not affect the total biomass produced in the microcosms but, in the case of saturated soil only, it did have a significant effect on the total richness of the community. The total biomass of *Illecebrum* was significantly higher in saturated soil and lower in flooded and dry soils. The addition of *Illecebrum* seeds combined with the saturated soil treatment led to a reduction in total species richness due to the high growth rate of the *Illecebrum* competing with the community species. This finding is in agreement with the results of Brewer et al. (1997) and Lenssen et al. (1999), who found that competition only plays an important role in the rarely flooded parts of wetland habitats. Conversely, in case of flooding or drought, the production of the community is low. Consequently, the growth of seed-added species remains limited and does not affect total richness or competition intensity.

These results suggest that fluctuations in the reproductive success of species in Mediterranean temporary pools (simulated experimentally by seed addition) only affect the richness of communities during the course of the subsequent cycle if the hydrological conditions are favourable (saturated).

Implications for the conservation of temporary pools

The results obtained in this study confirm that in Mediterranean temporary pools, as in wetland in general, hydrology is the primordial factor that structures and selects the species of the community. The hydrology modifies any disturbance effects and influences competition intensity through its impact on

Table 5 Comparison (Kruskal–Wallis) of the effects of the mechanical disturbance treatment (*D* disturbed; *C* control) in each of the three hydrological conditions (wet, flooded and dry) on the abundance of individual species

	Freq.	Wet			Flooded			Dry		
		χ^2	*P*	Comparison	χ^2	*P*	Comparison	χ^2	*P*	Comparison
Glyceria fluitans	13.58	4.66	<0.05	$D < C$	0.17	>0.05	ns	0.38	>0.05	ns
Pulicaria arabica	14.2	4.02	<0.05	$D < C$	0.27	>0.05	ns	1.25	>0.05	ns
Bolboschoenus maritimus	2.01	2.66	>0.05	ns	6.43	0.01	$D < C$	0	>0.05	ns

Only species with significantly different abundances between treatments are presented in the table; for each analyse df = 1

ns Not significant

primary production. Competition only plays a role in the structuring of communities when production conditions are favourable (saturation of the sediment by water). Such conditions in the pools are transitory and their duration varies considerably from year to year, thus reducing the role of this factor in the long term.

The results of these experiments show that local disturbance (simulating that caused by herbivores) does not seem to affect the richness of temporary pool communities. This can be explained by a low degree of seed bank stratification resulting from the high frequency of this type of disturbance. Although surface seeds are buried, the previously buried seeds thus exposed still conserve high germinating power. This result does not mean that disturbance by herbivores, especially during the plants' growing season, is not likely to have a considerable impact on the vegetation and its richness, but rather that its impact can be modulated.

Acknowledgements We thank Deirdre Flanagan for help in English, Dr S. D. Muller (University of Montpellier 2) for supporting the project, Florence Daubigney for her logistical and technical support and two anonymous referees for constructive comments which helped in a significant improvement of manuscript. This project has been achieved with the financial support of the Egide Volubilis programme AI N° MA/07/172 and was partly funded by the Fondation Tour du Valat and Fondation MAVA.

References

Barrat-Segretain, M. H. & G. Bornette, 2000. Regeneration and colonization abilities of aquatic plant fragments: effects of disturbance seasonality. Hydrobiologia 421: 31–39.

Bisigato, A. J., P. E. Villagra, J. O. Ares & B. E. Rossi, 2008. Vegetation heterogeneity in Monte Desert ecosystems: a multi-scale approach linking patterns and processes. Journal of Arid Environments 73: 182–191.

Bliss, S. A. & P. A. Zedler, 1998. The germination process in vernal pools: sensitivity to environmental conditions and effects on community structure. Oecologia 113: 67–73.

Bonis, A., P. Grillas, C. Van Wijck & J. Lepart, 1993. The effect of salinity on the reproduction of coastal submerged macrophytes from experimental communities. Journal of Vegetation Science 4(4): 461–468.

Bonis, A., P. Grillas & J. Lepart, 1995. Seed bank dynamics and coexistence of annual macrophytes in temporary and variable habitat. Oikos 74: 81–92.

Bornette, G., C. Amoros & N. Lamouroux, 1998. Aquatic plant diversity in riverine wetlands: the role of connectivity. Freshwater Biology 39: 267–283.

Bornette, G., E. Tabacchi, C. Hupp, S. Puijalon & J. C. Rostan, 2008. A model of plant strategies in fluvial hydrosystems. Freshwater Biology 53: 1692–1705.

Brewer, J. S., J. M. Levin & M. D. Bertness, 1997. Effects of biomass removal and elevation on species richness in a New England salt marsh. Oikos 80: 333–341.

Brock, M. A., 1986. Adaptation to fluctuations rather than to extremes of environmental parameters. In de Deckker, P. & W. D. Williams (eds), Limnology in Australia. CSIRO/ Dr W. Junk, Melbourne/Netherlands: 131–140.

Brock, M. A. & K. H. Rogers, 1998. The regeneration potential of the seed bank of an ephemeral floodplain in South Africa. Aquatic Botany 61: 123–135.

Capon, S. J. & M. A. Brock, 2005. Flooding, soil seed bank dynamics and vegetation resilience of a hydrologically variable desert floodplain. Freshwater Biology 51(2): 206–223.

Chambers, J. C., 1995. Relationship between seed fates and seedling establishment in an alpine ecosystem. Ecology 76: 2124–2133.

Deil, U., 2005. A review on habitats, plant traits and vegetation of ephemeral wetlands – a global perspective. Phytocoenologia 35: 533–705.

Devictor, V., J. Moret & N. Machon, 2007. Impact of ploughing on soil seed bank dynamics in temporary pools. Plant Ecology 192(1): 45–53.

Elderd, B. D. & D. F. Doak, 2006. Comparing the direct- and community-mediated effects of disturbance on plant population dynamics: flooding, herbivory and *Mimulus guttatus*. Journal of Ecology 94(3): 656–669.

Fahrig, L., D. P. Coffin, W. K. Lauenroth & H. H. Shugart, 1994. The advantage of long-distance clonal spreading in highly disturbed habitats. Evolutionary Ecology 8(2): 172–187.

Fennane, M., M. Ibn Tattou, J. Mathez, A. Ouyahya & J. El Oualidi (eds), 1999. Flore pratique du Maroc. Manuel de détermination des plantes vasculaires, Vol. 1. Travaux de l'Institut Scientifique, Série Botanique No. 36, Rabat.

Fennane, M., M. Ibn Tattou, A. Ouyahya & J. El Oualidi (eds), 2007. Flore Pratique du Maroc, Manuel de détermination des plantes vasculaires, Vol. 2. Travaux de l'Institut Scientifique, Série Botanique No. 38. Rabat.

Foster, B. L. & T. Dickson, 2004. Grassland diversity and productivity: the interplay of resource availability and propagule pools. Ecology 85: 1541–1547.

Foster, B. L. & D. Tilman, 2003. Seed limitation and the regulation of community structure in Oak Savanna grassland. Journal of Ecology 91: 999–1007.

Fraga i Arguimbau, P., 2008. Vascular flora associated to Mediterranean temporary ponds on the island of Minorca. Anales del Jardín Botánico de Madrid 65(2): 393–414.

Gallet, S. & F. Rozé, 2001. Resistance of Atlantic heathlands to trampling in Brittany (France): influence of vegetation type, season and weather conditions. Biological Conservation 97: 189–198.

Gaudet, C. L. & P. A. Keddy, 1995. Competitive performance and species distribution in shoreline plant communities: a comparative approach. Ecology 76: 280–291.

Grillas, P. & G. Battedou, 1998. Effects of the date of flooding on the biomass, species composition and seed production of submerged macrophyte beds in temporary marshes in the Camargue (S. France). Proceedings of the Intecol Conference, Perth, September 1996. In Wetlands for the Future (McComb A.J. & J.A. Davis, eds), INTECOL'S V International Wetland Conference, pp: 207–218.

Grillas, P., V. Boy & C. Van Wijck, 1992. Transferring sediment containing seed bank: a method for studying plant community ecology. Hydrobiologia 228(1): 29–36.

Grillas, P., P. Gauthier, N. Yavercovski & C. Perennou, 2004. Mediterranean temporary pools: issues relating to conservation functioning and management, Vol. 1. Tour du Valat, Arles.

Grime, J. P., 2001. Plant Strategies, Vegetation Processes, and Ecosystem Properties. Wiley, Chichester.

Keddy, P. A. & L. H. Fraser, 2000. Four general principles for the management and conservation of wetlands in large lakes: the role of water levels, nutrients, competitive hierarchies and centrifugal organization. Lakes and Reservoirs: Research and Management 5: 177–185.

Keddy, P. A. & A. A. Reznicek, 1986. Great Lakes vegetation dynamics: the role of fluctuating water levels and buried seeds. Journal of Great Lakes Research 12: 25–36.

Kneitel, J. M. & J. M. Chase, 2004. Trade-offs in community ecology: linking spatial scales and species coexistence. Ecology Letters 7: 69–80.

Kobayashi, T., Y. Hori & N. Nomoto, 1997. Effects of trampling and vegetation removal on species diversity and micro-environment under different shade conditions. Journal of Vegetation Science 8: 873–880.

Koning, C., 2005. Vegetation patterns resulting from spatial and temporal variability in hydrology, soils and trampling in a northern basin marsh, New Hampshire, USA. Wetlands 25(2): 239–251.

Leck, M. A. & R. L. Simpson, 1995. Ten years seed bank and vegetation dynamics of tidal freshwater marsh. American Journal of Botany 82: 1547–1557.

Lenssen, J. P. M., F. B. J. Menting, W. H. van der Putten & C. W. P. M. Blom, 1999. Control of plant species richness and zonation of functional groups along a freshwater flooding gradient. Oikos 86: 523–534.

Lepori, F. & N. Hjerdt, 2006. Disturbance and aquatic biodiversity: reconciling contrasting views. Bioscience 56: 809–818.

Liddle, M. J., 1975. A selective review of the ecological effects of human trampling on natural ecosystems. Biological Conservation 7: 17–36.

Liddle, M. J., 1997. Recreation Ecology. Chapman & Hall, London.

Maire, R., 1952–1987. Flore de L'Afrique du Nord, Vol. 16. Paul Lechevalier, Paris.

Médail, F., H. Michaud, J. Molina, G. Paradis & R. Loisel, 1998. Conservation de la flore et de la végétation des mares temporaires dulçaquicoles et oligotrophes de France Méditerranéenne. Ecologia Medediterranea 24(2): 119–134.

Mitchell, D. S. & K. H. Rogers, 1985. Seasonality/aseasonality of aquatic macrophytes in Southern hemisphere inland waters. Hydrobiologia 125: 137–150.

Mitsch, W. J. & J. G. Gosselink, 2000. Wetlands, 3rd edn. Wiley, New York, NY, USA.

Mouquet, N., P. Leadley, J. Meriguet & M. Loreau, 2004. Immigration and local competition in herbaceous plant communities: a three-year seed-sowing experiment. Oikos 104: 77–90.

Rhazi, L., 2001. Etude de la végétation des mares temporaires et l'impact des activités humaines sur la richesse et la conservation des espèces rares au Maroc. Thèse de Doctorat, Université Hassan II Faculté des Sciences Ain Chock, Casablanca.

Rhazi, M., P. Grillas, F. Médail & L. Rhazi, 2005. Consequences of shrub dynamics on the richness of aquatic vegetation in oligotrophic seasonal pools in Southern France. Phytocoenologia 35: 489–510.

Rhazi, L., M. Rhazi, P. Grillas & D. El Khyari, 2006. Richness and structure of plant communities in temporary pools from Western Morocco: influence of human activities. Hydrobiologia 570: 197–203.

Siemann, E. & W. E. Rogers, 2003. Reduced resistance of invasive varieties of the alien tree *Sapium sebiferum* to a generalist herbivore. Oecologia 135: 451–457.

Tabacchi, E. & A. M. Planty-Tabacchi, 2005. Exotic and native plant community distributions within complex riparian landscapes: a positive correlation. Ecoscience 12: 423–434.

Thiéry, A., 1991. Multispecies coexistence of branchiopods (Anostraca, Notostraca & Spinicaudata) in temporary ponds of Chaouia plain (Western Morocco): sympatry or syntopy between usually allopatric species. Hydrobiologia 212: 117–136.

Tilman, D., 1997. Community invisibility, recruitment limitation, and grassland biodiversity. Ecology 78: 81–92.

Trémolières, M., 2004. Plant response strategies to stress and disturbance: the case of aquatic plants. Journal of Bioscience 29: 461–470.

Van der Valk, A. G. & C. B. Davis, 1978. The role of seed banks in the vegetation dynamics of prairie glacial marshes. Ecology 59: 322–335.

Van Eck, W. H. J. M., H. M. Van de Steeg, C. W. P. M. Blom & H. de Kroon, 2005. Recruitment limitation along disturbance gradients in river floodplains. Journal of Vegetation Science 16: 103–110.

Wilsey, B. J. & H. W. Polley, 2003. Effects of seed additions and grazing history on diversity and productivity of subhumid grasslands. Ecology 84: 920–931.

Winkler, E. & M. Fischer, 2001. The role of vegetative spread and seed dispersal for optimal life histories of clonal plants: a simulation study. Evolutionary Ecology 15: 281–301.

Zedler, P. H., 1987. The ecology of Southern California vernal pools: a community profile. U.S. Fish and Wildlife Service Biological Report 85: 7–11.

Zobel, M., M. Otsu, J. Tiira, M. Moora & T. Mols, 2000. Is small-scale species richness limited by seed availability or microsite availability? Ecology 81: 3274–3282.

Hydrobiologia (2009) 634:87–95
DOI 10.1007/s10750-009-9884-8

POND CONSERVATION

Restoring ponds for amphibians: a success story

R. Rannap · A. Lõhmus · L. Briggs

Published online: 29 July 2009

Abstract Large-scale restoration of quality habitats is often considered essential for the recovery of threatened pond-breeding amphibians but only a few successful cases are documented, so far. We describe a landscape-scale restoration project targeted at two declining species—the crested newt (*Triturus cristatus* Laur.) and the common spadefoot toad (*Pelobates fuscus* Wagler)—in six protected areas in southern and southeastern Estonia. The ponds were restored or created in clusters to (i) increase the density and number of breeding sites at local and landscape levels; (ii) provide adjacent ponds with differing depths, hydroperiods and littoral zones and (iii) restore an array of wetlands connected to appropriate terrestrial habitat. In only 3 years, where 22 of the 405 existing ponds (5%) were restored and 208 new ponds (51%) created, the number of ponds occupied by the common spadefoot toad increased 6.5 times. Concerning the crested newt and the moor frog (*Rana arvalis* Nilsson), the increase was 2.3 and 2.5 times, respectively. The target species had breeding attempts in most of the colonised ponds—even more frequently than common species. Also, the amphibian species richness was higher in the restored than in the untreated ponds. The crested newt preferably colonised ponds that had some submerged vegetation and were surrounded by forest or a mosaic of forest and open habitats. The common spadefoot toad favoured ponds having clear and transparent water. Our study reveals that habitat restoration for threatened pond-breeding amphibians can rapidly increase their numbers if the restoration is implemented at the landscape scale, taking into account the habitat requirements of target species and the ecological connectivity of populations. When the remnant populations are strong enough, translocation of individuals may not be necessary.

Keywords Estonia · Aquatic habitat management · Threatened species · *Pelobates fuscus* · *Triturus cristatus*

Guest editors: B. Oertli, R. Cereghino, A. Hull & R. Miracle
Pond Conservation: From Science to Practice. 3rd Conference of the European Pond Conservation Network, Valencia, Spain, 14–16 May 2008

R. Rannap (✉) · A. Lõhmus
Institute of Ecology and Earth Sciences, University of Tartu, Vanemuise 46, 51014 Tartu, Estonia
e-mail: riinu.rannap@envir.ee

R. Rannap
Ministry of the Environment, Narva Road 7A, 15172 Tallinn, Estonia

L. Briggs
Amphi Consult, International Sciencepark Odense, Forskerparken 10, 5230 Odense M, Denmark

Introduction

Ponds—small, isolated freshwater bodies—are essential habitat supporting considerably more species, more unique species and more scarce species than rivers, ditches and streams, thus playing a central role in maintaining high regional biodiversity (Williams et al., 2003). Despite their significant ecological values, pond ecosystems are threatened by a number of human activities: infilling, stocking with fish, pollution, mismanagement, desiccation etc. (Brönmark & Hansson, 2005; Oertli et al., 2005); these are typically related to the loss of ponds' historical function and a changed land use. During the twentieth century, enormous numbers of ponds have vanished—in the European states often more than 50% and occasionally 90% (Hull, 1997)—and those remaining have often lost their quality and connectivity for biota.

Amphibians have the highest proportion of threatened species among higher taxa in the world (Stuart et al., 2004), and—together with dragonflies and aquatic plants—they represent a major pond-dependent taxon comprising numerous critically endangered species (Beebee, 1992; Oertli et al., 2005). Pond-breeding amphibians require both terrestrial and aquatic habitats during their life cycle, which makes them particularly vulnerable to a range of anthropogenic processes, such as landscape cultivation, intensification of agriculture, urban development and road building (Alford et al., 2001; Cushman, 2006). Fortunately, habitat alteration is potentially reversible. Thus, elucidating the factors critical for restoring or maintaining quality habitats (Semlitsch, 2002; Rannap et al., 2009) and, by necessity, large-scale restoration of both aquatic and terrestrial habitats (Fog, 1997; Stumpel & van der Voet, 1998), are essential for amphibian recovery efforts. The ultimate goal is to renew the ecological integrity of degraded wetlands and to create self-sustaining systems for long-term persistence of resident populations (Petranka & Holbrook, 2006), including their metapopulation structure (Semlitsch, 2002).

Despite the obvious necessity, only a few successful examples (Denton et al., 1997; Briggs, 1997, 2001; Petranka et al., 2007) are available for large-scale species-specific habitat restoration for threatened amphibians. Most special restoration has been small-scale and scattered (usually 1–3 ponds locally; Petranka & Holbrook, 2006; Petranka et al., 2007), while traditional approaches to create and restore wetlands may have even reduced regional amphibian diversity (Porej & Hetherington, 2005). Poor results often reflect inadequate planning and a failure to create wetlands suitable for amphibians (Petranka et al., 2007), for instance, due to inappropriate pond morphology, presence of fish or a lack of terrestrial habitat for juveniles and adults (Porej & Hetherington, 2005). For threatened species, it is important to identify particular areas and habitats where the restoration is expected to give the best results (Baker & Halliday, 1999; Nyström et al., 2007), but even then the species may not become established on their own (Pechmann et al., 2001). Thus, to successfully restore amphibian populations it is important to compile and implement biologically based management strategies (Semlitsch, 2002). Due to the complexity of the task and the lack of information, documenting the design and results of successful efforts (especially regarding rare and threatened species) is extremely valuable.

In this article, we describe a large-scale restoration project where 230 ponds were restored or created for two European-wide threatened species—the crested newt (*Triturus cristatus* Laur.) and the common spadefoot toad (*Pelobates fuscus* Wagler). In six protected areas in southern Estonia (Fig. 1), where the population density of these species is the highest in Estonia, we restored numerous clusters of ponds to halt the species' declines and to save from extinction the small and isolated populations of the common spadefoot toad at the northern edge of its distribution range. Most pond clusters comprised only a single remnant breeding pond of either species. We explored natural colonisation of the restored ponds by amphibians over 3 years to determine (1) the colonising species and (2) their speed of colonisation; (3) the efficacy of restoration (in terms of the total number and the number of new breeding ponds) and (4) the habitat characteristics influencing the probability of pond colonisation by target species.

Materials and methods

Study area

An extensive pre-restoration inventory in June 2005 and the following pond restoration were carried out in the two largest Landscape Protected Areas (LPA) of southern Estonia: the Haanja LPA (27°2′ E; 57°43′ N) and the Otepää LPA (26°25′ E; 58°5′ N). Additional ponds were restored/created in the same region in four smaller protected areas (Sadrametsa, 228 ha; Piusa, 53 ha; Hauka, 14 ha; Karste, 9 ha) where isolated

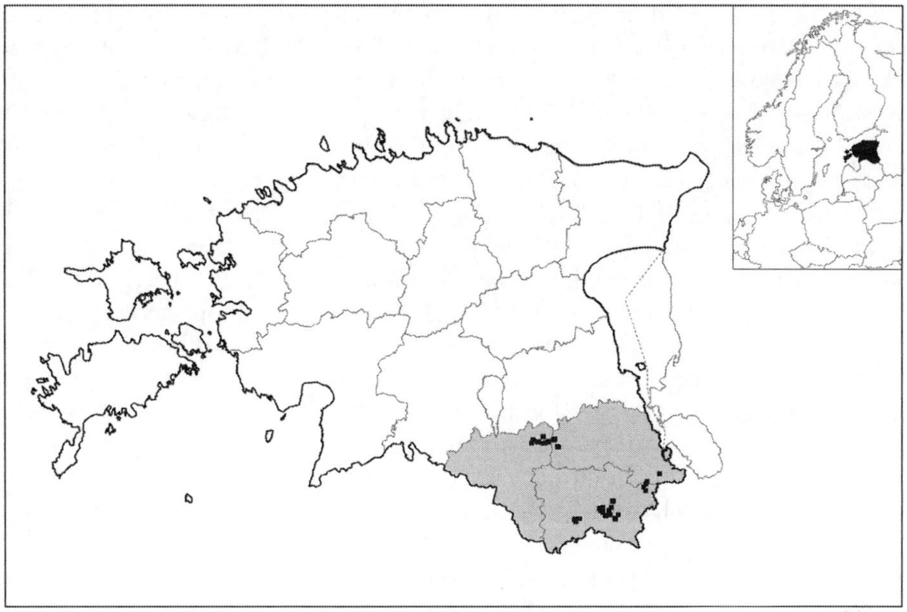

Fig. 1 The location of the study areas with constructed ponds in Estonia

populations of the crested newt and/or the spadefoot toad occurred in 2005. The hilly (altitudes 200–318 m a.s.l.) moraine landscape of the Haanja LPA (16,900 ha) represents a mosaic of forests (45%), grasslands (21%) and small extensively used fields and farmlands. Lakes, ponds, swamps and small bogs are situated in the depressions and valleys between the hills. The Otepää LPA (22,430 ha; 42% forest) also has a varied hilly relief that rises over 100 m above the surrounding plains, but the fields are generally larger than in Haanja though intensive farming practices are not in use (Evestus & Turb, 2002). Both areas have a great number of ponds of very diverse origin—created by natural processes (e.g. glaciation) or human activities (e.g. mineral extraction, cattle watering, water storage). The man-made ponds are situated mainly near human settlements.

Study species

Of the 11 Estonian amphibian species, 8 are found in the southern and southeastern part of the country. The smooth newt (*Triturus vulgaris* L.), the crested newt, the pool frog (*R. lessonae* Camerano) and the edible frog (*R.* kl. *esculenta* L.) are mainly aquatic species, while the common spadefoot toad, the common toad (*Bufo bufo* L.), the common frog (*Rana temporaria* L.) and the moor frog (*R. arvalis* Nilsson) are largely terrestrial. The target species in our study were the crested newt and the common spadefoot toad, which are listed in the Annexes of the EU Habitats Directive (92/43/EEC) encouraging the Member States to achieve favourable conservation status of those species. Also, the moor frog is an Annex IV species of the Habitats Directive, and though it was monitored in this study, there was no necessity for special habitat restoration—the moor frog is one of the most widespread and numerous amphibians in Estonia (Pappel & Rannap, 2007).

Of the target species, the crested newt has declined in most of its range countries (Edgar & Bird, 2006), while the common spadefoot toad has decreased dramatically within its northern distribution range (Fog, 1997; Nyström et al., 2002, 2007), including the range edge in Estonia. Both species strongly depend on permanent fish-free ponds (Joly et al., 2001; Nyström et al., 2002, 2007; Skei et al., 2006) surrounded by suitable terrestrial habitat: for the crested newt—forest or mosaic of forest and (semi)natural grasslands (Joly et al., 2001; Skei et al., 2006; Danoël & Ficetola, 2008; Rannap et al., 2009), for the common spadefoot toad—natural or semi-natural grasslands, small-scale extensively managed vegetable fields or gardens on sandy soils (Nyström et al., 2002, 2007; Stumpel, 2004). The proposed factors for the declines are habitat-related: the loss of ponds, habitat fragmentation, decreased connectivity between ponds, lack of pond management, introduction of fish, changes in agricultural systems etc. (Joly et al., 2001; Nyström et al., 2002; Stumpel, 2004; Edgar & Bird, 2006; Skei et al., 2006). In a crested newt population in southeastern Sweden (climatically similar to Estonia), population modelling has highlighted the importance of pond restoration and increased pond density due to the crucial role of the early life-cycle stages (Karlsson et al., 2007). Yet, for the crested newt only limited conservation work has taken place, so far (Edgar & Bird, 2006). For the common spadefoot toad, pond restoration and creation have been carried out at different scales in Denmark (Fog, 1997; Briggs et al., 2008), Sweden (Nyström et al., 2007), and the Netherlands (Stumpel, 2004), but the toad's reproductive success (Nyström et al., 2007) or colonisation rate (Stumpel, 2004; Briggs et al., 2008) has remained low.

Field methods and habitat restoration techniques

During the pre-restoration inventory in June 2005, 12 herpetologists from seven European countries checked 405 natural and man-made ponds, including natural depressions, beaver ponds, cattle ponds, garden ponds, sauna ponds and ponds historically used for flax soaking. Data collection was carefully standardised and simplified: we used a standard dip-netting of larvae (Skei et al., 2006) as the main method for detecting amphibians, and the absence of a species was only concluded after 10 min of dip-netting. In each pond, the dip-net sweeps covered all important microhabitats for amphibians. In addition, eggs of newts and egg-clusters of the 'green frogs' (the pool frog and the edible frog) were searched for. Due to the single visit to each pond, random effects in the number of caught individuals were still probably large and we used only presence–absence for analyses (reminding that 'absence' may include undetected presence in some cases). For each pond, we estimated the presence of fish and the pond quality for amphibian breeding. A pond was considered of high-quality if no extensive negative effects were observed, such as overgrowing (complete cover of bushes or tall vegetation such as *Typha latifolia* L.), eutrophication or silting (water unclear and full of algae, a thick mud layer) or shade (more than 80% of the water table). The presence of fish was established using the combined data of visual observation, the dip-netting and information from local people.

Based on the results of the pond inventory, in autumns 2005–2007 (after the reproductive period of most water organisms), 27 clusters with a total of 230 ponds (120 in Haanja, 74 in Otepää and 36 in the four smaller protected areas) were constructed for the target species, including 22 ponds restored and 208 new ponds created (of these, 73 were created at the places of old, vanished ponds; Fig. 1). Nineteen pond clusters (153 ponds) were designed for both species; six clusters (46 ponds) for the crested newt in Piusa, Karste and Hauka areas, and in isolated sites in Otepää and Haanja; and two clusters (31 ponds) for the common spadefoot toad in Otepää and Haanja. The clusters were designed for one target species only if the other was absent at the site in 2005. Bulldozers (for large shallow water bodies) or excavators (in smaller or wetter areas to restore an old pond or to create a deeper one) were used for digging, sometimes combining their use.

For pond construction (restoration or creation), we followed four principles:

(1) to increase colonisation probabilities and preserve the existing populations (Semlitsch, 2000; Petranka & Holbrook, 2006; Petranka et al., 2007), we constructed ponds in clusters (4–26 ponds in each), with distances between ponds no more than 500 m (on average, 116 m ± 8.9 SE; range 6–479 m) and at least one constructed pond within 200 m of an existing breeding pond of a target species. Land cover within 50 m from any constructed pond was mainly to consist of a mosaic of forest and (semi)natural grassland (for the crested newt) and (semi)natural grasslands and small extensively used potato fields or vegetable gardens (for the common spadefoot toad);
(2) to assure different hydroperiods (Semlitsch, 2002; Petranka et al., 2003), improve the ponds' quality for amphibians and to fit them into the landscape various treatments were applied in each cluster. Notably, we constructed ponds of various depths (0.4–2.5 m), sizes (12–5000 m^2), slopes (3°–90°; the mean: 24°), shapes and widths of shallow littoral zone (0.2–10 m). In case of existing ponds, we cleaned the ponds

from bushes and high dense vegetation (*Typha latifolia*), extracted the mud down to the mineral soil (mostly clay), to assure the quality and transparency of water as well as to eliminate the fish (for that purpose, the ponds were also pumped dry) and enlarged very small ponds and levelled the banks to create shallow littoral zones with warm water;

(3) none of the constructed ponds was allowed a connection to running water (ditch, stream, river) to avoid fish introduction or sedimentation (Semlitsch, 2000, 2002); by necessity, existing ditches were blocked for that purpose;

(4) as each pond construction was unique (depending on the relief, soil, hydrology, presence of drainage system, surrounding habitats etc.), it was guided in the field by experienced amphibian experts.

After construction, the ponds filled with rainwater, and allowed colonisation and succession to take their course. The post-restoration monitoring took place over 3 years (2006–2008). Each pond was visited once and examined in 10 min using visual counting of adults, dip-netting of larvae and searching for eggs of the newts and 'green frogs'. Breeding attempt was ascertained by the presence of eggs and/or larvae. For each pond, seven aquatic and one terrestrial habitat features were described in the field in June 2008 (Table 1), and its distance from the nearest pond occupied by the target species (source pond) was measured from the Estonian base map.

Data analyses

In order to detect habitat determinants of the pond colonisation by target species (presence–absence in 2008), multiple logistic regression models were built according to the procedure of Hosmer & Lemeshow (1989): (1) performed univariate analyses for each of the eight independent variables, (2) built preliminary multivariate models, which included the potentially important variables according to the univariate analyses and (3) omitted non-significant and/or redundant variables from the multivariate model considering their biological meaning and large differences in univariate significance levels. In the first two steps, the significance level was set at $P < 0.15$ (to retain the variables that could gain significance while in combination with other variables); in the final step, $P < 0.05$ was used. Performance of the final multivariate models was assessed by comparing observed versus expected presence/absence using the breakpoint at 0.5 for the expected values. The analyses were performed using STATISTICA 7.0 software.

Results

During the pre-restoration inventory, we recorded seven amphibian species. The only regionally present species, not found, was the edible frog. However, this species and the pool frog form mixed populations in Estonia and their field identification by egg-clusters

Table 1 Univariate relationships (likelihood-ratio tests of logistic regression) between the incidence of the crested newt (*T. cri.*) and the common spadefoot toad (*P. fus.*) and the aquatic and terrestrial habitat variables of the 230 restored ponds in June 2008

Variable	*N* *T. cri.*	*N* *P. fus.*	Mean ± SD for occupied ponds		*P*	
			T. cri.	*P. fus.*	*T. cri.*	*P. fus.*
Pond area (m^2)	127	29	419.4 ± 606.2	556.0 ± 901.9	0.451	0.121
Maximum water depth (m)	117	28	1.3 ± 0.4	1.3 ± 0.4	0.006	0.416
Mean width of shallow (up to 30 cm) littoral zone in the pond (m) measured from four cardinal edges	107	24	1.0 ± 1.1	1.2 ± 0.6	0.302	0.723
Mean slope (°) of the four cardinal banks of the pond	111	24	23.6 ± 11.4	22.7 ± 9.7	0.360	0.469
Water colour or transparency (four types)	127	29	–	–	0.086	0.041
Main land cover within 50 m (seven types)	127	29	–	–	<0.001	0.470
% Pond area occupied by floating vegetation	118	28	10.0 ± 17.5	9.4 ± 14.4	0.157	0.782
% Pond area occupied by submerged vegetation	118	28	12.8 ± 18.5	12.6 ± 14.7	0.004	0.394

Table 2 The occurrence of amphibian species in the 405 existing ponds in Haanja LPA and Otepää LPA in June 2005; in the constructed ponds over 3 years after restoration; and the number of constructed ponds occupied by amphibians in 2008

Species	Ponds occupied in 2005		Post-restoration colonisation of constructed ponds (%)			Ponds occupied in 2008	
	N	%	I year $N = 230$	II year $N = 193$	III year $N = 111$	*N*	Breeding attempt (%)
T. vulgaris	149	36.8	35.7	65.8	82.0	156	68.7
T. cristatus	94	24.2	16.1	54.9	71.2	127	98.4
P. fuscus	8	2.0	5.2	15.0	15.3	29	96.6
B. bufo	86	21.2	23.9	30.1	41.4	76	65.8
R. temporaria	90	22.2	25.7	37.3	44.1	95	86.3
R. arvalis	62	15.3	17.8	22.8	40.5	85	87.1
'*Green frogs*'	236	58.3	19.1	55.4	82.0	144	54.2

Table 3 The occurrence of amphibians in ponds with ($N = 194$) and without fish ($N = 211$) in June 2005

Species	Presence (%) in ponds		The effect of fish presence	
	With fish	Without fish	χ^2	*P*
T. vulgaris	52	97	15.7	<0.001
T. cristatus	27	67	18	<0.001
P. fuscus	0	8	–	–
B. bufo	50	36	4.6	0.032
R. temporaria	30	60	9.5	0.002
R. arvalis	16	46	14.3	<0.001
'*Green frogs*'	119	117	1.4	0.23

or tadpoles is complicated. Collectively, these 'green frogs' were the most frequent amphibians in the area, while the rarest species was the common spadefoot toad (Table 2). Importantly, only 22% of the 405 ponds were of high-quality for amphibian breeding. Forty-eight percent of the examined ponds were stocked with fish, mainly with crucian carp (*Carassius auratus gibelio* Bloch), which is an alien species in Estonia. All amphibian species, except 'green frogs', avoided ponds with fish, and the common spadefoot toad was never found in such ponds (Table 3). Fifteen percent of the examined ponds were completely overgrown with dense vegetation and/or bushes, 10% were eutrophicated or silted up, and 5% were completely in shade.

During the first post-restoration survey (June 2006), all the seven amphibian species (and no fish) were detected in the constructed ponds already (Table 2). Presence of the edible frog remained uncertain (no adults or juveniles were found). The breeding attempts of the crested newt were detected in the constructed ponds of 5 clusters of the 13 restored (38.5%), and the common spadefoot toad in 3 of 10 clusters (30%). By 2008, the breeding attempts of the crested newt had been recorded in 23 of 25 clusters (92%), and of the common spadefoot toad in 17 of 21 clusters (81%).

Altogether, in only 3 years when 22 of the 405 existing ponds (5%) were restored and 208 new ponds (51%) created, the number of ponds occupied by the common spadefood increased 6.5 times (from 2 to 13%). Concerning the crested newt and the moor frog (another (non-target) species listed in the EU Habitats Directive), the number of occupied ponds increased 2.3 (from 24 to 55%) and 2.5 times (from 15 to 37%; Table 2). In 2008, the 230 ponds constructed for amphibians hosted, on average, 3.1 ± 0.1 SE amphibian species per pond, while the 383 non-restored ponds had 1.8 ± 0.07 SE amphibian species (*t*-test: $t = 11.2$; $P < 0.001$).

The constructed ponds situated close to the source pond were colonised more quickly than ponds that were further away both in the case of the crested newt: (Kruskal–Wallis ANOVA: $\chi^2 = 17.6$; df = 3; $P < 0.001$) and the common spadefoot toad ($\chi^2 = 10.6$; df = 3; $P = 0.014$; Fig. 2). In terms of pond characteristics, the crested newt presence in the 230 constructed ponds in 2008 was explained (model log-likelihood = −129.0; $\chi^2 = 26.0$; $P = < 0.001$) by the land cover within 50 m (log-likelihood = −138.0; $\chi^2 = 18.1$; $P = 0.006$) and a higher percentage of submerged vegetation in the pond (estimate: 0.03 ± 0.01; SE; log-likelihood = −131.4; $\chi^2 = 4.8$; $P = 0.028$). The most favourable land cover type around the pond was forest (all six forest ponds being colonised), while the

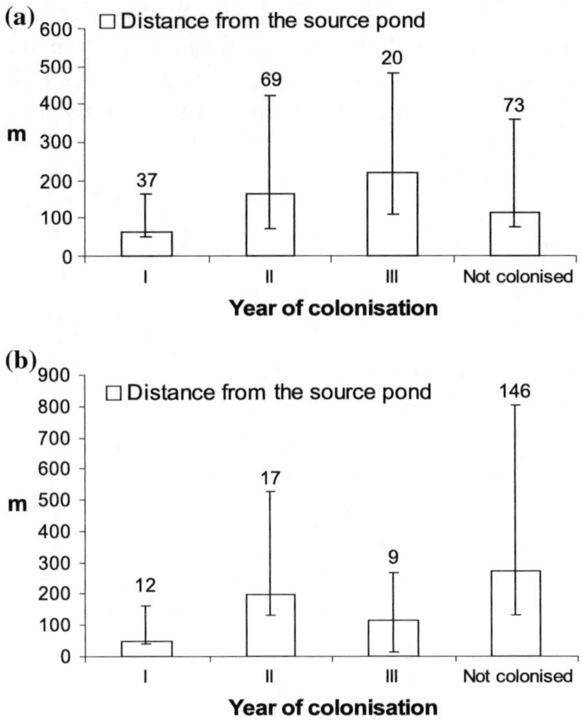

Fig. 2 Relationship between the year of colonisation and the distance (median and quartiles) of the constructed pond from the source pond in the crested newt (**a**) and the common spadefoot toad (**b**). The numbers above bars are sample sizes

29 ponds on meadows had the lowest colonisation rate (44.4%). The presence of forest, in combinations with meadows and farms, increased the suitability of the pond (70.4%; $N = 115$). Altogether, the multivariate model correctly classified 67% of the observations (81% for the presence, 49% for the absence of the species). At the univariate stage also depth of the pond appeared significant ($P = 0.005$; Table 3), but lost its significance in the final model. Pond colonisation by the common spadefoot toad was explained by the transparency and colour of water only (log-likelihood = −77.9; $\chi^2 = 8.3$; $P = 0.04$): transparent or clear but brownish water were favoured (96.6% of such ponds being colonised) and the unclear and muddy or algae-green water were avoided.

Discussion

The amphibian conservation management described in this study provides one of the rare success stories of its kind—given the rapid spontaneous colonisation of the constructed ponds and overall population increases of the two specifically targeted threatened species as well as the general increase in local amphibian diversity. Most of the previous pond restoration for amphibians has attempted to improve the local breeding conditions for amphibians, in general, with common species having benefited most (Lehtinen & Galatowitsch, 2001; Pechmann et al., 2001; Stumpel, 2004; Petranka et al., 2007). The success for threatened species has often remained low (Pechmann et al., 2001; Stumpel, 2004), even when these have been specifically targeted (Nyström et al., 2007; Briggs et al., 2008). Importantly, the few successful cases of habitat restoration for declining amphibians have always been carried out at the landscape scale, taking into account the particular terrestrial and aquatic habitat requirements of the target species (Denton et al., 1997; Briggs, 1997, 2001).

Before discussing the key factors for the success, two major limitations of the study need to be considered. First, only the short-term efficacy of habitat restoration (colonisation) was considered, while long-term monitoring will be needed to understand the viability of the populations (see also Petranka et al., 2003), which most likely depends on the succession in the restored ponds and the effects of annual fluctuations on the recovering, but small, populations of the threatened species. According to the management schemes in our project, the state of each of the 230 ponds will be monitored at 2- to 3-year intervals by local site managers. The emphasis is on the necessary management actions (bush cutting on the banks, removal of dense aquatic vegetation, mowing or grazing in the vicinity, fish elimination etc.), some of which (together with the restrictions for fish release) are included in special contracts with land owners. The potentially high risks for local extinctions from demographic or environmental stochasticity (Marsh, 2001) was addressed by selecting the areas where population densities of the two target species were the highest in Estonia and by restoring a variety of ponds in each cluster; however, the local populations were initially small and isolated (often a single extant breeding pond) and the long-term effects of fluctuations cannot be predicted. Second, the main effort was directed to aquatic habitats, although terrestrial habitats were also considered (and these appeared highly relevant for the crested newt). In degraded land areas, also a specific terrestrial habitat restoration may have to be considered as in successful population recovery cases of the fire-bellied toad (*Bombina bombina* L.) in Denmark (Briggs, 1997) and the natterjack toad (*Bufo calamita* Laur.) in England (Denton et al., 1997).

There were probably five general factors that contributed most to the success of the project. First, the restoration areas and habitats were carefully selected, as suggested by Nyström et al. (2007). The areas hosted the strongest, not the weakest, remnant populations in the region and protected areas with a high forest cover and a low-intensity agriculture were chosen to improve the long-term perspectives. Second, we restored ponds in clusters, taking into account the relatively limited dispersal abilities of the target species (Jehle, 2000; Kupfer & Kneitz, 2000; Nyström et al., 2002) and the preference of breeding adults to return to natal ponds (Berven & Grudzien, 1990). Indeed, the constructed ponds were significantly more rapidly colonised, when closer to source ponds, and the actual distances observed in the crested newt resemble the 400-m upper limit reported by Baker and Halliday (1999) for this species. Therefore, the clustering was apparently an effective way to increase the density and number of breeding sites both at the local population and at the landscape level. Third, pond quality was considered to be at least as important as pond availability (Danoël & Ficetola, 2008) and, as the exact requirements of the species are not precisely known and pond quality may fluctuate (e.g. depending on rainfall), a variety of ponds were created in each cluster. This also allows using natural pond drying to prevent and eliminate fish predation (Semlitsch, 2000), which was the fourth key consideration. In accordance with similar findings in many amphibian species, both our target species avoided ponds with fish (see also Joly et al., 2001; Skei et al., 2006; Nyström et al., 2007). Finally, we suggest that the participation of experienced experts in the field was essential for achieving good results.

The important technical details of pond reconstruction were species-specific. The shallow littoral zone of submerged vegetation, which can provide suitable egg laying, foraging and refugium sites for amphibians (Semlitsch, 2002; Porej & Hetherington, 2005), influenced colonisation of restored ponds by the crested newt. In addition to the habitat model, this was apparent in the increase of the colonisation rate after the submerged vegetation was established (Table 2). For the common spadefoot toad, the transparency of water was essential, which may indicate a high concentration of oxygen as favoured by this species (Nyström et al., 2002). Surprisingly, the presence of vegetation lacked significant effects to the common spadefoot toad, although other studies (Hels, 2002; Nyström et al., 2002) have reported its preference for ponds in late successional or eutrophic stages.

Conclusion

Habitat restoration for pond-breeding amphibians, especially for threatened species, can be successful if it is biologically based, implemented at the landscape scale, taking into account the habitat requirements of target species and the ecological connectivity of populations. The key considerations for short-term (colonisation) success highlighted by the study were: (i) at least some of the constructed (restored or created) ponds should be located in the close vicinity of existing source ponds; (ii) the ponds should be constructed in clusters, and each cluster should include a variety of ponds; (iii) the constructed ponds should be separated from running water to avoid fish introduction, sedimentation or pollution; (iv) the ponds have to be surrounded with terrestrial habitats suitable for target species; (v) the field guidance by experienced experts in the restoration is strongly advisable. In addition, long-term monitoring of the constructed ponds is necessary to assess the viability of the target populations and for adaptive management in the future.

Acknowledgements We thank L. C. Adrados, J. Foster, L. Iversen, J. Janse, T. Jairus, J. Kielgast, I. Lepik, M. Linnamägi, M. Markus, R. Novitsky, P. Pappel, O. Reshetylo, W. de Vries and V. Vuorio for help in the field. EU LIFE-Nature project LIFE04NAT/EE/000070 provided financial support for the project. Compilation of the paper was supported by the European Regional Development Fund (Centre of Excellence FIBIR), the Estonian Ministry of Education and Science (target-financing project 0180012s09) and the Estonian Science Foundation (grant 7402).

References

Alford, R. A., P. M. Dixon & J. H. K. Pechmann, 2001. Global amphibian population declines. Nature 412: 499–500.

Baker, J. M. R. & T. R. Halliday, 1999. Amphibian colonization of new ponds in an agricultural landscape. Herpetological Journal 9: 55–63.

Beebee, T. J. C., 1992. Amphibian decline. Nature 335: 120.

Berven, K. A. & T. A. Grudzien, 1990. Dispersal in the wood frog (*Rana sylvatica*): implications for genetic population structure. Evolution 44: 2047–2056.

Briggs, L., 1997. Recovery of *Bombina bombina* in Funen County, Denmark. Memoranda Societatis pro Fauna et Flora Fennica 73: 101–104.

Briggs, L., 2001. Conservation of temporary ponds for amphibians in northern and central Europe. In Rouen, K. (ed.), European Temporary Ponds: A Threatened Habitat. Freshwater Forum 17: 63–70.

Briggs L., R. Rannap & F. Biebelriether, 2008. Conservation of *Pelobates fuscus* as a result of breeding site creation. In Krone, A. (ed.), Die Knoblauchkröte (*Pelobates fuscus*) Verbreitung, Biologie, Ökologie und Schutz. RANA 5: 181–192.

Brönmark, C. & L. A. Hansson, 2005. The Biology of Lakes and Ponds. Oxford University Press, Oxford.

Cushman, S. A., 2006. Effects of habitat loss and fragmentation on amphibians: a review and prospectus. Biological Conservation 128: 321–340.

Danoël, M. & G. F. Ficetola, 2008. Conservation of newt guilds in an agricultural landscape of Belgium: the importance of aquatic and terrestrial habitats. Aquatic Conservation: Marine and Freshwater Ecosystems 18: 714–728.

Denton, J. S., S. P. Hitchings, T. J. C. Beebee & A. Gent, 1997. A recovery program for the natterjack toad (*Bufo calamita*) in Britain. Conservation Biology 11: 1329–1338.

Edgar, P. & D. Bird, 2006. Action Plan for the Conservation of the Crested Newt *Triturus cristatus* Species Complex in Europe. Strasbourg, France.

Evestus, T. & M. Turb (eds), 2002. The Management Plan of Otepää LPA. Estonian Ministry of the Environment, Tallinn.

Fog, K., 1997. A survey of the results of pond projects for rare amphibians in Denmark. Memoranda Societatis pro Fauna et Flora Fennica 73: 91–100.

Hels, T., 2002. Population dynamics in a Danish metapopulation of spadefoot toads *Pelobates fuscus*. Ecography 25: 303–313.

Hosmer, D. W. & S. Lemeshow, 1989. Applied Logistic Regression. Wiley, New York.

Hull, A., 1997. The Pond Life Project: a model of conservation and sustainability. In Boothby, J. (ed.), British Pond Landscape, Proceedings of the UK Conference of the Pond Life Project. Liverpool: 101–109.

Jehle, R., 2000. The terrestrial summer habitat of radio-tracked great crested newts *Triturus cristatus* and marbled newts *Triturus marmoratus*. Herpetological Journal 10: 137–142.

Joly, P., C. Miaud, A. Lehmann & O. Grolet, 2001. Habitat matrix effects on pond occupancy in newts. Conservation Biology 15: 239–248.

Karlsson, T., P. E. Betzholtz & J. C. Malmgren, 2007. Estimating viability and sensitivity of the great crested newt *Triturus cristatus* at a regional scale. Web Ecology 7: 63–76.

Kupfer, A. & S. Kneitz, 2000. Population ecology of the great crested newt in an agricultural landscape: dynamics, pond fidelity and dispersal. Herpetological Journal 10: 165–172.

Lehtinen, R. M. & S. M. Galatowitsch, 2001. Colonization of restored wetlands by amphibians in Minnesota. The American Midland Naturalist 145: 388–396.

Marsh, D. M., 2001. Fluctuations in amphibian populations: a meta-analysis. Biological Conservation 101: 327–335.

Nyström, P., L. Birkedal, C. Dahlberg & C. Brönmark, 2002. The declining spadefoot toad *Pelobates fuscus*: calling site choice and conservation. Ecography 25: 488–498.

Nyström, P., J. Hansson, J. Månsson, M. Sundstedt, C. Reslow & A. Broström, 2007. A documented amphibian decline over 40 years: possible causes and implications for species recovery. Biological Conservation 138: 399–411.

Oertli, B., J. Biggs, R. Cereghino, P. Grillas, P. Joly & J.-P. Lachavanne, 2005. Conservation and monitoring of pond biodiversity: introduction. Aquatic Conservation: Marine and Freshwater Ecosystems 15: 535–540.

Pappel, P. & R. Rannap, 2007. Monitoring of amphibians and reptiles in Estonia. In Sammul, M. (ed.), The 84th Yearbook of the Estonian Naturalists' Society. Tartu: 218–226.

Pechmann, J. H. K., R. A. Estes, D. E. Scott & J. W. Gibbons, 2001. Amphibian colonization and use of ponds created for trial mitigation of wetland loss. Wetlands 21: 93–111.

Petranka, J. W. & C. T. Holbrook, 2006. Wetland restoration for amphibians: should local site be designed to support metapopulations or patchy populations. Restoration Ecology 14: 404–411.

Petranka, J. W., C. A. Kennedy & S. S. Murray, 2003. Response of amphibians to restoration of a southern Appalachian wetlands: a long-term analysis of community dynamics. Wetlands 23: 1030–1042.

Petranka, J. W., E. M. Harp, C. T. Holbrook & J. A. Hamel, 2007. Long-term persistence of amphibian populations in a restored wetland complex. Biological Conservation 138: 371–380.

Porej, D. & T. E. Hetherington, 2005. Designing wetlands for amphibians: the importance of predatory fish and shallow littoral zones in structuring of amphibian communities. Wetlands Ecology and Management 13: 445–455.

Rannap R., A. Lõhmus & L. Briggs, 2009. Niche position, but not niche breadth, differs in two coexisting amphibians having contrasting trends in Europe. Diversity and Distributions 15: 692–700.

Semlitsch, R. D., 2000. Principles for management of aquatic-breeding amphibians. Journal of Wildlife Management 64: 615–631.

Semlitsch, R. D., 2002. Critical elements for biologically based recovery plans of aquatic-breeding amphibians. Conservation Biology 16: 619–629.

Skei, J. K., D. Dolmen, L. Rønning & T. H. Ringsby, 2006. Habitat use during the aquatic phase of the newts *Triturus vulgaris* (L.) and *T. cristatus* (Laurenti) in central Norway: proposition for a conservation and monitoring area. Amphibia-Reptilia 27: 309–324.

Stuart, S. N., J. S. Chanson, N. A. Cox, B. E. Young, A. S. L. Rodrigues, D. L. Fischman & R. W. Waller, 2004. Status and trends of amphibian declines and extinctions worldwide. Science 306: 1783–1786.

Stumpel, A. H. P., 2004. Reptiles and amphibians as targets for nature management. Alterra Scientific Contribution 13: 75–94.

Stumpel, A. H. P. & H. van der Voet, 1998. Characterizing the suitability of new ponds for amphibians. Amphibia-Reptilia 19: 125–142.

Williams, P., M. Whitfield, J. Biggs, S. Bray, G. Fox, P. Nicolet & D. Sear, 2003. Comparative biodiversity of rivers, streams, ditches and ponds in an agricultural landscape in Southern England. Biological Conservation 115: 329–341.

Hydrobiologia (2009) 634:97–105
DOI 10.1007/s10750-009-9892-8

POND CONSERVATION

High diversity of *Ruppia* meadows in saline ponds and lakes of the western Mediterranean

Ludwig Triest · Tim Sierens

Published online: 13 August 2009

Abstract Saline inland and coastal waterbodies are valuable habitats that deserve attention for the protection of their unique submerged macrophyte beds that render the water clear, stabilize sediments and provide a habitat for high biomasses of invertebrates as food for waterfowl. The 'continental seagrass' *Ruppia* has the widest salinity tolerance among the submerged macrophytes and occurs in a wide variety of saline saltmarsh pond and lagoon systems. Although two cosmopolitan species *Ruppia maritima* and *Ruppia cirrhosa* are recognized in Europe and *Ruppia drepanensis* in the western Mediterranean, their diversity and distribution are not well known. This previously held traditional idea that there are only two widespread *Ruppia* species suggests a uniform and very homogenized population structure following the hypothesis of long-distance-dispersal through strong bird-mediated dispersal events. Therefore, the *Ruppia* chloroplast DNA diversity was investigated along a more than 1,000 km transect of the Iberian Peninsula. We studied 492 individuals from 11 wetland areas (17 ponds) and sequenced a 1,753-bp length of seven chloroplast introns. Eight haplotypes represented at least four distinct groups or taxa which is higher than commonly accepted. Six wetland areas contained more than one haplotype and within-pond diversity occurred within distances as small as 30 m (5 out of 17 cases). This underlines the importance of single waterbodies for harbouring haplotypic diversity in *Ruppia*. Unique haplotypes were observed in four wetland areas and *R. maritima* was detected only from a low salinity pond, suggesting the species might be more rare than previously accepted. The present results tend to minimize an overall effect of strong bird-mediated dispersal. This emphasizes the role of regional pond habitat diversity for the preservation of *Ruppia* taxa and their unique haplotype diversity in extreme saline habitats.

Keywords Chloroplast DNA · Phylogeography · Seagrasses · Coastal · Conservation

Guest editors: B. Oertli, R. Cereghino, A. Hull & R. Miracle
Pond Conservation: From Science to Practice. 3rd Conference of the European Pond Conservation Network, Valencia, Spain, 14–16 May 2008.

L. Triest (✉) · T. Sierens
Research Group 'Plant Science and Nature Management', Vrije Universiteit Brussel, Pleinlaan 2, 1050 Brussels, Belgium
e-mail: ltriest@vub.ac.be

Introduction

Saline inland and coastal waterbodies are unique habitats. Very few submerged macrophytes can grow and reproduce under extreme conditions of high salinity, poorly oxygenated or anoxic sediments, large fluctuations in water level and shallow wind-exposed waters, which are often ephemeral. Among the genera that tolerate high salinity, such as *Althenia* Petit, *Lepilaena* J. Drumm. ex Harv., *Zannichellia* L., *Stuckenia* Börner. and *Ruppia* L., the latter survives the highest salinity, ranging from a few g/l to 230 g/l (Brock, 1982) which is more than for the related seagrasses. Ruppiaceae were phylogenetically closely associated with the other seagrass families such as Posidoniaceae and Cymodoceaceae (Les et al., 1997). *Ruppia*-dominated waters are clear and shallow with nutrient concentrations ranging from low to high (Verhoeven, 1979a). Large lagoons that are regularly influenced by seawater typically have lower nutrients than, for example, ponds in saltwater marshes accompanied with cattle or horse grazing. In saline waters, *Ruppia* forms monospecific beds and the genus can be regarded as a 'continental seagrass' because it grows in sheltered shallow non-tidal estuaries and lagoons as well as in coastal or inland waterbodies. In brackish waters, only a limited number of macrophytes species co-exist with *Ruppia*. Its vegetation often forms very dense monospecific beds over large surfaces and are inhabited by an abundant but species-poor fauna (Verhoeven, 1980a). The value of *Ruppia* seeds and habitat associated invertebrates as food for birds is especially known for coot, wigeon and flamingo (Verhoeven, 1980b). *Ruppia* beds deliver conditions for high quality food in sufficient quantities to support other wildlife.

Such saline pond and lagoon habitats are critical ecosystems that deliver a high biomass in relation to planktonic-based saltwater communities. The role of below ground biomass of the root and rhizome system is highly important in binding sediments, whereas the shoot system is providing a stable layer above the benthos and reducing the resuspension of sediments (Green & Short, 2003). *Ruppia* rhizomes can colonize rapidly and seeds persist under dry and hypersaline conditions (Verhoeven, 1979). *Ruppia* populations can show annual or perennial growth cycles (Malea et al., 2004) and much work on the ecology, biomass, productivity and ecophysiology of *Ruppia* was performed. However, the species ecotypic and genotypic variation is only partly understood (Triest & Symoens, 1991).

Ruppia has a cosmopolitan, but discontinuous distribution and is found on all continents, including many isolated islands from tropical to subarctic regions, as north as the White Sea and Iceland (Green & Short, 2003). It is one of the least known seagrasses and generally four species are recognized with *Ruppia cirrhosa* (Petagna) Grande and *Ruppia maritima* L. as widespread, nearly cosmopolitan taxa and both *R. megacarpa* Mason and *R. tuberosa* Davis & Tomlinson from Australia. Globally, seagrass diversity is the highest in tropical Asia and Western Australia and among the lowest in Europe and the Mediterranean. Unfortunately, species range maps for *Ruppia* could not be provided for any continent by the UNEP World Conservation Monitoring Centre as they faced insufficient data (Green & Short, 2003). In Europe, the EUNIS database for reporting on the distribution and status of species from protected areas (Natura 2000, Corine habitats and Biogenetic Reserves) considers only *R. maritima* and *R. cirrhosa*, following the treatment of Dandy (1980) in Flora Europaea. Many studies confirmed the existence of *Ruppia drepanensis* Tineo [or as the variety *R. cirrhosa* (Petagna) Grande var. *drepanensis* (Tineo) Symoens] on the Iberian peninsula, but they did not reach the attention at international level for monitoring and conservation purposes. Among these, we wish to mention morphological studies (Aedo & Fernandez Casado, 1988; Cirujano & Garcia-Murillo, 1990), cytotaxonomical investigations (Van Vierssen et al., 1981; Cirujano, 1986; Talavera et al., 1993) and finally autoecological and isozyme polymorphism studies (Triest & Symoens, 1991). Since the western Mediterranean may harbour African elements in its flora (Rivas-Martínez et al., 2003) and other submerged macrophytes such as *Zannichellia* also harbour a larger species diversity in the Iberian peninsula (Talavera et al., 1986; Triest & Vanhecke, 1991; Triest et al., 2007) we aimed to investigate the genetic diversity of *Ruppia* populations along a more than 1,000 km long transect of the western Mediterranean and several inland waterbodies of the Iberian Peninsula.

The objective of this study was to characterize the haplotypic chloroplast diversity for each species and to estimate the between- and within-pond diversity. This will permit a targeted evaluation and better conservation and protection of *Ruppia*-dominated ponds.

Materials and methods

Study sites and plant materials

Seventeen waterbodies from 11 wetland areas in the Iberian peninsula were investigated for their *Ruppia* diversity (Table 1). In each site we collected up to 30 individual shoots along a 30-m transect. Leaves were dried on silica gel and a reference herbarium for each population was deposited at the laboratory. Conductivity, salinity, temperature, water depth, vegetation composition and GPS coordinates were taken at each site.

DNA extraction, amplification and sequencing

Genomic DNA extractions were performed on dry material stored in silica gel (15–20 mg) using the E.Z.N.A. SP Plant DNA Mini Kit (Omega bio-tek). Three cpSSR primer pairs (Ccmp 2, Ccmp3 and Ccmp 10) derived from the complete sequence of tobacco (*Nicotiana tabacum*) chloroplast genome were used (Weising & Gardner, 1999) together with six primer pairs derived from the *Acorus calamus* chloroplast genome (Goremykin et al., 2005). The PCR amplification was carried out in 25 μl of reaction mixture containing 0.1 μl of genomic DNA, 2.5 μl 10× PCR buffer, 0.2 mM of each dNTP, 1.6 mM $MgCl_2$, 200 nM of the forward and reverse primer, 80 μg ml^{-1} bovine serum albumin (BSA) and 1 U Taq DNA polymerase. The PCR reactions were performed in a thermal cycler (MJ research PTC- 200 and Bio-Rad MyCycler) and started with 2 min at 94°C, followed by 30 cycles of 45 s at 92°C, 1 min at 50°C for the reactions with ccmp-primers and 1 min at 55°C for the reactions with Acorus-primers, 2 min at 72°C, and a final extension at 72°C for 5 min. The amplification products were separated on 5% non-denaturing polyacrylamide gels (acryl-bisacrylamide 19:1, 7 mM urea) with ethidium-bromide detection (Gel Scan 2000, Corbett research) and visualized with One-Dscan software (Scanalytics). Amplicon length was estimated using the GeneRuler 50 bp DNA ladder (Fermentas). Amplicon sequencing (in both forward and reverse direction) was performed by Macrogen Inc. (Seoul, South Korea).

A cladogram estimation (statistical parsimony) was performed for pairwise differences (probability >0.95) of the eight haplotypes using TCS1.21 (Clement et al., 2000). Gaps in a sequence were considered as a fifth state and duplications were considered as single events for constructing the haplotype network.

Results

Ruppia meadow and pond characteristics (Table 2)

The *Ruppia*-dominated vegetations mostly consisted of only *Ruppia* (12 out of 17 ponds) or were codominant with *Chara* species (2 ponds) or *Stuckenia pectinata* (L.) Börner (2 ponds). These *Ruppia* beds occurred at highly brackish to hypersaline conditions with conductivities (salinities) ranging from 18.21 mSiemens/cm (11‰ salinity) to 76.70 mSiemens/cm (53.7‰). Above 40 mSiemens/cm (19.9‰ salinity) *Ruppia* was the only submerged macrophyte. One *Stuckenia pectinata*-dominated vegetation contained *Zannichellia palustris* s.l. and *R. maritima* with a low cover and abundance at a lower conductivity (9.14 mSiemens/cm) and salinity (5.1‰) level.

The examined *Ruppia* populations were from various sizes of pond habitats and lagoons. Despite the close proximity to the Mediterranean Sea of 13 ponds, there was no permanent connection between the *Ruppia* sites and the sea. The following inland saltwater habitat types could be distinguished for *Ruppia*: lake, temporary wide ditch systems, temporary small drinking pool for cattle. The coastal wetland habitats containing *Ruppia* were shallow lagoons, temporary and isolated pools and ditches in (abandoned) salinas; apparently permanent pools connected with large wetland system, a former estuary and even newly created waterbodies in abandoned hotel construction zones, e.g. 'Marina' lagoon near Albufera NP and near Estartit.

Both small and large-sized waterbodies were characterized by dense *Ruppia* beds, mostly as large patches filling the small waterbodies and covering most of the lake surface as hectare sized homogenous patches. There was no relationship between *Ruppia* abundance, waterbody size and salinity.

Ruppia haplotype diversity

A total of eight haplotypes were detected in the 11 wetlands (Table 2; Figs. 1, 2). *Ruppia drepanensis* showed two haplotypes (A1 and A2) in a centrally located inland lake (Manjavacas) and one haplotype

Table 1 Characteristics of 17 ponds in 11 wetland areas of Spain (SP)

Pop. code	N.	Area	Locality	N	W–E	Conductivity mSiemens cm^{-1}	Salinity ‰	Habitat type	Other characteristics
SP1	30	La Mancha	Laguna de Manjavacas	39°24,691	W2°51,676	39.4	25.2	Shallow lake	With *Chara* sp.
SP2A	32	Doñana	Valverde	37°04,278	W6°16,333	18.2	11	Ditch	Disturbed by trampling of cows/horses
SP2B	30	Doñana	Valverde	37°04,278	W6°168,072	51	33.6	Ditch	Undisturbed ditch
SP2C	30	Doñana	Valverde	37°04,564	W6°23,024	39.3	25.3	Small cattle pool	Slightly disturbed drinking pool for cattle
SP3	30	Golfo de Almeria	Almerimar, Lago Victoria	36°42,571	W2°49,611	17.9	10.7	Coastal lagoon enclosed in urban area	With *Stuckenia pectinata*; slightly disturbed with stony debris
SP4A	30	Golfo de Almeria	Roquetas de Mar	36°42,780	W2°38,760	31.7	19.9	Shallow marshy coastal pool	With *Stuckenia pectinata*
SP4B	30	Golfo de Almeria	Roquetas de Mar	36°42,242	W2°39,667	69.5	48.5	Ditch connecting salinas	Undisturbed ditch
SP5A	30	Alicante	Santa Pola (opposite Torre de Tamarit)	38°11,073	W0°36,807	14.8	8.8	Coastal canal	Flowing water
SP5B	30	Alicante	Santa Pola (opposite Torre de Tamarit)	38°11,073	W0°36,807	76.7	53.7	Coastal marsh	Undisturbed saline marsh
SP5C	30	Alicante	Santa Pola (Torre de Tamarit)	38°11,073	W0°36,807	58.9	39.6	Coastal marsh	With *Chara* sp.; undisturbed marsh
SP6	30	Golfo de Valencia	Devesa and Albufera of Valencia NP, Lago de El Saler	39°20,767	W0°18,936	69.2	47.7	Man-made coastal lagoon	Man made harbour for boating facilities, but abandoned construction zone; spontaneous vegetation
SP7	30	Golfo de Valencia	Marjal dels Moros, Puçol	39°36,981	W0°15,498	19.9	12.1	Coastal lagoon	Nature reserve behind beach
SP8	15	Golfo de Valencia	Prat de Cabanes, Torreblanca NP	40°11,107	E0°12,581	9.1	5.1	Ditch from peat exploitation ponds to sea	With *Zannichellia palustris* and *Stuckenia pectinata;* nature reserve
SP9	25	Delta del Ebro	Illa de Buda	40°38,244	E0°44,781	48.2	32.1	Coastal marsh	Shallow marsh
SP10A	30	Costa Brava	L'Estartit, els Griells	42°01,887	E03°11,567	28.8	18	Newly created pond	Very shallow, abandoned construction zone
SP10B	30	Costa Brava	L'Estartit, els Griells	42°01,762	E03°11,444	32.4	20.4	Former coastal river branch	Continuous population
SP11	30	Costa Brava	Empuriabrava, Aiguamolls del Ampurdà	42°15,556	E03°08,766	65.4	44.8	Coastal marsh	Shallow marsh in *Salicornia* vegetation

Table 2 Distribution of chloroplast haplotypes and fragment length variants in 11 wetland areas and 17 ponds of Spain (SP)

Pop. code	% ind.	Haplotypes	ccmp2	ccmp3	ccmp10	acp1	acp2	acp4	acp5	acp6	acp7
SP1	50	A1	201	157	166a	175	169a	228a	222b	202a	194a
	50	A2	201	157	166b	175	169a	228c	222b	202a	194a
SP2A	100	A1	201	157	166a	175	169a	228a	222b	202a	194a
SP2B	100	A1	201	157	166a	175	169a	228a	222b	202a	194a
SP2C	100	B1	202	157	166a	175	169a	228a	222a	202a	194a
SP3	33	B1	202	157	166a	175	169a	228a	222a	202a	194a
	67	B2	204	157	173	175	169a	228a	222a	202a	194a
SP4A	24	B1	202	157	166a	175	169a	228a	222a	202a	194a
	76	B2	204	157	173	175	169a	228a	222a	202a	194a
SP4B	100	C1	210	157	166a	175	169a	228a	222a	202a	194a
SP5A	100	C1	210	157	166a	175	169a	228a	222a	202a	194a
SP5B	4	B1	202	157	166a	175	169a	228a	222a	202a	194a
	20	C1	210	157	166a	175	169a	228a	222a	202a	194a
	72	C2	210	157	174	175	169a	228a	222a	202a	194a
	4	C3	211	157	166a	175	169a	228a	222a	202a	194a
SP5C	100	C1	210	157	166a	175	169a	228a	222a	202a	194a
SP6	100	B1	202	157	166a	175	169a	228a	222a	202a	194a
SP7	100	B1	202	157	166a	175	169b	228b	222c	202b	194b
SP8	100	D1	203	152	166b	178	169a	228a	222a	202a	194a
SP9	56	C1	210	157	166a	175	169a	228a	222a	202a	194a
	44	C3	210	157	174	175	169a	228a	222a	202a	194a
SP10A	100	C1	210	157	166a	175	169a	228a	222a	202a	194a
SP10B	100	C1	210	157	166a	175	169a	228a	222a	202a	194a
SP11	100	C1	210	157	166a	175	169a	228a	222a	202a	194a

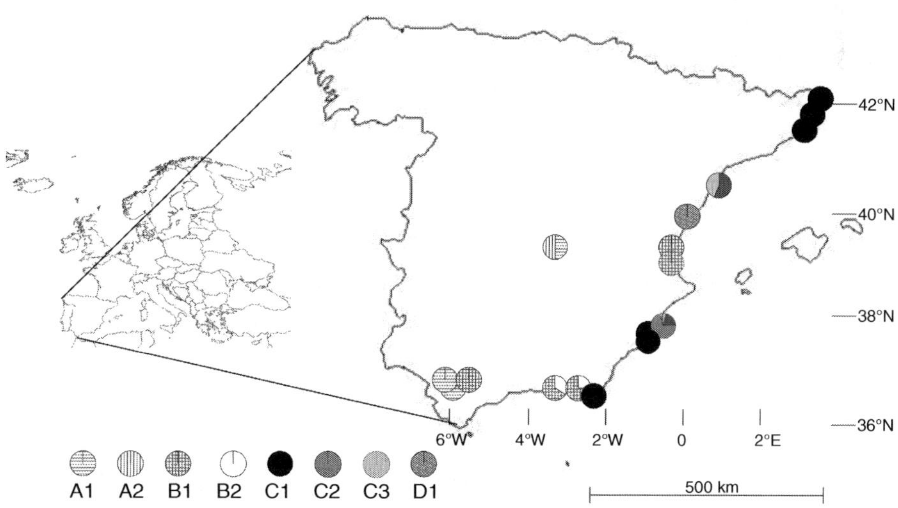

Fig. 1 Distribution of eight chloroplast haplotypes in *Ruppia* of 17 ponds and lakes

(A1) in the Doñana region. The haplotypes A1 and A2 differed in only two substitutions (for ccmp10 and acp4) and can be distinguished from all other haplotypes (ccmp2 with substitutions, indels or T-repeat; acp4 and acp5 with substitutions). *Ruppia cirrhosa* and some *R. maritima*-like populations

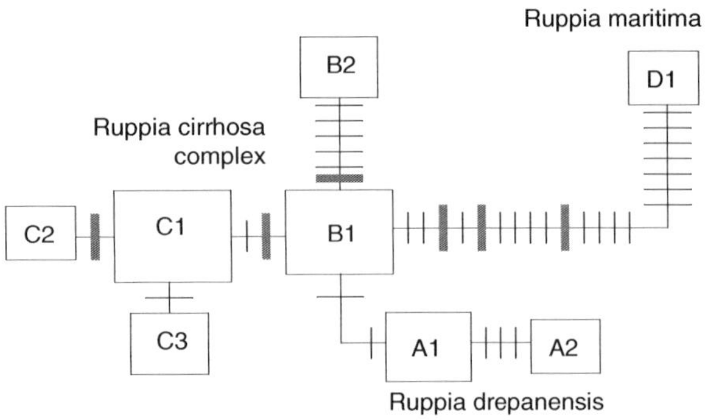

Fig. 2 Haplotype network of eight chloroplast haplotypes with sizes proportional to the number of ponds. *Thin linear bars* represent a substitution or repeat; the *thicker grey bars* represent indels

showed various haplotypes (B1, B2, C1, C2 and C3) differing in one or more substitutions and indels. C-haplotypes share an 8 bp duplication in ccmp2. The *R. maritima* haplotype D1 differs strongly from all other haplotypes (i.e. more than 15 substitutions, an A-repeat and 3 indels in ccmp2 and ccmp3).

From the 11 wetland areas, six showed more than one haplotype (Table 2; Fig. 1). Within a wetland area, different haplotypes may occur in separate waterbodies (e.g. *R. drepanensis* and *R. cirrhosa* in Doñana) but most were present as mixtures within the same pond or lagoon (B1 and B2 in southernmost lagoons of Almerimar-Roquetas de Mar; C1 and C3 in the Ebro delta; B1, C1, C2 and C3 in salinas near St. Pola). These mixtures of two to four haplotypes occurred within a transect distance of only 10–30 m of 20 cm shallow waters, indicating that diversity can be locally maintained. Four wetland areas harboured unique haplotypes within the studied zone, e.g. A2 for *R. drepanensis* in central Spain, B2 in Almerimar-Roquetas de Mar (which is situated south of Sierra Nevada); C2 in salinas near St. Pola and D1 in Tor de Sal. It was difficult to assign the B and C haplotypes to either *R. cirrhosa* or another taxon because of several non-fruiting populations and intermediate characteristics of the peduncle showing one or more coils.

Along the western Mediterranean coastline, haplotype C was 100% abundant in the northern half (from Aiguemolls to the Ebro delta), whereas haplotypes B and C were equally abundant in the southern half (from Puçol to Costa Almeria). Haplotype B2 was confined to the southernmost lagoons (Costa Almeria) and could represent another taxon as there were five substitutions and one duplication (Fig. 2).

Discussion

Although the genus *Ruppia* is studied in all parts of the world, there remain different interpretations of its species diversity. For conservation purposes at European level, the widespread *R. maritima* and *R. cirrhosa* are recognized (Dandy, 1980; Casper & Krausch, 1980) and hence not considered as rare taxa, vulnerable or threatened at regional level. However, additional taxa such as *R. drepanensis* (*R. cirrhosa* var. *drepanensis*), *R. maritima* var. *brevirostris* (Agardh) Aschers. & Graebn. and *R. maritima* var. *longipes* Hagström have been considered by many authors but there are no clear diagnostic features in their morphology and karyotype (Triest & Symoens, 1991).

Our results indicate that for two species there is indeed a larger genetic variability at the cpDNA level. *Ruppia maritima* is evolutionarily very distant from all other *Ruppia* haplotypes or taxa from the western Mediterranean. For another submerged brackish to freshwater macrophyte *Zannichellia*, the chloroplast haplotypes were not numerous, but corresponded very well with the known taxa and showed a biogeographical pattern (Triest et al., 2007).

In southern Europe, two cytodemes of *Ruppia* are found. The diploid $2x = 20$ was recorded in *R. maritima* populations in north western Spain (Aedo & Fernandez Casado, 1988) and Sicily (Marchioni Ortu, 1982) and in *R. drepanensis* in Central Spain (Cirujano, 1986). The tetraploid $2x = 40$ was recorded in *R. maritima* from Central Spain (Cirujano, 1986) and *R. maritima* var. *longipes* from south western Spain (Van Vierssen et al., 1981). These chromosome counts should be re-evaluated in the context of the cpDNA haplotypic variation because the morphological distinction of either *R. maritima* or *R. cirrhosa* could be problematic. Although a worldwide evaluation of the genus is needed before taxonomic questions can be answered, we emphasize that the European *Ruppia* are a good starting point for examining genetic diversity at local, regional and continental level for conservation purposes. It is evident that more than two *Ruppia* taxa should be considered for the interpretation of ecological studies in the Mediterranean.

A former study on isozymes at European and Mediterranean level revealed a low level of polymorphism with *R. maritima* being a separate taxon; a slight difference based on alcohol dehydrogenase between *R. cirrhosa* var. *cirrhosa* and *R. cirrhosa* var. *drepanensis* was found (Triest & Symoens, 1991). The genetic structure of each population was uniform, and no multiclonal populations were observed with isozymes, most likely due to the poorer sampling of genes. Our results with cpDNA confirmed the identity of *R. maritima* but additionally revealed a high diversity of haplotypes in the '*R. cirrhosa*-complex' in the southern part of the western Mediterranean. Eight haplotypes were observed instead of an expected three on the basis of previous isozyme studies (*R. maritima*, *R. cirrhosa* and *R. drepanensis*). This fact together with the possibility of a regional pattern (e.g. haplotype B2 only south of the Sierra Nevada) and rare and local populations (e.g. D1) enhances the conservation value of *Ruppia*-dominated ponds and lagoons.

The previously held traditional idea that there are only two widespread *Ruppia* species suggests a uniform and very homogenized population structure across the continent and evidence of long distance dispersal in other water plants (Santamaria, 2002). This would suggest strong bird-mediated dispersal events, but the present results tend to minimize such an overall effect. This is because the morphospecies of *Ruppia* do not exactly correspond with the haplotypes that are much better representing the evolutionary significant units. At least for what we considered here as a '*Ruppia cirrhosa* complex' there seems not to be an overall wide distribution of each haplotype. The cpDNA reflects the dispersal and spread through seeds and rhizomes which for *Ruppia* can be considered as more important than pollen-mediated gene dispersal. *Ruppia* species have a mixed reproduction system and can reproduce by seeds and persist through rhizomes. *Ruppia maritima* and *R. drepanensis* usually produce more flowers and seeds than *R. cirrhosa*. The pollen flow will mostly be restricted to a single waterbody as mature pollen grains are detached from the plants and float on the water surface. The female flower parts are situated on a peduncle and stigmas reach the water surface such that pollination depends on water movements through wind and wave action (Triest & Symoens, 1991).

In seagrass species such as *Zostera* L. spp., local patches were reported to be composed of a single genet (Reusch, 2001) or showing genet coexistence within a small spatial scale of up to 40 m (Araki & Kunii, 2006). Major colonists are supposed to be seeds rather than vegetative shoots, followed by sometimes extensive seagrass bed development as a result of local clonal growth and spread. However, the establishment of a seagrass population is an interplay between species identity, reproductive success and adaptation to local physical conditions (Green & Short, 2003). Chloroplast sequences are not indicative of similarity but for dissimilarity of genets, and we observed that coexistence of different *Ruppia* haplotypes can occur within a small spatial scale. Another explanation for haplotype diversity that cannot be ruled out is their generation via somatic mutations in plants capable of vegetative reproduction; however, this is difficult to document.

There are a few possible hypotheses to explain the occurrence of several haplotypes in the same waterbody, whereas neighboring waterbodies contain only a single *Ruppia* haplotype. We can assume that the haplotype recruitment may not be very frequent or even be the result of just a single introduction and that local dispersal of propagules by water currents is restricted, as well as the local dispersal of seeds, because either fruits are rare or because waterfowl are not foraging in a particular area. An alternative hypothesis is that after introduction of various haplotypes, the local spread among waterbodies was not limited by seed dispersal but that the habitat selects those genotypes adapted to the environmental condition of a waterbody. After dispersal over either short or long distances, the genotypes will be selected by the pond environmental conditions. Not much is known about the relationship between pond habitat characteristics and macrophyte gene diversity of but, e.g. in the context of a river system, a non-random distribution of *Callitriche* L. clones was observed in relation to alkalinity (Triest & Mannaert, 2006). Single ponds within a lagoon system can show different physical properties, salinities, inundation periods, sediment oxygenation, etc. Therefore, the pond habitat diversity of saltwater and brackish water lagoon systems should be regarded as important units for the conservation of *Ruppia* diversity.

Conclusion

This study of *Ruppia* populations along a transect of the western Mediterranean clearly showed that there is more chloroplast DNA variability at the level of species than could be expected from a traditional opinion of only two widespread taxa (*R. maritima* and *R. cirrhosa*) and a more restricted *R. drepanensis*. Eight haplotypes represented at least four distinct groups or taxa. Several wetland areas contained more than one haplotype and within-pond diversity was shown at small distances. Single waterbodies therefore may harbour substantial haplotypic diversity in *Ruppia*. Unique haplotypes were observed in four wetland areas and *R. maritima* was detected only from a low salinity pond, suggesting that the species might be more rare in that part of Europe than those previously accepted. It is hypothesized that either haplotype recruitment and local dispersal of *Ruppia* propagules by water currents is restricted, or that the habitat selects those genotypes adapted to the environmental condition of a waterbody. Both scenarios emphasize the role of regional pond habitat diversity for the preservation of *Ruppia* taxa and their unique haplotype diversity in extreme saline environments.

Acknowledgments This project was financed by the Fund for Scientific Research Flanders (KN 1.5.124.03, research projects G.0056.03, G.0076.05, Sabbatical leave contract for L. Triest), the Vrije Universiteit Brussel (OZR 1172, OZR1189BOF). Jordi Figuerola (Estación Biológica Doñana, Sevilla) and Xavier Quintana (Univ. Girona) kindly helped with the indication of suitable sites to collect plant shoots.

References

Aedo, C. & M. Fernandez Casado, 1988. The taxonomic position of *Ruppia* populations along the Cantabrian coast. Aquatic Botany 32: 187–192.

Araki, S. & H. Kunii, 2006. Allozymic implications of the propagation of eelgrass *Zostera japonica* within a river system. Limnology 7: 15–21.

Brock, M. A., 1982. Biology of the salinity tolerant genus *Ruppia* L. in saline lakes in South Australia II. Population ecology and reproductive biology. Aquatic Botany 13: 249–268.

Casper, S. & H.-D. Krausch, 1980. Süsswasserflora von Mitteleuropa 23. Gustav Fisher, Stuttgart.

Cirujano, S., 1986. El género *Ruppia* L. (Potamogetonaceae) en la Mancha (Espana). *Boletim da Sociedade Broteriana*, Série 2, 59: 293–303.

Cirujano, S. & P. Garcia-Murillo, 1990. Ruppiaceae. Fontqueria 28: 159–165.

Clement, M., D. Posada & K. Crandall, 2000. TCS: a computer program to estimate gene genealogies. Molecular Ecology 9: 1657–1660.

Dandy, J., 1980. *Ruppia*. In Tutin, T. G., H. Heywood, N. A. Burges, D. M. Moore, D. H. Valentine, S. M. Walters & D. A. Webb (eds), Flora Europaea. Vol V (1st ed.) Alismataceae to Orchidaceae. Oxford, Cambridge University Press: 11.

Goremykin, V. V., B. Holland, K. I. Hirsch-Ernst & F. H. Hellwig, 2005. Analysis of *Acorus calamus* chloroplast genome and its phylogenetic implications. Molecular Biology and Evolution 22: 1813–1822.

Green, P. & F. T. Short, 2003. World atlas of seagrasses. UNEP-WCMC, University of California Press, Berkeley and Los Angeles: 298.

Les, D. H., M. Cleland & M. Waycott, 1997. Phylogenetic studies in Alismatidae, II: evolution of marine angiosperms (Seagrasses) and hydrophily. Systematic Botany 22: 443–463.

Malea, P., T. Kevrekidis & A. Mogias, 2004. Annual versus perennial growth cycle in *Ruppia maritima* L.: temporal variation in population characteristics in Mediterranean lagoons (Monolimni and Drana lagoons, Northern Aegean Sea). Botanica Marina 47: 357–366.

Marchioni Ortu, A., 1982. Numeri cromosomici per la flora italiana: 873-876. Informatore Botanico Italiano 14: 234–237.

Reusch, T. B., 2001. Fitness-consequences of geitonogamous selfing in a clonal marine angiosperm (*Zostera marina*). Journal of Evolutionary Biology 14: 129–138.

Rivas-Martínez, S., A. Asensi, B. Díez-Garretas, J. Molero & F. Valle, 2003. Biogeographical synthesis of Andalusia (southern Spain). Journal of Biogeography 24: 915–928.

Santamaria, L., 2002. Why are most aquatic plants widely distributed? Dispersal, clonal growth and small-scale heterogeneity in a stressful environment. Acta Oecologia 23: 137–154.

Talavera, S., P. Garcia-Murillo & H. Smit, 1986. Sobre el genero *Zannichellia* L. (Zannichelliaceae). Lagascalia 14: 241–271.

Talavera, S., P. Garcia-Murillo & J. Herrera, 1993. Chromosome numbers and a new model for karyotype evolution in *Ruppia* L. (Ruppiaceae). Aquatic Botany 45: 1–13.

Triest, L. & A. Mannaert, 2006. The relationship between *Callitriche* L. clones and environmental variables using genotypes. Hydrobiologia 570: 73–77.

Triest, L. & J. J. Symoens, 1991. Isozyme variation in populations of the submerged halophyte *Ruppia* (Ruppiaceae). Opera Botanica Belgica 4: 115–132.

Triest, L. & L. Vanhecke, 1991. Isozymes in European and Mediterranean *Zannichellia* (Zannichelliaceae)

populations: a situation of predominant inbreeders. Opera Botanica Belgica 4: 133–166.
Triest, L., V. Tran Thi & T. Sierens, 2007. Chloroplast microsatellite markers reveal *Zannichellia* haplotypes across Europe using herbarium DNA. Belgian Journal of Botany 140: 109–120.
Van Vierssen, W., R. Van Wijk & J. Vander Zee, 1981. Some additional notes on the cytotaxonomy of *Ruppia* taxa in Western Europe. Aquatic Botany 11: 297–301.
Verhoeven, J., 1979. The ecology of *Ruppia*-dominated communities in Western Europe. I. Distribution of Ruppia representatives in relation to their autoecology. Aquatic Botany 6: 197–268.
Verhoeven, J., 1980a. Synecological classification, structure and dynamics of the macroflora and macrofauna communities. Aquatic Botany 8: 1–85.
Verhoeven, J., 1980b. Aspects of production, consumption and decomposition. Aquatic Botany 8: 209–253.
Weising, K. & C. Gardner, 1999. A set of conserved PCR primers for the analysis of simple sequence repeat polymorphisms in chloroplast genomes of dicotyledonous angiosperms. Genome 42: 9–19.

Hydrobiologia (2009) 634:107–114
DOI 10.1007/s10750-009-9886-6

POND CONSERVATION

Gravel pits support waterbird diversity in an urban landscape

F. Santoul · A. Gaujard · S. Angélibert ·
S. Mastrorillo · R. Céréghino

Published online: 29 July 2009

Abstract We assessed the benefit of 11 gravel pits for the settlement of waterbird communities in an urbanized area lacking natural wetlands. Gravel pits captured 57% of the regional species pool of aquatic birds. We identified 39 species, among which five were regionally rare. We used the Self-Organizing Map algorithm to calculate the probabilities of presence of species, and to bring out habitat conditions that predict assemblage patterns. The age of the pits did not correlate with assemblage composition and species richness. There was a positive influence of macrophyte cover on waterbird species richness. Larger pits did not support more species, but species richness increased with connectivity. As alternative wetland habitats, gravel pits are attractive to waterbirds, when they act as stepping stones that ensure connectivity between larger natural and/or artificial wetlands separated in space.

Keywords Artificial wetlands · Probability of presence · Rare species · Species richness · Self-Organizing Maps · Waterbirds

Guest editors: B. Oertli, R. Cereghino, A. Hull & R. Miracle
Pond Conservation: From Science to Practice. 3rd Conference of the European Pond Conservation Network, Valencia, Spain, 14–16 May 2008

F. Santoul · S. Mastrorillo · R. Céréghino (✉)
EcoLab, UMR 5245, Université Paul Sabatier, 118 Route de Narbonne, 31062 Toulouse Cedex 9, France
e-mail: cereghin@cict.fr

A. Gaujard
Fédération des Chasseurs de Haute-Garonne, 17 Avenue Jean Gonord, BP 5861, 31506 Toulouse Cedex 5, France

S. Angélibert
Department of Nature Management, University of Applied Sciences of Western Switzerland – EIL, 150 Route de Presinge, 1254 Jussy, Geneva, Switzerland

Introduction

While human activity has resulted in the destruction of natural wetlands (Hull, 1997), artificial pools, such as farm ponds, rice fields, etc., became important alternative habitats for the pond biota (Declerck et al., 2006; Céréghino et al., 2008). It also becomes increasingly accepted that man-made ecosystems are likely to support biodiversity while they provide resources that have economic values, calling for more attention on the importance of these ecosystems to both wildlife and people (Odling-Smee, 2005). Throughout the world, gravel pits contribute to local economies while they constitute new wetlands for species of conservation interest, notably dragonflies, amphibians and birds (Frochot & Godreau, 1995). In France, gravel pits cover an area of about 90,000 ha, and about 5,000 ha are still created each year to satisfy the demands of the construction trade (Barnaud & Le Bloch, 1998). These new wetlands are colonized by waterbirds, creating a situation which raises new management concerns. We performed a test to verify the hypothesis that gravel pits contribute to waterbird diversity in urban areas. We recorded the species occurring in 11 gravel pits in the suburbs of the city of Toulouse (SW France), and the regional species pool of aquatic birds (gamma diversity, SW France 57,000 km^2) was used to assess the benefit of gravel pits for the diversity of waterbird communities. In order to maximize the information extracted from "simple" presence–absence data, we used the Self-Organizing Map (SOM, neural network) to calculate the probabilities of the presence of each species in the various clusters of pits. Subsequently, environmental variables were introduced into the SOM trained with biological variables to interpret the variability of waterbird communities with respect to habitat features.

Methods

Several gravel pits were excavated in the River Garonne floodplain around the city of Toulouse, SW France (Fig. 1), a highly populated area (over 1 million inhabitants) from which natural wetlands were largely eliminated by drainage during the nineteenth and twentieth centuries. Eleven gravel pits were characterized with four environmental variables: age (years), surface area (ha), % macrophyte cover (estimated from point 30 samples using an Eckman grab, each year at the end of the spring period), and an index of connectivity C, calculated as $C = \sum_{i=1}^{n} S_i \times D_i$ where S (surface) and D (distance) are divided into classes defined in Oertli et al. (2000). The use of simple variables was intended to keep models broadly applicable for management applications. All gravel pits had a similar mean depth of 3–4 m.

Bird censuses were carried out weekly from October 1996 to October 1998 using binoculars (8 × 30) and telescopes (20 × 60). Waterbirds were selected according to Gillier et al. (2000). The adequacy of sampling was assessed by plotting the cumulative frequency of species against sampling effort (sample-rarefaction curve with 500 randomizations) (Colwell et al., 2004). The regional species pool (after Maurel et al., 2004) consists of 68 waterbird species occurring in SW France (see Table 1).

A full description of the SOM modellng procedure was given in Céréghino et al. (2008). The structure of the SOM consisted of two layers of neurons connected by weights (connection intensities): the input layer constituted by 39 neurons (one by species) connected to the 11 gravel pits, and the output layer constituted by 15 neurons (visualized as hexagonal cells) organized on an array with five rows and three columns. The SOM plots the similarities of the data in a 2D grid, by grouping similar data items together through an iterative learning process. At the end of the training, the connection intensity between input and output layers calculated during the learning process can be considered as the probability of occurrence of each species in the area concerned (see Céréghino et al., 2005). The occurrence probabilities of each species were visualized on the SOM map in grey scale, and allowed us to analyse the effect of each species on the patterning input dataset (sites). The SOM was clustered using Ward's algorithm (Leroy et al., 2009). In order to bring out relationships between biological and environmental variables, we introduced the environmental variables into the SOM previously trained with bird occurrences

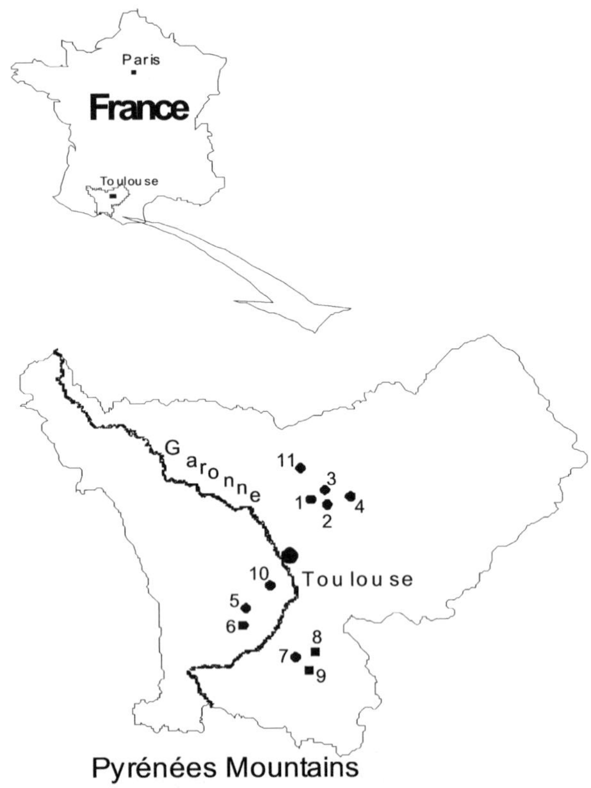

Fig. 1 Location of the 11 gravel pits (1–11) around the city of Toulouse, SW France

Table 1 List of the waterbird species occurring at the regional scale (SW France)

		Gravel pit										
Regional species pool	Regional distribution	1	2	3	4	5	6	7	8	9	10	11
***Actitis hypoleucos* (Linnaeus, 1758)**	Common	+			+				+			
***Anas acuta* Linnaeus, 1758**	Common	+			+			+	+			
***Anas clypeata* Linnaeus, 1758**	Common	+		+	+	+	+	+	+	+		
***Anas crecca* Linnaeus, 1758**	Common	+	+	+	+	+	+	+	+		+	+
***Anas penelope* Linnaeus, 1758**	Common	+			+			+	+	+		
***Anas platyrhynchos* Linnaeus, 1758**	Common	+	+	+	+	+	+	+	+	+	+	+
***Anas querquedula* Linnaeus, 1758**	Common	+			+			+	+	+		
***Anas strepera* Linnaeus, 1758**	Common	+	+		+	+	+	+	+	+	+	
Anser anser (Linnaeus, 1758)	Common											
***Ardea cinerea* Linnaeus, 1758**	Common	+	+	+	+	+	+	+	+	+	+	+
***Ardea purpurea* Linnaeus, 1758**	Rare	+		+	+		+	+	+	+		
Ardeola ralloides (Scopoli, 1769)	Rare											
***Aythya ferina* (Linnaeus, 1758)**	Common	+		+	+	+	+		+	+		
***Aythya fuligula* (Linnaeus, 1758**	Rare	+		+	+	+	+		+	+		
***Aythya marila* (Linnaeus, 1758)**	Rare	+			+							
Aythya nyroca (Gûld, 1769)	Rare											
Bubulcus ibis (Linnaeus, 1758)	Common		+	+	+	+	+	+	+	+		+
Bucephala clangula (Linnaeus, 1758)	Common											
***Calidris alpina* (Linnaeus, 1758)**	Common				+							
Calidris minuta (Leisler, 1812)	Common											
***Charadrius dubius* Scopoli, 1786**	Common				+		+		+		+	
Chlidonias hybridus (Pallas, 1811)	Common											
Chlidonias niger (Linnaeus, 1758)	Common											
Egretta alba (Linnaeus, 1758)	Common											
***Egretta garzetta* (Linnaeus, 1758)**	Common	+	+	+	+	+	+	+	+	+	+	+
Eudromias morinellus (Linnaeus, 1758)	Rare											
***Fulica atra* Linnaeus, 1758**	Common	+	+	+	+	+	+	+	+	+		+
***Gallinago gallinago* (Linnaeus, 1758)**	Common				+				+		+	+
***Gallinula chloropus* (Linnaeus, 1758)**	Common		+	+		+		+	+	+	+	+
Gelochelidon nilotica (Gmelin, 1789)	Rare											
Glareola pratincola Linnaeus, 1766	Common				+							
***Grus grus* (Linnaeus, 1758)**	Common				+				+	+		
***Himantopus himantopus* (Linnaeus, 1758)**	Common				+			+	+			
Ixobrychus minutus (Linnaeus, 1766)	Rare											
***Larus cachinnans* (Pallas, 1826)**	Common	+		+	+	+	+	+	+	+	+	+
Larus canus Linnaeus, 1758	Common											
Larus fuscus Linnaeus, 1758	Common											
Larus minutus Pallas, 1776	Common											
***Larus ridibundus* Linnaeus, 1758**	Common	+	+		+			+	+	+		
***Limosa lapponica* (Linnaeus, 1758)**	Common				+							
Limosa limosa (Linnaeus, 1758)	Common											
Lymnocryptes minimus (Brûnn, 1764)	Common											
***Mergus albellus* Linnaeus, 1758**	Rare				+							

Table 1 continued

Regional species pool	Regional distribution	Gravel pit										
		1	2	3	4	5	6	7	8	9	10	11
Mergus merganser Linnaeus, 1758	Rare											
Mergus serrator Linnaeus, 1758	Common											
***Netta rufina* (Pallas, 1773)**	Common	+			+				+			
Numenius arquata (Linnaeus, 1758)	Rare											
***Nycticorax nycticorax* (Linnaeus, 1758)**	Common	+		+	+	+		+	+			+
***Phalacrocorax carbo* (Linnaeus, 1758)**	Common	+	+	+	+	+	+	+	+	+	+	+
Philomachus pugnax (Linnaeus, 1758)	Common											
***Platalea leucorodia* Linnaeus, 1758**	Rare				+							
Pluvialis apricaria (Linnaeus, 1758)	Common											
***Podiceps auritus* (Linnaeus, 1758)**	Common	+			+							
***Podiceps cristatus* (Linnaeus, 1758)**	Common	+	+	+	+	+	+	+	+	+	+	+
Podiceps nigricollis Brehm, 1831	Common											
Porzana porzana (Linnaeus, 1758)	Rare											
Rallus aquaticus Linnaeus, 1758	Common											
Recurvirostra avosetta Linnaeus, 1758	Common											
Sterna albifrons Pallas, 1764	Common											
***Sterna hirundo* Linnaeus, 1758**	Common				+			+				
***Tachybaptus ruficollis* (Pallas, 1764)**	Common	+	+	+	+			+				
***Tadorna tadorna* (Linnaeus, 1758)**	Common	+			+				+	+		
Tringa erythropus (Pallas, 1764)	Common											
Tringa glareola Linnaeus, 1758	Common											
***Tringa nebularia* (Gunnerus, 1767)**	Common				+				+			
Tringa ochropus Linnaeus, 1758	Common											
***Tringa totanus* (Linnaeus, 1758)**	Common				+				+			
***Vanellus vanellus* (Linnaeus, 1758)**	Common				+	+						

The 39 species recorded in our 11 gravel pits (Toulouse suburbs) appear in bold, + = presence. Information on rarity and commonness at the regional scale is given (after Maurel et al., 2004)

(see Park et al., 2003). All mean values of environmental variables assigned on the SOM map were visualized in grey scale.

Results

Fifty-seven percent of the regional species pool was captured by our 11 gravel pits (Table 1). The number of species per pit ranged from 12 to 38. Among the 39 species recorded, five species were regionally rare, and the remaining ones were common. Accumulation of new species reached its asymptote (Fig. 2); we could thus consider that our sampling was satisfactory. After training, the SOM with species occurrences, Ward's algorithm helped to derive three clusters of sites (gravel pits) according to waterbird assemblages (Fig. 3a). The ordinate of the SOM represented the number of species, from low (top areas of the SOM) to high (bottom) (Fig. 3b). Eleven species (*Podiceps auritus*, *Platalea leucorodia*, *Aythya marila*, *Netta rufina*, *Mergus albellus*, *Glareola pratincola*, *Tringa totanus*, *Tringa nebularia*, *Actitis hypoleucos*, *Limosa lapponica*, and *Calidris alpina*) only occurred in cluster C (Fig. 4). Five species (*Podiceps cristatus*, *Phalacrocorax carbo*, *Ardea cinerea*, *Egretta garzetta*, and *Anas platyrhynchos*) occurred in all the gravel pits, and the remaining 23 species occurred in two clusters of sites. When environmental variables were introduced into the SOM (Fig. 5), the ordinate on the SOM showed a gradient of connectivity and macrophyte cover [from low (top area) to high (bottom)].

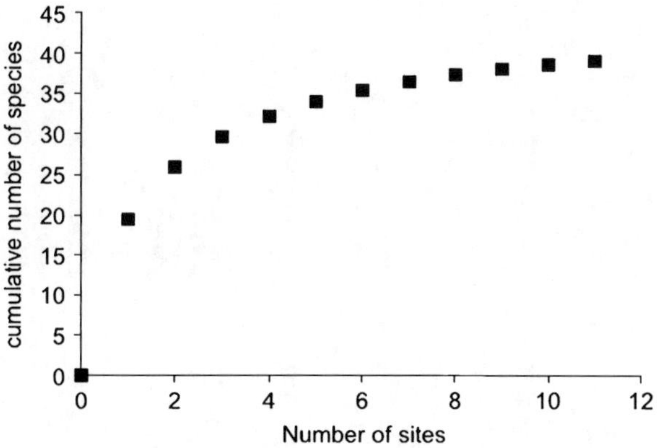

Fig. 2 Sample-based rarefaction curves representing the number of species accumulated by sampling the 11 gravel pits

Site 4, which had the highest index of connectivity and a macrophyte cover of 90%, hosted 38 waterbird species. Other variables under consideration (age and surface area) did not show clear patterns and were thus not considered as structuring variables.

When clusters of gravel pits were compared to their distribution on the geographical map of the study area, the species hosted by some pits within the same sub-areas tended to be similar (gravel pits 7 and 9 in cluster B; 5, 6 and 10, and 2, 3 and 11 in cluster A, 1 and 4 in cluster C, see Fig. 1) and those characteristics tended to differ when sites belonged to more distinct areas. However, gravel pits from distinct areas also had very similar assemblages (e.g., 8 and 4; 3 and 5–6; 11 and 10), thus suggesting that local factors (e.g., macrophyte cover) interacted with connectivity to shape bird assemblages.

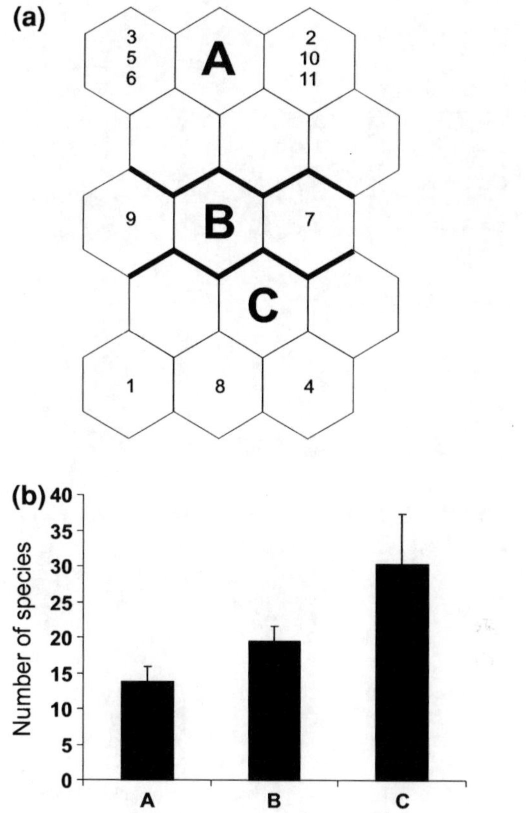

Fig. 3 **a** Distribution of gravel pits on the self-organizing map (SOM) according to the presence or absence of 39 waterbird species. *Solid lines* show the cluster boundaries (i.e., for clusters A, B, C), delineated according to Ward's algorithm. Gravel pits that are neighbors within clusters are expected to have similar waterbird assemblages. Codes (1–11) correspond to gravel pits. **b** Mean number of species (±SE) per cluster

Discussion

While the potential pool of colonists in our study region was made of 68 waterbird species, 11 gravel pits allowed the presence of 39 species in the Toulouse city suburbs. Our observations therefore highlight how a small set of artificial wetlands may sustain an important fraction (57%) of the larger regional species pool in landscapes where natural wetlands are lacking. Other observations in France and UK showed increases in population densities of several waterbirds during the past decades (*Aythya fuligula*, *Anas strepera*, *Phalacrocorax carbo*, *Podiceps cristatus*, *Sterna hirundo*), as a result of the creation of gravel pits and reservoirs (Frochot et al., 2008; Fuller & Ausden, 2008).

Species richness and the presence of rare species are frequently cited criteria for site selection by conservationists (Myers et al., 2000). If man-made habitats only attract the common (widespread) species, one may argue that they do not make a significant contribution to biodiversity. Conversely, rare species are of special interest to environmental managers (Rey-Benayas et al., 1999), and it was recently demonstrated that areas which carry rare species may also concentrate an important fraction of the regional biodiversity (Cucherousset et al., 2008). For instance, some species were exclusive to gravel pit 4, which also hosted the highest species richness. Such artificial wetlands might therefore benefit from

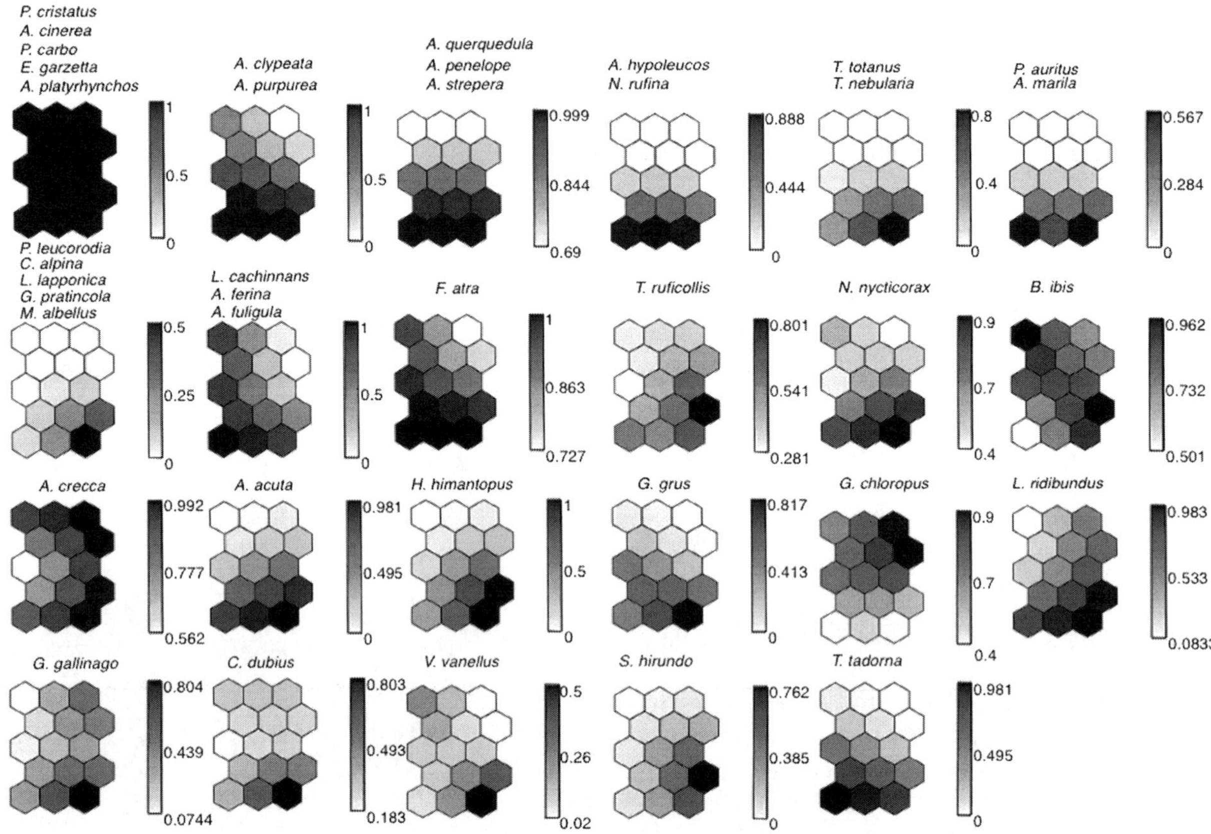

Fig. 4 Gradient analysis of the probability of occurrence of each waterbird species on the trained SOM (see Fig. 3a), with visualization in shading scale (*dark* = high probability of occurrence, *light* = low probability of occurrence). Each map is superimposed on the map representing the distribution of sites presented in Fig. 3a. See Table 1 for full species names and authorities

higher management priorities. We thus support the idea that urban landscapes containing man-made wetlands make a significant contribution to freshwater biodiversity. Larger gravel pits did not support more waterbird species. This absence of species–area relationship suggests that larger gravel pits were not more easily colonized by immigrants, and/or were not likely to show a higher diversity of ecological niches facilitating the coexistence of a larger number of species. Smaller but well-connected gravel pits ($\approx$4 ha) had the greatest susceptibility to host more taxa, and more rare species. They potentially had higher conservation value for waterbirds than larger gravel pits, although small surface may become a limiting factor if the carrying capacity becomes insufficient for waterbirds. Thus, gravel pits were certainly attractive to waterbirds when they acted as stepping stones that ensured connectivity between larger natural and/or artificial wetlands separated in space (Bournaud et al., 1982). Species-poor assemblages were subsets of richer assemblages, suggesting nested patterns of waterbird assemblages. We assume that such patterns would be colonization-driven because most waterbirds did not live at the gravel pits throughout the year. The geographical location of the study area near the Pyrenees mountainous barrier makes the region important as a stop-over for migrant birds (Hoyer, 1994). Rare species were specifically present during the stop-over period, and therefore they probably preferred those connected sites which allowed them to find quieter sites in case of disturbance. There was a positive influence of the extent of the macrophyte cover on the number of waterbird species present at a site. Aquatic plants over large areas are attractive to many birds, such as charophytes, and the many invertebrates living on

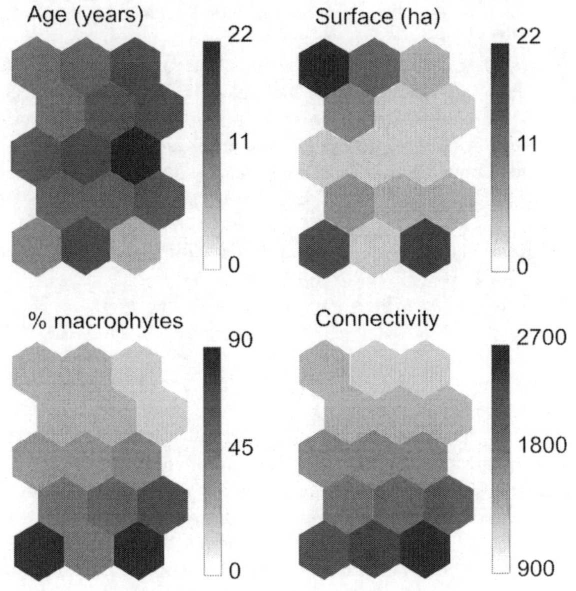

Fig. 5 Visualization of environmental variables on the SOM trained with waterbird species. The mean value of each variable was calculated in each output neuron of the trained SOM. *Dark* represents a high value, while *light* is low

them provide food for ducks and other waterbirds (Knapton & Petrie, 1999; Blindow et al., 2000). In turn, aquatic birds may have a negative impact on macrophyte abundance, and may be important determinants of the aquatic system dynamics (reviewed in Lodge et al., 1998). The age of the gravel pit was not a structuring factor for waterbird assemblages, at least within the range of 3–22 years considered in this study (no gravel pits older than 22 years in the area). Nevertheless, Frochot & Godreau (1995) emphasized that gravel pits over 30 years are less attractive for waterbirds, due to their more homogeneous and dense habitats, although some species such as herons may benefit from the development of trees.

In conclusion, gravel pits should be seen as a network of habitats integrated within the broader network of natural and artificial wetlands. We must enlarge the spatial scale needed to manage/survey these particular habitats, and we should primarily pay attention to wetland networks rather than attempting to target some species and/or bodies of water for particular management actions. The lack of fundamental knowledge needed to implement management plans usually limits the conservation potential of gravel pits. Restoration is generally carried out to transform gravel pits into recreational areas or fishing lakes, and only rarely into habitats favorable to waterbirds. Owing to the continuing loss of natural wetlands, there is a need to enhance the contribution of artificial wetlands such as gravel pits for future conservation of waterbirds, and probably other taxa such as amphibians, insects, or plants. Therefore, further understanding of the distribution of biological diversity in non-natural systems may facilitate the adoption of positive solutions for wildlife, with limited costs for human activities.

References

Barnaud, G. & F. Le Bloch, 1998. Entre terre et eau, agir pour les zones humides. Dossier d'Information du Ministère de l'Aménagement du Territoire et de l'Environnement (France).

Blindow, I., A. Hargeby & G. Andersson, 2000. Long-term waterfowl fluctuations in relation to alternative states in two shallow lakes. In Comín, F. A., J. A. Herrera & J. Ramírez (eds), Limnology and Aquatic Birds. Monitoring, Modelling and Management. Proceedings of the 2nd International Symposium on Limnology and Aquatic Birds. Universidad Autónoma de Yucatán, Mérida: 165–175.

Bournaud, M., J. P. Ledant, J. P. Broyer & M. Richoux, 1982. L'espace étang dans ses rapports avec l'avifaune en période de nidification. Bulletin d'Ecologie 13: 125–144.

Céréghino, R., F. Santoul, A. Compin & S. Mastrorillo, 2005. Using self-organizing maps to investigate spatial patterns of non-native species. Biological Conservation 125: 459–465.

Céréghino, R., A. Ruggiero, P. Marty & S. Angélibert, 2008. Biodiversity and distribution patterns of freshwater invertebrates in farm ponds of a southwestern French agricultural landscape. Hydrobiologia 597: 43–51.

Colwell, R. K., C. X. Mao & J. Chang, 2004. Interpolating, extrapolating, and compared incidence-based species accumulation curves. Ecology 85: 2717–2727.

Cucherousset, J., F. Santoul, J. Figuerola & R. Céréghino, 2008. How do biodiversity patterns of river animals emerge from the distributions of common and rare species? Biological Conservation 141: 2984–2992.

Declerck, S., T. De Bie, D. Ercken, H. Hampel, S. Schrijvers, J. Van Wichelen, V. Gillard, R. Mandiki, B. Losson, D. Bauwens, S. Keijers, W. Vyverman, B. Goddeeris, L. De Meester, L. Brendonck & K. Martens, 2006. Ecological characteristics of small farmland ponds: associations with land use practices at multiple spatial scales. Biological Conservation 131: 523–532.

Frochot, B. & V. Godreau, 1995. Intérêt écologique des carrières, terrils et mines. Natures Sciences Sociétés 3 (special issue): 66–76.

Frochot, B., V. Godreau & J. Roché, 2008. L'expansion récente des oiseaux d'eau. Alauda 76: 279–286.

Fuller, R. J. & M. Ausden, 2008. Birds and habitat change in Britain. Part I: a review of losses and gains in the twentieth century. British Birds 101: 644–675.

Gillier, J. M., R. Mahéo & F. Gabillard, 2000. Les comptages d'oiseaux d'eau hivernant en France: actualisation des connaissances, effectifs moyens, critères numériques d'importance internationale et nationale. Alauda 68: 45–54.

Hoyer, F., 1994. Rôle des gravières et des lacs colinéaires. Le journal du chasseur 39: 6–15.

Hull, A., 1997. The pond life project: a model for conservation and sustainability. In Boothby, J. (ed.), British Pond Landscape, Proceedings from the UK Conference of the Pond Life Project. Pond Life Project, Liverpool: 101–109.

Knapton, R. W. & S. A. Petrie, 1999. Changes in distribution and abundance of submerged macrophytes in the inner bay at Long Poing, Lake Erie: implications for foraging waterfowl. Journal of Great Lakes Research 25: 783–798.

Leroy, C., M. Guéroult, N. S. Wahyuni, J. Escoute, R. Céréghino, S. Sabatier & D. Auclair, 2009. Morphogenetic trends in morphological, optical, biochemical features of phyllodes in *Acacia mangium* Willd (Mimosaceae). Trees – Structure and Function 23: 37–49.

Lodge, D. M., G. Cronin, E. Van Donk & A. J. Froelich, 1998. Impact of herbivory on plant standing crop: comparisons among biomes, between vascular and among freshwater herbivory taxa. In Jeppensen, E., M. Sondergaard & K. Christoffersen (eds), The structuring role of submerged macrophytes in lakes. Springer-Verlag, New York: 149–174.

Maurel, C., C. Arthur, G. Bechard, J. F. Bousquet, M. Fily, V. Heaulme, J. Joachim, E. Menoni & H. Redon, 2004. Liste de collemboles déterminants. In Durand C., L. Pontcharraud & A. Bertrand (eds), Modernisation de l'inventaire des Zones Naturelles d'Intérêt Ecologique, Faunistique et Floristique (Znieff) en Midi-Pyrénées. Listes préliminaires d'espèces et cortèges de faune déterminants. Conservatoire Régional des Espaces Naturels de Midi-Pyrénées, DIREN Midi-Pyrénées, European Union: 17–29.

Myers, N., R. A. Mittermeier, C. G. Mittermeier, G. A. B. da Fonseca & J. Kent, 2000. Biodiversity hotspots for conservation priorities. Nature 403: 853–858.

Odling-Smee, L., 2005. Dollars and sense. Nature 437: 614–616.

Oertli, B., D. Auderset Joye, E. Castella, R. Juge & J. B. Lachavanne, 2000. Diversité biologique et typologie écologique des étangs et petits lacs de Suisse. Technical Report, LEBA, Université de Genève: 434 pp.

Park, Y. S., R. Céréghino, A. Compin & S. Lek, 2003. Applications of artificial neural networks for patterning and predicting aquatic insect species richness in running waters. Ecological Modelling 160: 265–280.

Rey-Benayas, J. M., S. M. Scheiner, M. Garcia Sanchez-Colomer & C. Levassor, 1999. Commonness and rarity: theory and application of a new model to Mediterranean montane grasslands. Conservation Ecology 3: 5.

Hydrobiologia (2009) 634:115–124
DOI 10.1007/s10750-009-9887-5

POND CONSERVATION

Competition in microcosm between a clonal plant species (*Bolboschoenus maritimus*) and a rare quillwort (*Isoetes setacea*) from Mediterranean temporary pools of southern France

Mouhssine Rhazi · Patrick Grillas · Laïla Rhazi · Anne Charpentier · Frédéric Médail

Published online: 5 August 2009

Abstract *Bolboschoenus maritimus*, a clonal species, is locally invasive in Mediterranean temporary pools where it threatens endangered rare plant species such as *Isoetes setacea*. The combination of management modifications (grazing) and of the progressive accumulation of fine sediments in the pools contributed to the establishment of competitive perennial plants such as *B. maritimus*. The competitive advantage of *B. maritimus* on *I. setacea* has been studied in controlled conditions. The goal of this experiment was to assess the role of environmental conditions in the output of the competition between *Bolboschoenus* and *Isoetes*, notably hydrology and soil richness. For this purpose, *Isoetes* was cultivated alone (three individuals/pot) and with *Bolboschoenus* (three individuals of both species). The experiment was run with five replicates on six types of sediment (gradient of richness in sand/silt/clay) combined with three hydrological treatments (flooded, wet. and dry). The competitive advantage of *Bolboschoenus* was measured as the ratio of the production of *Isoetes* in mixture versus monoculture. The results showed that *Isoetes* was always outcompeted by *Bolboschoenus*. However, the competitive advantage of *Bolboschoenus* on *Isoetes*, was more related to hydrology than to soil richness. The competitive advantage of *Bolboschoenus* was very high in wet and flooded conditions and very low in dry conditions. This situation may lead to the extinction, medium-term, of the populations of *I. setacea*. The introduction of ovine grazing or of cut back practices in temporary pools could reduce the *B. maritimus* biomass and help toward the conservation of *I. setacea* populations.

Keywords Competition · *Bolboschoenus maritimus* · *Isoetes setacea* · Temporary pools · Hydrology · Soil

---

Guest editors: B. Oertli, R. Cereghino, A. Hull & R. Miracle
Pond Conservation: From Science to Practice. 3rd Conference of the European Pond Conservation Network, Valencia, Spain, 14–16 May 2008

M. Rhazi
Department of Biology, Faculty of Sciences and Techniques, Moulay Ismail University, BP 509, Boutalamine, Errachidia, Morocco
e-mail: rhazimouhssine@yahoo.fr

P. Grillas (✉)
Tour du Valat, Le Sambuc, 13200 Arles, France
e-mail: grillas@tourduvalat.org

L. Rhazi
Laboratory of Aquatic Ecology and Environment, Hassan II University, BP 5366, Maarif Casablanca, Morocco

A. Charpentier
CEFE, Montpellier II, University, Montpellier, France

F. Médail
Mediterranean Institute of Ecology and Paleoecology (IMEP, UMR CNRS-IRD 6116), Aix-Marseille University (University Paul Cézanne), Europôle Méditerranéen de l'Arbois, BP 80,
13545 Aix-en-Provence Cedex 04, France

Introduction

Competitive exclusion is often advanced as a hypothesis to explain the regression or the disappearance of rare and threatened species (e.g., Gaston, 1994; Vilà & Weiner, 2004). For wetlands, the development of invasive and competitive species can result from many causes such changes in the management (e.g., eutrophication) or in ecological processes (e.g., sedimentation). The success and impacts of the competitive colonizing species depend on the life history traits of the invaders, the environmental characteristics of the colonized ecosystem and the biotic interactions with the other species of the receiving community (e.g., Connel, 1975; Tilma, 1988; Vilà & Weiner, 2004). Competitor plants are characterized by particular life history traits (e.g., Grime, 2001): the clonal vegetative multiplication and the varied modes of dispersal are key factors in the persistence and diffusion of this group. The intensity of the competition, and therefore, the exclusion potential generally increases with the productivity (Keddy, 1989; Wisheu & Keddy, 1992; Twolan-Strutt & Keddy, 1996).

Nevertheless, in spite of the numerous works performed on the conceptual and applied aspects of interspecific competition, few experimental studies have confronted the level of interference between a threatened plant versus a competitor plant (but see Huenneke & Thomson, 1995; Walck et al., 1999). In the Mediterranean region, where water stress levels are often high, the analysis of functional traits of endemic species, which are often rare and threatened plants, indicates that the "competitive" strategy (*sensu* Grime, 2001) is under-represented with 3–7% of the endemic flora of the south-eastern France (Médail & Verlaque, 1997). If we consider the Leaf–Height–Seed (L–H–S) *sensu* Westoby (1998), endemic plants possess generally a weak competitive capacity, but they only differ from their widespread congeners by their smaller stature (Lavergne et al., 2003).

Several methods have been developed for assessing the importance of interspecific competition (e.g., Mead & Wille, 1980; Connoll, 1986; Akey et al., 1991; Goldberg et al., 1999). These methods generally confront the relative performances of the species (density and/or productivity) in mixed culture and in monoculture. The analysis of the ratio of biomass in mixed versus pure culture (Goldberg et al., 1999) has been frequently used as it provides robust results and is easy to undertake.

In the Nature Reserve of Roque-Haute (43°18′ N; 3°22′ E; Hérault, Southern France), *Bolboschoenus maritimus* recently colonized only a few pools (Trabaud, 1998; Rhazi, 2005) among the 205 pools that it contains. This recent situation probably results from a combination of management modifications (grazing) and of the progressive accumulation of fine sediments in the pools. The colonization of *B. maritimus* could have important consequences in terms of conservation as it could have a negative impact on the Mediterranean Temporary Pools type of vegetation, habitat of European importance in the Habitats Directive 92/43/EEC (Gaudillat et al., 2002), and on the numerous rare and protected species associated to the temporary pools. Among these species, *Isoetes setacea*, an endemic of the Western Mediterranean, is abundant in the pools and finds its main location among the few places where the species exists in France (Rhazi et al., 2004). The on-going replacement of *I. setacea* by *B. maritimus* was suggested by the abundance of the macrospores of *Isoetes setacea* that were found in seed banks within the patches of *B. maritimus*, contrasting with the absence of *Isoetes* in the extant vegetation (Grillas et al., 2004a). The eutrophication and the decrease of the drought stress in summer, both resulting from sediment accumulation in the pools, are two likely mechanisms that could explain the success of *B. maritimus* in the pools of Roque-Haute. Therefore, the extension of *B. maritimus* in more pools is expected, if we consider the ongoing process of global change and increase of dryness in the Mediterranean region. However, at the current pioneer stage of colonization, the dispersal of *B. maritimus* could be limited by a low seed production resulting from a low number of genotypes (clones) which reduces the success of sexual reproduction of this strictly allogamous species (Charpentier et al., 2000).

The objective of this work was to assess the range of environmental conditions in which *I. setacea* would not be displaced by *B. maritimus*. With this perspective, an experiment was conducted to (1) test the impact of hydrological regime and the type of substratum on the output of competition and the sexual reproduction of *Isoetes setacea* and *Bolboschoenus maritimus*; (2) estimate the survival chances of *Isoetes setacea* facing this new competitor.

Materials and methods

Site study

Roque-Haute Nature Reserve (43°18′ N; 3°22′ E; Hérault, Southern France) contains 205 temporary pools on basaltic substratum wherein are found many rare and protected species of plants and amphibians (Molina, 1998; Médail et al., 1998; Jakob et al., 1998). Most of the pools have an artificial origin, as a result of the extraction of the basalt during the nineteenth and beginning of the twentieth century (Crochet, 1998). The Nature Reserve was created in 1975 to protect four rare ferns characteristics of the Mediterranean temporary pools: *Isoetes setacea*, *I. durieui*, *Pilularia minuta,* and *Marsilea strigosa*. The vegetation of the Mediterranean temporary pools is dominated by annual and geophyte species with often short periods of growth within the annual cycle and a weak vegetative development (Braun-Blanquet, 1936; Deil, 2005). These traits are interpreted as adaptations to the high unpredictability and low productivity of the habitat (Médail et al., 1998). Since its creation, the vegetation of the pools of Roque-Haute has shown important modifications, notably the colonization by helophytes such as *Bolboschoenus maritimus* and by shrubs (*Ulmus minor* or *Fraxinus angustifolia*) (Trabaud, 1998; Rhazi et al., 2004). These modifications probably result from management modifications (ending of sheep grazing), and of a primary succession dynamics due to the progressive accumulation of fine sediments in some pools after the ending of basalt extraction.

The species

Isoetes setacea

Isoetes setacea (Isoetaceae) is a heterosporous Lycopodiophyta whose height varies between 3 and 40 cm. It is a perennial amphibious (bulbous geophyte) species, characteristic of temporary pools (Braun-Blanquet, 1936). It presents cylindrical sporophylls, disposed in rosettes, with megasporangia (external sporophylls) or microsporangia (internal sporophylls) at the base. It has a haploid–diploid digenetic life cycle with a very short gametophytic stage (Prelli, 2002). *Isoetes setacea* is a western Mediterranean species, occurring in Portugal, Spain, France, and Morocco (Quézel, 1998; Titolet & Rhazi, 1999). In France, it has a national protection status (Olivier et al., 1995), currently occurring in only four locations: in the Hérault (Roque–Haute plateau and Béziers plain) and in the Pyrénées–Orientales (around the Saint-Estève pool and on the Rodès plateau) (Grillas et al., 2004b).

Bolboschoenus maritimus

Bolboschoenus maritimus is a perennial Cyperaceae with an average height of 1.20 m. It is often found in shallow, freshwater, or brackish swamps (Kantru, 1996) and considered a widespread species in France. *B. maritimus* has a modular development, with an aerial stem developing a tuber at its base and a system of rhizomes (1–3) that produce new aerial stems (Charpentier et al., 1998). Therefore, during an annual cycle, a seedling can be at the origin of several tens of stems, interconnected by the rhizomes. At the end of the summer, when the aerial parts of the plant die, the tubers and rhizomes remain dormant in the soil. In the spring, only some stems form the aerial apical inflorescences with several spikelets to hermaphrodite flowers. The seeds are produced after anemophilous pollination and remain dormant for some years in the soil (Clevering, 1995). Their germination is dependent on the amount of light and seedlings have a higher chance of surviving when the water depth is low (Clevering, 1995).

Competition experiments between *Bolboschoenus maritimus* and *Isoetes setacea*

Two experiments were carried out during this study. The first compared the development and production of *Isoetes setacea* monocultures in different substrata and in three distinct hydrological situations. The second experiment was to assess the competition between *Bolboschoenus maritimus* and *Isoetes setacea* in the same conditions of substratum and hydrology that were analyzed in the first experiment.

Bulbs of *Isoetes setacea* were harvested in the summer of 2001, in a pool (pool 51) in the Roque–Haute Nature Reserve (43°18′ N; 3° 22′ E). In the laboratory, the bulbs were sorted and kept in paper bags at room temperature before being used for the two experiments. Tubers of *Bolboschoenus maritimus* were also harvested in the summer of 2001, in a pool (pool 66) in the Reserve. In the laboratory, the tubers were covered with slightly humid sand and kept in a refrigerator at 5°C, until the beginning of the experiment in the following spring.

Effect of the hydrology and the granulometry of the substratum on Isoetes setacea in pure culture

For the experiment, 90 pots, with 16 cm in height and 20 cm in diameter, were used. Each one was filled with one of the following six types of substratum made up of sand (S), silt (I), and clay (C): (IS: 0% C, 75% I, 25% S); (SIC: 25% C, 25% I, 50% S); (CIS: 50% C, 25% I, 25% S); (CI: 75% C, 25% I, 0% S); (CS: 90% C, 0% I, 10% S) and (C: 100% C, 0% I, 0% S). These substrata were sterilized at 100°C.

Three randomly chosen bulbs of *Isoetes setacea* were weighed (min./max.: 1.63/2.2 g) and placed in each pot. The effect of hydrology was tested on each of the six types of substrata (five replicates/substratum) according to three hydrological treatments:

- "*Flooded*": each pot was completely flooded (top layer of soil was under 6 cm of water) (the percentage of water saturation was of 100%).
- "*Wet*": the pots were watered with a frequency of 5 min/h (percentage of water saturation after watering was of 47%; SD: 2.9%)
- "*Dry*": the pots were manually and weakly watered two times a week (percentage of water saturation after watering was of 18%; SD: 7.5%).

The experiment, carried out in a greenhouse, started on April 15, 2002 and finished on June 30, 2002 (76 days). The number and length of the sporophylls of *I. setacea* were measured every 2 weeks. At the end of the experiment, the bulbs were harvested and their weight measured. The number of microsporangia, megasporangia, and macrospores per macrosporangia were counted; the weight of 40 macrospores and that of a microsporangia was measured for each plant.

Competition between Bolboschoenus maritimus and Isoetes setacea

A series of 90 pots with the same type of treatments ("hydrology" and "substratum") and the same number of replicates were prepared. In each pot, three bulbs of *Isoetes setacea* and three tubers of *Bolboschoenus maritimus*, randomly chosen and previously weighed (min./max.: 1.732/2.36 g and min./max.: 1.613/18.42 g, respectively) were placed according to a constant alternate disposition to facilitate their location. The pots used in these experimentations have the same characteristics and offer a sufficient space for the growth of the plants of both species.

The experiment, carried out in a greenhouse, started on April 12, 2002 and finished on June 30, 2002 (79 days). During this experiment, the plants of both *Isoetes* and *Bolboschoenus* were measured every 2 weeks:

- For *Isoetes setacea*, the same measurements were taken as those carried out in the first experiment.
- For *Bolboschoenus maritimus*, the following measurements were taken for each pot: number of vegetative and reproductive stems, mean stem length, mean number of leafs per stem and number of ears and spikelets of each reproductive stem. At the end of the experiment, the number of tubers produced per individual was counted and weighed. The spikelets were harvested and the number of seeds produced counted. The length of the seeds, as well as the weight of 5 seeds per individual, was recorded.

Retained model of competition

Several models have been created to study the competition between species. Among these models, the "Absolute competition intensity" (ACI), the "Relative competition intensity" (RCI) (Grace, 1995), as well as the "log Response Ratio" (logRR) (Goldberg et al., 1999), are often used. The use of the ACI or the RCI models can lead to contradictory conclusions, creating the problem of choosing the more suitable model (Grace, 1995). No qualitative difference has been found between the RCI and the "logRR" models (Weigelt et al., 2002). Hedges et al. (1999) and Weigelt et al. (2002) encourage the use, mainly for statistical reasons, of "logRR". Moreover, this model enables the linearization of the measurements and the normalization of the data distribution. Besides, Goldberg et al. (1999) believe that the "logRR" model can provide more appropriate results for the analysis of competition interactions than the RCI model. Finally, the "logRR" model is symmetrical for the competition interactions between species and does not impose a limit on the possible maximum of competition intensity. The model used in this study, therefore, draws inspiration from the one established by Goldberg et al. (1999).

Data analysis

The morphological variables measured, for both species, were strongly correlated. In order to reduce the number of redundant variables, a Correspondence analysis (CA) per species was carried out, using all the variables, and a limited number of them were retained. The variables retained were, for *I. setacea*, the number and length of the sporophylls, the weight gained by the bulbs, the number of megasporangia and microsporangia, and the weight of the 40 spores. The stem length, the weight gained by the tubers, and the number and weight of the seeds were the variables retained for *B. maritimus*.

The two-factors analyses of variance (ANOVA-2) on the variables measured for both *Isoetes setacea* and *Bolboschoenus maritimus*, identified a significant effect due to the "hydrological" treatment but not the "substratum" treatment or the interaction between these two factors. The "substratum" variable has therefore been suppressed. After the verification of the data distribution, analyses of variance (ANOVA-1), followed by mean comparisons (Tukey–Kramer test), were used to test the effect of the "hydrological" treatment on the different variables of both species. These tests were also used to analyze the initial biomass and weight gain ratios of "*Bolboschoenus* tubers/*Isoetes* bulbs". When the distribution of the data was not normal (number of seeds per ear and weight of five seeds, both *B. maritimus* measurements), the non-parametric test of Kruskall–Wallis was used.

The effect of *B. maritimus*, in the three "hydrological" treatments, on the different variables of vegetative production and sexual reproduction of *I. setacea*, was measured as the ratio: P_{mix}/P_{pure}, where P_{pure} represents the performance of *Isoetes* in monoculture and P_{mix} its performance in mixture. This model is similar to that of Goldberg et al. (1999) (log Response Ratio = log (P_{mix}/P_{pure}). The presence of the logarithm in the model of Goldberg et al. (1999) enables the normalization of the data distribution (Hedges et al., 1999). In the model used, the distribution of the vegetative production data was normal (Shapiro–Wilk test W, $P > 0.05$) but the sexual reproduction data was not, even after transformation. The differences in the ratios (P_{mix}/P_{pure}) between the three "hydrological" treatments was tested by an analysis of variance, followed by a means comparison per pairs (Tukey–Kramer test).

Results

Effect of hydrology on *Isoetes setacea* in monoculture

The hydrological factor had a significant effect on all the variables measured, with significantly lower values for the dry treatment (Table 1). The length of the sporophylls and the number of microsporangia per plant were significantly higher in the flooded treatment than in the wet one. Conversely, the number of sporophylls and the number of megasporangia per plant were significantly higher in the wet treatment (Table 1). The biomass gain per bulb and the weight of the macrospores did not differ between the wet and flooded treatments.

Effect of hydrology on *Bolboschoenus maritimus* and *Isoetes setacea* in competition

Bolboschoenus maritimus

The hydrological treatment had a significant effect on the stem length and biomass gain of the tubers of *B. maritimus*. The averages for both these variables were significantly different between the three treatments, with maximal values recorded in the flooded treatment and minimal values in the dry treatment (Table 1). No sexual reproduction was observed in the dry treatment. The number of seeds per spikelet and the mean seed weight did not differ significantly between the flooded and wet treatments (Table 1).

Isoetes setacea

When *Isoetes setacea* was grown with *B. maritimus*, its response to the different hydrological treatments was similar to the response when grown in monoculture. The results of the comparison of the variables of vegetative production between the three treatments were analogous to those found when this species was grown on its own.

The number of megasporangia produced per plant was significantly higher in the flooded treatment than in the wet and dry treatments (Table 1). The number of microsporangia produced per plant and the weight of the

Table 1 The results of the analysis of variances between the three hydrological treatments of the pure culture and the mixture of *Isoetes setacea* (for the number and the length of the sporophylls, bulbs production, number of microsporangia and macrosporangia, and the weight of 40 spores) as well as the means comparison (*F* flooded treatment, *W* wet treatment, and *D* dry treatment. *Different letters* indicate a significant difference between treatments $P < 0.05$) and the results of the analysis of variances between the three hydrological treatments (flooded, wet, and dry) of *Bolboschoenus maritimus* for the length of the individuals, the tubers' production, and the results of the Kruskal–Wallis test between the flooded and wet treatments for the number of seeds by ears and the weight of five seeds/ears, as well as the medians

	ANOVA			Means ± standard errors			Means comparison
	d*F*	*F*	*P*	Flooded	Wet	Dry	
Isoetes in pure culture							
Sporophylls number	2	426.51	***	9.91 ± 0.32	14.47 ± 0.30	2.92 ± 0.19	$W^1F^2D^3$
Sporophylls length (cm)	2	457.33	***	15.58 ± 0.52	7.97 ± 0.19	1.60 ± 0.13	$F^1W^2D^3$
Bulbs production (g)	2	233.85	***	0.09 ± 0.01	0.09 ± 0.00	0.01 ± 0.00	$F^1W^1D^2$
Microsporangia number	2	373.00	***	5.92 ± 0.18	5.36 ± 0.10	0.73 ± 0.15	$F^1W^2D^3$
Macrosporangia number	2	257.51	***	11.32 ± 0.57	13.08 ± 0.38	0.93 ± 0.18	$W^1F^2D^3$
Weight of 40 spores (g)	2	156.56	***	0.005 ± 0.00	0.005 ± 0.00	0.001 ± 0.00	$F^1W^1D^2$
Isoetes in mixture							
Sporophylls number	2	369.37	***	6.98 ± 0.21	12.37 ± 0.31	2.93 ± 0.19	$W^1F^2D^3$
Sporophylls length (cm)	2	258.22	***	11.58 ± 0.48	5.72 ± 0.20	1.60 ± 0.12	$F^1W^2D^3$
Bulbs production (g)	2	220.09	***	0.08 ± 0.00	0.08 ± 0.00	0.01 ± 0.00	$F^1W^1D^2$
Microsporangia number	2	112.85	***	3.32 ± 0.17	3.33 ± 0.08	0.73 ± 0.14	$F^1W^1D^2$
Macrosporangia number	2	177.41	***	7.20 ± 0.52	11.00 ± 0.35	0.93 ± 0.18	$W^1F^2D^3$
Weight of 40 spores (g)	2	147.28	***	0.004 ± 0.00	0.004 ± 0.00	0.001 ± 0.00	$F^1W^1D^2$
B. maritimus in mixture							
Individuals length (cm)	2	83.81	***	27.13 ± 1.16	23.65 ± 1.14	7.72 ± 0.71	$F^1W^2D^3$
Tubers production (g)	2	290.22	***	10.05 ± 0.37	5.97 ± 0.23	0.91 ± 0.16	$F^1W^2D^3$
	d*F*	χ^2	*P*	Flooded	Wet	Means comparison	
B. maritimus in mixture							
Seeds number	1	0.75	ns	10.16	10.00	F^1W^1	
Weight of seeds (g)	1	3.36	ns	2.16	2.88	F^1W^1	

ns non-significant

* $P < 0.05$; ** $P < 0.01$; *** $P < 0.001$

40 spores were significantly higher in both the flooded and wet treatments than in the dry one (Table 1).

In relation to the previous experiment, a reduction in the number and length of the sporophylls, bulb weight gain, number of megasporangia and microsporangia, as well as in the macrospore weight, was found in the flooded and wet treatments but not in the dry treatment (Table 1).

Results of the competition between *Bolboschoenus maritimus* and *Isoetes setacea*

The ratios of the initial biomasses (initial weight of the *Bolboschoenus* tubers/initial weight of the *Isoetes* bulbs) were generally low, ranging between 2.32 and 3.16. By chance, this ratio differed weakly, but significantly, between the treatments in the beginning of the experiment ($F = 9.61$, d$F = 2$, $P = 0.0002$). It was weaker for the flooded treatment than for the wet and dry treatments (Fig. 1). At the end of the experiment, the ratio of the final biomasses (weight gained by the *Bolboschoenus* tubers/weight gained by the *Isoetes* bulbs) was higher than the initial biomasses and significantly different between treatments ($F = 4.79$, d$F = 2$, $P = 0.01$). The final biomass ratio was significantly different between the flooded and wet treatments, but not between these two treatments and the dry one, that presented an intermediate value (Fig. 1).

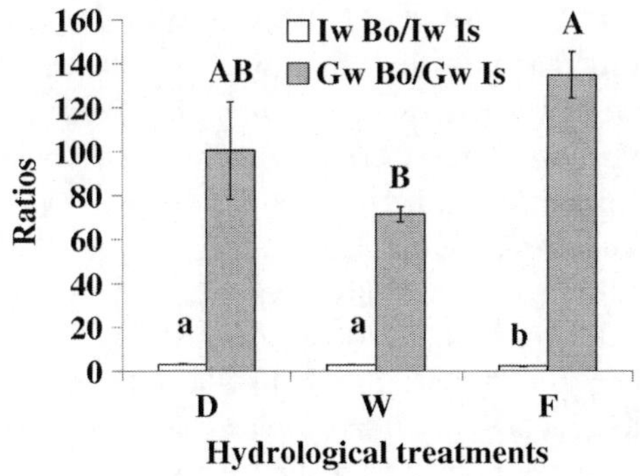

Fig. 1 Results of the analysis of variances on the comparison between the three hydrological treatments: dry (D), wet (W), and flooded (F), the ratios of the initial underground biomasses (Initial weight of tubers of *Bolboschoenus* "Iw Bo"/Initial weight of bulbs of *Isoetes* "Iw Is") and of the produced underground biomasses (Gw Bo/Gw Is). The comparison of ratios between pairs of hydrological treatments has been achieved by the Tukey–Kramer test. *Different letters* (a, b) and (A, B) on the diagram indicate a significant difference of the ratios, respectively, (Iw Bo/Iw Is) and (Gw Bo/Gw Is) between the three hydrological treatments ($P < 0.05$)

The ratio (P_{mix}/P_{pure}) for the number of sporophylls per plant was significantly different between the treatments ($F = 564.63$, d$F = 2$, $P < 0.0001$), with the highest ratio occurring in the dry treatment and the lowest in the flooded treatment (Fig. 2α). Regarding the length of the sporophylls, this ratio was significantly higher in the dry treatment ($F = 408.06$, d$F = 2$, $P < 0.0001$) than in the other two, which did not differ for this variable (Fig. 2α). The ratio for the weight gained by the bulbs was significantly lower in the flooded treatment than in the other two ($F = 182.59$, d$F = 2$, $P < 0.0001$) (Fig. 2α).

The ratios (P_{mix}/P_{pure}) calculated for the number of megasporangia and microsporangia varied significantly between the flooded, wet, and dry treatments (respectively, $X^2 = 65.47$, d$F = 2$, $P < 0.0001$; $X^2 = 47.52$, d$F = 2$, $P < 0.0001$). The highest ratios occurred in the dry treatment and the lowest in the flooded one (Fig. 2β). The ratio calculated for the weight of the macrospores was significantly lower in the flooded treatment than in the other two ($X^2 = 53.32$, d$F = 2$, $P < 0.0001$) (Fig. 2β).

Discussion

In these experiments, several species traits of *Bolboschoenus maritimus* and *Isoetes setacea*, were found to

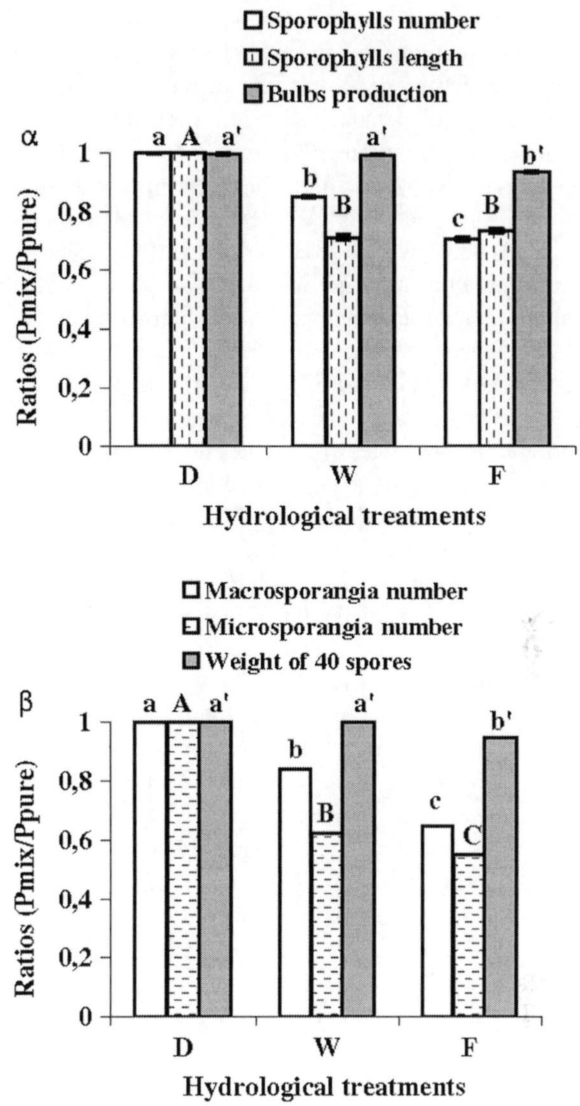

Fig. 2 Results of the parametric analysis of variance (**α**) and of the non-parametric Kruskal–Wallis test (**β**) on the comparison of the ratios (P_{mix}/P_{pure}), established between the number of sporophylls, their length, the bulbs production, the number of macrosporangia and microsporangia, and the weight of 40 spores of *Isoetes setacea* in mixture with *Bolboschoenus maritimus* and their corresponding control, between the three hydrological treatments: dry (D), wet (W), and flooded (F). The comparison between pairs of treatments has been achieved by the Tukey–Kramer test for the vegetative production and by a simple comparison between flooded and wet treatments by the Kruskal–Wallis test for the sexual reproduction. *Different letters* (a, b, c); (A, B, C), and (a′, b′, c′) on the diagram indicate a significant difference between the three hydrological treatments ($P < 0.05$) of the ratios established, respectively, for the number of sporophylls, their length, and the bulbs production (**α**), and, respectively, for the number of macrosporangia and microsporangia and the weight of 40 spores (**β**)

be much affected by the hydrological regime, but they were not modified by the nature of the substratum. Their vegetative development, sexual reproduction, and competitive capacity varied according to the hydrological situations. The two species developed weakly in the dry conditions but were highly productive in the other two treatments, in particular, the flooded one. Clevering & Hundscheid (1998) recorded maximal elongation in *B. maritimus* in a flooded situation, and the same result was obtained by Rhazi (2001) for another amphibious species of the Moroccan temporary pools. For Jackson (1985) and Ridge (1987), the increase of the water levels leads the young shoots of macrophytes to elongate and emerge from the water according to the depth accommodation process. This adaptive strategy seems to have been adopted by the two species studied in this experiment. The aerial vegetative biomass produced by the plants in the flooded situation, distinctly influences the produced underground biomass that seems to be very important in the same hydrological situation.

In the dry condition *B. maritimus*, in contrast to *I. setacea*, was unable to perform a sexual reproduction. However, in flooded and wet conditions, the performance of *B. maritimus* was better than that of *I. setacea*, revealing a high capacity to increase its vegetative productivity. The productivity of biomass has often been used as an indicator of competitive strength and invasive potential (e.g., Claridge & Franklin, 2002).

The results obtained from the soil enrichment experiment were not statistically significant. The abandonment of the exploitation of the basalt in the Roque–Haute was followed by an on-going process of accumulation of fine sediment in the pools originated from the catchments (Grillas et al., 2004a). This accumulation of sediment increases the water retention capacity of the soil, thus decreasing the intensity of drought stress during the summer and hence facilitating the survival of perennial competitors such as *Bolboschoenus maritimus*.

Competitive advantage of *Bolboschoenus maritimus* on *Isoetes setacea*

The competitive advantage of *B. maritimus* over *I. setacea* was very important in flooded and wet situations. A ratio (P_{mix}/P_{pure}) lower than 1 represents a strong interspecific competition between the plants species (Vilà & Weiner, 2004). This competition process involved not only the vegetative growth parameters but also, in the case of *I. setacea*, those linked to sexual reproduction. Under these two hydrological conditions (flooded and wet), the production of the *B. maritimus* distinctly surpassed that of *I. setacea*. The maximal size recorded for B. maritimus (height = 65 cm) was four times the maximal size recorded for the quillwort (height = 17.25 cm). Similarly, the maximal underground biomass of *B. maritimus* (biomass = 13.9 g) was 87 times higher than that of the quillwort (biomass = 0.16 g). The reduction of the *I. setacea* performance in the mixture, in comparison to its performance in monoculture, is explained by the strong competition imposed by *B. maritimus* under these two hydrological conditions. Several authors place emphasis on the narrow relationship between the intensity of competition and the productivity (Dutoit et al., 2001; Weigelt et al., 2002). Therefore, the distinctive competitive advantage of *B. maritimus* over the quillwort may lead to the displacement of the populations of *I. setacea*; our results thus support the replacement hypothesis of the *I. setacea* populations by those of *B. maritimus* (Grillas & Tan Ham, 1998).

Under dry conditions, the competitive advantage of *B. maritimus* strongly decreased, with ratios of approximately 1. Indeed, under drought, the productivity of the *B. maritimus* was very low and sexual reproduction was not recorded. However, in spite of the low productivity of *B. maritumus*, *I. setacea* did not reveal any competitive advantage over its competitor. The biological attributes of *I. setacea* (Table 1) under dry conditions reveals that it is unlikely that this species can maintain itself throughout an extended period of dry climate. The absence of competitive advantage of *Isoetes* over *Bolboschoenus* is tied to the low productivity of this fern in the presence of *Bolboschoenus* that still exhibits, under these adverse conditions, an advantage in terms of size and underground biomass production (Table 1).

Chances of survival of *Isoetes setacea*

The results of this study emphasized the weak competitive performance of *I. setacea* in temporary pools recently colonized by a clonal competitive plant. The current survival of this rare and threatened Lycopodiophyta present in isolated and fragmented populations could be related to its strong phenotypic plasticity which is mainly reflected by the high flexibility of its development cycle (Rhazi et al., 2004) and its high tolerance to desiccation. Indeed, *I. setacea* has an early (vernal) development and quickly completes its cycle, often adopting an ephemerophyte life strategy (Barbero et al., 1982). The intense but short drought to which *I. setacea* was exposed did not seem to affect the survival of its populations, but in the short term, these extreme conditions were not sufficient for this species to gain competitive advantage over *B. maritimus*. Nevertheless, the hypothesis that *Isoetes* could win a competitive advantage on *B. maritimus* under more intense and prolonged drought conditions cannot be excluded. Severe droughts strongly limit the biomass of plants (Grillas et al., 1993) and consequently limit the processes of competitive exclusion (Keddy, 1989; Vilà and Weiner, 2004). Indeed, such a drought would lead to the absence of sexual reproduction of *B. maritimus* and probably to the regression of its populations in the long term despite the local persistence of clones. Moreover, after keeping for 6 months, 90 *Bolboschoenus* tubers and 90 *Isoetes* bulbs in paper sachets, in a closed cardboard box at room temperature, none of the *Bolboschoenus* tubers germinated whereas all of the *Isoetes* bulbs did (Rhazi, unpublished data).

From a conservation perspective, the introduction of sheep grazing or of cut back practices in temporary pools could also help toward the maintenance of *I. setacea* populations in these habitats (Rhazi et al., 2004) through the reduction of the *Bolboschoenus maritimus* biomass (Grillas et al., 2004a).

Acknowledgments The English text was corrected and improved by Deirdre Flanagan, and Frank Torre (IMEP) has provided helpful insights in statistics. We thank the two anonymous reviewers for their remarks that improved the quality of this manuscript.

References

Akey, W. C., T. W. Jurik & J. Dekker, 1991. A replacement series evaluation of competition between velvetleaf (*Abutilon theophrasti*) and soybean (*Glycine max*). Weed Research 31: 63–72.

Barbero, M., J. Giudicelli, R. Loisel, P. Quézel & E. Terzian, 1982. Etude des biocénoses des mares et ruisseaux temporaires à éphémérophytes dominants en région méditerranéenne francaise. Bulletin d'Ecologie 13: 387–400.

Braun-Blanquet, J, 1936. Un Joyau floristique et phytosociologique, l'*Isoetion* méditerranéen. SIGMA, Communication 42.

Charpentier, A., F. Mesleard & J. D. Thompson, 1998. The effects of rhizome severing on the clonal growth and clonal architecture of *Scirpus maritimus*. Oikos 83: 107–116.

Charpentier, A., P. Grillas & J. D. Thompson, 2000. The effect of population size limitation on fecundity in mosaic populations of the clonal macrophyte *Scirpus maritimus* (Cyperaceae). American Journal of Botany 87: 502–507.

Claridge, K. & S. B. Franklin, 2002. Compensation and plasticity in an invasive plant species. Biological Invasion 4: 339–347.

Clevering, O., 1995. Germination and seedling emergence of *Scirpus lacustris* L. and *Scirpus maritimus* L. with special reference to the restoration of wetlands. Aquatic Botany 50: 63–78.

Clevering, O. & M. P. J. Hundscheid, 1998. Plastic and non-plastic variation in growth of newly established clones of *Scirpus* (*Bolboschoenus*) *maritimus* L. grown at different water depths. Aquatic Botany 62: 1–17.

Connell, J. H., 1975. Some mechanisms producing structure in natural communities: a model and evidence from field experiments. In Cody, M. L. & J. M. Diamond (eds), Ecology and Evolution of Communities. The Belknap Press of Harvard University Press, Cambridge, Massachusetts, USA: 460–490.

Connolly, J., 1986. On difficulties with replacement series methodology in mixture experiments. Journal of Applied Ecology 23: 125–137.

Crochet, J. Y., 1998. Le cadre géologique de la Réserve Naturelle de Roque-Haute. Ecologia Mediterranea 24: 179–183.

Deil, U., 2005. A review on habitats plant trait and vegetation of ephemeral wetlands-a global perspective. Phytocoenologia 35: 533–705.

Dutoit, T., E. Gerbaud, J. M. Ourcival, M. Roux & D. Alard, 2001. Recherche prospective sur la dualité entre caractéristiques morphologiques et capacités de compétition des végétaux: le cas des espèces adventices et du blé. Comptes Rendus de l'Académie des Sciences, série III, Sciences de la vie 324: 261–272.

Gaston, K. J., 1994. Rarity. Chapman & Hall, London: 205 pp.

Gaudillat, V., J. Haury, B. Barbier & F. Peschadour, 2002. Connaissance et gestion des habitats et des espèces d'intérêt Communautaire. Tome 3 Habitats Humides. La Documentation Française, Paris: 457 pp.

Goldberg, D. E., T. Rajaniemi, J. Gurevitch & A. Stewart-Oaten, 1999. Empirical approaches to quantifying interaction intensity competition and facilitation along productivity gradients. Ecology 80: 1118–1131.

Grace, J. B., 1995. On the measurement of plant competition intensity. Ecology 76: 305–308.

Grillas, P. & L. Tan Ham, 1998. Dynamique intra- et inter-annuelles de la végétation dans les mares de la Réserve Naturelle de Roque-Haute: programme d'étude et résultats préliminaires. Ecologia Mediterranea 24: 215–222.

Grillas, P., C. Van Wijck & A. Bonis, 1993. The effect of salinity on the dominance-diversity of experimental communities of coastal submerged macrophytes. Journal of Vegetation Science 4: 453–460.

Grillas, P., P. Gauthier, N. Yavercovski & C. Perennou, 2004a. Mediterranean Temporary Pools Volume 1, Issues Relating

to Conservation, Functioning and Management. Station Biologique de la Tour du Valat, Arles: 120 pp.
Grillas, P., P. Gauthier, N. Yavercovski & C. Perennou, 2004b. Mediterranean Temporary Pools. Vol. 2, Species Information Sheets. Station Biologique de la Tour du Valat, Arles: 127 pp.
Grime, J. P., 2001. Plant Strategies, Vegetation Processes, and Ecosystem Properties, 2nd ed. Wiley, Chichester: 417 pp.
Hedges, L. V., J. Gurevitch & P. S. Curtis, 1999. The meta-analysis of response ratios in experimental ecology. Ecology 80: 1150–1156.
Huenneke, L. F. & J. K. Thomson, 1995. Potential interference between a threatened endemic thistle and an invasive non native plant. Conservation Biology 9: 416–425.
Jackson, M. B., 1985. Ethylene and responses of plants to water logging and submergence. Annual Review of Plant Physiology 36: 145–174.
Jakob, C., M. Veith, A. Seitz & A. J. Crivelli, 1998. Données préliminaries sur la communauté d'amphibiens de la Réserve Naturelle de Roque-Haute dans le sud de la France. Ecologia Mediterranea 24: 235–240.
Kantrud, H.A, 1996. The alkali (*Scirpus maritimus* L) and salt marsh (*S. robustus* Pursh) bulrushes: a literature review. U.S. National Biological Service Information and Technology report 6: 77 pp.
Keddy, P. A., 1989. Competition, Population and Community Biology Series. Chapman and Hall, London: 202 pp.
Lavergne, S., E. Garnier & M. Debussche, 2003. Do rock endemic and widespread plant species differ under the leaf-height-seed plant ecology strategy scheme? Ecology Letters 6: 398–404.
Mead, R. & R. W. Willey, 1980. The concept of a "land equivalent ratio" and advantages in yields from intercropping. Experimental Agriculture 16: 217–228.
Médail, F. & R. Verlaque, 1997. Ecological characteristics and rarity of endemic plants from southeast France and Corsica: implications for biodiversity conservation. Biological Conservation 80: 269–271.
Médail, F., H. Michaud, J. Molina, G. Paradis & R. Loisel, 1998. Conservation de la flore et de la végétation des mares temporaires dulçaquicoles et oligotrophes de France méditerranéenne. Ecologia Mediterranea 24: 119–134.
Molina, J., 1998. Typologie des mares de Roque-Haute. Ecologia Mediterranea 24: 207–213.
Olivier, L., J. P. Galland & H. Maurin, 1995. Livre rouge de la flore menacée de France. Muséum National d'Histoire Naturelle, Institut d'Ecologie et de Gestion de la Biodiversité. Service du Patrimoine Naturel. Collection Patrimoines Naturels, Vol. 20, série Patrimoine Génétique, tome 1: Espèces prioritaires. 486 p+ annexes.
Prelli, R., 2002. Les fougères et plantes alliées de France et d'Europe occidentale. Belin, Paris: 432 pp.
Quézel, P., 1998. La végétation des mares transitoires à Isoetes en région méditerranéenne, intérêt patrimonial et conservation. Ecologia Mediterranea 24: 111–117.
Rhazi, L, 2001. Etude de la végétation des mares temporaires et l'impact des activités humaines sur la richesse et la conservation des espèces rares au Maroc. Thèse Doctorat d'Etat Es Sciences, Université Hassan II, Casablanca. 191 pp.
Rhazi, M, 2005. Ecologie de la restauration de la diversité végétale et des espèces rares dans les mares temporaires méditerranéennes (sud France). Thèse de Doctorat ès Sciences. Université Paul Cézanne, Aix-Marseille III. 162 p + annexes.
Rhazi, M., P. Grillas, A. Charpentier & F. Médail, 2004. Experimental management of Mediterranean temporary pools for conservation of the rare quillwort *Isoetes setacea*. Biological Conservation 118: 675–684.
Ridge, I., 1987. Ethylene and growth control in amphibious plants. In Crawford, R. M. M. (ed.), Plant Life in Aquatic and Amphibious Habitats. Blackwell, Oxford: 53–76.
Tilman, D., 1988. Plant Strategies and the Dynamics and Structure of Plant Communities. Princeton University Press, Princeton, New Jersey, USA.
Titolet, D. & L. Rhazi, 1999. Intérêt patrimonial d'un milieu associé aux suberaies: les mares temporaires des rives gauches et droites de l'oued Cherrat. Integrated Protection in Oak Forests, IOBC Bulletin 22: 189–194.
Trabaud, L., 1998. Historique de la création de la Réserve Naturelle de Roque-Haute et sa végétation. Ecologia Mediterranea 24: 173–177.
Twolan-Strutt, L. & P. A. Keddy, 1996. Above-and below-ground competition intensity in two contrasting wetland plant communities. Ecology 77: 259–270.
Vilà, M. & J. Weiner, 2004. Are invasive plant species better competitors than native plant species?- evidence from pair-wise experiments. Oikos 105: 229–238.
Walck, J. L., J. M. Baskin & C. C. Baskin, 1999. Effects of competition from introduced plants on establishment, survival, growth and reproduction of the rare plant *Solidago shortii* (Asteraceae). Biological Conservation 88: 213–219.
Weigelt, A., T. Steinlein & W. Beyschlag, 2002. Does plant competition intensity rather depend on biomass or on species identity? Basic and Applied Ecology 3: 85–94.
Westoby, M., 1998. A leaf–height–seed (LHS) plant ecology strategy scheme. Plant Soil 199: 213–227.
Wisheu, I. C. & P. A. Keddy, 1992. Competition and centrifugal organization of plant communities: theory and tests. Journal of Vegetation Science 3: 147–156.

Hydrobiologia (2009) 634:125–135
DOI 10.1007/s10750-009-9888-4

POND CONSERVATION

Restoration potential of biomanipulation for eutrophic peri-urban ponds: the role of zooplankton size and submerged macrophyte cover

Anatoly Peretyatko · Samuel Teissier ·
Sylvia De Backer · Ludwig Triest

Published online: 5 August 2009

Abstract Eight hypereutrophic phytoplankton dominated ponds from the Brussels Capital Region (Belgium) were biomanipulated (emptied with fish removal) to restore their ecological quality and reduce the risk of cyanobacterial bloom formation. Continuous monitoring of the ponds before and after the biomanipulation allowed the effects of the management intervention on different compartments of pond ecosystems (phytoplankton, zooplankton, submerged vegetation and nutrients) to be assessed. Fish removal resulted in a drastic reduction in phytoplankton biomass and a shift to the clear-water state in seven out of eight biomanipulated ponds. The reduction in phytoplankton biomass was associated with a marked increase in density and size of large cladocerans in six ponds and a restoration of submerged macrophytes in five ponds. The phytoplankton biomass in the ponds with extensive stands of submerged macrophytes was less affected by planktivorous fish recolonisation of some of the ponds later in the summer. The two non-vegetated ponds as well as one pond with sparse submerged vegetation showed a marked increase in phytoplankton biomass associated with the appearance of fish. Phytoplankton biomass increase coincided with the decrease in large Cladocera density and size. One pond lacking submerged macrophytes could maintain very low phytoplankton biomass owing to large Cladocera grazing alone. The results of this study confirmed the importance of large zooplankton grazing and revegetation with submerged macrophytes for the maintenance of the clear-water state and restoration success in hypereutrophic ponds. They also showed that large Cladocera size is more important than their number for efficient phytoplankton control and when cladocerans are large enough, they can considerably restrain phytoplankton growth, including bloom-forming cyanobacteria, even when submerged vegetation is not restored. The positive result of fish removal in seven out of eight biomanipulated ponds clearly indicated that such management intervention can be used, at least, for the short-term restoration of ecological water quality and prevention of noxious cyanobacterial bloom formation. The negative result of biomanipulation in one pond seems to be related to the pollution by sewage water.

Keywords Biomanipulation · Ponds · Fish removal · Submerged macrophytes · Phytoplankton · Large Cladocera size

Guest editors: B. Oertli, R. Cereghino, A. Hull & R. Miracle
Pond Conservation: From Science to Practice. 3rd Conference of the European Pond Conservation Network, Valencia, Spain, 14–16 May 2008

A. Peretyatko (✉) · S. Teissier · S. De Backer · L. Triest
Plant Science and Nature Management, Department of Biology, Vrije Universiteit Brussel, Pleinlaan 2, 1050 Brussels, Belgium
e-mail: Anatoly.Peretyatko@vub.ac.be

Introduction

During the past decades, eutrophication has become a serious threat to many European lakes and ponds. Increased productivity caused a considerable degradation of ecological water quality often resulting in the development of potentially toxic cyanobacterial blooms (Jeppesen et al., 1990, 1997; Willame et al., 2005) or profuse growth of filamentous green algae (Irfanullah and Moss, 2005).

Many approaches have been used to mitigate the effects of eutrophication. Substantial reduction of nutrient loading can be useful, particularly in cases when nutrients come from point sources (Jeppesen et al., 2005, 2007), but might not produce the desired effect or such effect can be considerably delayed in cases of accumulation of phosphorus in the sediment or diffuse external nutrient sources (Lauridsen et al., 2003; Carpenter, 2005; Søndergaard et al., 2007).

A number of studies have demonstrated, however, that substantial reduction in fish densities shifts ponds or lakes from phytoplankton to submerged vegetation dominance, thus improving the water quality and reducing the risk of cyanobacterial blooms (Shapiro, 1990; Van Wichelen et al., 2007).

Addition of piscivores or removal of all or most of the fish community is referred to as biomanipulation (Moss et al., 1996). It allows phytoplankton control through increase in large zooplankton grazing released from fish predation. Because of its cost effectiveness and quick response to the intervention as compared to nutrient reduction measures (Lammens, 1999; Lauridsen et al., 2003), biomanipulation has been a primary method of lake and pond restoration (Shapiro, 1990; Van Wichelen et al., 2007).

Given their small size, ponds respond faster than large lakes to changes in their trophic structure, and therefore are easier to biomanipulate. Ponds can also be completely emptied thus insuring that, at least for a short term, plankti-benthivorous fish populations are very small or absent. Beside the drastic reduction in phytoplankton biomass, biomanipulation often results in the recovery of submerged vegetation (macrophytes or filamentous green algae), which can stabilize the clear-water state through a number of associated mechanisms (Moss et al., 1996; Søndergaard and Moss, 1998; Irfanullah and Moss, 2005).

Although it is well established that large zooplankton grazing, resulting from the release of fish predation pressure, is the primary cause of the drop in phytoplankton biomass, the importance of large zooplankton size is rarely elucidated as in most studies zooplankters are not measured. Besides, the majority of published case studies show biomanipulation effects on lakes. Examples of pond biomanipulation are relatively rare.

We have collected phytoplankton, zooplankton, submerged vegetation and environmental data from eight Brussels ponds differentially affected by eutrophication before and after biomanipulation through complete emptying with fish removal. The objective was to assess the restoration potential of biomanipulation for eutrophic peri-urban ponds, with particular focus on the role of large cladocerans and submerged macrophytes in phytoplankton control.

Methods

Eight ponds from the Brussels Capital Region, Belgium, were sampled before/after biomanipulation (complete emptying with fish removal) carried out in winter/early spring 2007 to assess the restoration potential of such management intervention.

The ponds are all artificial, created by the damming of small low order streams in the twentieth century or earlier. They are shallow (maximum depth $<$1.5 m) and flat-bottomed and range in surface area from 0.1 to 2.3 ha. All the ponds are mainly fed by small rivulets and ground water seepage. Before biomanipulation, they were populated by fish communities typical of northern Europe. All of them were overstocked with fish ($>$500 kg ha^{-1}; mainly carp).

Before biomanipulation, the ponds studied were sampled in May, July and August; VKn2 and WPk1 in 2005, the rest of the ponds in 2006. The data from 2003 to 2004 for VKn2 and WPk1 (not presented here) and visual observations in 2006 support the idea that 2005 data are representative of the before manipulation state of these ponds. After biomanipulation, all the ponds were sampled monthly, from May to September 2007.

Quantitative samples of phytoplankton, zooplankton, main nutrients (total phosphorus—TP, soluble reactive phosphorus—SRP, NO_3 and NH_4), chlorophyll *a* (Chl *a*) as well as conductivity, pH, temperature, and water transparency (Secchi depth) data were collected according to the standard limnological procedures (see Peretyatko et al., 2007b for details).

Water samples preserved with Lugol's solution, sodium thiosulfate and buffered formalin (Kemp et al., 1993) were used for phytoplankton identification (genus level) and counting with an inverted microscope (a modified Utermöhl sedimentation technique; Hasle, 1978). Biovolumes were calculated using the approximations of cell shapes to simple geometrical forms (Wetzel and Likens, 1990).

For zooplankton, 10 sub-samples of 1 l were combined in the field, filtered through a 64-μm mesh net and preserved in 4% formaldehyde final concentration before being identified and counted using an inverted microscope. Different levels of identification were used: cladocerans were identified to genus level; copepods were divided into cyclopoids, calanoids and nauplii; rotifers were not discriminated. For the analyses, cladocerans were divided into two groups: 'large' (*Daphnia* spp., *Eurycercus* spp., *Sida* spp. and *Simocephalus* spp.) and 'small' (*Acroperus* spp., *Bosmina* spp., *Ceriodaphnia* spp., *Chydorus* spp., *Moina* spp. and *Pleuroxus* spp.) (Moss et al., 2003). Predator cladocerans, *Leptodora* spp. and *Polyphemus* spp., which feed mainly on other zooplankters (Reynolds, 2006), were not included in the group of large cladocerans. The length of large Cladocera species was measured and taken as an indicator of grazing intensity and size-selective predation (Pinel-Alloul, 1995; Carpenter et al., 2001).

Surface cover of aquatic vegetation was mapped visually from a boat during each field visit. After biomanipulation, visual assessment of fish presence was also done during each field visit. Morphometric variables of the ponds were measured in the field (depth), or using GIS software (area; digitized map, ArcView 3.2). Hydraulic retention time was estimated on the basis of the outlet discharge and the corresponding pond volume once a year in 2006 and monthly in 2007.

Standard univariate and multivariate statistical tests were used for the analysis of the data acquired. Because phytoplankton and zooplankton data were not normally distributed, non-parametric Mann–Whitney *U* test was used for statistical comparisons of the before and after biomanipulation situations. Redundancy analysis (RDA; ter Braak et al., 2002) based on averaged value per year of phytoplankton, and environment data were used to elucidate the relationships between phytoplankton and environmental factors operating in the ponds studied. Phytoplankton data were aggregated to division level.

Results

All biomanipulated ponds were hypereutrophic (mean $TP > 0.1$ mg l^{-1}; Table 1, Fig. 4). Before biomanipulation, they were phytoplankton dominated, lacking any submerged vegetation. Phytoplankton biomass, however, ranged from less than 20 to more than 100 mm^3 l^{-1} (Fig. 1). Three ponds (Leyb-a, Leyb-b and WPk1) were prone to persistent cyanobacterial blooms. Although often present in the other five ponds during the study, cyanobacteria in these ponds were generally less abundant than chlorophytes and diatoms.

After biomanipulation, all the ponds except PRB2 showed a considerable drop in phytoplankton biomass accompanied by a shift in zooplankton community towards larger zooplankters in general (Fig. 2), and a considerable increase in the number and size of large cladocerans in particular (Fig. 1). Biomanipulation also resulted in the recovery of submerged macrophytes (mainly *Potamogeton* spp., *Chara* spp., *Ceratophyllum demersum*) in five out of eight ponds (Beml, Leyb-a, Leyb-b, Sbsk and VKn2). The degree of the recovery of submerged macrophytes varied substantially from pond to pond. VKn2 was virtually packed with *Ceratophyllum demersum*, leaving very little open water (<10% of the pond surface) whereas Leyb-a showed only sparse growth of *Potamogeton* spp. In Beml, Leyb-b and Sbsk, submerged vegetation cover exceeded 60%. Submerged macrophytes were generally covered with a thick epiphytic growth. Leyb-a and WPk1 developed large patches of benthic filamentous green algae.

Despite the drastic decrease in phytoplankton biomass (Fig. 1) and increase in water transparency, no submerged macrophytes were observed in Dens. This can probably be explained by the small size (<0.5 ha) and low depth (mostly <0.5 m) of this pond combined with large populations of herbivorous birds feeding in it.

PRB2 became more turbid after biomanipulation (Fig. 1). It should be noted that it was polluted by sewage water soon after biomanipulation as a result of an overflow into the pond during particularly heavy rain. Some small fish were observed in this pond during the samplings suggesting a quick recolonisation. This is supported by the lack of increase in large Cladocera densities and size (Fig. 1). It is also possible that the polluted water

Table 1 General characteristics of the ponds studied (mean May–September values before and after biomanipulation)

Site	TP ($mgP\ l^{-1}$)	SRP ($mgP\ l^{-1}$)	NH_4 ($mgN\ l^{-1}$)	NO_3 ($mgN\ l^{-1}$)	Chl *a* ($\mu g\ l^{-1}$)	Cond ($\mu S\ cm^{-1}$)	pH	SD (m)	MD (m)	RT (day)	*T* (°C)	SM (%)
Before biomanipulation												
Beml	0.673	0.248	0.055	0.080	52	748	7.9	0.7	1.0	>300	20.9	0
Dens	0.351	0.039	0.045	0.047	88	422	8.4	0.4	0.7	220	22.2	0
Leyb-a	0.506	0.011	0.034	0.187	470	536	9.0	0.3	0.6	11	21.5	0
Leyb-b	0.407	0.007	0.028	0.214	349	557	8.8	0.3	0.8	21	21.2	0
Sbsk	0.426	0.025	0.246	0.146	83	781	8.4	0.6	0.7	47	21.4	0
VKn2	0.171	0.006	0.048	0.334	42	562	7.5	0.8	1.0	5	14.6	0
WPk1	0.191	0.001	0.012	0.315	41	901	7.7	0.7	1.0	30	17.3	0
PRB2	0.428	0.015	0.204	0.153	40	735	8.0	0.6	0.8	>300	21.2	0
After biomanipulation												
Beml	0.247	0.158	0.627	0.010	29	935	7.7	Bottom	1.0	>300	19.1	20
Dens	0.191	0.054	1.047	0.099	18	433	7.9	Bottom	0.6	217	17.6	0
Leyb-a	0.517	0.434	0.344	0.041	20	634	8.3	Bottom	0.5	13	17.5	38
Leyb-b	0.213	0.145	0.132	0.027	26	661	8.0	Bottom	0.9	20	17.5	31
Sbsk	0.196	0.154	0.692	0.048	7	711	7.8	Bottom	0.8	55	18.7	36
VKn2	0.100	0.019	0.032	0.070	7	525	7.6	Bottom	1.1	4	15.8	78
WPk1	0.131	0.040	0.385	0.311	15	924	7.8	Bottom	1.1	31	18.6	0
PRB2	0.324	0.028	0.190	0.039	151	624	8.0	0.4	0.8	>300	18.7	0

Abbreviations used: *TP* total phosphorus, *SRP* soluble reactive phosphorus, *Chl a* chlorophyll a, *Cond* conductivity, *SD* Secchi depth, *MD* maximum depth, *RT* hydraulic retention time, *T* temperature, *SM* submerged macrophyte cover

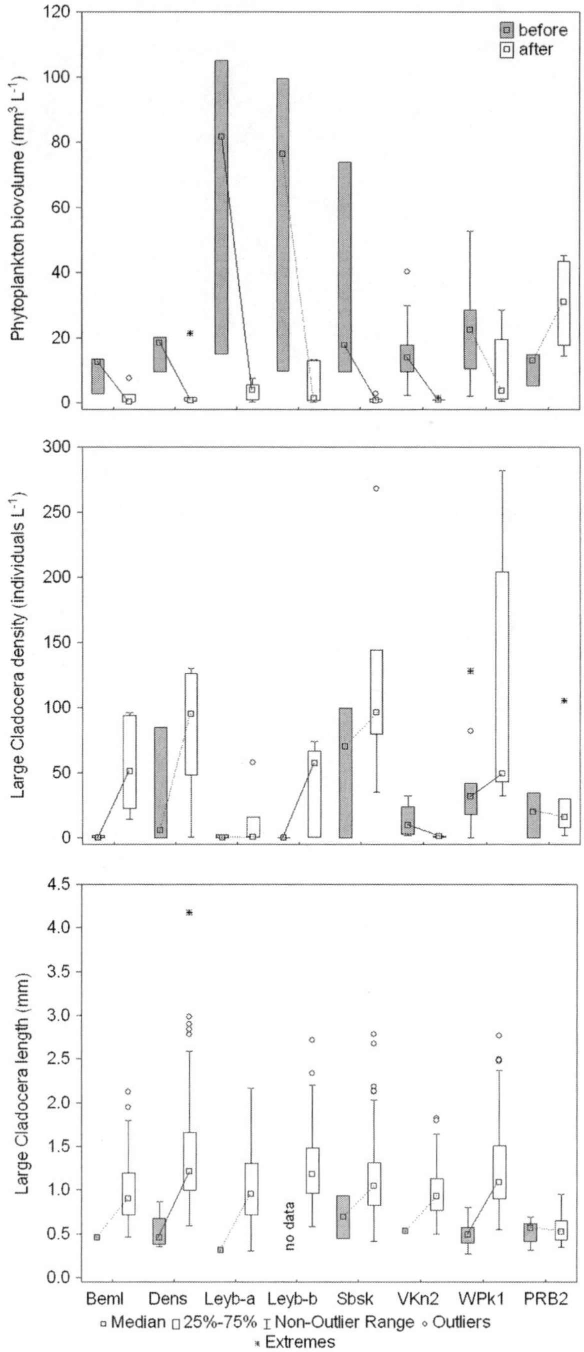

Fig. 1 Phytoplankton biovolume and large Cladocera density and length before and after biomanipulation. *Solid lines* indicate significant difference, *dotted lines*—non-significant difference (Mann–Whitney *U* test; $P < 0.05$)

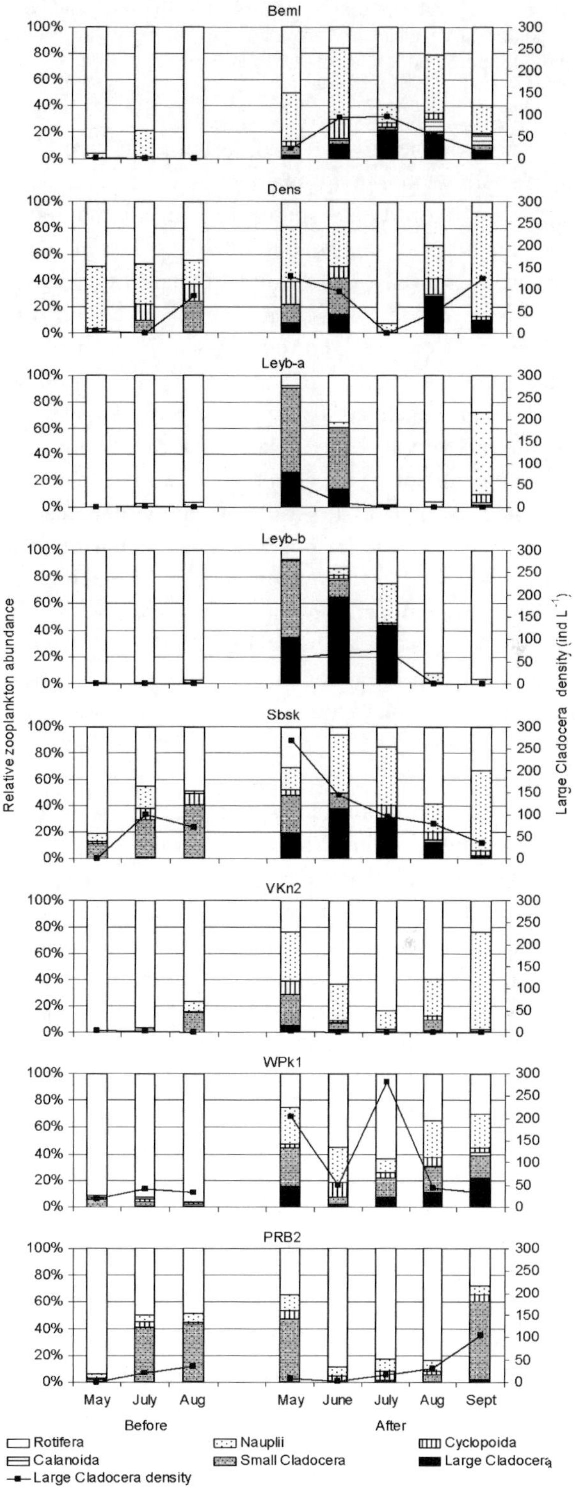

Fig. 2 Temporal variation in zooplankton community structure and large Cladocera densities in the ponds studied before and after biomanipulation

was poorly suitable for large cladocerans. Schools of small planktivorous fish were also observed in Leyb-a, Leyb-b, VKn2 and WPk1. From the latter, the fish

Fig. 3 Temporal variation in phytoplankton biomass and large Cladocera size (mean and standard deviation) in the ponds studied before and after biomanipulation

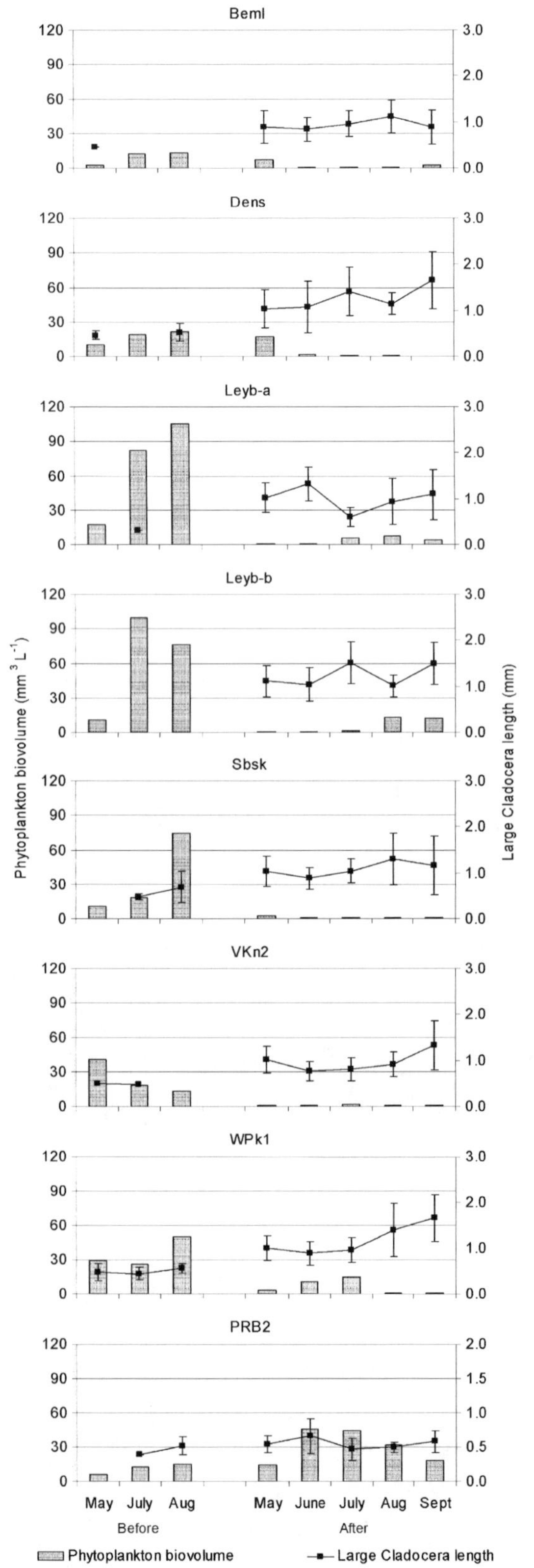

were not completely removed, as some survived in a large shallow pool that remained after the drawdown.

Although drastically reduced after biomanipulation in all the ponds but PRB2, phytoplankton biomass showed contrasting dynamics (Fig. 3) depending on the presence and coverage of submerged macrophytes and recolonisation by planktivorous fish. The ponds that were not recolonised by fish, either vegetated (Beml, Sbsk) or not (Dens), maintained very low phytoplankton biomass throughout the summer. The effect of planktivorous fish presence differed in function of submerged macrophyte cover. VKn2, a pond with an extensive and dense stand of *Ceratophyllum demersum*, maintained very low phytoplankton biomass despite very low density of large Cladocerans (Figs. 2, 3). In Leyb-a, a pond with sparse growth of *Potamogeton* spp., marked phytoplankton biomass increase was observed soon after the appearance of numerous small fish in early June. As Leyb-a and Leyb-b are connected, fish colonised both ponds. Nevertheless, phytoplankton biomass increase in Leyb-b was observed only after the collapse of the *Potamogeton* spp. stands, probably as a result of epiphytic overgrowth. Phytoplankton biomass increase in both ponds coincided with a marked reduction in large Cladocera density and size (Figs. 2, 3). This supports the idea that phytoplankton biomass increase was caused by fish predation on large zooplankton. WPk1, a pond where submerged macrophytes were not restored, showed a marked fluctuation in phytoplankton biomass and large Cladocera density (Figs. 2, 3).

Biomanipulation also affected nutrient dynamics. TP concentrations decreased in most of the ponds, whereas SRP concentrations increased (Fig. 4). DIN concentrations also increased, except for the ponds with extensive cover of submerged macrophytes or profuse growth of filamentous green algae, where DIN was generally depleted during the intense plant growth (Beml, Leyb-a, Leyb-b, VKn2; Fig. 4). SRP concentrations also varied in function of submerged vegetation presence but to a lesser degree than DIN, probably owing to internal phosphorus loading. In the non-vegetated ponds (Dens, WPk1 and PRB2) DIN

Fig. 4 Temporal variation in TP, SRP and DIN concentrations ▶ in the ponds studied before and after biomanipulation. * Indicates macrophyte cover whenever relevant: * <30%, ** 30–60%, *** >60%

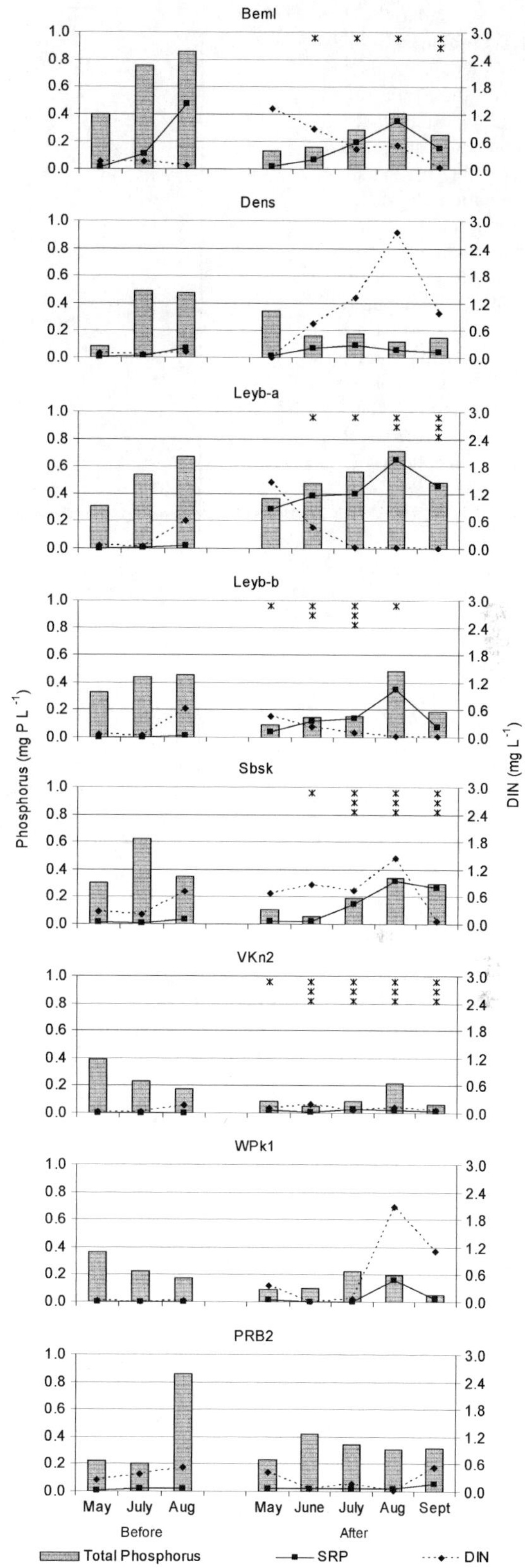

minimums corresponded to the phytoplankton biomass maximums.

The RDA results show that the factors having the strongest relationship with phytoplankton are large Cladocera length, pH and submerged vegetation cover (Fig. 5). The first two showed significant relationship with phytoplankton biomass and the latter marginally insignificant (Table 2). These three variables explained 63% of the variation in the phytoplankton data out of 87% explained by all the variables used in the model (Table 3). Large Cladocera length alone explained 40%. Nutrients and large Cladocera density showed poor relationship with the phytoplankton and explained a minor part of the variation in it.

Discussion

All the biomanipulated ponds were nutrient-rich phytoplankton dominated systems, overstocked with plankti-benthivorous fish before biomanipulation. Some of them developed persistent cyanobacterial blooms posing serious public health concerns. Fish removal alone was sufficient to shift seven out of the eight biomanipulated ponds from the turbid to the clear-water state. The main factor responsible for this shift is undoubtedly phytoplankton grazing by large cladocerans. This is supported by the drastic phytoplankton biomass decrease associated with the large Cladocera increase in numbers and size, the results of the RDA and a large number of previous studies (Carpenter et al., 1985; Shapiro, 1990; Christoffersen et al., 1993; Brönmark and Hansson, 2005).

Some authors, however, nurture doubt about the ability of large cladocerans to control phytoplankton when bloom-forming cyanobacteria are present (Benndorf et al., 2002). Others argue that large Cladocera grazing can favour grazing-resistant cyanobacteria by selectively removing other phytoplankters competing with cyanobacteria for light and nutrients (Gliwicz, 1990). Because large cladocerans are rarely measured, it is difficult to compare other study cases with these results. They clearly show,

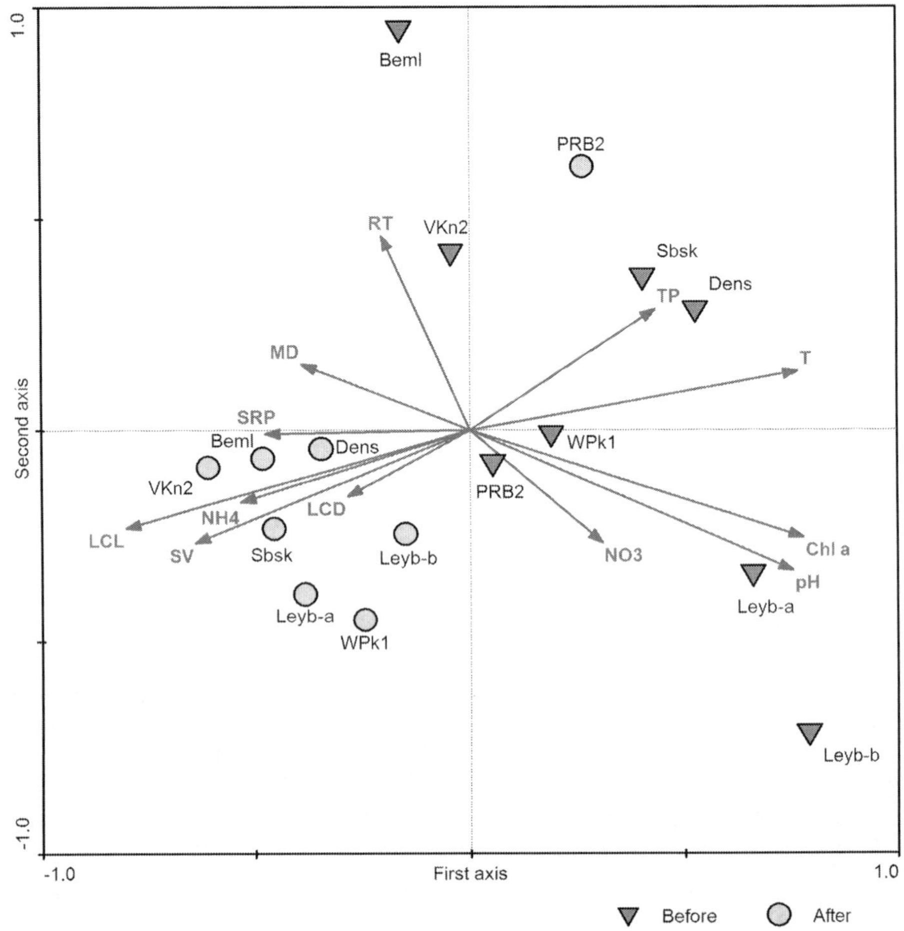

Fig. 5 Redundancy analysis biplot (sites and environmental variables) based on the averaged phytoplankton and environment data before and after biomanipulation. Abbreviations used: *Chl a* chlorophyll a, *LCL* large Cladocera length, *LCD* large Cladocera density, *MD* maximum depth, *RT* hydraulic retention time, *SD* Secchi depth, *SV* submerged vegetation (macrophytes and filamentous green algae) cover, *SRP* soluble reactive phosphorus, *T* temperature

however, that large Cladocera size is crucial for phytoplankton control. Significant reduction in phytoplankton biomass and low concentrations of cyanobacteria in Leyb-a and Leyb-b, these two ponds prone to massive cyanobacterial blooms before biomanipulation, also suggest that when cladocerans are sufficiently large they can considerably restrain cyanobacterial growth and prevent noxious bloom formation. Recolonisation of these two ponds by planktivorous fish in mid-summer caused a sharp drop in the numbers and size of large cladocerans. It seems that submerged vegetation (*Potamogeton* spp. and filamentous green algae), established in both ponds by that time, could buffer the effect of sharp reduction in zooplankton grazing through a number of associated mechanisms (Søndergaard and Moss, 1998; van Donk and van de Bund 2002; Peretyatko et al., 2007a) and prevent the formation of cyanobacterial blooms. Difference in submerged macrophyte cover and density between these two ponds was reflected on large Cladocera densities and size, as well as on phytoplankton biomass. Extended *Potamogeton* spp. stands in Leyb-b appear to have provided refugia to large cladocerans, as they could maintain high density and size in the presence of fish and, in combination with submerged macrophytes, maintained the water clear. The collapse of *Potamogeton* spp. stands in Leyb-b was rapidly followed by a drop in large Cladocera numbers and size associated with a marked phytoplankton biomass increase. Rapid decline in large Cladocera density and size in Leyb-a suggests that sparse *Potamogeton* spp. growth and patches of filamentous green algae provide poor protection from fish predation. Nevertheless, phytoplankton biomass in this pond remained much lower than before biomanipulation suggesting the involvement of other mechanisms of phytoplankton control than large Cladocera grazing, such as competition for nutrients, watercolumn stabilisation and/or allelopathy (Søndergaard and Moss, 1998;

Table 2 RDA forward selection results

Marginal effects		Conditional effects		P
Variable	Lambda1	Variable	LambdaA	
LCL	0.40	LCL	0.40	0.004
Chl a	0.39	pH	0.16	0.004
pH	0.37	SV	0.07	0.056
T	0.36	TP	0.06	0.090
SV	0.26	NH_4	0.04	0.252
NH_4	0.18	SRP	0.03	0.432
SRP	0.15	NO_3	0.02	0.474
TP	0.15	Chl a	0.02	0.500
MD	0.10	RT	0.02	0.522
LCD	0.09	T	0.03	0.664
NO_3	0.07	MD	0.02	0.696
RT	0.06	LCD	0.01	0.832

Marginal effects show the variance explained by each environmental variable alone (Lambda1); conditional effects show the significance of the addition of a given variable (P) and the additional variance explained at the time the variable was included into the model (LambdaA)

Gross, 1999; Gross et al., 2003; Peretyatko et al., 2007a).

The importance of submerged macrophytes for phytoplankton control was confirmed in other ponds where extensive stands of submerged macrophytes were also associated with low phytoplankton biomass and clear water despite the appearance of numerous planktivorous fish. Owing to a dense stand of *Ceratophyllum demersum* covering most of the pond surface, VKn2 maintained the water clear throughout the summer 2007 despite very low density of large cladocerans. This supports the idea that once established, submerged vegetation can considerably restrain phytoplankton growth even at low level of zooplankton grazing (Blindow et al., 2000; Peretyatko et al., 2007a) and underlines the importance of submerged macrophytes for biomanipulation success. It should be noted that excessive growth of submerged macrophytes can make a pond unsuitable for recreational use.

The example of Dens, the pond that maintained very low phytoplankton biomass despite the fact that submerged vegetation was not restored, shows that large cladocerans alone are able to control phytoplankton provided there are no fish. This situation is very unstable, as recolonisation by fish can rapidly shift the system back to the turbid state (Moss et al., 1996).

Conclusions

Although permanent effects of restoration can only be achieved if the external nutrient loading is reduced to sufficiently low levels and sediment phosphorus is depleted or removed, short-term improvement of ecological quality can be achieved by the food web manipulation even in nutrient rich ponds. At lower nutrient loading, there is a greater chance of biomanipulation success. When nutrient level is too high, as was the case of PRB2 polluted by sewage water, food web manipulation might not bring the desired effect.

The most important factor for phytoplankton control in the biomanipulated ponds is phytoplankton grazing by large cladocerans. For efficient grazing, cladocerans must be sufficiently large to tackle colonial and filamentous phytoplankters. Large cladocerans can restrain the growth of grazing resistant phytoplankters including bloom-forming cyanobacteria and thus prevent bloom formation. In ponds lacking submerged vegetation, this is only possible when there are no fish. Ponds harbouring extensive

Table 3 Summary of the RDA results

Axes	1	2	3	4	Total variance
Eigenvalues	0.602	0.102	0.072	0.063	1.000
Species–environment correlations	0.986	0.903	0.954	0.868	
Cumulative percentage variance					
Of species data	60.2	70.4	77.6	83.9	
Of species–environment relation	68.6	80.2	88.4	95.6	
Sum of all eigenvalues					1.000
Sum of all canonical eigenvalues					0.877

stands of submerged vegetation can maintain low phytoplankton biomass even in the presence of fish. The effect of fish recolonisation on phytoplankton biomass depends on the density and surface cover of submerged macrophytes. Restoration of submerged macrophytes seems, therefore, to be crucial for the longer-term success of biomanipulation.

Acknowledgements This study was supported by the Brussels Institute of Environment (BIM/IBGE), the Research in Brussels Action 2006 and the Belgian Science Policy, and is part of the project "B-Bbloom2": "Cyanobacterial blooms: toxicity, diversity, modelling and management"; contract SD/TE/01A.

References

Benndorf, J., W. Boing, J. Koop & I. Neubauer, 2002. Top-down control of phytoplankton: the role of time scale, lake depth and trophic state. Freshwater Biology 47: 2282–2295.

Blindow, I., A. Hargeby, B. M. A. Wagner & G. Andersson, 2000. How important is the crustacean plankton for the maintenance of water clarity in shallow lakes with abundant submerged vegetation? Freshwater Biology 44: 185–197.

Brönmark, C. & L.-A. Hansson, 2005. The Biology of Lakes and Ponds. Oxford University Press, Oxford.

Carpenter, S. R., 2005. Eutrophication of aquatic ecosystems: bistability and soil phosphorus. PNAS 102: 10002–10005.

Carpenter, S. R., J. F. Kitchell & J. R. Hodgson, 1985. Cascading trophic interactions and lake productivity: Fish predation and herbivory can regulate lake ecosystems. Bioscience 35: 634–639.

Carpenter, S. R., J. J. Cole, J. R. Hodgson, J. F. Kitchell, M. L. Pace, D. Bade, K. L. Cottingham, T. E. Essington, J. N. Houser & D. E. Schindler, 2001. Trophic cascades, nutrients, and lake productivity: whole-lake experiments. Ecological Monographs 71: 163–186.

Christoffersen, K., B. Riemann, A. Klysner & M. Sondergaard, 1993. Potential role of fish predation and natural populations of zooplankton in structuring a plankton community in eutrophic lake water. Limnology and Oceanography 38: 561–573.

Gliwicz, Z. M., 1990. Why do cladocerans fail to control algal blooms? Hydrobiologia 200: 83–97.

Gross, E. M., 1999. Allelopathy in benthic and littoral areas: case studies on allochemicals from benthic cyanobacteria and submersed macrophytes. In Inderjit, K. M. M. Dakshini & C. L. Foy (eds), Principles and Practices in Plant Ecology: Allelochemical Interactions. CRC Press/Begell House, New York: 179–199.

Gross, M. E., D. Erhard & E. Iványi, 2003. Allelopathic activity of *Ceratophyllum demersum* L. and *Najas marina* ssp. intermedia (Wolfgang) Casper. Hydrobiologia 506–509: 583–589.

Hasle, G. R., 1978. The inverted-microscope method. In Sournia, A. (ed.), Phytoplankton Manual. UNESCO, Paris: 88–96.

Irfanullah, H. M. & B. Moss, 2005. A filamentous green algae-dominated temperate shallow lake: Variations on the theme of clearwater stable states? Archiv Fur Hydrobiologie 163: 25–47.

Jeppesen, E., J. P. Jensen, P. Kristensen, M. Søndergaard, E. Mortensen, O. Sortkjaer & K. Olrik, 1990. Fish manipulation as a lake restoration tool in shallow, eutrophic, temperate lakes. 2. Threshold levels, long-term stability and conclusions. Hydrobiologia 200: 219–227.

Jeppesen, E., J. P. Jensen, M. Søndergaard, T. Lauridsen, L. J. Pedersen & L. Jensen, 1997. Top-down control in freshwater lakes: the role of nutrient state, submerged macrophytes and water depth. Hydrobiologia 342: 151–164.

Jeppesen, E., J. P. Jensen, M. Søndergaard & T. L. Lauridsen, 2005. Response of fish and plankton to nutrient loading reduction in eight shallow Danish lakes with special emphasis on seasonal dynamics. Freshwater Biology 50: 1616–1627.

Jeppesen, E., M. Sondergaard, M. Meerhoff, T. L. Lauridsen & J. P. Jensen, 2007. Shallow lake restoration by nutrient loading reduction—some recent findings and challenges ahead. Hydrobiologia 584: 239–252.

Kemp, P. F., B. F. Sherr, E. B. Sherr & J. J. Cole, 1993. Handbook of Methods in Aquatic Microbial Ecology. Lewis Publishers, Boca Raton.

Lammens, E. H. H. R., 1999. The central role of fish in lake restoration and management. Hydrobiologia 395(396): 191–198.

Lauridsen, T. L., J. P. Jensen, E. Jeppesen & M. Sondergaard, 2003. Response of submerged macrophytes in Danish lakes to nutrient loading reductions and biomanipulation. Hydrobiologia 506: 641–649.

Moss, B., J. Madgwick & G. Phillips, 1996. A Guide to the Restoration of Nutrient-enriched Shallow Lakes. Broads Authority.

Moss, B., D. Stephen, C. Alvarez, E. Becares, W. Van de Bund, S. E. Collings, E. Van Donk, E. De Eyto, T. Feldmann, C. Fernandez-Alaez, M. Fernandez-Alaez, R. J. M. Franken, F. Garcia-Criado, E. M. Gross, M. Gyllstrom, L. A. Hansson, K. Irvine, A. Jarvalt, J. P. Jensen, E. Jeppesen, T. Kairesalo, R. Kornijow, T. Krause, H. Kunnap, A. Laas, E. Lille, B. Lorens, H. Luup, M. R. Miracle, P. Noges, T. Noges, M. Nykanen, I. Ott, W. Peczula, E. Peeters, G. Phillips, S. Romo, V. Russell, J. Salujoe, M. Scheffer, K. Siewertsen, H. Smal, C. Tesch, H. Timm, L. Tuvikene, I. Tonno, T. Virro, E. Vicente & D. Wilson, 2003. The determination of ecological status in shallow lakes—a tested system (ECOFRAME) for implementation of the European Water Framework Directive. Aquatic Conservation-Marine and Freshwater Ecosystems 13: 507–549.

Peretyatko, A., J.-J. Symoens & L. Triest, 2007a. Impact of macrophytes on phytoplankton in eutrophic peri-urban ponds, implications for pond management and restoration. Belgian Journal of Botany 140: 83–99.

Peretyatko, A., S. Teissier, J.-J. Symoens & L. Triest, 2007b. Phytoplankton biomass and environmental factors over a gradient of clear to turbid peri-urban ponds. Aquatic Conservation-Marine and Freshwater Ecosystems 17: 584–601.

Pinel-Alloul, B., 1995. Impacts des prédateurs invertébrés sur les communautés aquatiques. In Pourriot, R. & M. Meybeck (eds), Limnologie Générale. Masson, Paris: 628–686.

Reynolds, C. S., 2006. Ecology of Phytoplankton. Cambridge University Press, Cambridge.

Shapiro, J., 1990. Biomanipulation: the next phase—making it stable. Hydrobiologia 200–201: 13–27.

Søndergaard, M. & B. Moss, 1998. Impact of submerged macrophytes on phytoplankton in shallow freshwater lakes. In Jeppessen, E., M. Sondergaard & K. Christoffersen (eds), The Structuring Role of Submerged Macrophytes in Lakes. Springer, New York: 115–133.

Søndergaard, M., E. Jeppesen, T. L. Lauridsen, C. Skov, E. H. Van Nes, R. Roijackers, E. Lammens & R. Portielje, 2007. Lake restoration: successes, failures and long-term effects. Journal of Applied Ecology 44: 1095–1105.

ter Braak, C. J. F. & P. Smilauer, 2002. CANOCO reference manual and user's guide to Canoco for Windows: software for canonical community ordination (version 4.5). Microcomputer Power, Ithaca.

van Donk, E. & W. J. van de Bund, 2002. Impact of submerged macrophytes including charophytes on phyto- and zooplankton communities: allelopathy versus other mechanisms. Aquatic Botany 72: 261–274.

Van Wichelen, J., S. Declerck, K. Muylaert, I. Hoste, V. Geenens, J. Vandekerkhove, E. Michels, N. De Pauw, M. Hoffmann, L. De Meester & W. Vyverman, 2007. The importance of drawdown and sediment removal for the restoration of the eutrophied shallow Lake Kraenepoel (Belgium). Hydrobiologia 584: 291–303.

Wetzel, G. R. & E. G. Likens, 1990. Limnological Analyses. Springer-Verlag, New York.

Willame, R., T. Jurczak, J. F. Iffly, T. Kull, J. Meriluoto & L. Hoffmann, 2005. Distribution of hepatotoxic cyanobacterial blooms in Belgium and Luxembourg. Hydrobiologia 551: 99–117.

Hydrobiologia (2009) 634:137–151
DOI 10.1007/s10750-009-9896-4

POND CONSERVATION

Spatial and temporal patterns of pioneer macrofauna in recently created ponds: taxonomic and functional approaches

A. Ruhí · D. Boix · J. Sala · S. Gascón · X. D. Quintana

Published online: 13 August 2009

Abstract Man-made ponds are often created to compensate for the loss and degradation of wetlands, but little is known about the processes taking place in these artificial environments, especially at a community level. The macrofaunal assemblage and water chemistry of newly created ponds in three nearby areas in the NE Iberian Peninsula were studied during the first year of life of these ponds in order to (i) detect if any invertebrate assemblage structure change was taking place, (ii) evaluate the effect of local factors on the invertebrate assemblage in each site, and (iii) compare the information obtained by taxonomic and functional approaches. Although invertebrate colonization was rapid, no relevant changes in assemblage parameters were related to time, implying that more time may be needed to detect successional changes in invertebrate assemblages. Local factors—especially those related to hydrological stability—produced notable differences both in the assemblage parameters and in the taxonomic and functional compositions of the invertebrate fauna. Finally, information provided by the functional approach was redundant with respect to that obtained by the classical taxonomic approach: in these newly created systems, the high dominance of a small number of taxa makes the functional approach a simple biological traits analysis of the few dominant species.

Keywords Mediterranean ponds · Succession · Colonization · Assemblage structure · Functional approach · Hydrological stability

Electronic supplementary material The online version of this article (doi:10.1007/s10750-009-9896-4) contains supplementary material, which is available to authorized users.

Guest editors: B. Oertli, R. Cereghino, A. Hull & R. Miracle
Pond Conservation: From Science to Practice. 3rd Conference of the European Pond Conservation Network, Valencia, Spain, 14–16 May 2008

A. Ruhí (✉) · D. Boix · J. Sala · S. Gascón · X. D. Quintana
Institute of Aquatic Ecology, University of Girona, Campus de Montilivi, Facultat de Ciències, 17071 Girona, Catalonia, Spain
e-mail: albert.ruhi@udg.edu

Introduction

Artificial ponds are very valuable to society, since they are often created for purposes such as water supply, floodwater retention, recreation and education, or wildlife management and research (Oertli et al., 2005). They are also often the result of mitigative measures to compensate for habitat destruction and the subsequent loss of species (National Research Council, 1992), but there is still little knowledge of the ecological function of these habitats, especially at the invertebrate community level (e.g., Gee et al., 1997; Herrmann et al., 2000). In this sense, a deeper knowledge of the biodiversity hosted in these environments is needed to evaluate if newly created ponds are appropriate management tools for biological conservation. Although biodiversity analyses have often been based only on species richness, it is important to take into account, among others, aspects concerning taxonomic relatedness (Warwick & Clarke, 1995). Thus, assemblages comprising only taxonomically related species should be regarded as less diverse than others that host more distantly related species (Abellán et al., 2006).

In general, newly created ponds are rapidly colonized by plants and invertebrates, particularly, when they are close to, or connected to, other wetlands (Gee et al., 1997; Fairchild et al., 2000). Any successional process includes not only this initial rapid colonization but also subsequent changes in structure and organization of the assemblage (Fisher et al., 1983). Hence, although it has been reported that the main increase in species richness takes place over the first year, assemblage structuring patterns—e.g., decreasing of the dominance values, shift from predators toward grazers and detritivores—can take longer (Barnes, 1983; Friday, 1987). Therefore, a lack of structure changes in assemblages of newly created ponds during their first year of life is hypothesized in this study.

Local factors, such as fish stocking, pond use, egg banks, water regime, hydroperiod length, or habitat heterogeneity, can relevantly influence the community structure (e.g., Schneider & Frost, 1996; Della Bella et al., 2005; Gascón et al., 2005). Even in highly interconnected ponds, local environmental constraints can be strong enough to prevail over regional homogenizing forces and structure local communities (Cottenie et al., 2003). In this study, we hypothesize that local factors will greatly influence the resulting assemblage.

Since ecosystem-level processes are affected by the functional characteristics of the organisms involved rather than by their taxonomic identity (Hooper et al., 2002; Verberk et al., 2008), functional analyses, taking into account both functional parameters and biological traits, have proved to be useful methods (Higgins & Merritt, 1999; Gascón et al., 2008). Moreover, functional approaches can provide information that is complementary to taxonomical data (e.g., Rodríguez & Magnan, 1993; Boix et al., 2004; Brucet et al., 2005). Therefore, in this study, taxonomic and functional approaches are expected to provide complementary, not redundant, information.

The aims of this study were: (i) to analyze changes that take place in macrofaunal assemblages of newly created habitats during the very early phase of colonization and succession, along the first year of creation of these ponds; (ii) to study the importance of local factors and to establish whether they can explain differences in the pond assemblage structure among nearby ponds; and (iii) to compare results obtained by taxonomic and functional approaches, and check if these procedures are complementary or redundant.

Materials and methods

Study site

The study was carried out in the NE Iberian Peninsula, in three different but nearby plain areas (Fig. 1): Pla de l'Estany (hereafter, PE), Plana del Baix Ter (hereafter, BT), and Plana de la Selva (hereafter, PS). In each site, several man-made ponds for habitat and species recovery and mitigative measures were constructed, and three ponds in each site were selected for this study (Table 1). All ponds at all sites were shallow (depth <2 m), were flooded for the first time in the summer of 2006 and were monitored during their first year of life, from September 2006 until September 2007. The main water supplies were different in the three sites (Table 1): ponds in PE were fed by a karstic stream, ponds in BT by a coastal aquifer, and ponds in PS by rainfall. Ponds located in this last site did not complete their first hydroperiod, drying up in summer due to a strong drought (Fig. 2). The study sites presented different abundances of the Eastern mosquitofish (*Gambusia holbrooki*) and the red swamp crawfish (*Procambarus clarkii*), as well as the percentage of the surface covered by aquatic vegetation, estimated by visual analysis in the field (Table 1). Some natural ponds were selected for having similar characteristics and being situated near the constructed ones, acting as controls to compare their invertebrate assemblages: two natural ponds were selected near PE (adding up 6 samples), one natural pond was selected near BT (13 samples), and another one near PS (5 samples). This data was obtained in the framework of a project focused on water quality assessment (Boix et al., 2005).

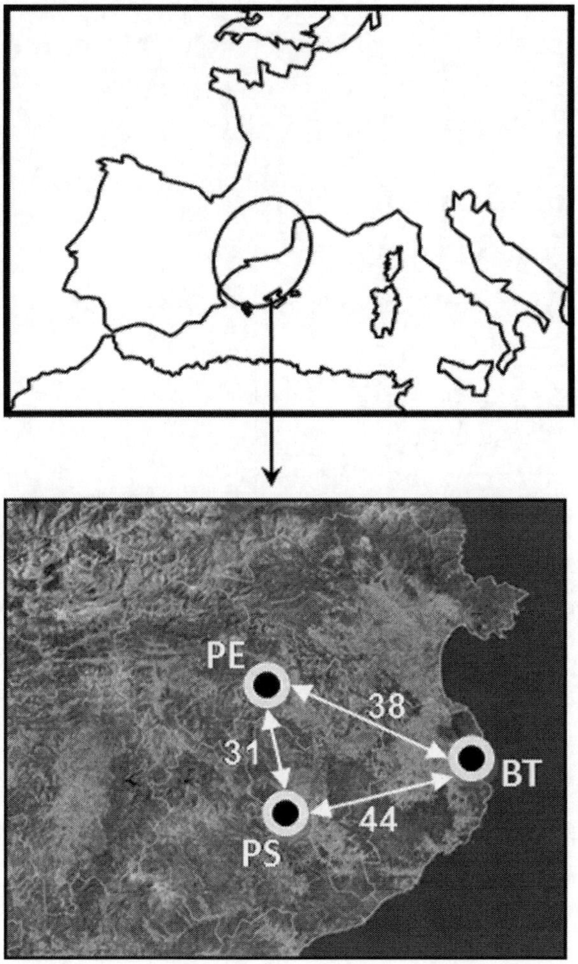

Fig. 1 Map of the study area with the location of the three studied sites, and distances (in km) between them

Sampling and sample processing

Newly created ponds were sampled quarterly, during the same week in all three areas. Physical and chemical water parameters and water level were measured in situ using a Crison 524 conductivity meter (for conductivity), an EcoScan ph6 (for pH), a Hach HQ10 Portable LDO meter (for temperature and dissolved oxygen), and a graduated gauge (for water level). Water samples were collected on every sampling day, filtered through GF/C Whatman filters, and frozen upon arrival at the laboratory. Analyses of dissolved inorganic nutrients (ammonium, NH_4^+; nitrite, NO_2^-; nitrate, NO_3^-; and soluble reactive phosphorus, PO_4^{3+}) from filtered samples were carried out according to Grasshoff et al. (1983). According to Talling & Driver (1963), chlorophyll-a content was later extracted from filters using 80% methanol and measured. Aquatic macrofauna was sampled using a 250 μm mesh size dip-net of 20 cm in diameter, and the sampling procedure was based on 20 dip-net sweeps in rapid sequence, covering all different microhabitats. This methodology was performed identically in the newly created and in the natural ponds. Samples were preserved in situ in 4% formalin. Subsequently, the invertebrate fauna was sorted, identified to the lowest taxonomic level possible, and 25 individuals of each taxa were randomly chosen and measured. Biomass was estimated as dry mass, which was calculated from individual lengths using existing equations for macroinvertebrates (Smock, 1980; Meyer, 1989; Lindegaard, 1992; Traina & Ende, 1992; Montes et al., 1993; Smit et al., 1993; Quintana, 1995; Benke et al., 1999; Cabral & Marques, 1999; Koutrakis & Tsikliras, 2003; Boix et al., 2004).

Assemblage parameters: taxonomic and functional approaches

The analyzed taxonomic parameters of the assemblage were species richness (S), Shannon–Wiener diversity (H′), and complementary measures of taxonomic distinctness. Three taxonomic distinctness indices were chosen for this study: taxonomic distinctness (Δ^*), average taxonomic distinctness (Δ^+), and variation in taxonomic distinctness (Λ^+). The first metric is based on quantitative data, whereas the last two are based on presence/absence data. Both the first (Δ^*) and the second (Δ^+) represent the average path length in the phylogenetic tree connecting two random species of those collected, but are not always highly correlated, suggesting that they capture different aspects of relatedness (Clarke & Gorley, 2006). Λ^+ measures the variance in pairwise lengths between each pair of species and reflects the unevenness of the taxonomic tree (Clarke & Warwick, 2001b). Neither total taxonomic distinctness ($S\Delta^+$) nor taxonomic diversity (Δ) was used $-S\Delta^+$ as it is redundant with the species richness, and Δ as it closely tracks traditional diversity measures, depending highly on the dominance of species (Clarke & Warwick, 2001b). The taxonomic levels taken into account for this analysis were species, genus, family, order, class, subphylum, and phylum, and the same branch length was weighted for each taxonomic level. Three assemblage functional parameters were also

Table 1 General characteristics of the ponds created at each site and mean values (standard deviation, in *italics*) of the water and environmental variables measured in each pond

Site	Situation	Water supply	Hydroperiod	Vegetation cover	*P. clarkii* and *G. holbrooki*	Pond	T (°C)	Ox. (mg/l)	Conductivity (μS cm^{-1})	pH	Chl a (μg/l)	DIN (mg/l)	SRP (mg/l)	Water level (cm)	Distance to the nearest natural pond (m)
Pla de l'Estany (PE)	Inland plain	Karstic stream	>12 months	High (>60%)	Present	PE1	18.7	9.60	**1145.0**	8.67	**2.304**	0.034	0.002	**102**	**2614**
							6.8	*4.83*	*150.2*	*1.10*	*3.503*	*0.014*	*0.001*	*42*	–
						PE2	16.9	6.83	**1236.0**	8.22	**32.087**	0.048	0.008	**61**	**2603**
							7.9	*2.38*	*661.3*	*0.32*	*66.392*	*0.034*	*0.006*	*36*	–
						PE3	19.5	8.58	**1153.8**	8.37	**3.396**	0.103	0.004	**96**	**2464**
							7.1	*1.49*	*337.6*	*0.42*	*2.896*	*0.103*	*0.004*	*47*	–
Plana del Baix Ter (BT)	Coastal plain	Coastal aquifer	>12 months	Medium (30–60%)	Present	BT1	17.0	7.60	**1393.6**	8.30	**3.918**	0.040	0.002	**119**	**3042**
							6.8	*0.50*	*226.3*	*0.68*	*1.753*	*0.024*	*0.002*	*9*	–
						BT2	17.2	8.42	**2834.0**	8.57	**12.194**	0.042	0.009	**112**	**3114**
							6.7	*2.17*	*940.1*	*1.41*	*12.663*	*0.033*	*0.015*	*37*	–
						BT3	16.3	6.52	**718.6**	8.14	**7.464**	0.093	0.012	**86**	**3097**
							6.8	*1.21*	*327.4*	*0.59*	*4.942*	*0.089*	*0.015*	*12*	–
Plana de la Selva (PS)	Inland plain	Rainfall	6–9 months	Low (<30%)	Absent	PS1	17.6	6.86	**309.0**	7.64	**3.243**	0.341	0.001	**25**	**3256**
							4.6	*0.53*	*455.8*	*0.54*	*4.587*	*0.367*	*0.001*	*7*	–
						PS2	16.4	5.89	**63.0**	7.51	**2.583**	0.375	0.014	**20**	**3236**
							4.5	*1.69*	*36.5*	*0.24*	*0.960*	*0.271*	*0.023*	*7*	–
						PS3	16.0	7.19	**63.8**	7.46	**3.200**	0.344	0.001	**24**	**3260**
							5.2	*1.31*	*28.5*	*0.30*	*3.127*	*0.185*	*0.000*	*3*	–

Variables significantly related to the invertebrate assemblage, according to the CCA and the forward selection, are shown in *bold* ($P < 0.05$)

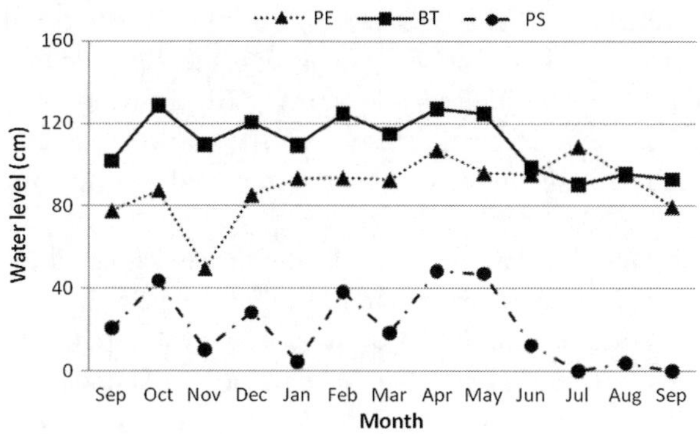

Fig. 2 Water level (cm) of the ponds studied in (PE) Pla de l'Estany, (BT) Plana del Baix Ter, and (PS) Plana de la Selva, during the first complete year after flooding of these ponds (September 2006–September 2007)

chosen: total numerical abundance (N), total biomass (B), and size diversity (μ). For size diversity (μ), the method recently described by Quintana et al. (2008) was followed.

Assemblage composition: taxonomic and functional approaches

The assemblage composition was studied both from a taxonomic point of view (at species and order levels) and using a functional approach. Two functional groups were created, one according to functional feeding groups and the other to life-history strategy groups. Following Merritt & Cummins (1996), six feeding-type groups were created: predators (PR), filterers (FI), scrapers (SC), piercing-herbivores (PH), collectors (CO), and shredders (SH). Moreover, each taxon was assigned to a life-history strategy group based on Wiggins et al. (1980). Groups 1, 2, and 3 include species that are dormant during the unfavorable season, but differ in the dispersion capacity and the timing of oviposition. Group 1 includes species with passive dispersal, group 2 includes species with active dispersal that need water for oviposition, and group 3 includes species with active dispersal that do not need water for oviposition. Group 4 includes species that cannot remain in the basin during the unfavorable season. A fifth group, added to the original classification, included taxa which need water to disperse and reproduce and cannot survive desiccation, as described by other authors (Hillman & Quinn, 2002; Gascón et al., 2008). Assignations of found species to the corresponding functional feeding and life-history group are shown in the Appendix—Supplementary Material.

Data analysis

The extent to which study sites differ in their hydrological stability was studied using the method described by Brownlow et al. (1994), which implies cluster analysis of the water level frequencies for each pond (group average as a conglomeration method and Manhattan distance as a similarity measure). In order to analyze the importance of local environmental variables among sites and over time, a principal component analysis (PCA) was conducted. The variables included in the environmental matrix were: conductivity, pH, temperature, dissolved oxygen, water level, concentration of dissolved inorganic nitrogen (DIN) as the sum of concentrations of NH_4^+, NO_2^- and NO_3^-, soluble reactive phosphorus (SRP), and concentration of chlorophyll-a (Chla). All environmental variables, except pH, were log-transformed using $\log_{10}$ (Var + 1). ANOVA tests were performed to check the significance of the positions of samples (using sample scores as dependent variable) classified by site or time (used as factors). Bonferroni post-hoc tests were also used to compare similarity among sites in ANOVA results. The environmental variables listed above, plus the distance to the nearest natural pond, were related to species variability by means of a canonical correspondence analysis (CCA) of the species matrix. A forward selection of the environmental variables was performed using Monte Carlo permutation tests (999 random permutations), retaining the variables with $P < 0.05$. This was computed from the 'forward.sel' function available in the R 'packfor' library (Dray, 2004). In order to quantify the "locality effect", a variation partitioning was performed using the function 'varpart' in the vegan library, written in R-language (Oksanen et al., 2005). Thus, "locality effect" was assessed not only considering the variance explained by the locality itself but also the shared variability between locality and environmental variables. In order to perform this analysis, three matrices were used: the species matrix, the locality matrix, and the significant environmental variables matrix. The species matrix contained the abundance of taxa by samples, the locality matrix included three dummy variables identifying samples from the same site (PE, BT, and PS), and the environmental variables matrix was the result of the selection of the significant environmental variables performed by the forward selection procedure previously described.

Changes in the assemblage parameters among sites and over time were tested by a mixed linear model, with site and time as fixed effects and time as a random effect, nested within the site. The inclusion of time in the random part of the model allows for the control of pseudoreplication problems due to sampling the same pond over time. The tested taxonomic parameters of the assemblage were species richness (S), Shannon–Wiener diversity (H′), average taxonomic distinctness (Δ^{+}), variation in taxonomic distinctness (Λ^{+}), and taxonomic distinctness (Δ^{*}). The tested functional parameters of the assemblage were total numerical abundance (N), total biomass (B), and size diversity (μ). Numerical abundance, biomass, and species richness were log-transformed to ensure a better fit of errors to a normal distribution.

Changes in the assemblage composition among sites and over time were tested with an analysis of similarities (ANOSIM) test. This type of test operates on a resemblance matrix and is similar to a standard univariate ANOVA, but does not need either normality or homoscedasticity of data. The chosen option was the two-way nested layout, where the two factors were hierarchic (time factor nested within site factor), and the results showed a global R that oscillated from 0 and 1 and a *P*-value expressed in percentage. When R was 0 or close to 0, similarities within groups (i.e., in samples within the same site) and between groups (i.e., in samples between different sites) were equivalent. In contrast, when higher values of R were obtained, more samples within the group resembled each other than they did between groups, differentiating groups that would correspond to samples from different sites (Clarke & Warwick, 2001b). The distance matrix was built with Bray–Curtis similarity, based on a rectangular, original standardized, and square-root transformed matrix, and the ANOSIM test operated 999 permutations in all cases. This test was applied to eight compositional matrices, which were the result of two approaches (taxonomic and functional), two levels for each approach (species and orders for the taxonomic approach; functional feeding groups and life-history strategy groups for the functional approach) and two matrices for each level (numerical abundance and biomass). A similarity percentages (SIMPER) test was used to analyze and detect the characteristic species of a site. Specifically, the test performed was a two-way SIMPER analysis with the numerical abundance species composition matrix, with site and time as factors and a cut off for low contributions at 90%.

Finally, a non-parametric multidimensional scaling (MDS) was performed to compare the assemblage structure of samples from the newly created ponds to those of natural ponds situated nearby. Abundances were square-root transformed and Bray–Curtis was selected as a similarity distance. Subsequently, a two-way nested ANOSIM (natural/artificial factor nested within the site factor) was used to test differences when comparing assemblages among sites and pond types.

The cluster analysis, the ANOSIM and SIMPER tests, the MDS and the calculations of the assemblage parameters were performed with PRIMER v. 6.0 for Windows. The PCA was performed with CANOCO 4.5, and R v. 2.8.1 was used for the CCA, the forward selection of the environmental variables and the variation partitioning test. ANOVA tests, Bonferroni post-hoc tests, and mixed linear models were performed with the software package SPSS 15.0.

Results

Pond characterization

In accordance with the temporary behavior of PS ponds, the analysis of the hydrological stability resulted in a cluster that separated PS ponds from the rest (Fig. 3). When studying the site factor, the position of the samples in the PCA (Fig. 4) showed significant differences among sites for axis 1 (ANOVA, $F_{2,37} = 21.701$; $P < 0.001$). Again, PS was different from the other sites with the post-hoc test ($P < 0.001$), while BT and PE were not distinguishable ($P = 1.000$). Low levels of conductivity and water column depth were characteristic of PS, as well as high levels of dissolved nitrogen in comparison with ponds in PE and BT, more related to higher conductivities and water levels (Table 1). Ponds in PE and BT also had more variability within axis 2, associated with the concentration of chlorophyll values. Nevertheless, no significant differences in sample positions were detected for the second axis

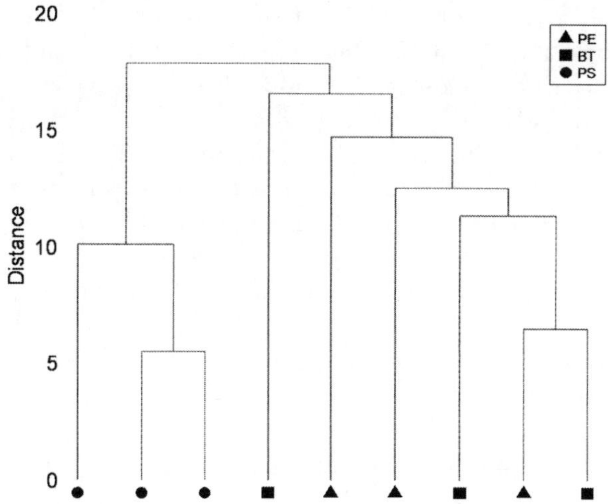

Fig. 3 Clustering of the distances of the newly created ponds according to their hydrological stability. *PE* Pla de l'Estany, *BT* Plana del Baix Ter, and *PS* Plana de la Selva

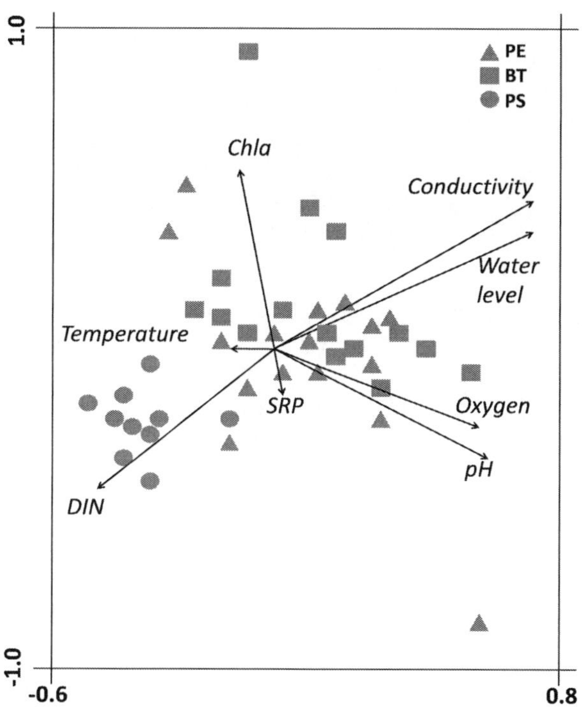

Fig. 4 Sample position in relation to water and environmental variables in the space created by the first two axes of the PCA. Axis 1 explains 54.3% of the variability; axis 2 explains 24.2%. *DIN* concentration of dissolved inorganic nitrogen, *SRP* soluble reactive phosphorus, *Chl a* concentration of Chlorophyll-a, *PE* Pla de l'Estany, *BT* Plana del Baix Ter, and *PS* Plana de la Selva

(ANOVA, $F_{2,37} = 2.14$; $P > 0.100$). On the other hand, water and environmental characteristics did not differ over time since the position of the samples in the PCA did not show significant differences either for axis 1 (ANOVA, $F_{4,35} = 2.088$; $P > 0.100$), or for axis 2 (ANOVA, $F_{4,35} = 1.871$; $P > 0.100$). According to the forward selection, the environmental variables which appeared to significantly affect the assemblage were distance to the nearest natural pond ($F = 5.705$; $P = 0.001$), conductivity ($F = 2.465$; $P = 0.010$), concentration of chlorophyll ($F = 2.300$; $P = 0.024$), and water level ($F = 2.291$; $P = 0.014$). These four variables alone explained 10% of the variance, whereas the shared variability between these variables and the site explained 14% and the site by itself explained only 1%. The high percentage of shared variability implies that each locality had characteristic environmental conditions.

Pioneer assemblages: composition and abundance

A total of 87 species were detected, of which 57.5% were exclusively found at one site and 19.5% of them were found at all sites. Species richness was, however, concentrated in a few taxonomic groups, as only three families were responsible for more than a third of the total richness. Chironomidae (18 taxa), Dytiscidae (7 taxa), and Corixidae (5 taxa), and the corresponding orders (Diptera, 27 taxa; Coleoptera, 18 taxa; and Heteroptera, 12 taxa) were the richest groups within Insecta. Moreover, insects alone, with 66 species, represented more than 75% of the macrofaunal richness. Although the three main dominating orders in number of individuals were the same at all sites (Ephemeroptera, Heteroptera, and Diptera), the dominant one was different for each site: Ephemeroptera for PE, Heteroptera for BT, and Diptera for PS. In contrast, none of these groups dominated in biomass in the corresponding site. In PE and BT, a small number of individuals of Malacostraca (more than 99% of them were red swamp crawfish) represented 73% (PE) and 48% (BT) of the total macrofaunal biomass. In contrast, in PS insects represented practically the whole biomass. It is remarkable that two out of the three areas (PE and BT) were colonized early (during the first year of life) by the red swamp crawfish and the Eastern mosquitofish, whereas in PS these invasive species were not detected (Figs. 5A, B).

When analyzing functional feeding strategies, collectors represented the main feeding group in PE, both in biomass and in numerical abundance, collectors and predators were the dominant feeding groups in BT, and predators dominated in PS (Fig. 5C). Analyzing life-history strategies, results for abundance and biomass were only similar in PS, where the fourth strategy was dominant, followed by the second strategy. The fourth strategy comprises species active in dispersal, which need water to reproduce and can leave the pond during a dry phase; and taxa comprised in the second strategy show the same behavior but remain during an eventual desiccation of the pond, often as an egg or immature stage. In PE and BT, the dominant strategy for biomass was the fifth, which contained the crawfish. In contrast, when analyzing numerical abundance, the main strategy was the second in PE and the fourth in BT (Fig. 5D).

Analyzing the effect of time on species richness, it was observed that during the first sampling period (September 2006), 51% of the cumulative species richness (44 out of 87) had already arrived at least in one site. A close look showed that this percentage differed among sites, ranging from around 40% in PE and PS to 50% in BT. Nevertheless, the increase in species richness obtained during the first period (June to September 2006) was the highest at all sites.

Assemblage structure among sites and over time

Assemblage parameters: taxonomic and functional approaches

Log-transformed species richness ($F_{2,11} = 6.163$, $P = 0.006$) and variation in taxonomic distinctness ($F_{2,11} = 6.696$, $P = 0.005$) were the only taxonomic parameters of the assemblage significantly related to site factor (Table 2). These variables showed their highest values in PE and their lowest in PS, indicating that sites with higher richness also had a higher unevenness in their taxonomic tree. In contrast, none

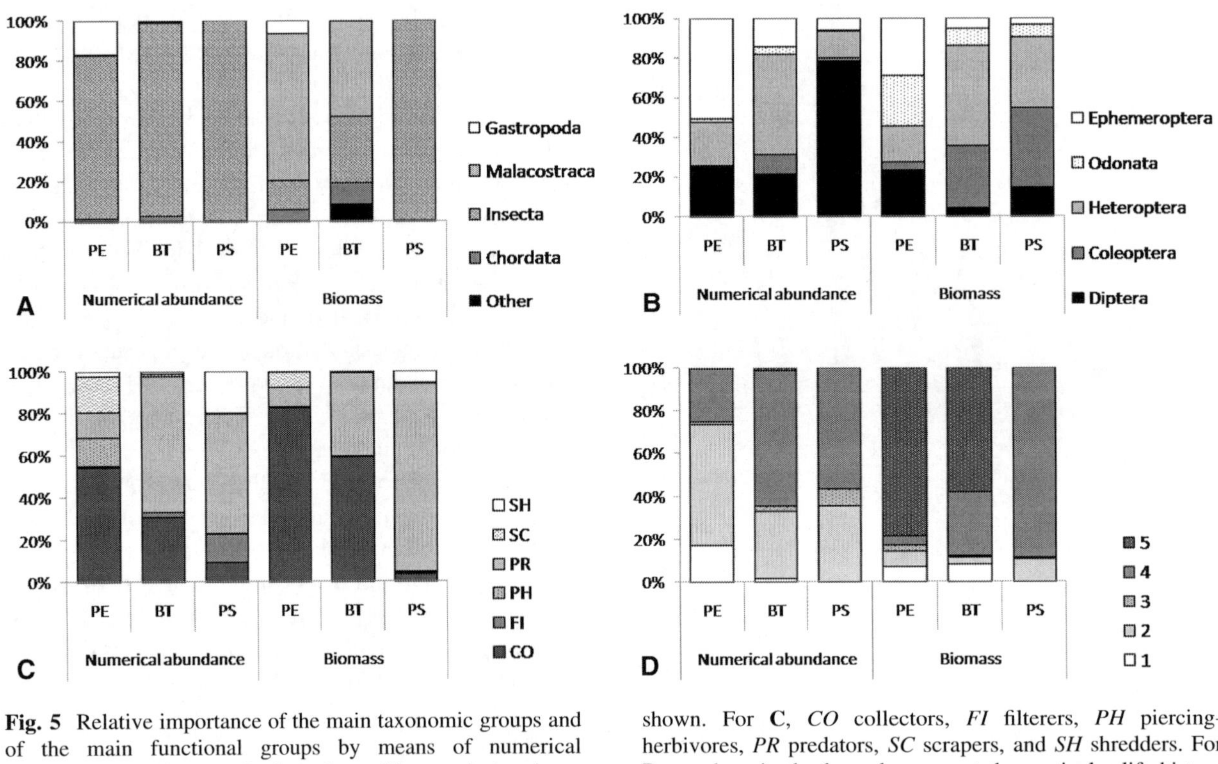

Fig. 5 Relative importance of the main taxonomic groups and of the main functional groups by means of numerical abundance and biomass. In **A**, orders with a total abundance of less than 50 individuals and a total biomass lower than 1 g have been included in the "other" category (including the orders of the following classes: Leptolida, Oligochaeta, and Arachnida), except for Chordata. In **B**, only Insecta orders are shown. For **C**, *CO* collectors, *FI* filterers, *PH* piercing-herbivores, *PR* predators, *SC* scrapers, and *SH* shredders. For **D**, numbers in the legend represent the particular life-history strategy group (see text for the biological characteristics of the five life-history strategy groups). *PE* Pla de l'Estany, *BT* Plana del Baix Ter, and *PS* Plana de la Selva

of the five assemblage taxonomic parameters showed significant differences when related to the time factor, and only one was marginally significant: species richness ($F_{2,11} = 2.155$, $P = 0.053$). Analyzing the functional parameters of the assemblage, both total numerical abundance ($F_{2,11} = 7.313$, $P = 0.003$) and total biomass ($F_{2,11} = 5.722$, $P = 0.009$), presented significant differences for the site factor. The highest values of total numerical abundance were obtained in PE, while lower values of total biomass were measured in PS than at the other two sites. Total biomass was the only parameter related to the time factor ($F_{2,11} = 4.718$, $P = 0.001$), showing its highest values at the end (Fig. 6); and total numerical abundance gave a marginally significant result ($F_{2,11} = 2.140$, $P = 0.054$). At BT and PS, adult Coleoptera and Heteroptera contributed highly to biomass at the beginning, with a relevant decrease during autumn and winter, and an increase in spring. In contrast, in PE Coleoptera and Heteroptera were only relevant in biomass at the beginning, and other groups dominated during winter and spring, mainly crawfish from June onwards.

Assemblage composition: taxonomic and functional approaches

When analyzing the site factor, the two-way nested ANOSIM tests showed significant differences for all levels of both approaches (taxonomic and functional), testing either the numerical abundance matrix or the biomass matrix (Table 3). However, only at species level were all pairwise comparison tests significant. In this case, all R values were high (>0.8), indicating that the species composition was different in each area, either analyzing the numerical abundance or analyzing the biomass matrix. For the rest of the levels (orders, functional feeding groups, and life-history strategy groups), although global tests were significant ($P < 0.01$), pairwise comparisons were not always significant. Differences were higher with numerical abundance matrices (higher values of R and a higher number of significant pairwise comparison tests) than with biomass matrices; and only comparisons between two sites (PE vs. BT) were

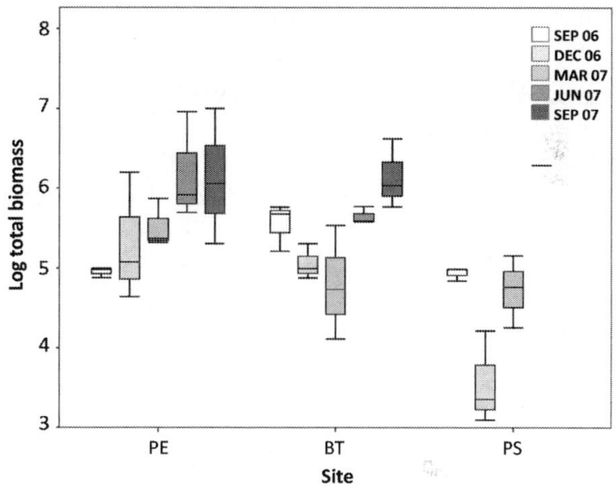

Fig. 6 Box-plots showing the total biomass over time. *PE* Pla de l'Estany, *BT* Plana del Baix Ter, and *PS* Plana de la Selva

Table 2 Assemblage parameters compared among sites and over time

Assemblage parameter	Site		Time	
	$F_{2,11}$	P	$F_{2,11}$	P
Taxonomic				
Species richness (S)	**6.163**	**0.006**	2.155	0.053
Shannon–Wiener diversity (H′)	1.300	0.290	0.557	0.845
Average taxonomic distinctness (Δ^+)	0.138	0.872	1.289	0.285
Variation in taxonomic distinctness (Λ^+)	**6.696**	**0.005**	0.981	0.487
Taxonomic distinctness (Δ^*)	0.790	0.464	1.470	0.202
Functional				
Total numerical abundance	**7.313**	**0.003**	2.140	0.054
Total biomass	**5.722**	**0.009**	**4.718**	**0.001**
Size diversity (μ)	3.017	0.066	1.562	0.169

The results of the linear mixed models (site = fixed factor, time = random factor nested inside site) are shown. Significant results are shown in *bold* ($P < 0.05$)

Table 3 Taxonomic and functional assemblage parameters compared among sites and over time

	Site		Time	
	N. ab.	Biom.	N. ab	Biom.
Taxonomic approach				
Species				
Global *R*	**0.871**	**0.811**	ns ($P > 0.01$)	
Global *P*	**0.001**	**0.001**		
R_{PE-BT}	**0.874**	**0.805**	–	
R_{PE-PS}	**0.963**	**0.825**		
R_{BT-PS}	**0.881**	**0.881**		
Orders				
Global *R*	**0.756**	**0.656**	ns ($P > 0.01$)	
Global *P*	**0.001**	**0.001**		
R_{PE-BT}	**0.840**	**0.790**	–	
R_{PE-PS}	**0.831**	0.594		
R_{BT-PS}	0.681	**0.669**		
Functional approach				
Feeding groups				
Global *R*	**0.799**	**0.663**	ns ($P > 0.01$)	
Global *P*	**0.001**	**0.001**		
R_{PE-BT}	**0.976**	**0.772**	–	
R_{PE-PS}	**0.963**	0.719		
R_{BT-PS}	**0.550**	0.600		
Life-history strategy groups				
Global R	**0.743**	**0.631**	ns ($P > 0.01$)	
Global *P*	**0.001**	**0.001**		
R_{PE-BT}	**0.896**	**0.752**	–	
R_{PE-PS}	**0.813**	0.525		
R_{BT-PS}	0.594	0.744		

The results of the two-way nested ANOSIM test (time nested within site factor) are shown. For pairwise comparisons, *PE* Pla de l'Estany, *BT* Plana del Baix Ter, and *PS* Plana de la Selva

N. ab. numerical abundance matrix, *Biom.* = biomass matrix

Significant results (global tests and pairwise comparisons) are shown in bold ($P < 0.01$)

significant for all levels and matrices. The time factor was not significantly related to changes in the taxonomic or functional assemblage composition. On the other hand, the results of the two-way SIMPER analysis with numerical abundance of species composition allowed us to identify the characteristic species of each site (Table 4). Thus, although, some species were characteristic in samples of all sites—*Cloeon inscriptum* and *Sigara lateralis* contributed highly to similarity among samples for all three sites—others were particularly common only in one site, such as *Physella acuta* and *Procladius choreus* in PE, *Anisops sardeus* and *Psectrocladius* gr. *sordidellus* in BT, or *Chaoborus flavicans* and *Anopheles* gr. *maculipennis* in PS.

Comparison of newly created and natural ponds' assemblages

The MDS showed that samples from newly created ponds appeared separated from the assemblages of the corresponding natural ponds in each site (Fig. 7). The ANOSIM analysis detected significant differences between newly created and natural ponds ($R = 0.347$; $P = 0.001$), but not among sites ($R = 0.222$; $P = 0.867$). This implies that, when comparing natural to newly created ponds, the large differences between the typology of the pond (natural–artificial) reduces the existent site differences until make them non-significant.

Discussion

Assemblage structure changes

Succession and colonization have traditionally been considered as equivalent and inseparable processes, but succession includes colonization and the subsequent changes in structure and organization of the assemblage (Velasco et al., 1993). The rapid colonization of the studied ponds was fairly predictable, as it is described as being common in newly created ponds (Gee et al., 1997; Fairchild et al., 2000), and it is also known in other aquatic environments, such as newly created streams or channels (e.g., Malmqvist et al., 1991), experimentally flooded lakes and wetlands (e.g., Herrmann et al., 2000), or impounded reservoirs (e.g., Bass, 1992). In our case, about 50% of the species found in this study were already captured at the first sampling after flooding (1–2 month period). Studies going deeper into colonization mechanisms show that newly created, small and isolated ponds can be compared to oceanic islands, with a colonization rate that results from species immigration and extinction, and tends to be quicker, when dispersal distances are short and active

Table 4 Characteristic species (two-way SIMPER analysis, factors = site and time)

PE (32.7% average similarity)		BT (25.0% average similarity)		PS (59.9% average similarity)	
Species	%	Species	%	Species	%
Cloeon inscriptum	24.7	*Anisops sardeus*	31.5	*Chaoborus flavicans*	47.8
Sigara lateralis	16.4	*Sigara lateralis*	24.5	*Cloeon inscriptum*	17.4
Physella acuta	9.3	*Cloeon inscriptum*	10.7	*Anopheles* gr. *maculipennis*	15.7
Procladius choreus	8.6	*Psectrocladius* gr. *sordidellus*	4.9	*Sigara lateralis*	12.0

The first four species in % of similarity have been selected

dispersal mechanisms are efficient (Barnes, 1983; Havel & Shurin, 2004; Csabai & Boda, 2005).

Although the colonization of newly created ponds was quick, only one assemblage parameter (total biomass) proved to be related to time during their first year of life. Since there was no significant increase in the total numerical abundance of individuals, the increasing biomass can probably be explained by secondary production inside the pond (individual growth), and the arrival of larger-bodied species. Despite a biomass shift at two sites (BT and PS) during colder months (December 2006 and March 2007), when biomass was a bit lower, no more relevant changes in assemblage parameters were related to time. Other findings are consistent with these results and have demonstrated that pioneer communities usually last longer (Barnes, 1983; Friday, 1987). Fairchild et al. (2000) found that young ponds were colonized early by large predatory beetles and acquired beetle assemblages comparable

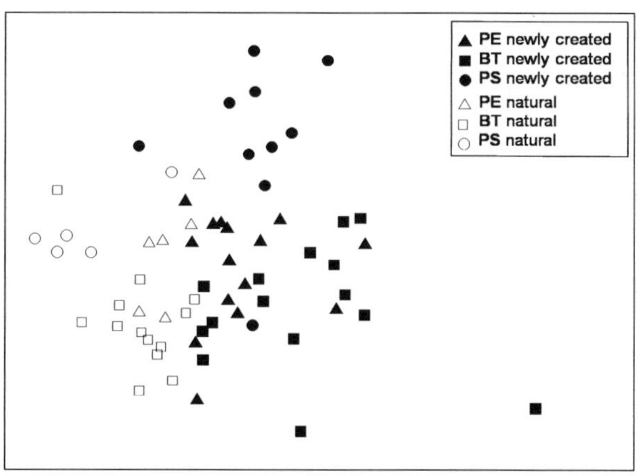

Fig. 7 MDS plot of the invertebrate assemblages of newly created ponds and those of nearby natural ponds (stress = 0.19). *PE* Pla de l'Estany, *BT* Plana del Baix Ter, and *PS* Plana de la Selva

in species richness to those of older ponds, but taxa dependent on particular plant species occurred later. This led to the conclusion that several years are necessary after site construction to host these rarer specialist taxa. The lack of these specialist taxa in the newly created ponds, together with the great dominance of a small number of pioneer colonizers, may explain the distance between assemblages of these ponds and those in natural ones.

Importance of local factors

Conductivity and water level are two covarying local factors: low conductivity values and low water column levels at PS are due to the fact that ponds in this site depended exclusively on rainfall, and the year during which this study took place was very dry. Higher conductivity values in BT and PE are attributed to the coastal influence and the karstic stream, respectively, whereas more regular and higher water column depths are related to their small dependence on rainfall.

Although salinity has been widely described as a community constraint (e.g., Williams & Williams, 1998; Boix et al., 2007), in this case, values did not reach 5 mS/cm, described as a threshold for invertebrate communities (Frey, 1993; Boix et al., 2005). Instead, differences found in assemblage parameters (species richness (S), variation in taxonomic distinctness (Λ^+), total numerical abundance, and total biomass) could be related to the hydrological stability of each site. Being true that, in some cases, temporary ponds certainly contain a smaller number of taxa than permanent ponds (Della Bella et al., 2005). Williams (2006) has called for a review of assumptions that temporary environments support lower biodiversity. In some studies, comparing species richness of temporary and permanent environments, water permanence has not appeared to have a relevant effect on species richness (Boix et al., 2008), and some other authors have pointed out that a large number of species are well-adapted and associated with intermittent drying (e.g., Williams, 1996). On the other hand, hydroperiod lengths have been described as one of the main environmental factors influencing macrofaunal assemblages (Schneider & Frost, 1996), but the predictability of these hydroperiods has proven to be even more relevant than the length of hydroperiods themselves, with regard to the ability of animal communities to predict water fluctuations (Williams, 2006). Hydrological stability, including drying out and water level fluctuations, summarizes the physical factors that ultimately determine benthic fauna (Gascón et al., 2005). This stability seems to be related to differences between the two permanent sites and the least hydrologically stable site. Thus, hydrological stability has also probably altered the attraction and probability of colonization for macroinvertebrates, as described by other authors (Williams, 2006).

It is interesting to observe that higher levels of hydrological stability in PE and BT were not reflected in more phylogenetic evenness in the arriving taxa. A high variation in taxonomic distinctness (Λ^{+}) indicates that although more species arrived, they were not only equally distributed among taxonomic groups but also concentrated in a few taxonomic levels, each of which tended to be relatively more species-rich (Clarke & Warwick, 2001a). This may be due to the fact that short term (one year) macroinvertebrate colonization processes in newly created ponds are mainly driven by a few taxonomic groups, which is in agreement with the observed high dominance in numerical abundance of a few taxa. Whereas in PE collectors dominated, PS and BT were dominated by predators, *Chaoborus flavicans* and *Anisops sardeus*, respectively. Both predators have been described as having an important effect on community structure (Luecke & Litt, 1987; Lindholm & Hessen, 2007). The dominance of predators has been related to the first months of the hydroperiod in natural temporary ponds and also in newly created ones (Friday, 1987; Layton & Voshell, 1991; Fairchild et al., 2000), supporting the idea that assemblages in these two sites were more pioneering than those in PE. The life-history strategy group analysis is coherent with this statement: although dominant strategies presented active dispersion at all sites (which is expected for young ponds, e.g., Velasco et al., 1993; Fairchild et al., 2000), in PE, the first strategy (species with passive dispersal) had a significant presence, probably as the superficial stream made it possible for passive dispersers to reach this site.

Trophic state (using concentration of chlorophyll as a surrogate) and connectivity are factors which have also appeared to affect pioneering assemblages. The relationship between trophic state and macrofaunal assemblages' structure in the Mediterranean area has not always been strong (e.g., García-Criado et al., 2005; Boix et al., 2008). On the other hand, connectivity to natural ponds is known to be important, since colonization and extinction of aquatic insects in a water body is mainly explained by pond and landscape characteristics (e.g., Jeffries, 2005). Nevertheless, both connectivity and trophic state covaried with other previously commented factors, making the importance by themselves difficult to establish. For example, the temporary and least hydrologically stable site was also the most isolated, with less chlorophyll content and lower conductivity values. Therefore, it is interesting to remark that covariation is expressed by the high shared variability found in the variation partitioning analysis between the environmental variables and the locality factor, which would imply that each site presented several characteristic conditions that affected colonization processes differently.

Finally, the vegetation cover and the abundance of invasive species are other local factors that may determine the assemblage structure. In our case, on one hand, the high aquatic vegetation cover, important for providing refuge, and structural heterogeneity (e.g., Savage et al., 1998; Della Bella et al., 2005), must have had a strong connection with the great dominance of *Cloeon inscriptum* in the well-vegetated site, since it is an herbivore–detritivore larva (Merritt & Cummins, 1996). On the other hand, the low hydrological stability of PS may have been a constraint for the colonization of the red swamp crawfish and the Eastern mosquitofish. This has probably had consequences, since the red swamp crawfish, the main invasive species in terms of biomass, causes negative impacts on the trophic web and on assemblage biodiversity (Geiger et al., 2005; Gherardi & Acquistapace, 2007); and the introduced Eastern mosquitofish also profoundly alters the functioning of the ecosystem through trophic cascades (Blanco et al., 2004).

Taxonomic and functional approaches

Despite the literature defending the idea that taxonomic and functional approaches are complementary (Rodríguez & Magnan, 1993; Brucet et al., 2005), in this study, only a single functional parameter (total biomass) presented a significantly different response when relating it to a factor (time), while the results for the rest of the assemblage parameters coincided with the taxonomic approach. A similar result was obtained in the case of taxonomic/functional composition approaches. Although the use of functional groups has also become popular to study ecosystem processes (e.g., Gladden & Smock, 1990; Gascón et al., 2008), results provided by both approaches were redundant in our case. The functional assignations have become biased by the great dominance of a small number of species, and the functional analysis has lost a large part of its meaning by really only showing the roles of the most dominant taxa. This explanation can be added to other drawbacks associated with assigning species to groups, and with the arbitrary scale on which differences between species qualify as functionally significant (Hooper et al., 2002; Wright et al., 2006).

Conclusions

In conclusion, although a quick colonization has been observed, subsequent successional changes in the assemblage structure have not yet manifested. In addition, local factors—especially those related to hydrological stability—have highly affected both the assemblage parameters and the composition. Finally, although analyses from a functional point of view have proved to be interesting tools for studying aquatic communities, they are not appropriate in short term studies of newly created environments since the dominance of a few taxa causes them to provide practically redundant information with respect to the traditional taxonomic approach.

Acknowledgements This study was supported by a Ph.D. grant and a Scientific Research grant (CGL2008 05778/BOS) from the Ministerio de Ciencia y Tecnología of the Spanish Government. Also, the authors would like to thank the anonymous reviewers for valuable comments, Dawn Egan for the English revision, and the following field collaborators for their help: Francesc Canet, Jordi Compte, Cristina Conchillo, Helena Dehesa, Núria Pla, Martí Queralt, and Maria Mercè Vidal.

References

Abellán, P., D. T. Bilton, A. Millán, D. Sánchez-Fernández & P. M. Ramsay, 2006. Can taxonomic distinctness assess anthropogenic impacts in inland waters? A case study from a Mediterranean river basin. Freshwater Biology 51: 1744–1756.

Barnes, L. E., 1983. The colonization of ball-clay ponds by macroinvertebrates and macrophytes. Freshwater Biology 13: 561–578.

Bass, D., 1992. Colonization and succession of benthic macroinvertebrates in Arcadia Lake, a South-Central USA reservoir. Hydrobiologia 242: 123–131.

Benke, A. C., A. D. Huryn, L. A. Smock & J. B. Wallace, 1999. Length–mass relationships for freshwater macroinvertebrates in North America with particular reference to the Southeastern United States. Journal of North American Benthological Society 18: 308–343.

Blanco, S., S. Romo & M. J. Villena, 2004. Experimental study on the diet of mosquitofish (*Gambusia holbrooki*) under different ecological conditions in a shallow lake. International Review of Hydrobiology 89: 250–262.

Boix, D., J. Sala, X. D. Quintana & R. Moreno-Amich, 2004. Succession of the animal community in a Mediterranean temporary pond. Journal of the North American Benthological Society 23: 29–49.

Boix, D., S. Gascón, J. Sala, M. Martinoy, J. Gifre & X. D. Quintana, 2005. A new index of water quality assessment in Mediterranean wetlands based on crustacean and insect assemblages: the case of Catalunya (NE Iberian peninsula). Aquatic Conservation: Marine and Freshwater Ecosystems 15: 635–651.

Boix, D., J. Sala, S. Gascón, M. Martinoy, J. Gifre, S. Brucet, A. Badosa, R. López-Flores & X. D. Quintana, 2007. Comparative biodiversity of crustaceans and aquatic insects from various water body types in coastal Mediterranean wetlands. Hydrobiologia 584: 347–359.

Boix, D., S. Gascón, J. Sala, A. Badosa, S. Brucet, R. López-Flores, M. Martinoy, J. Gifre & X. D. Quintana, 2008. Patterns of composition and species richness of crustaceans and aquatic insects along environmental gradients in Mediterranean water bodies. Hydrobiologia 597: 53–69.

Brownlow, M. D., A. D. Sparrow & G. G. Ganf, 1994. Classification of water regimes in systems of fluctuating water level. Australian Journal of Marine and Freshwater Research 45: 1375–1385.

Brucet, S., D. Boix, R. López-Flores, A. Badosa, R. Moreno-Amich & X. D. Quintana, 2005. Zooplankton structure and dynamics in permanent and temporary Mediterranean salt marshes: taxon-based and size-based approaches. Archiv für Hydrobiologie 162: 535–555.

Cabral, J. A. & J. C. Marques, 1999. Life history, population dynamics and production of eastern mosquitofish, *Gambusia holbrooki* (Pisces, Poeciliidae), in rice fields of the lower Mondego River Valley, western Portugal. Acta Oecologica 20: 607–620.

Clarke, K. R. & R. N. Gorley, 2006. PRIMER v6: User Manual/Tutorial. PRIMER-E, Plymouth, UK.

Clarke, K. R. & R. M. Warwick, 2001a. A further biodiversity index applicable to species lists: variation in taxonomic distinctness. Marine Ecology Progress Series 216: 265–278.

Clarke, K. R. & R. M. Warwick, 2001b. Changes in Marine Communities: An Approach to Statistical Analysis and Interpretation, 2nd ed. PRIMER-E, Plymouth, UK.

Cottenie, K., E. Michels, N. Nuytten & L. De Meester, 2003. Zooplankton metacommunity structure: regional vs local processes in highly interconnected ponds. Ecology 84: 991–1000.

Csabai, Z. & P. Boda, 2005. Effect of the wind speed on the migration activity of aquatic insects (Coleoptera, Heteroptera). Acta Biologica Hungarica 13: 37–42.

Della Bella, V., M. Bazzanti & F. Chiarotti, 2005. Macroinvertebrate diversity and conservation status of Mediterranean ponds in Italy: water permanence and mesohabitat influence. Aquatic Conservation: Marine and Freshwater Ecosystems 15: 583–600.

Dray, S., 2004. Packfor R package v. 0.0–7. Available from: http://cran.r-project.org (accessed Jaunary 2009).

Fairchild, G. W., A. M. Faulds & J. F. Matta, 2000. Beetle assemblages in ponds: effects of habitat and site age. Freshwater Biology 44: 523–534.

Fisher, S. G., 1983. Succession in streams. In Barnes, J. R. & G. W. Minshall (eds), Stream Ecology: Application and Testing of General Ecological Theory. Plenum Press, New York.

Frey, D. G., 1993. The penetration of cladocerans into saline waters. Hydrobiologia 267: 233–248.

Friday, L. E., 1987. The diversity of macroinvertebrate and macrophyte communities in ponds. Freshwater Biology 18: 87–104.

García-Criado, F., C. Bécares, C. Fernández-Aláez & M. Fernández-Aláez, 2005. Plant-associated invertebrates and ecological quality in some Mediterranean shallow lakes: implications for the application of the EC Water. Aquatic Conservation: Marine and Freshwater Ecosystems 15: 31–50.

Gascón, S., D. Boix, J. Sala & X. D. Quintana, 2005. Variability of benthic assemblages in relation to the hydrological pattern in Mediterranean salt marshes (Empordà wetlands, NE Iberian Peninsula). Archiv für Hydrobiologie 163: 163–181.

Gascón, S., D. Boix, J. Sala & X. D. Quintana, 2008. Relation between macroinvertebrate life strategies and habitat traits in Mediterranean salt marsh ponds (Empordà wetlands, NE Iberian Peninsula). Hydrobiologia 597: 71–83.

Gee, J. H. R., B. D. Smith, K. M. Lee & S. W. Griffiths, 1997. The ecological basis of freshwater pond management for biodiversity. Aquatic Conservation: Marine and Freshwater Ecosystems 7: 91–104.

Geiger, W., P. Alcorlo, A. Baltanás & C. Montes, 2005. Impact of an introduced Crustacean on the trophic webs of Mediterranean wetlands. Biological Invasions 7: 49–73.

Gherardi, F. & P. Acquistapace, 2007. Invasive crayfish in Europe: the impact of *Procambarus clarkii* on the littoral community of a Mediterranean lake. Freshwater Biology 52: 1249–1259.

Gladden, J. E. & L. A. Smock, 1990. Macroinvertebrate distribution and production on the floodplains of two lowland headwater streams. Freshwater Biology 24: 533–545.

Grasshoff, K., M. Ehrhardt & K. Kremling, 1983. Methods of Seawater Analysis. Verlag Chemie, Weinheim.

Havel, J. E. & J. B. Shurin, 2004. Mechanisms, effects and scales of dispersal in freshwater zooplankton. Limnology and Oceanography 49: 1229–1238.

Herrmann, J., A. Boström & I. Bohman, 2000. Invertebrate colonization into the man-made Kalmar Dämme wetland dam system. Verhandlungen/Internationale Vereinigung für theoretische und angewandte Limnologie 27: 1653–1656.

Higgins, M. J. & R. W. Merritt, 1999. Invertebrate seasonal patterns and trophic relationships. In Batzer, D., R. B. Rader & S. A. Wissinger (eds), Invertebrates in Freshwater Wetlands of North America. Wiley, NewYork: 279–297.

Hillman, T. J. & G. P. Quinn, 2002. Temporal changes in macroinvertebrate assemblages following experimental flooding in permanent and temporary wetlands in an Australian floodplain forest. River Research and Applications 18: 137–154.

Hooper, D. U., M. Solan, A. Symstad, S. Díaz, M. O. Gessner, N. Buchmann, V. Degrange, P. Grime, F. Hulot, F. Mermillod-Blondin, J. Roy, E. Spehn & L. van Peer, 2002. Species diversity, functional diversity and ecosystem functioning. In Loreau, M., S. Naeem & P. Inchausti (eds), Biodiversity and Ecosystem Functioning: Synthesis and Prespectives. Oxford University Press, Oxford: 195–208.

Jeffries, M., 2005. Local-scale turnover of pond insects: intrapond habitat quality and inter-pond geometry are both important. Hydrobiologia 543: 207–220.

Koutrakis, E. T. & A. C. Tsikliras, 2003. Length-weight relationships of fishes from three northern Aegean estuarine systems (Greece). Journal of Applied Ichthyology 19: 258–260.

Layton, R. J. & J. R. Voshell Jr., 1991. Colonization of new experimental ponds by benthic macroinvertebrates. Environmental Entomology 20: 110–117.

Lindegaard, C., 1992. Zoobenthos ecology of Thingvallavatn: vertical distribution, abundance, population dynamics and production. Oikos 64: 257–304.

Lindholm, M. & D. O. Hessen, 2007. Competition and niche partitioning in a floodplain ecosystem: a cladoceran community squeezed between fish and invertebrate predation. African Zoology 42: 158–164.

Luecke, C. & A. H. Litt, 1987. Effects of predation by *Chaoborus flavicans* on crustacean zooplankton of Lake Lenore, Washington. Freshwater Biology 18: 1185–1192.

Malmqvist, B., S. Rundle, C. Brönmark & A. Erlandsson, 1991. Invertebrate colonization of a new, man-made stream in southern Sweden. Freshwater Biology 26: 307–324.

Merritt, R. W. & K. W. Cummins, 1996. Aquatic Insects of North America. Kendal/Hunt Publishing Company, Dubuque, Iowa.

Meyer, E., 1989. The relationship between body length parameters and dry mass in running water invertebrates. Archiv für Hydrobiologie 117: 191–203.

Montes, C., M. A. Bravo-Utrera, A. Baltanás, C. Duarte & P. J. Gutierrez-Yurrita, 1993. Bases ecológicas para la gestión del cangrejo rojo en el Parque Nacional de Doñana. ICONA-Technical Report, Madrid, Spain.

National Research Council, 1992. Restoration of Aquatic Ecosystems. National Academy Press, Washington, DC.

Oertli, B., J. Biggs, R. Céréghino, P. Grillas, P. Joly & J.-B. Lachavanne, 2005. Conservation and monitoring of pond biodiversity: introduction. Aquatic Conservation: Marine and Freshwater Ecosystems 15: 535–540.

Oksanen, J., R. Kindt, P. Legendre & R. B. O'hara, 2005. Vegan: community ecology package, v. 1.7-81. Available from: http://cran.r-project.org (accessed Jaunary 2009).

Quintana, X. D., 1995. Relaciones entre el peso y la longitud en *Aedes*, *Culex* y *Gammarus*. Limnética 11: 15–17.

Quintana, X. D., S. Brucet, D. Boix, R. López-Flores, S. Gascón, A. Badosa, J. Sala, R. Moreno-Amich & J. J. Egozcue, 2008. A non-parametric method for the measurement of size diversity, with emphasis on data standardization. Limnology and Oceanography: Methods 6: 75–86.

Rodríguez, M. A. & P. Magnan, 1993. Community structure of lacustrine macrobenthos: do taxon-based and size-based approaches yield similar insights? Canadian Journal of Fisheries and Aquatic Sciences 50: 800–815.

Savage, A. A., J. H. Mathews & D. L. Beaumont, 1998. Community development in the benthic macroinvertebrate fauna of a lowland lake, Oak Mere, from 1994 to 1996. Archiv für Hydrobiologie 143: 295–305.

Schneider, D. W. & T. M. Frost, 1996. Habitat duration and community structure in temporary ponds. Journal of the North American Benthological Society 15: 64–86.

Smit, H., E. D. Van Heel & S. Wiersma, 1993. Biovolume as a tool in biomass determination of Oligochaeta and Chironomidae. Freshwater Biology 29: 37–46.

Smock, L. A., 1980. Relationships between body size and biomass of aquatic insects. Freshwater Biology 10: 375–383.

Talling, J. F., & D. Driver, 1963. Some problems in the estimation of chlorophyll a in phytoplankton. In Proceedings of a Conference on Primary Productivity Measurements, Marine and Freshwater, University of Hawaii, Honolulu, 1961. US Atomic Energy Commission TID-7633: 142–146.

Traina, J. A. & C. N. Ende, 1992. Estimation of larval dry weight of *Chaoborus americanus*. Hydrobiologia 228: 219–223.

Velasco, J., A. Millán & L. Ramírez-Díaz, 1993. Colonización y sucesión de nuevos medios acuáticos II. Variación temporal de la composición y estructura de las comunidades de insectos. Limnética 9: 73–85.

Verberk, W. C. E. P., H. Siepel & H. Esselink, 2008. Life-history strategies in freshwater macroinvertebrates. Freshwater Biology 53: 1722–1738.

Warwick, R. M. & K. R. Clarke, 1995. New 'biodiversity' measures reveal a decrease in taxonomic distinctness with increasing stress. Marine Ecology Progress Series 129: 301–305.

Wiggins, G. B., R. J. Mackay & I. M. Smith, 1980. Evolutionary and ecological strategies of animals in annual temporary pools. Archiv für Hydrobiologie supplement 58: 97–206.

Williams, D. D., 1996. Environmental constraints in temporary fresh waters and their consequences for insect fauna. Journal of the North American Benthological Society 15: 634–650.

Williams, D. D., 2006. The Biology of Temporary Waters. Oxford University Press, Oxford.

Williams, D. D. & N. E. Williams, 1998. Aquatic insects in an estuarine environment: densities, distribution and salinity tolerance. Freshwater Biology 39: 411–421.

Wright, J. P., S. Naeem, A. Hector, C. Lehman, P. B. Reich, B. Schmid & D. Tilman, 2006. Conventional functional classification schemes underestimate the relationship with ecosystem functioning. Ecology Letters 9: 111–120.

Hydrobiologia (2009) 634:153–165
DOI 10.1007/s10750-009-9900-z

POND CONSERVATION

Comparison of macroinvertebrate community structure and driving environmental factors in natural and wastewater treatment ponds

G. Becerra Jurado · M. Callanan · M. Gioria · J.-R. Baars · R. Harrington · M. Kelly-Quinn

Published online: 29 July 2009

Abstract Eutrophication still continues to be an issue of major concern for the protection of water quality, and accordingly, the European Union Water Framework Directive has set a minimum target for all waters where "good status" is defined as a slight departure from the biological community which would be expected in conditions of minimal anthropogenic impact. The use of constructed ponds for wastewater treatment aimed at achieving this target has shown to be an effective alternative to conventional systems in the farm landscape. Their applicability in these areas is of great interest since these ponds have the added potential to combine their wastewater treatment properties with that of biodiversity enhancement. This article focuses on exploring the community structure of both natural and constructed ponds used for wastewater treatment and the driving environmental factors. A total of 15 constructed and 5 natural ponds were sampled for aquatic macroinvertebrates and hydrochemistry in spring and summer 2006. Results showed that the most important factors responsible for the differences in the community structure between these two types of ponds were pH, vegetation structure and pollution levels. These gradients helped to structure a large proportion of the communities with some taxa being associated with the constructed ponds. These results highlight the potential contribution of constructed ponds used for wastewater treatment to the landscape biodiversity. The present findings also open the possibility for a more integrated management of water quality and biodiversity enhancement in farmland areas.

Keywords Macroinvertebrates · Community structure · Constructed ponds · Wastewater treatment · Farmland areas · Wetlands

Electronic supplementary material The online version of this article (doi:10.1007/s10750-009-9900-z) contains supplementary material, which is available to authorized users.

Guest editors: B. Oertli, R. Cereghino, A. Hull & R. Miracle
Pond Conservation: From Science to Practice. 3rd Conference of the European Pond Conservation Network, Valencia, Spain, 14–16 May 2008

G. Becerra Jurado (✉) · M. Callanan · M. Gioria · J.-R. Baars · M. Kelly-Quinn
Freshwater Biodiversity, Ecology and Fisheries (FreBEF), School of Biology and Environmental Science, University College of Dublin, Belfield, Dublin 4, Ireland
e-mail: gustavo.becerra@ucdconnect.ie

R. Harrington
Water and Policy Division, Department of Environment, Heritage and Local Government, Old Custom House, 106 The Quay, Waterford, Ireland

Introduction

The protection of water quality and the conservation of freshwater biodiversity have become increasingly important worldwide. The European Union Water

Framework Directive (EU 2000/86/EEC) has set a minimum target of good status for all water bodies to protect water resources and ecosystems. In order to achieve this objective, it is of crucial importance that effective measures are put in place to address both diffuse and point sources of pollution. Eutrophication from diffuse nutrient inputs is probably the key issue affecting water resources throughout the EU. In Ireland, the loss of nutrients from agricultural practices is responsible for the majority of diffuse pollution (Tunney et al., 1997; Withers et al., 2000; McGarrigle, 2005).

Constructed wetlands can be used for the treatment of various waste types. Among those built to date in Ireland, integrated constructed wetlands (ICWs) exhibit the singularity of being formed by a number of interconnected ponds (Scholz et al., 2007). These systems are based on the "small watershed technique" (Bormann & Likens, 1981) and have been proven to be effective at treating wastewater, mainly from farms (Dunne et al., 2005). Quantitative approaches have been used to construct these engineered systems, which depend on the presence of emergent vegetation for their functioning, to enhance their purifying capacity (Mitsch & Jorgensen, 1989). Furthermore, ICWs have the potential to fully integrate effluent treatment and management properties with that of aquatic macroinvertebrate diversity enhancement in the farmland landscape (Scholz et al., 2007).

In Europe, ponds have been found to significantly contribute to regional diversity (Brönmark & Hansson, 2002; Oertli et al., 2004; Williams et al., 2004) and may host more rare species than other waterbody types (e.g. Biggs et al., 1994; Davies et al., 2008). A number of studies have shown that the creation and maintenance of a wide range of ponds can be especially useful for the enhancement of gamma biodiversity (e.g. Biggs et al., 1994). One of the advantages of focusing on ponds to conserve and enhance biodiversity is linked with their relatively small area compared to other waterbodies in the landscape, which makes their protection more feasible (Davies et al., 2008). What is more, Céréghino et al. (2008) suggested that manmade, agricultural ponds can greatly contribute to landscape biodiversity. However, ponds treating agricultural wastewater were not considered.

Few studies have investigated the environmental variables structuring the aquatic communities of permanent natural ponds and the available results are variable. Biggs et al. (2005) showed that area, connectedness, pH and abundance of vegetation were the most important environmental variables correlated with a number of species in UK ponds. On the other hand, Oertli et al. (2008) suggested that connectivity was highly important in structuring the communities in alpine ponds. Other important factors that affect the abundance and number of species include the presence of fish, as shown by Fairchild et al. (2000). Despite the ecological and conservation role of ponds, there is a paucity of knowledge regarding how these and other environmental factors, such as size, hydrology, type of vegetation and physicochemical characteristics may structure aquatic macroinvertebrate diversity in constructed ponds. Little information is available on whether there are differences between macroinvertebrate communities in constructed and natural ponds, and on the factors determining such differences. To date, only a few studies have explored the macroinvertebrate communities of constructed wetlands and the driving environmental factors (Spieles & Mitsch, 2000; Fairchild et al., 2000; Balcombe et al., 2005). Becerra et al. (2008) suggested in a previous study on ICWs that vegetation structure, pond profile and pollution levels were the most important environmental factors within ICW ponds. This study showed that the last ponds of the systems were deeper and were characterised by lower pollution levels and less emergent vegetation cover. However, the study solely focused on ICW ponds, positioned at different stages of the water treatment process and did not make comparative analyses with natural ponds.

We investigate the effects of a number of environmental factors that may structure macroinvertebrate communities and taxon composition in both natural and constructed ponds used for treating agricultural wastewater. Understanding these driving factors will be of benefit to managers involved in enhancing freshwater biodiversity in ponds within similar agricultural landscapes.

Methods

Study area

A total of 15 ICWs and 5 natural ponds were studied in southeast Ireland (Fig. 1; Table 1). The ICW ponds sites were located within the Annestown River Catchment, Co. Waterford, Ireland. This catchment has an approximate area of 25 km^2. The geology is complex but mostly consists of igneous rock with metamorphic deposits. In this area, 75% of all farmyard dirty water generated within the watershed is treated by the ICWs. All ponds were created in 2000, and the sampled ponds correspond to the last three ponds in the chain. No large fish species are known to populate the ICW ponds.

The five natural ponds were located in the surrounding area. Their geology is more varied but is mostly formed by limestone and mudstone. These ponds were considered to be representative of the area and un-impacted by effluent inputs. Three of the ponds were spring fed (Fennor, Wexford and Hoare Abbey) and the other two were fed by rivers (Kildalton and Mobarnane House). The natural ponds were situated in pasture areas used for grazing cattle or sheep, except for the pond at Mobarnane House. This pond was created around one hundred years ago when a stream was diverted and the pond was used as a landscape feature for its ornamental and aesthetic values on the grounds of the estate. Unlike the ICWs, it is not known if large fish species inhabit these natural ponds. Both natural and constructed ponds had neutral to alkaline pH, ranging from 6.81 to 8.64.

Sampling methods and statistical analysis

Macroinvertebrate sampling was conducted both in spring and summer (March–April and July–August) 2006. All ponds, except the first one (one or two ponds, depending on the system), in each of the five ICW systems were sampled. The first ponds in each system had heavily polluted conditions that would have compromised health and safety standards due to the presence of untreated sewage and animal slurry. Three 3-min multihabitat sweep samples were collected from the ponds. A standard 1-mm pond net (frame size 20 × 25 cm) was used. Pond netting consisted of vigorous sweeping through the upper part of the water column. The 3-min period was proportionally allocated to the different mesohabitats according to the area they covered within each pond (Biggs et al., 1998). Sampling of the Hoare Abbey pond changed slightly between seasons. The area of this pond was significantly reduced in summer (<3 m in diameter) after an atypical severe drought. Sampling on that occasion was restricted to one 15-s pond net sample.

The sweep samples were placed in plastic bags and preserved with 70% industrial methylated spirits (IMSs). The macroinvertebrates from the activity trap samples were directly decanted into a 250-μm sieve. The contents of the sieve were then backwashed with

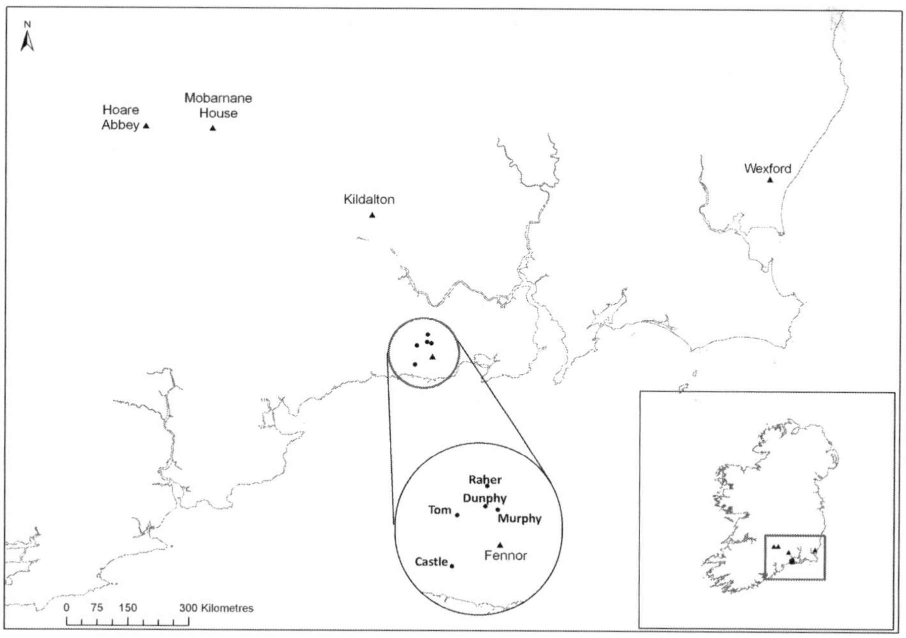

Fig. 1 Map of Ireland showing pond locations. *Filled circle* ICW ponds and *filled triangle* natural ponds

Table 1 Names, locations and underlying geology of natural and ICW ponds

Owner/site name	County	Code	Grid reference	Ponds sampled	Geology
ICWs					
Anthony Raher	Waterford	Raher	E252412, N105554	R1, R2[a] and R3[b]	Intermediate volcanics
Castle Farm	Waterford	Castle	E250309, N100578	C1, C2 and C3	Dark blue-grey, green and purple slates
Edmun Dunphy	Waterford	Dunphy	E252300, N104300	D1, D2[a] and D3	Rhyolitic volcanics, grey and brown slates
Tom Murphy	Waterford	Tom	E250600, N103750	T1[a], T2 and T3	Rhyolitic volcanics, grey and brown slates
Milo Murphy	Waterford	Milo	E253050, N104087	M1, M2 and M3	Rhyolitic volcanics, grey and brown slates
Natural ponds					
Kildalton College, Piltown	Waterford	Kildalton	E243200, N125600	1 pond[b]	Massive unbedded lime-mudstone
Mobarnane House, Fethard	Tipperary	Mobarnane House	E216700, N140090	1 pond[b]	Wackestone/packstone limestone
Fennor bog, Waterford	Waterford	Fennor	E253200, N101950	1 pond	Felsic volcanics
Ballinra, Wexford	Wexford	Wexford	E309600, N131850	1 pond	Grey-green greywackes and slates
Hoare Abbey, Cashel	Tipperary	Hoare Abbey	E206200, N140040	1 pond	Pale-grey bedded limestone with chert

Numbers for ponds refer to pond order along the treatment process, where 1 = first sampled pond, 3 = last pond in the treatment

[a] Pond dried out during summer

[b] River fed pond

the IMS solution into labelled plastic bag. In the laboratory, all macroinvertebrates were sorted from the samples, except for *Asellus* sp. which was subsampled due to the high numbers present. Macroinvertebrates were identified to species level where possible using the keys Macan & Cooper (1977), Elliott & Mann (1979), Friday (1988), Elliott et al. (1988), Savage (1989), Wallace et al. (1990), Edington & Hildrew (1995) and Nilsson (1996), (1997). Dipteran larvae were only identified to family level. Oligochaeta, Hydracarina, Aeshnidae, Coenagrionidae and Piralidae were considered as one taxon each.

The physico-chemistry in the ICW ponds was monitored for a few months prior to the macroinvertebrate sampling period, i.e. from August 2005 to August 2006. Natural ponds were sampled on two occasions during this period (9 March and 14 August, 2006). Standard analysis methods were applied as per APHA (1990).

A number of physico-chemical variables were included in the analyses (Table 2). Variables were subjected to the forward selection procedure using the program CANOCO with 999 permutations ($P < 0.2$) (ter Braak & Šmilauer, 2002). Canonical correspondence analysis (CCA) was undertaken using the selected variables to determine which environmental factors were responsible for the relationships between taxa composition and the physico-chemical variables (Seaby et al., 2004). The most important environmental variables were highlighted using correspondence analysis (CA) with 1,000 iterations. Further analyses were carried out to test the significance in community composition between the first and the last constructed ponds, as well as the last constructed ponds in the chain and the natural ponds. These analyses were conducted using PRIMER 6 (Clarke & Gorley, 2006) and PERMANOVA+B version (Anderson & Gorley, 2007). Similarity percentage analysis (SIMPER) (Clarke, 1993) was then used to identify the percentage contribution of each taxon to any observed differences in community composition.

Table 2 Physico-chemical parameters included in the CCA analyses

Environmental variable	Unit and/or definition
Easting	Six digit coordinate
Northing	Six digit coordinate
Distpond	Distance to closest pond (km)
Area	m^2
Water20	% area of pond with water depth between 0 and 20 cm
Water50	% area of pond with water depth between 20 and 50 cm
Water100	% area of pond with water depth greater than 50 cm
Water flow	Categorical values (still or flow)
TerrVeg	% perimeter of pond surrounded by shrubs and trees
Submerged	% area of pond covered with submerged vegetation
Emergent	% area of pond covered with emergent vegetation
Floating	% area of pond covered with floating vegetation
pH	Adimensional
Conductivity	$\mu S\ cm^{-1}$
BOD_5	$mg\ l^{-1}$ O_2 after 5-day incubation at 20°C
DO	$mg\ l^{-1}$ Dissolved oxygen
Ammonia	$mg\ l^{-1}$ N
Nitrate	$mg\ l^{-1}$ NO_3
Molybdate reactive phosphorus (MRP) Phosphorus	$mg\ l^{-1}$ P

Results

A total of 151 taxa were recorded from all ponds, of which 123 taxa were found in the ICW ponds and 115 taxa in the natural ponds. The same number of taxa (115) were found in the last ICW ponds. In both types of ponds (natural vs. constructed), the community was dominated by Coleoptera (35% in natural and 45% in constructed ponds) and Hemiptera (22 and 17%, respectively).

Results showed that, in general, pH and MRP seemed to explain the differences between natural and constructed ponds in terms of their community structure as indicated in the CCA biplot (Fig. 2). Using the full data set, the three most important variables highlighted and the variance explained by the analysis were: pH (14.5%), MRP (12.0%) and Wat50 (10.7%) (Table 3). The model generated using CA highlighted three variables: pH, MRP and emergent vegetation, together accounting for 31.9% of the variance. While the gradient in pH seemed to explain the differences between natural and constructed ponds, the gradient in MRP generally did so for the differences along the treatment process. It is important to note that the geographical distance between studied ponds did not seem to explain the differences between ponds. Also interestingly, the last ponds in the water treatment process were closely oriented to the natural ponds. In fact, pond T3 had similar environmental characteristics as the natural ponds.

In addition, the gradients in pH and MRP seemed to structure a large proportion of the community (Fig. 3). Some taxa were associated with the constructed ponds, which were characterised by relatively low pH and high MRP values. These were: *Asellus* sp., all Diptera taxa, Oligochaeta, Tricladida, the Mollusca *Pisidium/Sphaerium* spp., *Lymnaea stagnalis* (Linnaeus), *Physa acuta* (Draparnaud), *Succinea putris* (Linnaeus), *Zenitoides nitidus* (Müller) and *Aplexa hypnorum* Linnaeus, as well as Coleoptera larvae. On the other hand, the Trichoptera *Glyphotaelius pellucidus* (Retzius), *Halesus radiatus* (Curtis), *Triaenodes biocolor* (Curtis), *Adicella reducta* (McLachlan), *Agraylea sexmaculata* (Curtis) and *Agraylea multipunctata* (Curtis), as well as all Hemiptera, Ephemeroptera and Plecoptera were associated with natural ponds, characterised by relatively high pH and low MRP values.

Fig. 2 CCA biplot of environmental and pond relationships. *Filled circle* ICW ponds and *filled triangle* natural ponds. Variance accounted for by the canonical axes = 27.6%

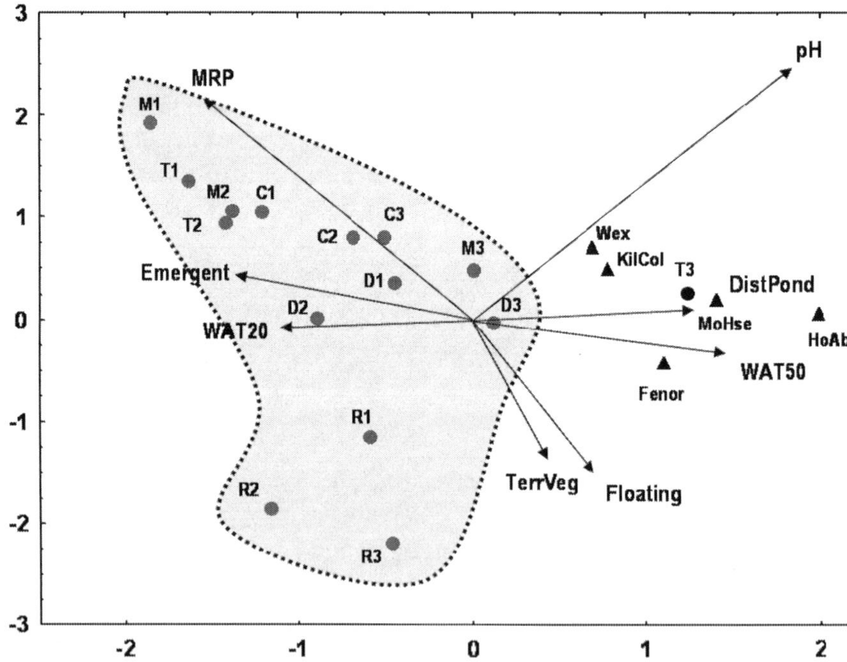

What is more, the community composition from the first and the last constructed ponds in the chain differed significantly (PERMANOVA, $F_{1,8} = 4.18$, $P < 0.01$), as well as that from the last constructed ponds and the natural ponds (PERMANOVA, $F_{1,8} = 2.55$, $P < 0.01$). The analysis of the contribution of each taxon to the observed differences among the constructed ponds showed that *Noterus clavicornis* (Degeer), Aeshnidae, Coenagrionidae, Tricladida, *Hydrometra stagnorum* (Linnaeus), *Gerris lacustris* (Linnaeus), Gerris sp. immature, *Helocentropus picicornis* (Stephens), *Nepa cinerea* Linnaeus and Psychodidae contributed most to the differences (16.49% contribution) (Appendix 1a—Supplementary Material). All but Psychodidae were associated with the last ponds. Besides, the analysis of the contribution of each taxon to the observed differences between the constructed ponds (last in the chain) and the natural ponds revealed that Piralidae, *H. stagnorum*, *Limnephilus marmoratus* Curtis, Psychodidae, *Hydroporus striola* (Gyllenhal), *Hydroporus angustatus Sturm*, *Sigara nigrolineata* (Fieber), *Hyphydrus ovatus* (Linnaeus), *Ilybius quadriguttatus* (Lacordaire and Boisduval), Corixidae immature and *Lymnaea peregra* (Müller) contributed most to the differences (15.37% contribution) (Appendix 1b—Supplementary Material). Among them, *H. stagnorum*, *L. marmoratus*, *H. striola*, *H. angustatus*, *I. quadriguttatus*, Corixidae immature and *L. peregra* were associated with the last ponds of the ICW systems.

Since the beetles were the most represented group, they were extracted from the full data set for analysis of the environment–species correlation. Results showed that the gradients in pH, as well as emergent vegetation and Wat20 generally structured these communities (Fig. 4). Here, also the pH gradient seemed to account for the differences between natural and constructed ponds, while the emergent vegetation and Wat20 gradient seemed to explain the differences along the water treatment process. However, this time the distance to the closest pond seemed to explain some of the variance. In fact, results showed that pH (12.0%), emergent vegetation (9.6%) and distance to closest pond (9.0%) were the variables that explained most of the variance (Table 4). On this occasion, the model generated using CA highlighted pH, emergent vegetation and conductivity, together accounting for 29.7% of the variance. The Coleoptera taxa associated with the constructed ponds were *Hydroporus* sp., *Cercyon* sp., *Laccobius* sp., *Anacaena* sp. and *Enochrus* sp. On the

Table 3 Canonical analysis (CA) results for the whole taxa

Variables included in the model	Variance	% of variance
Variable (all variables in model)		
pH	0.2338	14.4846
MRP	0.1945	12.0498
WAT50	0.1730	10.7178
Emergent	0.1680	10.4081
DistPond	0.1616	10.0116
TerrVeg	0.1392	8.6238
WAT20	0.1354	8.3884
Floating	0.1176	7.2857
Easting	0.1124	6.9635
Northing	0.1094	6.7776
The single best variable = pH		
Variable (with best variable in model = pH)		
MRP	0.3938	24.3970
TerrVeg	0.3711	22.9907
Emergent	0.3660	22.6747
Floating	0.3517	21.7888
WAT20	0.3492	21.6339
WAT50	0.3477	21.5410
DistPond	0.3438	21.2994
Easting	0.3336	20.6675
Northing	0.3315	20.5374
Two best variable = pH and MRP		
Variable (with 2 best variables in model = pH and MRP)		
Emergent	0.5158	31.9553
TerrVeg	0.5108	31.6455
WAT20	0.5104	31.6207
WAT50	0.4999	30.9702
DistPond	0.4994	30.9392
Northing	0.4983	30.8711
Easting	0.4959	30.7224
Floating	0.4868	30.1586
The 3 best variables = pH, MRP and Emergent		

Total variance = 1.6141

other hand, the Coleoptera taxa of the family Hydroporinae (other than *Hydroporus* sp.) and *Haliplus* sp. were associated with the natural ponds (Fig. 5).

As for the whole taxa, the Coleoptera community composition from the first and the last constructed ponds in the chain differed significantly (PERMANOVA, $F_{1,8} = 2.40$, $P < 0.05$). Likewise, significant differences were found between the last constructed ponds and the natural ponds (PERMANOVA, $F_{1,8} = 1.83$, $P < 0.05$). *N. clavicornis*, *Laccophilus minutus* (Linnaeus), *Porhydrus lineatus* (Fabricius), *Haliplus ruficollis* (Degeer) and *I. quadriguttatus* contributed the most to the observed differences between the first and the last constructed ponds (17.43% contribution) (Appendix 2a—Supplementary Material), while *H. striola*, *H. ovatus*, *H. angustatus*, *I. quadriguttatus* and *Hydroporus planus* (Fabricius) did so for the differences between the constructed and natural ponds (16.48% contribution) (Appendix 2b—Supplementary Material). All taxa but *H. ovatus* were associated with the last constructed ponds.

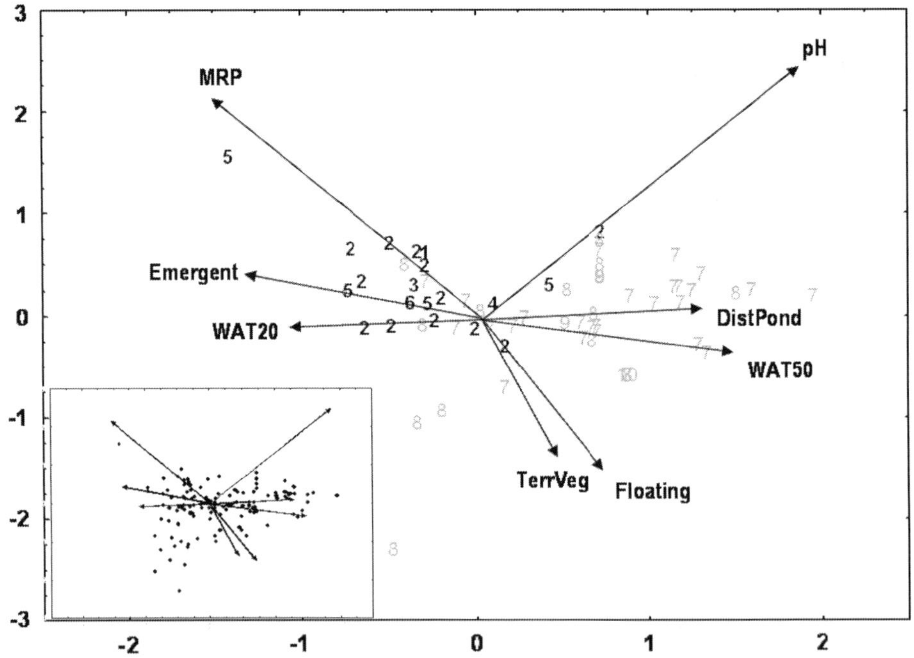

Fig. 3 CCA biplot of environmental and taxa relationships. Where 1 = Asellus sp., 2 = Diptera, 3 = Oligochaeta, 4 = Tricladida, 5 = Pisidium/Sphaerium sp., *Lymnaea stagnalis*, *Physa acuta*, *Succinea putris*, *Zenitoides nitidus* and *Aplexa hypnorum*, 6 = Coleoptera larvae 7 = Hemiptera, 8 = Trichoptera, 9 = Ephemeroptera and 10 = Plecoptera. Variance accounted for by the canonical axes = 27.6%

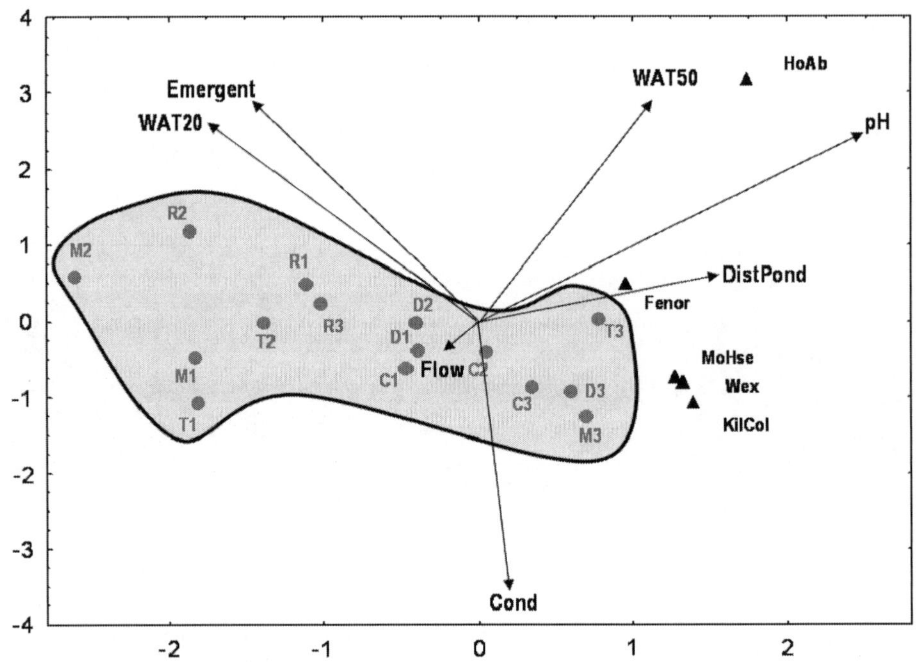

Fig. 4 CCA biplot of environmental and pond relationships for Coleoptera taxa. *Filled circle* ICW ponds and *filled triangle* natural ponds. Variance accounted for by the canonical axes = 24.3%

Discussion

The CCA showed that the constructed ponds were generally positioned in the plot according to their order in the water treatment process. The first sampled ponds were characterised by high MRP values and were generally shallower than the last constructed ponds. This pond sequence was followed in the plot by the natural ponds. The varied history and catchment characteristics of natural ponds generally give rise to a higher variability. Natural ponds normally have small catchments, and any variation in

Table 4 Canonical analysis (CA) results for the Coleoptera taxa

Variables included in the model	Variance	% of variance
Variable (all variables in model)		
pH	0.2291	11.9988
Emergent	0.1844	9.6577
DistPond	0.1729	9.0554
Cond	0.1559	8.1650
WAT20	0.1526	7.9922
WAT50	0.1435	7.5156
FLOW	0.1349	7.0652
The single best variable = pH		
Variable (with best variable in model = pH)		
Emergent	0.3998	20.9389
Cond	0.3881	20.3262
WAT20	0.3652	19.1268
FLOW	0.3633	19.0273
DistPond	0.3584	18.7707
WAT50	0.3491	18.2836
Two best variable: pH and emergent		
Variable (with two best variables in model: pH and Emergent)		
Cond	0.5677	29.7325
DistPond	0.5218	27.3285
FLOW	0.5193	27.1976
WAT50	0.5171	27.0824
WAT20	0.5037	26.3806
The three best variables: pH, Emergent and Cond		

Total variance 1.9094

their surrounding areas will be reflected in the character of these waterbodies (Biggs et al., 2005). For this reason, the ICW ponds were expected to have more uniform environmental characteristics compared to the natural ponds since their design is specific for water treatment (Scholz et al., 2007). However, results revealed that the characteristics exhibited by the ICW ponds varied widely. This may be explained by the fact that while the main function of the first ponds is to treat the wastewater load through microbial respiration, the last ponds in the system were essentially created to dilute the contaminants still present in the water column.

The CCA also revealed that the most important factors responsible for the differences in the whole community structure between natural and constructed ponds were pH, MRP, emergent vegetation, pond profile (Wat20 and Wat50) and the distance to the closest pond. Spieles & Mitsch (2000) also found a pH gradient along the treatment process in high- and low-nutrient constructed wetlands. These wetlands received wastewater that had previously been treated in a secondary water treatment plant. Results from this study showed that both soluble reactive phosphorus and pH were useful predictors of a number of invertebrate measures (e.g. % tolerant organisms). These results are highly comparable since the ponds sampled in the present study were the last three ponds in the chain of a total of 4 + and waters entering these three ponds generally exhibit lowered pollution values (Scholz et al., 2007). Even though the pH values for all ponds were within the circumneutral and alkaline range (6.81–8.64), it is fairly well known that pH (or the related measures such as alkalinity) strongly structures macroinvertebrate communities in freshwater habitats (e.g. Feldman & Connor, 1992) including lakes (e.g. WGCRA, 2005).

Despite the complex geology that can be found in southeast Ireland, all ICW ponds of the present study were situated in similar geological areas. One possible explanation for the pH gradient within the ICW ponds is that, as a consequence of microbial

Fig. 5 CCA biplot of environmental and Coleoptera taxa relationships. Where 1 = *Hydroporus* sp., 2 = *Cercyon* sp., 3 = *Laccobius* sp., 4 = *Anacaena* sp., 5 = *Enochrus* sp, 6 = *Helophorus* sp., 7 = *Gyrinus* sp., 8 = *Laccophilus* sp., 9 = Hydroporinae (other than *Hydroporus* sp.) and 10 = *Haliplus* sp. Variance accounted for by the canonical axes = 24.3%

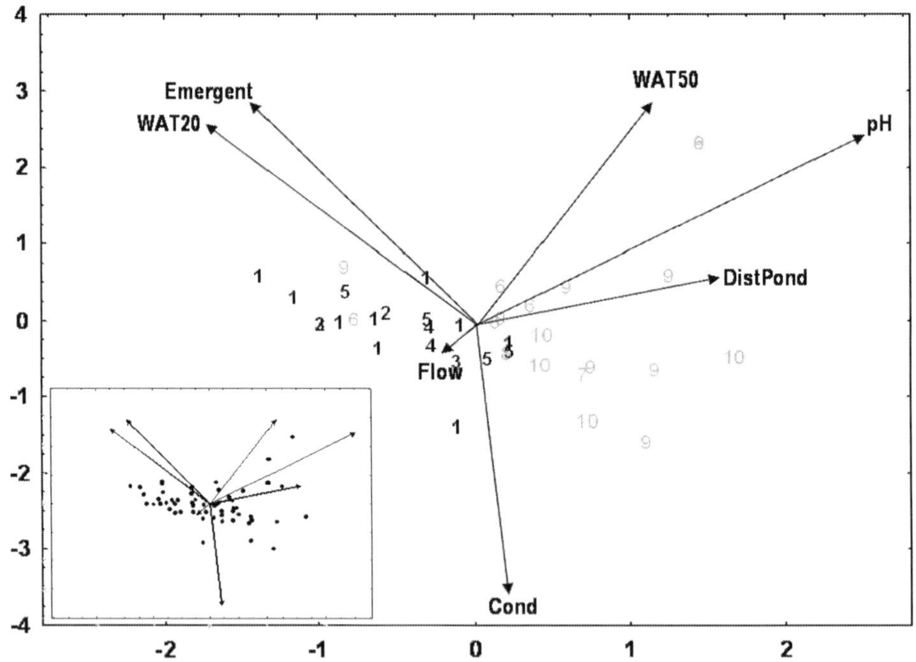

respiration processes, more carbon dioxide is added to the water in the first constructed ponds. This process is known to free more H^+ ions in the water leading to a lower pH (Brönmark & Hansson, 1998). Microbial respiration is likely to decrease from the first to the last pond and this is reflected in the slight increase in pH. Interestingly, when comparing natural and constructed ponds, the former recorded the highest pH values of all ponds. The fact that these ponds were generally underlain by a different local geology (limestone and mudstone) may explain the pH to be higher. In addition, some of the natural ponds are spring fed. This may have caused the pH to increase as water is in contact with the underlying geology for a longer period of time (WGG, 2004). A more comprehensive study would be necessary to accept or refute these hypotheses.

In terms of MRP, highlighted as differentiating the communities of the natural and constructed ponds, the results reflect the relatively higher pollution levels in the ICW ponds compared to natural ponds. Furthermore, MRP particularly differentiated the ICW pond communities along the water treatment process. Organic pollution is a factor that has been widely studied in rivers and there are numerous examples in the literature (e.g. Hawkes, 1998; Kelly-Quinn & Bracken, 2000; McGarrigle et al., 2002). These studies show that some sensitive taxa are not able to complete their life cycles under such conditions. However, it is worth noting that even though natural ponds have also been the subject of such studies and certain taxa were shown to respond negatively to organic pollution (e.g. Menetrey et al., 2008), some natural ponds may act as important natural sinks (Davies et al., 2008) with an increased capacity for receiving nutrients. In this respect, ICW ponds act as purposely built sinks of nutrients, particularly in the first ponds of the systems, where most emergent vegetation is planted. It is estimated that a mean rate of ~3 cm of sediments is accumulated per year and that sediment removal is required every 10 years from the first ponds. In contrast, it is estimated that sediment removal is required every 20–60 years from the subsequent ponds in the chain (Scholz et al., 2007).

Not surprisingly, emergent vegetation was also highlighted in the analyses and was closely related to the pond profile. A similar finding was reported by De Szalay & Resh (2008) as influencing macroinvertebrate colonisation in seasonal wetlands. In particular, a number of dipteran families were positively correlated with the amount of emergent cover in their study. On the other hand, Corixidae, Chironomidae and Hydrophilidae were negatively correlated. In this respect, Parsons & Matthews (1995) also showed associations between macroinvertebrates and macrophytes, and linked the types of these (submerged vs. emergent) with a number of macroinvertebrates. In the present study, emergent vegetation seemed to explain the differences in both the whole taxa and the beetle community composition. Ponds characterised by high emergent vegetation cover were also the shallowest and most gently sloped of all, as indicated by Wat20 (percentage of the pond with water depth <20 cm). These results are similar to those of Nilsson et al. (1994), who suggested that the abundance and species richness of Dytiscidae were correlated with the amount and complexity of vegetation, as well as shore slope in a number of Swedish lakes.

A factor that also had an effect on the results was pond connectivity. Indeed, the connectivity between ponds may have an important effect on pond community structure since some macroinvertebrates require the presence of other waterbodies in the surrounding area for their aerial dispersal (Anderson & Smith, 2004; Briers & Biggs, 2005). In general, newly constructed ponds are known to be quickly colonised (Balcombe et al., 2005; Williams et al., 2008), with an estimation of around 5 years to reach a plateau for macroinvertebrate colonisation (Hansson et al., 2005). The proximity of other ponds to the newly created ponds may greatly facilitate this (Nelson et al., 2000). It is worth noting that it was only when the Coleoptera data were analysed that this factor was highlighted in the results. It is, therefore, reasonable to suggest that existing natural ponds in the ICW study area, e.g. Fennor pond, may have acted as a source of Coleoptera colonisers. Moreover, the high connectivity within any ICW system may have an effect on structuring the Coleoptera communities.

Of particular interest are the taxa found in constructed farm ponds since they can greatly contribute to landscape biodiversity as shown by Céréghino et al. (2008). Results for ICW ponds showed that the community composition present in these ponds differed when compared to other ponds. In general, the coleopterans *Hydroporus* sp., *Cercyon* sp., *Laccobius* sp., *Anacaena* sp. and *Enochrus* sp, a number of Hemiptera, Trichoptera and Mollusca, as well as the Oligochaeta, *Asellus* sp. and all dipterans were mostly found in these ponds. The pH values obtained in the constructed ponds were within the range necessary to maintain a viable population for these taxa (Nilsson, 1996, 1997).

Among the taxa found in the constructed ponds, some were associated with the first ponds. This was the case of *Asellus* sp., Oligochaeta and the Dipteran family Phsycodidae. It is clear that the high organic pollution levels may provide the necessary conditions for these taxa to thrive. For example, the larvae of Phsycodidae normally live in organic sludge (drains and sewage pipes) (Nilsson, 1997), which are the conditions found at the early stage of the wastewater treatment.

On the other hand, other taxa were associated with the last ponds, when compared to both the first constructed ponds and the natural ponds. All the taxa found in the last ponds were relatively tolerant to organic pollution. However, apart from organic pollution, other factors seemed to explain why certain taxa were associated with last ponds of the chain. For example, the trichopteran species of these ponds are typically found in small waterbodies with no flow of water. This was the case of *H. picicornis* and *L. marmoratus* (Wallace et al., 1990). Some differences may also be explained by abundance of emergent vegetation. Some species are known to be closely associated with aquatic vegetation such as *N. clavicornis* (Friday, 1988), *N. cinerea* (Savage, 1989) and *S. putris* (Macan & Cooper, 1977). For instance, a number of specimens of this Molluscan species were observed to inhabit the aerial parts of the emergent vegetation present in ICW ponds. It is also important to note that the taxa associated with the last ponds require a deeper habitat. This was the case of taxa such as *H. stagnorum*, *G. lacustris* and Corixidae immature. Finally, other taxa, such as *H. ruficollis*, generally feed on algae, which was only present in the last constructed ponds.

This study shows that ICW ponds have the potential to provide a habitat that may be otherwise unavailable within the surrounding landscape. What is more, if the aim of mimicking natural conditions is intended, ICW pond systems may benefit from an extension of the treatment process to further cleanse the water. This would create an additional environment within the ICW systems, presumably characterised by higher pH values and more pristine conditions, which would be suitable for species only found in a subset of natural ponds. All in all, ICWs constitute a valid option in farmland areas since they have the potential to combine their wastewater treating properties with that of providing a suitable habitat for macroinvertebrates.

Acknowledgements This project has been funded through the European Union programme INTERREG IIIA. We gratefully acknowledge the help and enthusiasm of Waterford County Council and everyone in the Freshwater Biodiversity, Ecology and Fisheries (FreBEF) laboratory. We would also like to thank two anonymous referees for their valuable comments on a draft of this manuscript.

References

Anderson, J. T. & L. M. Smith, 2004. Persistence and colonization strategies of playa wetland invertebrates. Hydrobiologia 513: 77–86.

Anderson, M. J. & R. N. Gorley, 2007. PERMANOVA+ for PRIMER: 450 guide to software and statistical methods. PRIMER-E, Plymouth.

APHA (American Public Health Organization), 1990. Standard methods for the examination of water and wastewater, including bottom sediments and sludges, 18th ed. APHA, Washington.

Balcombe, C. K., J. T. Anderson, R. H. Fortney & W. S. Kordek, 2005. Aquatic macroinvertebrate assemblages in mitigated and natural wetlands. Hydrobiologia 541: 175–188.

Becerra Jurado, G., R. Matson, R. Harrington, J. R. Baars & M. Kelly-Quinn, 2008. Macroinvertebrate diversity in constructed ponds: community structure and driving environmental factors. Verhandlungen Internationale Vereinigung für theoretische und angewandte Limnologie 30: 489–492.

Biggs, J., A. Corfield, D. Walker, M. Whitfield & P. Williams, 1994. New approaches to the management of ponds. British Wildlife 5: 273–287.

Biggs, J., G. Fox, P. Nicolet, D. Walker, M. M. Whitfield & P. Williams, 1998. A guide to the methods of the national pond survey. Pond Action, Oxford.

Biggs, J., P. Williams, M. Whitfield, P. Nicolet & A. Weatherby, 2005. 15 years of pond assessment in Britain: results and lessons learned from the work of Pond Conservation. Aquatic Conservation: Marine and Freshwater Ecosystems 15: 693–714.

Bormann, F. H. & G. E. Likens, 1981. Pattern and process in a forested ecosystem. Springer, New York: 253.

Briers, R. A. & J. Biggs, 2005. Spatial patterns in pond invertebrate communities: separating environmental and distance effects. Aquatic Conservation: Marine and Freshwater Ecosystems. 15: 549–557.

Brönmark, C. & L. A. Hansson, 1998. The Biology of Lakes and Ponds. Oxford University Press, New York.

Brönmark, C. & L. A. Hansson, 2002. Environmental issues in lakes and ponds: current state and perspectives. Environmental Conservation 29: 290–307.

Céréghino, R., A. Ruggiero, P. Marty & S. Angelibert, 2008. Biodiversity and distribution patterns of freshwater invertebrates in farm ponds of a south-western French agricultural landscape. Hydrobiologia 597: 43–51.

Clarke, K. R., 1993. Non-parametric multivariate analysis of changes in community structure. Australian Journal of Ecology 18: 117–143.

Clarke, K. R. & R. N. Gorley, 2006. PRIMER (v6): User Manual/495 Tutorial. PRIMER-E, 496 Plymouth, UK.

Davies, B. R., J. Biggs, P. J. Williams, J. T. Lee & S. Thompson, 2008. A comparison of the catchment sizes of rivers, streams, ponds, ditches and lakes: implications for protecting aquatic biodiversity in an agricultural landscape. Hydrobiologia 597: 7–17.

De Szalay, F. A. & V. H. Resh, 2008. Factors influencing macroinvertebrate colonization of seasonal wetlands: responses to emergent plant cover. Freshwater Biology 45: 295–308.

Dunne, E. J., N. Culleton, G. O'Donovan, R. Harrington & A. E. Olse, 2005. An integrated constructed wetland to treat contaminants and nutrients from dairy farmyard dirty water. Ecological Engineering 24: 219–232.

Edington, J. M. & A. G. Hildrew, 1995. A Revised Key to the Caseless Caddis Larvae of the British Isles with Notes on Their Ecology. Freshwater Biological Association, Scientific Publication No. 53, Ambleside.

Elliott, J. M. & K. H. Mann, 1979. A Key to the British Freshwater Leeches. Freshwater Biological Association, Scientific Publication No. 40, Ambleside.

Elliott, J. M., U. H. Humpesch & T. T. Macan, 1988. Larvae of the British Ephemeroptera. Freshwater Biological Association, Scientific Publication No. 49, Ambleside.

European Union, 2000. Water Framework Directive (2000/86/EEC). European Union, Brussels.

Fairchild, G. W., A. M. Faulds & J. F. Matta, 2000. Beetle assemblages in ponds: effects of habitat and site age. Freshwater Biology 44: 523–534.

Feldman, R. S. & E. F. Connor, 1992. The relationship between pH and community structure of invertebrates in streams of the Shenandoah National Park, Virginia, USA. Freshwater Biology 27: 261–276.

Friday, L. E., 1988. A key to adults of British water beetles. Field Studies Council Publication 189.

Hansson, L.-A., C. Brönmark, P. Anders Nilsson & K. Åbjörnsson, 2005. Conflicting demands on wetland ecosystem services: nutrient retention, biodiversity or both? Freshwater Biology 50: 705–714.

Hawkes, H. A., 1998. Origin and development of the Biological Monitoring Working Party Score System. Water Research 32: 946–984.

Kelly-Quinn, M. & J. J. Bracken, 2000. Ephemeropteran assemblages in Ireland. Verhandlungen Internationale Vereinigung für theoretische und angewandte Limnologie 27: 963–969.

Macan, T. T. & R. D. Cooper, 1977. A Key to the British Fresh- and Brackish-Water Gastropods. Freshwater Biological Association, Scientific Publication No. 13, Ambleside.

McGarrigle, M. L., 2005. Water quality in Ireland: diffuse agricultural eutrophication, a key problem. In Dunne, E. J., R. Reddy & O. T. Carton (eds), Nutrient Management

in Agricultural Watersheds: A Wetlands Solution. Wageningen Academic Publishers, Wageningen: 15–17.
McGarrigle, M. L., J. J. Bowman, K. J. Clabby, J. Lucey, P. Cunningham, M. MacCárthaigh, M. Keegan, B. Cantrell, M. Lehane, C. Clenaghan & P. F. Toner, 2002. Water Quality in Ireland 1998–2000. Environmental Protection Agency.
Menetrey, N., B. Oertli, M. Sartori, A. Wagner & J. B. Lachavanne, 2008. Eutrophication: are mayflies (Ephemeroptera) good bioindicators for ponds? Hydrobiologia 597: 125–135.
Mitsch, W. J. & S. E. Jorgensen, 1989. Ecological Engineering: An Introduction to Ecotechnology. Wiley, New York: 203.
Nelson, M. S., R. A. Roline, J. S. Thullen, J. J. Sartoris & J. E. Boutwell, 2000. Invertebrate assemblage and trace element bioaccumulation associated with constructed wetlands. Wetlands 20: 406–415.
Nilsson, A. N., 1996. Aquatic insects of northern Europe. A taxonomic handbook, Vol. 1. Apollo Books, Stenstrup.
Nilsson, A. N., 1997. Aquatic insects of northern Europe. A taxonomic handbook, Vol. 2. Apollo Books, Stenstrup.
Nilsson, A. N., J. Elmberg & K. Sjöberg, 1994. Abundance and species richness patterns of predaceous diving beetles (Coleoptera, Dytiscidae) in Swedish lakes. Journal of Biogeography 21: 197–206.
Oertli, B., D. Auderset Joye, N. Indermuehle, R. Juge & J. B. Lachavanne, 2004. 1st European Pond Workshop: conservation and monitoring of pond biodiversity. Archives des Sciences 57: 69–71.
Oertli, B., N. Indermuehle, S. Angelibert, H. Hinden & A. Stoll, 2008. Macroinvertebrate assemblages in 25 high alpine ponds of the Swiss National Park (Cirque of Macun) and relation to environmental variables. Hydrobiologia 597: 29–41.
Parsons, J. K. & R. A. Matthews, 1995. Analysis of the associations between macroinvertebrates and macrophytes in a freshwater pond. Northwest Science 4: 265–275.
Savage, A. A., 1989. Adults of the British Aquatic Hemiptera Heteroptera: A Key with Ecological Notes. Freshwater Biological Association, Scientific Publication No. 50, Ambleside.
Scholz, M., R. Harrington, P. Carroll & A. Mustafa, 2007. The integrated constructed wetlands (ICW) concept. Wetlands 27: 337–354.
Seaby, M. R. H., P. A. Henderson & J. R. Prendergast, 2004. Community Analysis Package, Version 2.01. PISCES Conservation Ltd., Lymington, Hampshire.
Spieles, D. J. & W. Mitsch, 2000. Macroinvertebrate community structure in high- and low-nutrient constructed wetlands. Wetlands 20: 716–729.
ter Braak, C. J. F. & P. Šmilauer, 2002. CANOCO Reference Manual and CanoDraw for Windows User's Guide: Software for Canonical Community Ordination (version 4.5). Microcomputer Power, Ithaca, NY, USA.
Tunney, H., A. Beeuwsma & P. Withers, 1997. Fertilizer strategies: present and future. In Tunney, H., O. T. Carton, P. C. Brookes & A. E. Johnston (eds), Phosphorus Loss from Soil to Water. CAB (Centre of Agriculture and Biosciences) International, Oxon: 177–204.
Wallace, I. D., B. Wallace & G. N. Philipson, 1990. A Key to the Case-Bearing Caddis Larvae of Britain and Ireland. Freshwater Biological Association, Scientific Publication No. 51, Ambleside.
Williams, P., M. Whitfield, J. Biggs, S. Bray, G. Fox, P. Nicolet & D. Sear, 2004. Comparative biodiversity of rivers, streams, ditches and ponds in an agricultural landscape in Southern England. Biological Conservation 115: 329–341.
Williams, P., M. Whitfield & J. Biggs, 2008. How can we make new ponds biodiverse? A case study monitoried over 7 years. Hydrobiologia 597: 137–148.
Withers, P. J. A., I. A. Davidson & R. H. Foy, 2000. Prospects for controlling nonpoint phosphorus loss to water: a UK perspective. Journal of Environmental Quality 29: 167–176.
Working Group on Groundwater, 2004. The calcareous/non-calcareous ("siliceous") classification of bedrock aquifers in the Republic of Ireland. Guidance Document No. GW3. Available on line: http://www.wfdireland.ie.
Working Group on Characterisation and Risk Assessment, 2005. Summary note of Irish lake typology to be applied in Ireland's River Basin Districts. Surface Water Guidance Document. Available on line: http://www.wfdireland.ie.

Hydrobiologia (2009) 634:167–183
DOI 10.1007/s10750-009-9897-3

POND CONSERVATION

Inter- and intra-annual variations of macroinvertebrate assemblages are related to the hydroperiod in Mediterranean temporary ponds

Margarita Florencio · Laura Serrano · Carola Gómez-Rodríguez · Andrés Millán · Carmen Díaz-Paniagua

Published online: 5 August 2009

Abstract Macroinvertebrate assemblages of 22 temporary ponds with different hydroperiod were sampled monthly during a dry year (2005–2006) and a wet year (2006–2007). Coleopteran and Heteropteran adults were most abundant at the end of the hydroperiod, while Coleopteran larvae, mainly Dytiscidae, were mostly recorded in spring. Macroinvertebrate assemblages differed between study years. The shorter hydroperiod of ponds in the dry year constrained the length of the aquatic period for macroinvertebrates, and three distinct wet phases of community composition could be distinguished: filling phase, aquatic phase and drying phase. In the wet year, with a longer pond hydroperiod, five phases could be identified, with the aquatic phase differentiated into winter, early spring and late spring phases. Dispersers such as *Anisops sardeus*, *Berosus guttalis* or *Anacaena lutescens* were typical during the filling phase and *Corixa affinis* or *Enochrus fuscipennis* during the drying phase. The ponds with intermediate hydroperiod showed a similar composition (mainly dispersers) at the beginning and end of their wet period; this is not being seen in early drying or long hydroperiod ponds. A general pattern was detected, with similar variation between both years, which may be associated with the life histories of the macroinvertebrate taxa recorded.

Keywords Aquatic macroinvertebrates · Temporal variation · Wet phases · Hydroperiod · Community composition · Life cycle

Guest editors: B. Oertli, R. Cereghino, A. Hull & R. Miracle
Pond Conservation: From Science to Practice. 3rd Conference of the European Pond Conservation Network, Valencia, Spain, 14–16 May 2008

M. Florencio (✉) · C. Gómez-Rodríguez · C. Díaz-Paniagua
Doñana Biological Station-CSIC, P.O. Box 1056, 41080 Seville, Spain
e-mail: margarita@ebd.csic.es

C. Gómez-Rodríguez
e-mail: carola@ebd.csic.es

C. Díaz-Paniagua
e-mail: poli@ebd.csic.es

L. Serrano
Department of Plant Biology and Ecology, University of Seville, P.O. Box 1095, 41080 Seville, Spain
e-mail: serrano@us.es

A. Millán
Department of Ecology and Hydrology, University of Murcia, Campus Espinardo, 30100 Murcia, Spain
e-mail: acmillan@um.es

Introduction

Temporary ponds are optimal habitats for many macroinvertebrate species, being important for the conservation of their specialized fauna (Strayer, 2006). However, these ponds have been frequently neglected in conservation programmes that have traditionally considered protection of extensive wetlands but not of small water bodies, despite their high biodiversity (Collinson et al., 1995; Céréghino et al., 2008). Moreover, temporary ponds are highly suitable for ecological studies due to their wide environmental gradients of salinity, temperature, vegetation, pH or hydroperiod (Herbst, 2001; Batzer et al., 2004; Waterkeyn et al., 2008; Bilton et al., 2009).

Despite the fact that permanent ponds may contain many aquatic species (Bazzanti et al., 1996; Brooks, 2000; Serrano & Fahd, 2005; Della Bella et al., 2005), temporary ponds usually harbour exclusive species or large populations of species which are scarce in or absent from permanent waters (Collinson et al., 1995; Williams, 1997; Boix et al., 2001; Della Bella et al., 2005; Céréghino et al., 2008). While the dry period may exclude many aquatic organisms from temporary ponds, the absence of large predators, such as fish, is a critical factor that determines the presence of specialist taxa (Wellborn et al., 1996).

Many macroinvertebrate species require an aquatic phase to complete their complex life cycles for which different life history strategies have been reported. Among the most important challenges for the macroinvertebrates of temporary ponds is survival during the dry period. Some adaptations for living in temporary ponds are dispersal to more permanent waters, or resistance of eggs, larvae, or adults to desiccation (Wiggins et al., 1980). Physiological and behavioural mechanisms to survive desiccation have also been described in different aquatic invertebrates (Williams, 2006). Wiggins et al. (1980) segregated groups of macroinvertebrates according to their life history strategies, justifying the presence of specific fauna in different ponds. Differences in the life history strategies of species allow the identification of functional groups which appear at different times in the ponds (Gascón et al., 2008) or to differences in optimal habitats, being able to only complete their life cycles in ponds with a long hydroperiod, but not in ephemeral ponds (Schneider & Frost, 1996).

Annual and seasonal variations of macroinvertebrate assemblages have been reported in temporary ponds (Brooks, 2000) and have been associated with seasonal changes in environmental conditions during the wet phase (Boulton & Lake, 1992). Jeffries (1994) found differences in the macroinvertebrate assemblages of the same ponds in three different years, including a low rainfall year in which ponds did not fill. Different macroinvertebrate groups have been described as characteristics of different pond phases; usually classified as filling, aquatic and drying phases, out of which the aquatic phase could be further differentiated into three additional phases (Boulton & Lake, 1992; Bazzanti et al., 1996; Boix et al., 2004).

Our study has been carried out in an area in which more than 3000 water bodies support a robust network of aquatic habitats (Fortuna et al., 2006) that exhibit high conservation values and encompass a wide range of hydroperiod and environmental conditions (Gómez-Rodríguez et al., 2009). Several studies have focused on the limnology of these ponds (García Novo et al., 1991; Serrano & Toja, 1995; Serrano et al., 2006) and their use as amphibian breeding sites (Díaz-Paniagua, 1990; Díaz-Paniagua et al., 2005). In contrast, only preliminary data on macroinvertebrates (Agüesse, 1962; Bigot & Marazanof, 1966; Marazanof, 1967; Millán et al., 2005) and studies on abundance of Coleoptera, Heteroptera and Odonata (Montes et al., 1982) have been reported.

In this research, we have studied temporal variation in macroinvertebrate abundance and composition in temporary ponds, with the following specific aims: (1) detecting inter-annual variation; (2) detecting seasonal variation in relation to different phases of the wet period of ponds; (3) comparing monthly variation within ponds of different hydroperiod and (4) determining if there is a general pattern of temporal variation for all ponds in the study area.

Methods

The study was carried out in 22 ponds located in the Doñana Biological Reserve (Doñana National Park, Southwestern Spain, Fig. 1). This area is located between the Atlantic coast and the mouth of the Guadalalquivir River. It includes a high number of temporary ponds, appearing during autumn or winter, and two permanent ponds. The type of climate is Mediterranean sub-humid, with hot and dry summers, mild winters, and rainfall mainly falling in autumn and winter (see Siljeström et al., 1994, García-Novo & Marín, 2006 for a detailed description of the area).

Our study period was from October 2005 to July 2007. Annual rainfall was calculated as the amount of rainfall collected from 1st September to 31st August of the following year. This amounted to 468.3 mm in

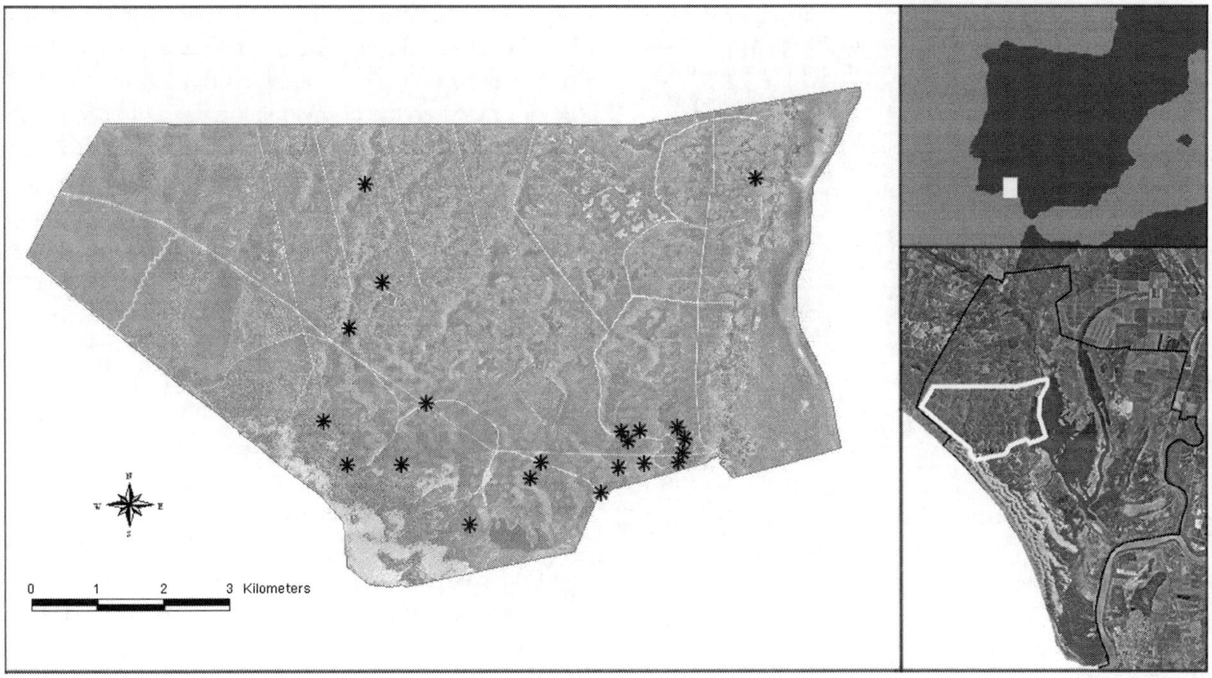

Fig. 1 Location of the 22 temporary ponds in the Doñana Biological Reserve, Doñana National Park (SW Spain)

the first year (hereafter referred to as the dry year), when we sampled 18 temporary ponds which usually dry out every summer and one semi permanent pond which only dries out in years of severe drought. As this pond was dry in 2005, prior to our study period, we considered it as a temporary pond. Ponds were selected to encompass the highest possible range of hydroperiods, being representative of the range of ponds found in the study area. In the second year, annual rainfall was 716.9 mm (hereafter referred to as the wet year) when a higher number of ponds with short hydroperiod were formed in the area. In order to assess the widest range of hydroperiod during the wet year, we sampled three of these new ponds, although the total number of ponds sampled was the same as the year before. In the dry year, most temporary ponds were wet from February to June and from October to July during the wet year, although the ponds with longest hydroperiod had water even during August in both years. A detailed description of the characteristics of Doñana temporary ponds, including most of our study ponds, is given by Gómez-Rodríguez et al. (2009). Hydroperiod and maximum depth of ponds during our study, as well as their basin areas, are shown in Table 1. Pond area was extracted from a pond cartography obtained in a moment of large inundation (see Gómez-Rodríguez et al., 2008). Vegetation in the ponds was mainly composed of meadow plants as *Mentha pulegium* L., *Illecebrum verticillatum* L. or *Hypericum elodes* L., in the littoral zone, while aquatic macrophytes were common in deeper areas, such as *Juncus heterophyllus* Dufour, *Myriophyllum alterniflorum* DC. in Lam & DC., *Potamogeton pectinatus* L. and *Ranunculus peltatus* Schrank (García Murillo et al., 2006).

Macroinvertebrates were sampled monthly in each pond by using a dip-net with a 1 mm mesh, netting a stretch of water of ~1.5 m length in each sampling unit. In the wet year, the four ponds with the shortest hydroperiod (including the three ponds only sampled during this year) were sampled every 15 days. In each pond, we sampled at different points along one or two transects from the littoral to the open water, the number of sampling points being proportional to pond size. We also took additional samples in microhabitats which were not represented in these transects. The maximum number of samples per pond ranged from 6 to 13 in the month of maximal inundation. As pond size decreased during the season, the number of samples taken was reduced accordingly. Most macroinvertebrates captured were identified in situ, being counted and released. Individuals of unidentified

Table 1 Hydroperiod and maximum depth of every pond (named with *three letters*) are shown for the dry year and wet year, and also the pond area calculated in a large inundation moment (hydroperiod is only given for the year in which each pond was sampled)

Pond	Hydroperiod (months)		Maximum depth (cm)		Pond area (m^2)
	Dry year	Wet year	Dry year	Wet year	Maximum inundation
Pol	3.1	7.2	33	50	1,200
Acm	3.3	6.4	34	44	50
Rp	2.1	7.2	24	64	4,075
Pg	3.1	7.2	31	54	3,925
Jim	2.2	7.2	9.5	86	39,900
Cam	2.7	7.2	23	55	2,200
Zah	3.8	9.1	47	69	48,189
Lve	6.1	12	104	132	3,300
Dul	8.9	12	142	165	122,672
Abe	2.3	6.9	18	43	50
Bre	3.4	7.9	47	85	2,150
Pp	3.1	7.2	42	82	875
Tej	2.8	7.2	22	67	150
Orf	4.3	9	80	82	850
Ant	1.4	6.8	15	45	5,131
Wou	3.4	–	31.5	–	14,375
Mor	3.2	–	25	–	14,725
Tar	4.4	–	55	–	81,250
Arm	–	4.2	–	21	25
Vac	0.4	6.2	–	51	25
Len	–	5.5	–	24	650
Tps	–	6.1	–	39	6,375

species were preserved in 70% ethanol for identification in the laboratory. Whenever possible, individuals were identified to species level, except for Diptera, which were identified to family. For Chironomidae and Ceratopogonidae, only presence–absence data were recorded. All recorded taxa with only presence–absence data were not included in analyses.

For the analysis of the macroinvertebrate assemblage composition, we estimated the relative abundance of each taxon, as the total number of individuals captured across all samples taken in a pond, divided by the total number of samples taken in that pond. In these analyses, we differentiated adults from larvae or nymphs, and considered these as different taxa (hereafter referred to as "taxa" for simplicity) in our data matrix. Relative abundance was log transformed ($X + 1$) to calculate the similarity matrix with the Bray–Curtis similarity index (Clarke & Warwick, 2001).

For each pond, we computed the Spearman correlations between the corresponding taxa of each pair of similarity matrices of relative abundances in different months, using the RELATE program (Primer v.6, Clarke & Warwick, 2001) to assess *monthly variation in the macroinvertebrate assemblages within ponds*. The Spearman correlation coefficient (ρ) was close to one when the monthly similarity matrices were highly corresponding. These analyses detected if the similarity among the composition of macroinvertebrates was higher in subsequent months (Serial RELATE) than in more distant months, such as the beginning and the end of each hydroperiod (Cyclic RELATE). Similarity distances among months were represented using non-metric multidimensional scaling (NMDS). As pond hydroperiod was relatively short in the dry year, these analyses of monthly variation of macroinvertebrate assemblages were performed only for the wet year.

In order to assess *seasonal variation* in macroinvertebrate assemblages, we used a NMDS representation of the similarity matrices of relative abundances of all ponds and months except for the February matrix of one pond in the dry year which had been previously filled. The different groups observed in the NMDS were used as grouping factor including three or five levels depending on number of observed groups in every case. We then tested differences among observed groups using one-way ANOSIM analyses (performed with 9999 number of permutations). The ANOSIM test statistic, R, is close to 1 when the levels of grouping factor are different; that is to say, all dissimilarities between levels of grouping factor are larger than any dissimilarity among samples in every level of grouping factor (Clarke & Warwick, 2001). An exploratory analysis (SIMPER) was used to detect those taxa with the highest contribution to the dissimilarity of each level of grouping factor versus all other levels for the same factor (Primer v.6, Clarke & Warwick, 2001).

In order to explore particular questions about the temporal variation of macroinvertebrate assemblages, we averaged the relative abundances of macroinvertebrates in different ways. (1) To analyse *inter-annual variation* between the dry year and wet year, we averaged the relative abundance of macroinvertebrate taxa every year by dividing by the numbers of months that every pond was sampled. These averaged matrices were represented in NMDS to observe whether both years corresponded to different groups. We tested if macroinvertebrate assemblages were different in two study years through one-way ANOSIM analysis, using the year as groping factor with two levels. SIMPER analysis detected those taxa making a higher contribution to dissimilarity between the two years (Primer v.6, Clarke & Warwick, 2001). We removed the macroinvertebrate assemblages of two ponds sampled in the dry year of these analyses and NMDS representation because they only were sample once, not being comparable with the rest of ponds in both years. (2) To analyse if *a general pattern of monthly variation* occurred in both study years, we averaged the relative abundance of individual taxa across all ponds every month by dividing by the number of sampled ponds per month. Then, we used only one averaged matrix of relative abundance of macroinvertebrates per month, representing a unique similarity value per month in a NMDS. The Spearman correlation between these monthly similarity values for the average matrix of relative abundance of macroinvertebrates across all ponds was calculated for each year through a Serial RELATE. The Spearman correlation coefficient value (ρ) would be 1 in case of maximum correlation. Prior to these analyses, we tested whether the variation among months was higher than among ponds within a month, using the complete relative abundance matrix of both study years through a one-way ANOSIM analysis where months in every year were the grouping factor with a total of 18 levels.

Monthly matrices of relative abundance of macroinvertebrates were not included in the analyses when any or very scarce abundances were detected in a pond (mainly during the initial stages of annual sampling). Some taxa had to be combined to compare between years, because some species were not identified during the first year (adults of all species of *Haliplus* were included in one taxon, as were adults of Corixidae, except for *Corixa affinis* Leach, 1817).

Results

Macroinvertebrate taxa and their monthly variation

The macroinvertebrates recorded in the Doñana ponds included 123 different taxa, including 97 species, and additionally unidentified species included in 6 genera, 16 families, 3 subfamilies and 1 order. The most abundant species were *C. affinis*, *Cloeon* spp. and *Anisops sardeus* Herrich-Schäffer, 1849, while other species such as *Coenagrion scitulum* (Rambur, 1842) appeared occasionally and with very low abundance (Table 2). Coleoptera, Heteroptera and Odonata were the orders that included the highest number of species and individuals during both years. The monthly variation of the average number of individuals caught in all the samples during both years is shown in Fig. 2. Adults of Coleoptera and Heteroptera showed the highest abundance both at the end of the wet period and at the beginning during the wet year. Dytiscidae and Hydrophilidae were the most abundant families of Coleoptera. Larvae of Coleoptera (mainly Dytiscidae) were found in the middle of the wet period, while the highest abundance of Coleoptera was reached by adults of Hydrophilidae at the end of the wet period, in July during the wet year and in May during the dry year, when ponds had shorter hydroperiod. Adults of Heteroptera (mainly Corixidae and Notonectidae) reached their highest abundance in ponds with longer hydroperiods in summer. *A. sardeus* and *C. affinis* were the most abundant heteropterans; *C. affinis* being much more abundant during the wet year than the dry year. Among Odonata, Libellulidae [mainly *Sympetrum fonscolombei* (Selys, 1841)] were found throughout the wet period, while Coenagrionidae [mainly *Ischnura pumilio* (Charp., 1825)] were especially abundant at the end of this.

Inter-annual variation

Macroinvertebrate compositions of every pond were segregated in two groups corresponding with both study years, although the dissimilarity between the dry year and the wet year was not very strong (ANOSIM: $R = 0.235$; $P = 0.02$) (Fig. 3). SIMPER analyses showed that adults of *Hydroglyphus geminus* (Fabricius, 1792) (13.67%), *Anacaena lutescens* (Stephens, 1829) (13.64%) and Notonectidae larvae (10.91%) mainly contributed to these differences in the dry year, while *Cloeon* spp. (11%) and adults of *A. sardeus* (10.81%) had a larger contribution in the wet year.

Seasonal variation

During the dry year, we observed three groups in the NMDS composed by different macroinvertebrate

Table 2 Taxa of macroinvertebrates recorded in the study ponds during both years

Taxa	Family	Average		Maximum	
		Adult	Larva	Adult	Larva
Acari					
Hydrachnellae	–	0.029		11	
Bassomatophora					
Physa spp.	Physidae	1.849		446	
Planorbidae	Planorbidae	1.486		259	
Coleoptera					
Donacia spp.	Chrysomelidae	[a]		[a]	
Bagous spp.	Curculionidae	0.017		3	
Bagous revelieri Tournier, 1884[b]	Curculionidae	[a]		[a]	
Bagous subcarinatus Gyllenhal, 1836[b]	Curculionidae	[a]		[a]	
Bagous vivesi González, 1967[b]	Curculionidae	[a]		[a]	
Dryops luridus (Erichson, 1847)	Dryopidae	[a]		[a]	
Dryops spp.	Dryopidae	0.602	0.01	54	2
Agabus bipustulatus (Linnaeus, 1767)	Dytiscidae	[a]		[a]	
Agabus conspersus (Marsham 1802)	Dytiscidae	0.026		5	
Agabus didymus (Olivier, 1795)	Dytiscidae	0.001		1	
Agabus nebulosus (Forster, 1771)	Dytiscidae	0.012		3	
Agabus spp.	Dytiscidae		0.397		19
Cybister (Scaphinectes) lateralimarginalis (De Geer, 1774)	Dytiscidae	0.014	0.025	3	2
Dytiscus circumflexus Fabricius, 1801	Dytiscidae	0.001	0.02	1	2
Eretes griseus (Fabricius, 1781)	Dytiscidae	0.001		1	
Graptodytes flavipes (Olivier, 1795)	Dytiscidae	0.004		1	
Hydaticus (Guignotites) leander (Rossi, 1790)	Dytiscidae	0.001	0.001	1	1
Hydroglyphus geminus (Fabricius, 1792)	Dytiscidae	0.516		92	
Hydroporus gyllenhali Schiödte, 1841	Dytiscidae	0.023		5	
Hydroporus lucasi Reiche, 1866	Dytiscidae	0.079		22	
Hygrotus confluens (Fabricius, 1787)	Dytiscidae	0.012		3	
Hygrotus lagari (Fery, 1992)	Dytiscidae	0.478		47	
Hydroporus spp. or *Hygrotus* spp.	Dytiscidae		0.151		12
Hyphydrus aubei Ganglbauer, 1892	Dytiscidae	0.009	0.033	2	5
Ilybius montanus (Stephens, 1828)	Dytiscidae	0.003		1	
Laccophiluis minutus (Linnaeus, 1758)	Dytiscidae	0.171	0.358	65	33
Liopterus atriceps (Sharp, 1882)	Dytiscidae	0.044		15	
Rhantus (Rhantus) hispanicus Sharp, 1882	Dytiscidae	0.063		5	
Rhantus (Rhantus) suturalis (McLeay, 1825)	Dytiscidae	0.02		6	
Colymbetes fuscus (Linnaeus, 1758)	Dytiscidae	0.063		19	
Rhantus spp. or *Colymbetes fuscus*	Dytiscidae		0.442		18
Gyrinus (Gyrinus) dejeani Brullé, 1832	Gyrinidae	0.007	0.007	1	1
Haliplus (Liaphlus) andalusicus Wehncke, 1874	Haliplidae	0.018		4	
Haliplus (Liaphlus) guttatus Aubé, 1836	Haliplidae	0.01		2	
Haliplus (Neohaliplus) lineatocollis (Marsham, 1802)	Haliplidae	0.008		2	
Haliplus spp.	Haliplidae		0.020		3
Helophorus spp.	Helophoridae	0.367	0.001	93	1

Table 2 continued

Taxa	Family	Average		Maximum	
		Adult	Larva	Adult	Larva
Helophorus (Trichohelophorus) alternans Gené, 1836	Helophoridae	[a]		[a]	
Helophorus (Rhopalohelophorus) longitarsis Wollaston, 1864	Helophoridae	[a]		[a]	
Hydraena (Hydraena) rugosa Mulsant, 1844	Hydraenidae	0.012		2	
Limnebius furcatus Baudi, 1872	Hydraenidae	0.001		1	
Ochthebius (Asiobates) dilatatus Stephens, 1829	Hydraenidae	0.018		9	
Ochthebius (Ochthebius) punctatus Stephens, 1829	Hydraenidae	0.004		1	
Ochthebius (Ochthebius) auropallens Fairmaire, 1879	Hydraenidae	0.060		17	
Hydrochus flavipennis Küster, 1852	Hydrochidae	0.029		12	
Anacaena (Anacaena) lutescens (Stephens, 1829)	Hydrophilidae	1.165		421	
Berosus (Berosus) affinis Brullé, 1835	Hydrophilidae	0.455		136	
Berosus (Enoplurus) guttalis Rey, 1883	Hydrophilidae	0.165		13	
Berosus (Berosus) signaticollis (Charpentier, 1825)	Hydrophilidae	0.201		12	
Berosus spp.	Hydrophilidae		0.084		5
Enochrus (Lumetus) bicolor (Fabricius, 1792)	Hydrophilidae	0.059		6	
Enochrus (Lumetus) fuscipennis (C.G. Thomsom, 1884)	Hydrophilidae	1.192		242	
Enochrus spp.	Hydrophilidae		0.007		1
Helochares (Helochares) lividus (Forster, 1771)	Hydrophilidae	0.029		14	
Hydrobius convexus Brullé, 1835	Hydrophilidae	[a]		[a]	
Hydrobius fuscipes (Linnaeus, 1758) & *Limnoxenus niger* (Zschach, 1788)	Hydrophilidae	0.369		40	
Hydrobius spp. or *Limnoxenus niger*	Hydrophilidae		0.084		21
Hydrochara flavipes (Steven, 1808)	Hydrophilidae	0.023	0.007	6	2
Hydrophilus (Hydrophilus) pistaceus (Laporte, 1840)	Hydrophilidae	0.001	0.007	1	2
Laccobius (Hydroxenus) revelierei Perris, 1864	Hydrophilidae	0.003		2	
Paracymus scutellaris (Rosenhauer, 1856)	Hydrophilidae	0.222		110	
Hygrobia hermanni (Fabricius, 1775)	Paelobiidae	0.029	0.107	12	8
Noterus laevis Sturm, 1834	Noteridae	0.019		11	
Hydrocyphon spp.	Scirtidae		0.019		4
Decapoda					
Procambarus clarkii (Girard, 1852)	Cambaridae		0.027		7
Ephemeroptera					
Cloeon spp.	Baetidae		6.179		394
Haplotaxida					
Lumbricidae & Sparganophilidae	Lumbricidae & Sparganophilidae		[a]		[a]
Tubificidae	Tubificidae		[a]		[a]
Heteroptera					
Corixa affinis Leach, 1817	Corixidae	8.879		2209	
Micronecta scholzi (Fieber, 1860)	Corixidae	0.001		1	
Paracorixa concinna (Fieber, 1848)	Corixidae	0.006		3	
Sigara (Vermicorixa) lateralis (Leach, 1817)	Corixidae	0.391		59	
Sigara (Vermicorixa) scripta (Rambur, 1840)	Corixidae	0.04		14	
Sigara (Halicorixa) selecta (Fieber, 1848)	Corixidae	0.003		1	

Table 2 continued

Taxa	Family	Average		Maximum	
		Adult	Larva	Adult	Larva
Sigara (Halicorixa) stagnallis (Leach, 1817)	Corixidae	0.037		6	
Trichocorixa verticalis (Fieber, 1851)	Corixidae	0.009		2	
Corixidae spp.	Corixidae		1.258		99
Gerris (Gerris) cf. *maculatus* Tamanini, 1946	Gerridae	0.002		1	
Gerris (Gerris) thoracicus Schummel, 1832	Gerridae	0.229		2	
Gerris spp.	Gerridae		0.282		12
Microvelia pygmaea (Dufour, 1833)	Microveliidae	0.011		2	
Naucoris maculatus Fabricius, 1798	Naucoridae	0.01	0.03	5	12
Nepa cinerea Linnaeus, 1798	Nepidae	0.008	0.009	5	6
Anisops sardeus Herrich-Schäffer, 1849	Notonectidae	3.704		272	
Notonecta glauca Linnaeus, 1758 ssp. *glauca*	Notonectidae	0.025		4	
Notonecta glauca Linnaeus, 1758 ssp. *meridionalis* Poisson, 1926	Notonectidae	0.039		4	
Notonecta maculata Fabricius, 1794	Notonectidae	0.011		3	
Notonecta viridis Delcourt, 1909	Notonectidae	0.029		6	
Notonectidae spp.	Notonectidae		1.219		91
Plea minutissima Leach, 1817	Pleidae	0.677	0.208	130	38
Saldidae	Saldidae	0.018		12	
Isopoda					
Asellus aquaticus (Linnaeus, 1758)	Asellidae	0.014		11	
Lumbriculida					
Lumbriculidae	Lumbriculidae	[a]		[a]	
Notostraca					
Triops mauritanicus (Ghigi, 1921)	Triopsidae	0.055		6	
Spinicaudata					
Cyzicus grubei Simon, 1886	Cyzicidae	[a]		[a]	
Maghrebestheria maroccana Thiéry, 1988	Leptestheriidae	[a]		[a]	
Anostraca					
Branchipus cortesi Alonso y Jaume, 1991	Branchipodidae	[a]		[a]	
Branchipus schafferi Fischer de Waldheim, 1834	Branchipodidae	[a]		[a]	
Tanymastix stagnalis (Linnaeus, 1758)	Tanymastigiidae	[a]		[a]	
Chirocephalus diaphanus Desmarest, 1823	Chirocephalidae	[a]		[a]	
Odonata					
Aeshna affinis Vander Linden, 1823	Aeshnidae		0.005		1
Aeshna mixta Latreille, 1805	Aeshnidae		0.012		2
Anax imperator Leach, 1815	Aeshnidae		[a]		[a]
Hemianax (Anax) ephippiger (Burmeister, 1839)	Aeshnidae		0.003		1
Coenagrion scitulum (Rambur, 1842)	Coenagrionidae		0.001		1
Ishnura elegans (Vander Linden, 1820)	Coenagrionidae		0.052		9
Ishnura pumilio (Charp., 1825)	Coenagrionidae		0.525		50
Lestes barbarus (Fabr., 1798)	Lestidae		0.028		16
Lestes dryas Kirby, 1890	Lestidae		0.008		2
Lestes macrostigma (Eversm., 1836)	Lestidae		0.001		1
Lestes virens (Charpentier, 1825)	Lestidae		0.002		1

Table 2 continued

Taxa	Family	Average		Maximum	
		Adult	Larva	Adult	Larva
Crocothemis erythrarea (Brullé, 1832)	Libellulidae		0.033		5
Sympetrum fonscolombei (Selys, 1841)	Libellulidae		0.252		9
Sympetrum meridionale (Selys, 1841)	Libellulidae		0.024		3
Sympetrum sanguineum (Müller, 1764)	Libellulidae		0.038		5
Sympetrum striotalum (Charpentier, 1840)	Libellulidae		0.048		4

Taxa	Family	Average		Maximum	
		Larva	Nymph	Larva	Nymph
Diptera					
Ceratopogoninae	Ceratopogonidae	[a]		[a]	
Chaoborus spp.	Chaoboridae	[a]		[a]	
Chironomidae sp.	Chironomidae	[a]	0.001	[a]	1
Chironomus plumosus	Chironomidae	[a]	0.001	[a]	1
Culicidae	Culicidae	0.567	0.132	253	8
Dixa spp.	Dixidae	0.01	0.004	2	2
Dolichopodidae	Dolichopodidae	0.016		11	
Ephydridae	Ephydridae	0.023	0.013	12	4
Orthocladiinae	Chironomidae	[a]		[a]	
Rhagionidae	Rhagionidae	0.009		2	
Scatophagidae	Scatophagidae	0.001		1	
Sciomyzidae	Sciomyzidae	0.001		1	
Syrphidae	Syrphidae	0.005	0.004	3	2
Tabanidae	Tabanidae	0.005		1	
Tanypodinae	Chironomidae	[a]	0.001	[a]	1
Thaumelidae	Thaumelidae		0.001		1
Tipulidae	Tipulidae	0.011	0.004	2	2

Average and maximum number of individuals per sample is shown for adults, larvae and nymphs

[a] Only presence was recorded

[b] New records for Doñana National Park

compositions detected in every pond and month, which corresponded to different wet phases of the ponds (Fig. 4): filling phase (February), aquatic phase (March and April) and drying phase (May–September) (ANOSIM, global $R = 0.615$, $P = 0.01$). In the aquatic phase, we also distinguished a weak segregation in two subgroups: early spring (March) and late spring (April) phases (ANOSIM, $R = 0.297$, $P = 0.02$). We identified the main taxa that contributed to the dissimilarity of the three phases with a SIMPER analysis: Adults of *Berosus affinis* Brullé, 1835 (17.32%), *Helophorus* spp. (17.24%), *A. lutescens* (15.74%), Corixidae (without *C. affinis*) (12.52%) and *Dryops* spp. (11.97%) in the filling phase; Notonectidae larvae (14.14%) and adults of *H. geminus* (13.69%) in the aquatic phase; adults of Corixidae (without *C. affinis*) (25.83%) and *A. sardeus* (19.18%) in the drying phase. During the wet year, we observed five consecutive groups of macroinvertebrates compositions of every pond and month in the NMDS that corresponded to different wet phases (Fig. 4): filling phase (November), winter (December and January), early spring (February and March), late spring (April) and drying phase (May–August). The wet phases presented different similarities according to an ANOSIM analysis (global $R = 0.538$, $P = 0.01$).

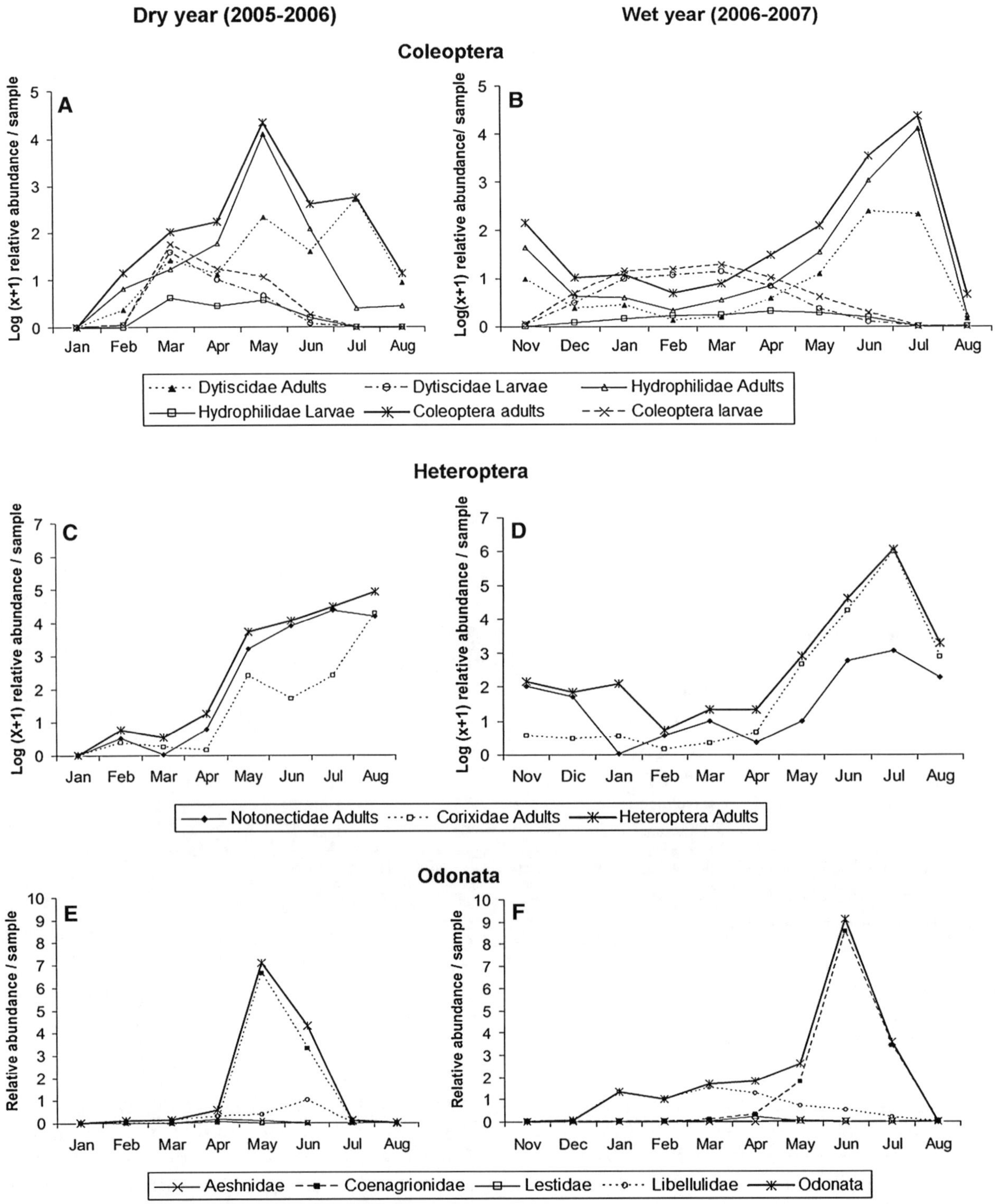

Fig. 2 Monthly variation in the relative abundance of individuals of different taxa of macroinvertebrates averaging data of all ponds in a dry year and a wet year: Coleoptera (**A**, **B**), Heteroptera (**C**, **D**) and Odonata (**E**, **F**)

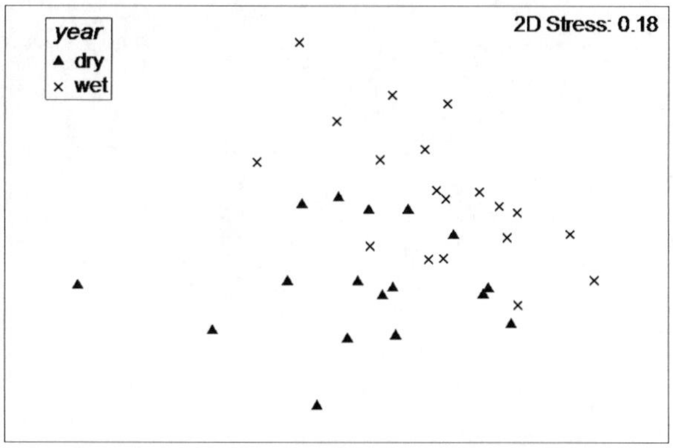

Fig. 3 NMDS ordination of the relative abundance of macroinvertebrates in every pond, showing their inter-annual variation. It is averaging the number of individuals per month in the dry year and the wet year

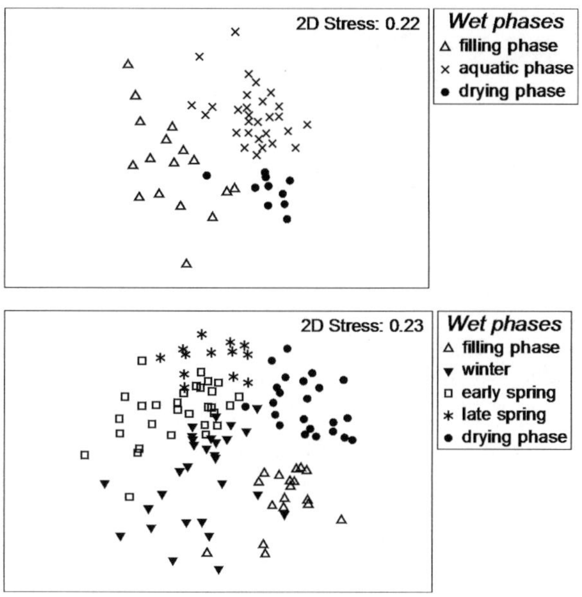

Fig. 4 NMDS ordination of the relative abundance of macroinvertebrates in different ponds and months during the dry year (*top*) and wet year (*bottom*). Different phases identified are indicated on the plot

The highest R value in the pairwise comparison of wet phases during the wet year was for filling phase versus early spring ($R = 0.826$; $P = 0.01$) and filling phase versus late spring ($R = 0.912$; $P = 0.01$). SIMPER analysis revealed that the taxa with highest contribution to global dissimilarity were: adults of *A. sardeus* (23.90%) and *Berosus guttalis* Rey, 1883 (12.30%) in the filling phase; *A. sardeus* (adults) (31.41%) and *Cloeon* spp. (19.34%) in the winter phase; *Cloeon* spp. (39.79%) in the early spring phase; *Gerris* spp. larvae (14.04%), Notonectidae larvae (11.94%) and *Cloeon* spp. (11.93%) in the late spring phase; and adults of *C. affinis* (25.15%) and *Enochrus fuscipennis* (C.G. Thomsom, 1884) (10.44%) in the drying phase.

Monthly variation of macroinvertebrate assemblages within ponds

In the wet year, the Spearman correlations comparing the similarity matrices of monthly macroinvertebrate assemblages in each pond tended to present higher ρ values in serial than in cyclic correlations in 13 ponds (Table 3; Fig. 5A). In contrast, in five ponds they tended to present higher cyclic correlations (Table 3; Fig. 5B). In one pond, the ρ value for these similarity

Table 3 The Spearman correlation coefficients (ρ) calculated among monthly macroinvertebrate assemblages in each pond (Serial and Cyclic RELATE analyses) during the wet year

Pond	Spearman's correlation (ρ)			
	Monthly		Every 15 days	
	Serial	Cyclic	Serial	Cyclic
Pol	**0.732** (7)**	0.491** (7)		
Acm	**0.627** (6)**	0.466* (6)		
Rp	**0.618** (7)**	0.472** (7)		
Pg	**0.736** (7)**	0.616** (7)		
Jim	**0.757** (7)**	0.478** (7)		
Cam	**0.641** (7)**	0.287** (7)		
Zah	**0.881** (9)**	0.658** (9)		
Lve	**0.653** (10)**	0.614** (10)		
Dul	**0.724** (12)**	0.665** (12)		
Abe	**0.593* (7)**	0.549** (7)		
Bre	0.241 (9)	**0.474** (9)**		
Pp	0.511* (7)	**0.636** (7)**		
Tej	0.555* (7)	**0.742** (7)**		
Orf	0.391* (9)	**0.471** (9)**		
Ant	0.35 (6)	0.086* (6)		
Arm	0.156 (6)	0.362 (6)	0.309 (8)	**0.428* (8)**
Vac	**0.724** (7)**	0.684** (7)	**0.553** (11)**	0.42** (11)
Len	0.736* (7)	0.736* (7)	**0.509** (10)**	0.358** (10)
Tps	**0.583** (6)**	0.265* (6)	**0.506** (10)**	0.291* (10)

For ponds sampled every 15 days, both monthly and 15-day analyses are shown. Number of samples in each correlation analysis is given in *brackets*. * $P < 0.05$; ** $P < 0.01$. The highest significant ρ value (cyclic or serial correlations) for each pond is marked in *bold*

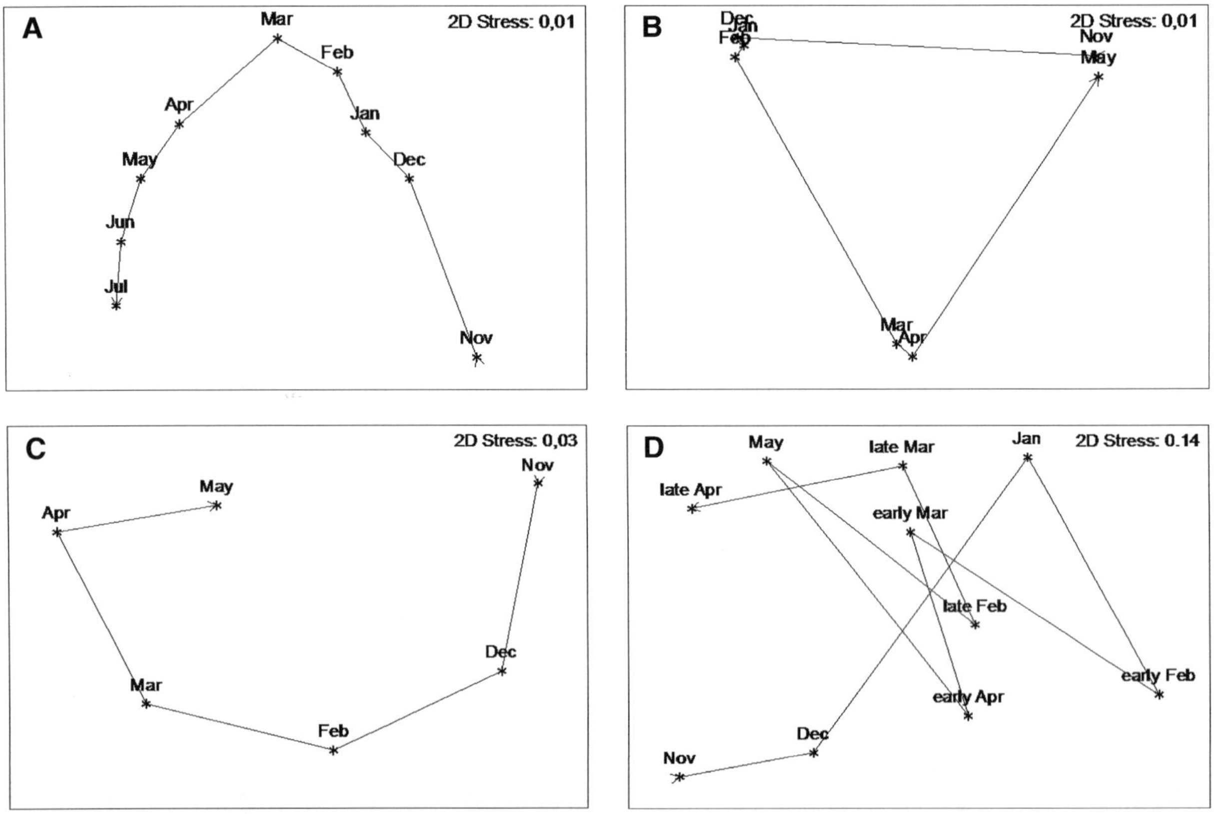

Fig. 5 Monthly variation of the macroinvertebrate assemblage in three different temporary ponds (**A**, **B**, **C**) represented in a NMDS during the wet year. Pond **A** (Zah) presented a higher serial correlation while **B** (Pp) presented a higher cyclic correlation. **C** and **D** show the same pond (Len) with monthly samples (**C**) and 15-day samples (**D**). Pond shown in **C** and **D** was occasionally dried in January when we could not record data for the monthly sample (**C**), but only for the 15-day samples (**D**)

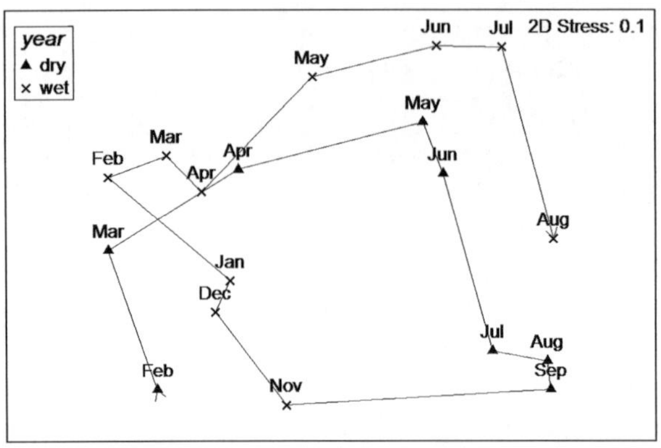

Fig. 6 NMDS ordination of the relative abundance of macroinvertebrates showing monthly variation after averaging across all the study ponds per month in the consecutive dry year and wet year

matrices was not significant in the case of serial correlation and was very low in the case of cyclic correlation (Table 3). The correlations were higher in three of the four ponds sampled monthly compared to every 15-day sampled ponds. In these three ponds, monthly macroinvertebrate assemblages presented higher serial than cyclic correlations, while in one pond they only presented a significant cyclic correlation for 15-day samples (Table 3). These ponds exhibited a high variability among 15-day samples in the NMDS representation, pointing out their fluctuating trajectory which was not detected among monthly samples (Fig. 5D, C).

General pattern of monthly variation in the macroinvertebrate community

We detected differences in the macroinvertebrate assemblages of all ponds and months using the sampling month in both study years as grouping factor, with an ANOSIM analysis (global $R = 0.475$, $P = 0.01$). It showed that the variation among months was higher than among ponds for each month. The monthly variation of the community showed a similar pattern in both years, observed in the NMDS representation. The highest similarity was found between the months of April of both years (Fig. 6). The monthly macroinvertebrate community presented a strong serial correlation (Serial RELATE) in both the dry year ($\rho = 0.916$; $P = 0.01$) and the wet year ($\rho = 0.788$; $P = 0.01$).

Discussion

Macroinvertebrate fauna of temporary ponds

Our temporary ponds had a rich macroinvertebrate fauna with similar or even higher richness than other temporary (Schneider & Frost, 1996; Bazzanti et al., 1996; Brooks, 2000; Boix et al., 2001) or permanent ponds (Heino, 2000; Della Bella et al., 2005). The high richness found in our study does not correspond to a single pond, but to a system of temporary ponds which allow movement and dispersal of individuals among ponds (Fortuna et al., 2006). In this kind of systems, the high connectivity and non-fragmentation area are very important factors to conserve their invertebrate biodiversity (Briers & Biggs, 2005; Van de Meutter et al., 2006). In the past, temporary ponds were usually excluded from conservation plans for wetlands, neglecting the diversity of their associated fauna due to their small size and temporal behaviour (Williams et al., 2001; Grillas et al., 2004; Williams, 2006; Zacharias et al., 2007). The high richness of macroinvertebrates in temporary ponds justifies the necessity of their conservation, this also being important since they include different fauna from permanent aquatic habitats, including many rare species (Collinson et al., 1995). These temporary habitats also allow the occurrence of many species which are vulnerable to predation and adapted to survive their characteristic dry phase (Wellborn et al., 1996; Williams, 2006).

Inter-annual variation

Temporary ponds are fluctuating habitats, and in this study we have detected significant changes in their macroinvertebrate composition. Many physical characteristics of temporary ponds are widely dependent on rainfall, with important variation from dry year to wet year. Consequently, macroinvertebrate assemblages may differ among wet year and dry year (Jeffries, 1994). Historical events, such as very dry years, may affect the macroinvertebrate community composition as much as site-specific abiotic differences among ponds (Boulton & Lake, 1992). Between our dry and wet study years, the same ponds differed in their hydroperiod, as well as in water depth and area, and accordingly we also found significant differences in the macroinvertebrate composition between years, despite these being consecutive. The shorter hydroperiod of the ponds in the dry year constrained the length of the aquatic period for macroinvertebrates. Thus, the occurrence of larvae of Coleoptera and Odonata was more concentrated and we detected differences in the peak of abundance of Coleoptera and Odonata that occurred in May in the dry year, 1 or 2 months earlier than in the wet year (June–July).

Seasonal variation

From filling to desiccation, temporary ponds experience large physicochemical variations (García Novo et al., 1991; Serrano & Toja, 1995; Gómez-Rodríguez et al., 2009), characterizing different phases according to the wet period (Bazzanti et al., 1996). Particular macroinvertebrate compositions have been described as characteristic of different wet phases of such ponds. They are explained as a consequence of the changes experienced in these aquatic habitats, which present optimal environmental conditions for different macroinvertebrates (Boulton & Lake, 1992; Boix et al., 2004; Culioli et al., 2006). In fact, different taxa of macroinvertebrates show wide differences in their life strategies, such as in reproduction, feeding, development or dispersal, and other particularities of their life cycle (Bilton et al., 2001; Williams, 2006; Verberk et al., 2008). The macroinvertebrate groups obtained in the NMDS representation of ponds and months also revealed this variation in macroinvertebrate assemblages (including adults, larvae and nymphs), changing through the different wet phases of the ponds. However, the shorter pond hydroperiod of the dry year also reduced the number of phases observed in this year relative to the wet year. During the dry year, only three distinct phases were identified: filling phase, aquatic phase and drying phase, while during the wet year, five phases were detected: filling phase, winter, early spring, late spring phases (aquatic phase) and drying phase. The reduction of the number of phases in years of low rainfall causes macroinvertebrates to synchronize their life histories (Wiggins et al., 1980; Nilsson, 2005b) concentrating biological processes into the short hydroperiod available. In ponds with very short hydroperiods, the number of phases may be even lower than three (Boix et al., 2004). In dry years, organisms with long life cycles may be the taxa most affected by short hydroperiods, as they cannot successfully complete their aquatic development (Schneider & Frost, 1996; Taylor et al., 1999).

The filling phase, just after pond formation, is characterized by the arrival of coleopterans and heteropterans through dispersal from other (more permanent) ponds (Wiggins et al., 1980). The taxa most characteristic of this phase did not coincide in our two study years, probably because the date of filling occurred in different seasons in both years, affecting the activity cycles of the species. We also found other macroinvertebrates taxa that usually spend the dry period in the mud, such as adults of *Berosus signaticollis* (Charpentier, 1825) (Boix et al., 2001) or adults of some species of Hydrophilidae which have a period of flight to colonize new habitats in newly filled ponds (Wiggins et al., 1980; Hansen, 2005; Williams, 2006).

The aquatic phase was longer in the wet year, and also the species characteristic of this or these phases were different among years, except for Notonectidae larvae which mostly appeared in the late spring phase of the wet year, having their peak abundance in the same month of the dry year. The environmental conditions of the wet year appear to have favoured particular species, such as *Cloeon* spp., which was very abundant in the wet year, being the only taxon characteristic of the three phases, winter, early and late spring that constituted the aquatic phase of this year. In contrast, its abundance was not high during the dry year.

The taxa most characteristics of the drying phases of both years did not coincide either in both study years, although adult corixids were characteristic of this phase in both the dry year and wet year. In the drying phase, adult heteropterans and coleopterans were the most common taxa in our study ponds, as described in other studies (Boulton & Lake, 1992; Culioli et al., 2006; Garrido & Munilla, 2008). Some beetles and almost all hemipterans possess excellent dispersal capabilities (Wiggins et al., 1980; Bilton et al., 2001). The high abundance of these taxa may be explained by the arrival of dispersers, moving from dry ponds to other ponds while dispersing to more permanent habitats to survive during dry periods (Wiggins et al., 1980; Higgins & Merrit, 1999; Bilton et al., 2001, Williams, 2006). We observed some dispersing individuals landing in some of our study ponds during the drying phase, such as *Colymbetes fuscus* (Linnaeus, 1758), *Gerris thoracicus* Schummel, 1832, and *C. affinis*. In the dry phase, as well as in the filling phase, the increase in number of species recorded in particular ponds were mainly due to dispersers, as reported for summer and autumn seasons by Verberk et al. (2005).

Monthly variation of macroinvertebrate assemblages within ponds

The variation of the macroinvertebrate composition in the ponds was not only attributable to differences between a wet and a dry year, or to the wet phases. Our study ponds had been chosen within a wide hydroperiod gradient, and while most of them filled in approximately the same month of each year, they clearly differed in the timing of desiccation, with some ponds drying earlier than others. As a consequence of the different desiccation times of the ponds, we observed different monthly variation in macroinvertebrate assemblages: some ponds showing cyclic correlation with similar assemblage composition at the beginning and the end of the hydroperiod, while others differed at these two phases, showing serial correlations instead. These differences may be explained in relation to the capability of many species to move between ponds via dispersal (Bilton et al., 2001; Rundle et al., 2002; Williams, 2006). At the end of hydroperiod, many Dytiscidae and Hydrophilidae suddenly leave the water, dispersing to more permanent waters (Nilsson, 2005a). Some adults and larvae can also leave the water and bury into the mud for pupation or as resistance stages (Hansen, 2005; Nilsson, 2005b, c, d) waiting for the next filling phase. Ponds with serial correlations would correspond to: (a) early drying ponds in which coleopterans and heteropterans were forced to move as desiccation progressed and (b) ponds with very long hydroperiod with a high abundance of heteropterans (mainly corixids) in summer. In contrast, ponds with cyclic correlations would correspond to intermediate hydroperiod ponds which still have water when the other ponds are drying and could act as intermediate sites during dispersal of organisms towards more permanent aquatic habitats, with similar taxa occurring in drying and filling phases. All the ponds with non-significant or weak correlation values had short hydroperiods, indicating that they were much more fluctuating than the other ponds. It was detected mainly in 15-day sampled pond compositions when compared with monthly samples of the same ponds. Richness and biodiversity have been related to the stability along of time (White, 2004), being invertebrate assemblages more stable in ponds with more permanence of water and highly fluctuating in ephemeral ponds (Shurin, 2007). We recommend increasing the frequency of samples along the time in ephemeral ponds with respect more permanent ponds to record all the variability of their macroinvertebrate assemblages, and maybe of other groups like macrophytes, amphibians and other invertebrates.

General pattern of monthly variation in the macroinvertebrate community

Despite differences in macroinvertebrate composition among ponds and in the same ponds at seasonal and inter-annual scales, a general pattern was detected, with similar variation between both years. This may be associated with the general development of the life cycle of many macroinvertebrates within the hydrological cycle of temporary ponds. From flooding to desiccation, we detected the successive appearance of adults, larvae and nymphs at different phases of the ponds. In long hydroperiod years, this general pattern may be extended from autumn to summer, while in short hydroperiod years it is concentrated. In our two study years, we detected a similar variation in macroinvertebrate composition through both wet phases. We also detected a similar composition between filling months although it occurred in February in the dry year and November in the wet year, as well as between the last month of the drying phases (August). This consistent general pattern revealed a high monthly correlation during both years, which was apparently repeated in the 2 years of the study.

Acknowledgements We are grateful to Azahara Gómez Flores, Alexandre Portheault and Carlos Marfil Daza for assistance in fieldwork. We are grateful to David Bilton for identifying *Bagous* spp. and reviewing the language, and the anonymous referees for comments which have improved the manuscript. We wish to thank the Spanish Ministry of Science and Innovation and EU Feder Funds (CGL2006-04458/BOS and Fellowship grants CSIC-I3P to M.F. and AP-2001-3475 to C.G.-R.) and Junta de Andalucía (Excellence Research Project 932). The fellowship grant to M.F. was funded by European Union Social Fund.

References

Agüesse, P., 1962. Quelques Odonates du Coto Doñana. Archivos del Instituto de Aclimatación de Almería 11: 9–12.

Batzer, D. P., B. J. Palik & R. Buech, 2004. Relationships between environmental characteristics and macroinvertebrate communities in seasonal woodland ponds of Minnesota. Journal of the North American Benthological Society 23: 50–68.

Bazzanti, M., S. Baldoni & M. Seminara, 1996. Invertebrate macrofauna of a temporary pond in Central Italy: composition, community parameters and temporal succession. Archiv für Hydrobiologie 137: 77–94.

Bigot, L. & F. Marazanof, 1966. Notes sur l'êcologie des colêoptêres aquatiques des marismas du Guadalquivir et premier inventaire des colêoptêres et lêpidoptêres du Coto Doñana (Andalucia). Annales de Limnologie 2: 491–502.

Bilton, D. T., J. R. Freeland & B. Okamura, 2001. Dispersal in freshwater invertebrates. Annual Review of Ecology and Systematics 32: 159–181.

Bilton, D. T., L. McAbendroth, P. Nicolet, A. Bedford, S. D. Rundle, A. Foggo & P. M. Ramsay, 2009. Ecology and conservation status of temporary and fluctuating ponds in two areas of Southern England. Aquatic Conservation: Marine and Freshwater Ecosystems 19: 134–146.

Boix, D., J. Sala & R. Moreno-Amichi, 2001. The faunal composition of Espolla pond (NE Iberian Peninsula): the neglected biodiversity of temporary waters. Wetlands 21: 577–592.

Boix, D., J. Sala, X. D. Quintana & R. Moreno-Amichi, 2004. Succession of the animal community in a Mediterranean temporary pond. Journal of the North American Benthological Society 23: 29–49.

Boulton, A. J. & P. S. Lake, 1992. The ecology of two intermittent streams in Victoria, Australia. II. Comparisons of faunal composition between habitats, rivers and years. Freshwater Biology 27: 99–121.

Briers, R. A. & J. Biggs, 2005. Spatial patterns in pond invertebrate communities: separating environmental and distance effects. Aquatic Conservation: Marine and Freshwater Ecosystems 15: 549–557.

Brooks, R. T., 2000. Annual and seasonal variation and the effects of hydroperiod on benthic macroinvertebrates of seasonal forest ("vernal") ponds in Central Massachusetts, USA. Wetlands 20: 707–715.

Céréghino, R., J. Biggs, B. Oertli & S. Declerck, 2008. The ecology of European ponds: defining the characteristics of a neglected freshwater habitat. Hydrobiologia 597: 1–6.

Clarke, K. R. & R. M. Warwick, 2001. Change in marine communities: an approach to statistical analysis and interpretation, 2nd ed. PRIMER-E, Plymouth.

Collinson, N. H., J. Biggs, A. Corfield, M. J. Hodson, D. Walker, M. Whitfield & P. J. Williams, 1995. Temporary and permanent ponds: an assessment of the effects of drying out on the conservation value of aquatic macroinvertebrate communities. Biological Conservation 74: 125–133.

Culioli, J. L., J. Foata, C. Mori, A. Orsini & B. Marchand, 2006. Temporal succession of the macroinvertebrate fauna in a corsican temporary pond. Viet et Milieu-Life and Environment 56: 215–221.

Della Bella, V., M. Bazzanti & F. Chiarotti, 2005. Macroinvertebrate diversity and conservation status of Mediterranean ponds in Italy: water permanence and mesohabitat influence. Aquatic Conservation: Marine and Freshwater Ecosystems 15: 583–600.

Díaz-Paniagua, C., 1990. Temporary ponds as breeding sites of amphibians at a locality in Southwestern Spain. Herpetological Journal 1: 447–453.

Díaz-Paniagua, C., C. Gómez-Rodríguez, A. Portheault & W. de Vries, 2005. Los Anfibios de Doñana. Organismo Autónomo de Parques Nacionales. Ministerio de Medio Ambiente, Madrid.

Fortuna, M. A., C. Gómez-Rodríguez & J. Bascompte, 2006. Spatial network structure and amphibian persistence in stochastic environments. Proceedings of the Royal Society B: Biological Sciences 273: 1429–1434.

García Murillo, P., R. Fernández Zamudio, S. Cirujano & A. Sousa, 2006. Aquatic macrophytes in Doñana protectes area (SW Spain): an overview. Limnetica 25: 71–80.

García-Novo, F. & C. Marín, 2006. Doñana. Water and biosphere. Doñana 2005, Confederación Hidrográfica del Guadalquivir. Spanish Ministry of Environment, Madrid.

García Novo, F., D. Galindo, J. A. García Sánchez, C. Guisande, J. Jáuregui, T. López, N. Mazuelos, J. C. Muñoz, L. Serrano & J. Toja, 1991. Tipificación de los ecosistemas acuáticos sobre sustrato arenoso del Parque Nacional de Doñana. Instituto Tecnológico GeoMinero de España 1: 165–177.

Garrido, J. & I. Munilla, 2008. Aquatic Coleoptera and Hemiptera assemblages in three coastal lagoons of the NW Iberian Peninsula: assessment of conservation value and response to environmental factors. Aquatic Conservation: Marine and Freshwater Ecosystems 18: 557–569.

Gascón, S., D. Boix, J. Sala & X. D. Quintana, 2008. Relation between macroinvertebrate life strategies and habitat traits in Mediterranean salt marsh ponds (Empordà wetlands, NE Iberian Peninsula). Hydrobiologia 597: 71–83.

Gómez-Rodríguez, C., J. Bustamante, S. Koponen & C. Díaz-Paniagua, 2008. High-resolution remote-sensing data in amphibian studies: identification of breeding sites and contribution to habitat models. Herpetological Journal 18: 103–113.

Gómez-Rodríguez, C., C. Díaz-Paniagua, L. Serrano, M. Florencio & A. Portheault, 2009. Mediterranean temporary ponds as amphibian breeding habitats: the importance of preserving pond network. Aquatic Ecology. doi 10.1007/s10452-009-9235-x.

Grillas, P., P. Gauthier, N. Yavercovski & P. Perennou, 2004. Mediterranean Temporary Pools, Vol. 1. Issues relating to conservation, functioning and management. Station Biologique de la Tour du Valat, Arles.

Hansen, M., 2005. Coleoptera Hydrophiloidea and Hydraenidae, water scavengers beetles. In Nilsson, A. (ed.), The Aquatic Insects of North Europe, Vol. 1. Apolo Books, Denmark: 146–173.

Heino, J., 2000. Lentic macroinvertebrate assemblage structure along gradients in spatial heterogeneity, habitat size and water chemistry. Hydrobiologia 418: 229–242.

Herbst, D. B., 2001. Gradients of salinity stress, environmental stability and water chemistry as a templet for defining habitat types and physiological strategies in inland salt waters. Hydrobiologia 466: 209–219.

Higgins, M. J. & R. W. Merrit, 1999. Invertebrate seasonal patterns and trophic relationships. In Batzer, D., R. B. Rader & S. A. Wissinger (eds), Invertebrates in Freshwater Wetlands of North America. Wiley, New York: 279–297.

Jeffries, M., 1994. Invertebrate communities and turnover in wetland ponds affected by drought. Freshwater Biology 32: 603–612.

Marazanof, F., 1967. Ostracodes, Cladocères, Hétéroptères et hydracariens noveaux pour les Marismas du Guadalquivir (Andalousie). Données écologiques. Annales de Limnologie 3: 47–64.

Millán, A., C. Hernando, P. Aguilera, A. Castro & I. Ribera, 2005. Los coleópteros acuáticos y semiacuáticos de Doñana: reconocimiento de su biodiversidad y prioridades de conservación. Boletín de la Sociedad Entomológica Aragonesa 36: 157–164.

Montes, C., L. Ramírez & A. Soler, 1982. Variación estacional de las taxocenosis de odonatos, coleópteros y heterópteros acuáticos en algunos ecosistemas del Bajo Guadalquivir durante un ciclo anual. Anales de la Universidad de Murcia 38: 19–100.

Nilsson, A. N., 2005a. Coleoptera, Introduction. In Nilsson, A. (ed.), The Aquatic Insects of North Europe, Vol. 1. Apolo Books, Denmark: 146–173.

Nilsson, A. N., 2005b. Coleoptera Dytiscidae, diving water beetles. In Nilsson, A. (ed.), The Aquatic Insects of North Europe, Vol. 1. Apolo Books, Denmark: 146–173.

Nilsson, A. N., 2005c. Coleoptera Gyrinidae, whirligig beetles. In Nilsson, A. (ed.), The Aquatic Insects of North Europe, Vol. 1. Apolo Books, Denmark: 146–173.

Nilsson, A. N., 2005d. Coleoptera Haliplidae, crawling water beetles. In Nilsson, A. (ed.), The Aquatic Insects of North Europe, Vol. 1. Apolo Books, Denmark: 146–173.

Rundle, S. D., A. Foggo, V. Choiseul & D. T. Bilton, 2002. Are distribution patterns linked to dispersal mechanism? An investigation using pond invertebrate assemblages. Freshwater Biology 47: 1571–1581.

Schneider, D. W. & T. M. Frost, 1996. Habitat duration and community structure in temporary ponds. Journal of the North American Benthological Society 15: 64–86.

Serrano, L. & K. Fahd, 2005. Zooplankton communities across a hydroperiod gradient of temporary ponds in the Doñana National Park (SW Spain). Wetlands 25: 101–111.

Serrano, L. & J. Toja, 1995. Limnological description of four temporary ponds in the Doñana National Park (SW, Spain). Archiv für Hydrobiologie 133: 497–516.

Serrano, L., M. Reina, G. Martín, I. Reyes, A. Arechederra, D. León & J. Toja, 2006. The aquatic systems of Doñana (SW Spain): watersheds and frontiers. Limnetica 25: 11–32.

Shurin, J. B., 2007. How is diversity related to species turnover through time? Oikos 116: 957–965.

Siljeström, P. A., A. Moreno, L. V. García & L. E. Clemente, 1994. Doñana National Park (south-west Spain): geomorphological characterization through a soil-vegetation study. Journal of Arid Environments 26: 315–323.

Strayer, D., 2006. Challenges for freshwater invertebrate conservation. Journal of the North American Benthological Society 25: 271–287.

Taylor, B. E., D. A. Leeper, M. A. McClure & A. E. DeBiase, 1999. Carolina Bays: ecology of aquatic invertebrates and perspectives on conservation. In Batzer, D., R. B. Rader & S. A. Wissinger (eds), Invertebrates in Freshwater Wetlands of North America. Wiley, New York: 167–196.

Van de Meutter, F., R. Stoks & L. De Meester, 2006. Lotic dispersal of lentic macroinvertebrates. Ecography 29: 223–230.

Verberk, W. C. E. P., H. H. Van Kleef, M. Dijkman, P. Van Hoek, P. Spienrenburg & H. Esselink, 2005. Seasonal changes on two different spatial scales: response of aquatic invertebrates to water body and microhabitat. Insect Science 12: 263–280.

Verberk, W. C. E. P., H. Siepel & H. Esselink, 2008. Life-history strategies in freshwater macroinvertebrates. Freshwater Biology 53: 1722–1738.

Waterkeyn, A., P. Grillas, B. Vanschoenwinkel & L. Brendonck, 2008. Invertebrate community patterns in Mediterranean temporary wetlands along hydroperiod and salinity gradients. Freshwater Biology 53: 1808–1822.

Wellborn, G., D. K. Skelly & E. E. Werner, 1996. Mechanisms creating community structure across a freshwater habitat gradient. Annual Review of Ecology and Systematics 27: 337–363.

White, E. P., 2004. Two-phase species-time relationships in North American land birds. Ecology Letters 7: 329–336.

Wiggins, G. B., R. J. Mackay & I. M. Smith, 1980. Evolutionary and ecological strategies of animals in annual temporary pools. Archiv für Hydrobiologie 58(Suppl): 97–206.

Williams, D. D., 1997. Temporary ponds and their invertebrate communities. Aquatic Conservation: Marine and Freshwater Ecosystems 7: 105–117.

Williams, D. D., 2006. The Biology of Temporary Waters. Oxford University Press, Oxford.

Williams, P., J. Biggs, G. Fox, P. Nicolet & M. Whitfield, 2001. History, origins and importance of temporary ponds. In Freshwater Biological Association (ed.), European Temporary Ponds: A Threatened Habitat. Freshwater Biological Association, Birmingham: 7–15.

Zacharias, I., E. Dimitrou, A. Dekker & E. Dorsman, 2007. Overview of temporary ponds in the Mediterranean region: threats, management and conservation issues. Journal of Environmental Biology 28: 1–9.

Hydrobiologia (2009) 634:185–194
DOI 10.1007/s10750-009-9893-7

POND CONSERVATION

Ten-year dynamics of vegetation in a Mediterranean temporary pool in western Morocco

Laïla Rhazi · Patrick Grillas · Mouhssine Rhazi · Jean-Christophe Aznar

Published online: 5 August 2009

Abstract The aim of this work was to test the hypotheses that the species composition of the vegetation of one pool in Morocco change continuously along with rainfall fluctuations, that among the vegetation can be recognized Pool species and Opportunistic species with distinct dynamics in time. We expected the Pool species to show lower inter-annual variation than the Opportunistic species. This hypothesis was tested in a 10-year study of the species composition of the vegetation along two permanent transects. The results showed high cumulative species richness (95 species) with large differences between years and a predominance of annual species (77). The proportion of Pool species during these 10 years was low (39%) when compared to opportunists (61%). In dry years the Opportunistic species were dominant and declined during wet years. The number of Pool species was correlated with the amount of rainfall. A large number of these species revealed a preference for wet years. No negative interaction between annuals/perennials and pools/non-pools species was found, suggesting that competition was not a major process during the survey. The intensity of the drought and flood stress, related to climate fluctuations, seems to be the main factors controlling the species composition of the vegetation of this unstable habitat. However, beyond the inter-annual fluctuation of the species composition of the vegetation a directional change was noticed. This directional change could result from a recovery process of the vegetation during the first years of the study after a severe flood which extirpated most of the Opportunistic species of the pool. In the last years this directional change of the species composition of the vegetation is less clear and random recruitment of the Opportunistic species from the surrounding forested habitats could contribute to explain inter-annual changes. The data collected over

Electronic supplementary material The online version of this article (doi:10.1007/s10750-009-9893-7) contains supplementary material, which is available to authorized users.

Guest editors: B. Oertli, R. Cereghino, A. Hull & R. Miracle
Pond Conservation: From Science to Practice. 3rd Conference of the European Pond Conservation Network, Valencia, Spain, 14–16 May 2008

L. Rhazi
Laboratory of Aquatic Ecology and Environment, Hassan II University, BP 5366 Maarif, Casablanca, Morocco
e-mail: rhazilaila@yahoo.fr

P. Grillas (✉)
Tour du Valat, Research Centre for the Conservation of Mediterranean Wetlands, Le Sambuc, 13200 Arles, France
e-mail: grillas@tourduvalat.org

M. Rhazi
Department of Biology, Faculty of Sciences and Techniques of Errachidia, Mouslay Ismail University, BP 509 Boutalamine, Errachidia, Morocco
e-mail: rhazimouhssine@yahoo.fr

J.-C. Aznar
Direction of the Forest Research, Québec, 2700, Einstein, Sainte-Foy, QC G1P 3W8, Canada

these 10 years led to the speculation of hypotheses on the consequences of climate change. The expected reduction of humid years and of rainfall regionally may lead to important changes in the species composition of the vegetation of the temporary pools in Morocco.

Keywords Temporary pools · Inter-annual dynamics · Plant community · Hydrology · Climate change · Morocco

Introduction

Temporary aquatic habitats are varied ecosystems extensively widespread on a world scale (Deil, 2005; Williams, 2006). They play an important role in the landscape such as controlling flooding, refilling aquifers, retaining toxic products, and recycling nutrients (Keddy, 2000; Williams, 2006). They also have an important use for the population (Williams et al., 2004; Biggs et al., 2005) and constitute a remarkable habitat for unique flora and fauna contributing to regional biodiversity (Williams et al., 2004; Biggs et al., 2005; Oertli et al., 2008).

Among these temporary humid habitats, special attention should be given to Mediterranean temporary pools that are present in the five regions of the world with Mediterranean climate (Europe: Braun-Blanquet, 1936; Barbéro et al., 1982; Quézel, 1998; Grillas et al. 2004; Australia: Jacobs & Brock, 1993; California: Barbour & Major, 1977; Zedler, 1987; Keeley & Zedler, 1998; South America: Bliss & Zedler, 1998; South Africa: Been et al., 1993; North Africa: Chevassut & Quézel, 1956; Nègre, 1956; Rhazi et al., 2006). Mediterranean temporary pools are very important habitats for biodiversity in particular for plants (Médail et al., 1998; Grillas et al., 2004; Rhazi et al., 2006), invertebrates (Thiéry, 1991; Waterkeyn et al., 2008), and amphibians (Jacob et al., 2003). They are extensively recognized as important zones in terms of conservation, because of their fast disappearance due to human activity in North Africa (urbanization, infilling, agriculture, etc.; Grillas et al., 2004; Saber, 2006).

Often occupying endoreic depressions, temporary pools show a high diversity of size, depth, shape, landscape context, use, and habitat characteristics (substrate, salinity, etc.; Grillas et al., 2004; Rhazi et al., 2006). In spite of this diversity, temporary pools share important common features related to the hydrologic cycle and to adaptations to the alternation of flooded and dry phases. These conditions encourage the abundance of short cycle species, adapted to the unpredictability of this habitat in terms of water regime (Thiéry, 1991; Médail et al., 1998; Grillas et al., 2004).

Temporary pools are sensitive habitats to environmental changes, in particular those affecting their hydrological functioning (e.g., Zedler, 1987; Keeley & Zedler, 1998). The local climate conditions, the topography, and the catchment size influence their hydrology (Brooks & Hayashi, 2002; Bauder, 2005; Brooks, 2005). The large intra- and inter-annual rainfall variability of the Mediterranean climate led to important differences between years in terms of water volume and the periods of setting in water (Metge, 1986) thus determining the specific composition of the communities (Bonis, 1993; Grillas & Battedou, 1998; Waterkeyn et al., 2008). The temporary flooded pools are the main habitat for some species including rare ones (e.g., *Pilularia minuta*, *Elatine brochonii*, and *Marsilea strigosa*). The pools also host more Opportunistic species which find their main habitat in the surrounding landscape. Due to their small size and simple community structure, temporary pools are often considered as early warning systems of the effects of the long-term changes to larger aquatic systems (for example, changes in the hydroperiod due to the global change; De Meester et al., 2005; Pyke, 2005; Hulsmans et al., 2008).

The climate changes for the Mediterranean region foresee a decrease of the rainfall, an increase in the evapotranspiration, and a higher frequency of severe droughts in the south of the Mediterranean basin (Bates et al., 2008; Elouali, 2008) as well as the amplification of the variance of the monthly and annual rainfall (Alibou, 2002). The consequences of climate changes could strongly influence the dynamics and the specific composition of plant communities due to the strong sensitivity of the pools to the variations of the hydrologic balance. The species with the highest requirements in terms of height, duration, and phenology of flooding could be negatively affected. In contrast, the terrestrial species would be favored as well as those with short and flexible life cycles (Thomas et al., 2004). Strong human pressure on the pools (drainage, infilling, agriculture, etc.) and its use as a freshwater source constitute supplementary effects that could accelerate these changes.

The aim of this work was to study the inter-annual dynamics of vegetation in relation to the rainfall during a 10-year period in a temporary pool in Morocco, while distinguishing the characteristic species of the pools from the opportunist ones and the annual species from the perennials. The initial hypotheses were the following:

(1) every year, only a fraction of the total richness of vegetation expresses itself and its expression is dependent on hydrologic conditions;
(2) the Opportunistic species of the pools constitute an important fraction of their biodiversity, and vary significantly according to hydrology cycles; and
(3) the characteristic species of temporary pools are more stable in time than Opportunistic species because the former have large permanent seed banks; in contrast, the Opportunistic species should not have well established populations, their presence putatively resulting from successful establishment from the seed rain from surrounding habitats during the short favorable periods of drought.

Materials and methods

The study site is in the province of Benslimane (western Morocco), located between Rabat and Casablanca. This region has a semi-arid Mediterranean climate, with a mean annual rainfall of 450 mm and large inter-annual fluctuations (range: 142–803 mm for the period of 1961–2006). This province has a remarkably high density of temporary pools (670) covering a total surface area of 1,994 ha, about 0.8% of the total surface of the area (Saber, 2006). These pools are very diverse in terms of size, depth, shape, landscape features, and use (Rhazi et al., 2006). They are grazed extensively by cattle and sheep; at the regional scale the pools outside forested areas are submitted to high human pressure mostly by urbanization and agriculture (Rhazi et al., 2001a; Saber, 2006).

Within this system of temporary pools, only one site (N 33°38,497′, W 007°05,242′; elevation: 259 m; area: 0.4 ha) located in the Benslimane cork oak forest, was selected for an inter-annual vegetation survey (Rhazi et al., 2001b). This pool is located on an underlying quartzitic sandstone rock and a silty-clayey, acidic soil (Rhazi et al., 2001b) and is representative of the forest pools of the area (Rhazi et al., 2001a).

The vegetation of this pool was studied from 1997 to 2006 with two sampling dates per year (March–April and June) in 79 permanent quadrats (0.3×0.3 m, divided into nine squares of 0.1×0.1 m) regularly distributed (every 2 m) along two permanent orthogonal transects ($T1 = 80$ m and $T2 = 74$ m) starting near the trees of cork oak and passing through the deepest part of the pool. The species frequency was measured in each quadrat in each visit as the number of squares in which it was found (0–9). The data of the two sampling date for each year were integrated into single annual values: the frequency of each species was the maximum value found in the two sampling dates, the species richness per quadrat was the cumulated number of species found at the two dates.

The water level was measured at each quadrat at the same date of the vegetation survey (March–April and June) and also at January when the pools are usually at maximum flooding conditions. Two groups of species were distinguished, i.e., (1) Pool species: aquatic and amphibious species (sensu lato), identified as characteristic species of Mediterranean temporary pools by Nègre (1956) and Médail et al. (1998) or more generally of wetlands and (2) Opportunistic: terrestrial species typically found outside wetlands. The annual or perennial trait was attributed following the flora of North Africa (Maire, 1952–1987) and Morocco (Fennane et al., 1999, 2007). The total richness and the richness of the specific groups (annuals, perennials, Pool, and Opportunistic) were calculated for each year. A correspondence analysis (CA; ADE-4 software) was performed on the 10-year data vegetation, considering for each species and each year the maximum value of frequency observed by quadrat. The CA was carried out with 78 species, excluding those that found in less than four quadrats (<0.5% of the total number of quadrats) over the 10 years. The barycenters of the distribution of the quadrats of each year were positioned on the ½ biplot. Variations in time of the total species richness, measured as the total number of species listed per year in all of the quadrats, as well as of its components ("Characteristic"/"Opportunistic", "Annual"/"Perennial") were studied by nonparametric correlations of Spearman. The difference in frequencies for each species between dry and wet years was tested by nonparametric variance analysis (Kruskal–Wallis). The species tested were those present in at least three out of the 10 years. The statistical analyses were conducted using JMP™ software.

Results

The period in study (1997–2006) was characterized by significant inter-annual rainfall fluctuations with four wet (1997, 2003, 2004, and 2006) and six dry years (1998, 1999, 2000, 2001, 2002, and 2005) (Fig. 1). The maximum levels of water recorded in the pool varied between 40 and 54 cm in the wet years against 5–17 cm in the dry years. The duration of the flooding periods ranged between 3 and 23 weeks. The average rainfall during the 10 years (439 mm) was slightly lower than that of the 45-year period (458 mm), but without significant difference ($P > 0.05$). The January (maximum) water levels were strongly correlated with the annual rainfall ($r^2 = 0.97$; $P < 0.0001$; $n = 10$).

A total of 95 species was found in the pool during the study period and, the annual richness varied between 28 (1997) and 68 species (2003). Of this total, 18 species occurred only 1 year, conversely 13 species were present every year. A total of 37 Pool species were found (39% of the total richness) and 58 were listed as Opportunistic (61%). The annuals (77 species) represented 81% of the accumulated total richness over the 10 years (perennials = 19%).

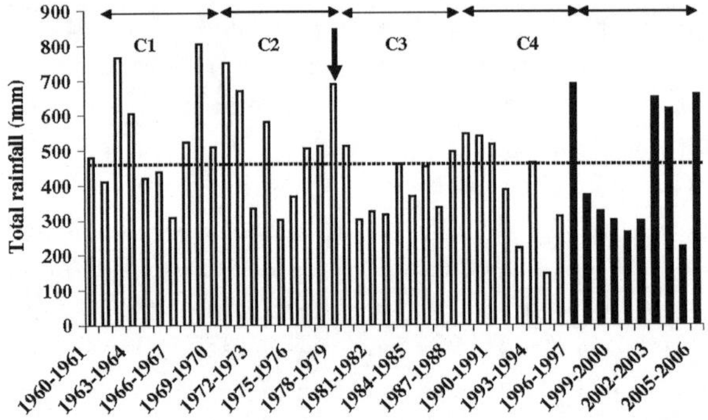

Fig. 1 Total rainfall during annual hydrological cycles (September–August) between 1960 and 2006 at Benslimane (in *black* are the years studied; *broken line*: average of 1961–2006; C1, C2, C3, and C4 correspond to cycles similar to the survey period)

The cumulated frequency of species over the 10 years was significantly higher for the Pool species than for the Opportunistic species (nonparametric test of Wilcoxon, one-way test, $\chi^2 = 25.38$; $P < 0.001$). However, some Opportunistic species occurred at high frequencies (found in 9 years out of 10), notably *Leontodon saxatilis* (218), *Narcissus viridiflorus* (189), *Scilla autumnalis* (177), and *Filago gallica* (154) (Supplementary material—Annex 1).

Temporal dynamics of the vegetation

Axis 1 (32% of variance) of the CA (Fig. 2) opposes terrestrial (*Sanguisorba minor*, *Cistus salviifolius*, *Lolium rigidum*...) and aquatic species (*Myriophyllum alterniflorum*, *Callitriche brutia*, *Ranunculus peltatus*...). Axis 2 (26% of total variance) opposes the forest terrestrial species (*Gaudinia fragilis*, *Stachys arvensis*, *Anagallis arvensis*...) and amphibious characteristic pool species (*Pilularia minuta*, *Damasonium stellatum*, *Exaculum pusillum*, *Elatine brochonii*, etc.).

The annual barycenter of the quadrats shows an important displacement on the ½ biplot of the CA (Fig. 2). The coordinates on axis 1 of the annual barycenter were significantly correlated with the maximum depth of water ($r^2 = 0.84$; $P < 0.001$; $n = 10$). This axis 1 opposes the wet (1997, 2003, 2004, and 2006) and the dry years (1998, 1999, 2000, 2001, 2002, and 2005). The coordinates on axis 2 of the barycenter of the distribution of the quadrats for each year, increased significantly with time (Spearman $\rho = 0.91$; $P < 0.001$) and they are also significantly correlated with the total species richness (linear regression, $r^2 = 0.68$; $P < 0.01$; $n = 10$).

Species richness

The total number of species found each year in the pool was the lowest (28) in 1997 (Fig. 3); it significantly increased during the survey period (Spearman $\rho = 0.91$; $P < 0.001$) reaching 68 species in 2003 (142% increase) and then fluctuated around 60 species for the following years. The number of Pool species found each year fluctuated between

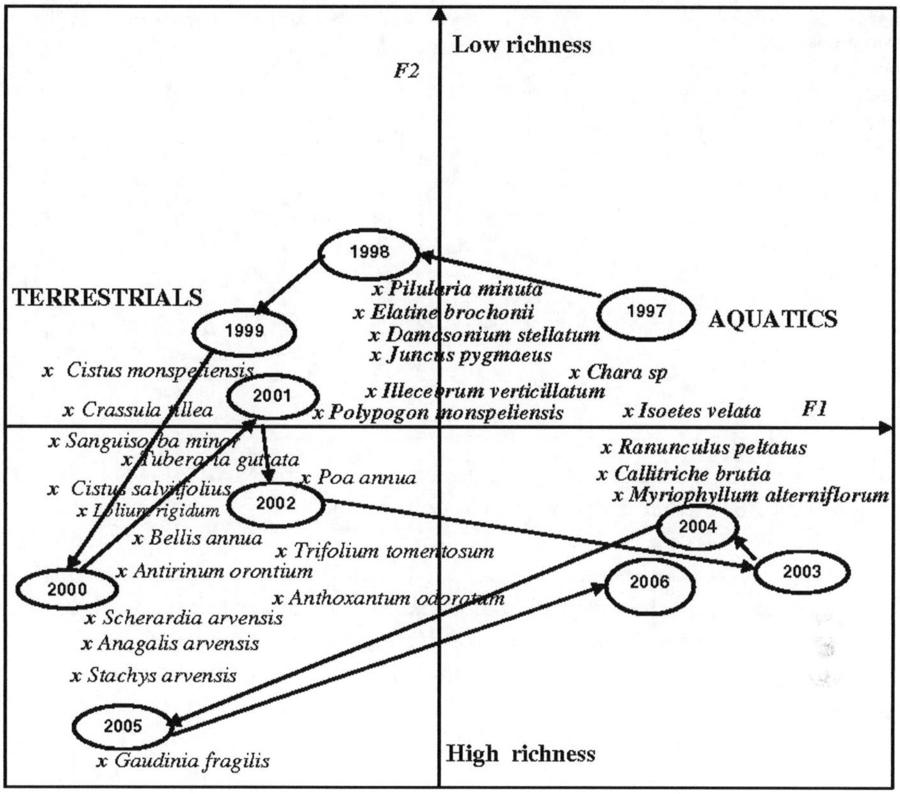

Fig. 2 Plot ½ of the CA vegetation (1997–2006) with positioning of the barycenters of the years (in *bold* characteristic pool species and in *non-bold* terrestrial species)

years between 20 and 35 species. The Pool species contributed only to 39% of the total species richness during the 10 years; they were, however, more frequent than the Opportunistic species. The Opportunistic species were numerous (58 species: 61% of the total number of species found in the pool), but generally less frequent over the 10 years. Some Opportunistic species were however found every year (*Cynodon dactylon*) or absent for only 1 year (Supplementary material—Annex 1). Among these two annual (*Leontodon saxatilis* and *Filago gallica*) and perennial species (*Narcissus viridiflorus* and *Scilla autumnalis*) (Supplementary material—Annex 1) were found 9 years out of 10 each of them occupying a cumulated number of quadrats higher than 150 (close to 20% or more of the maximum possible).

The number of Pool species was significantly correlated with rainfall ($r^2 = 0.61$; $P < 0.01$) and did not show a significant trend in time (Spearman $\rho = 0.34$; $P = 0.34$). In contrast, the annual number of Opportunistic species, was not correlated with rainfall ($r^2 = 0.21$; $P = 0.17$) and significantly increased during the survey period (Spearman $\rho = 0.66$; $P < 0.05$) from a very low value (three species) in 1997 to 39 in 2005. This increase showed two phases: until 2001, the number of Opportunistic species steadily increased from 3 to 32 species; from 2001 to 2006 it fluctuated around 30 species (Fig. 3). The contribution of the Opportunistic species to the flora of the pool increased from 10% of flora in 1997 to about 50% from 2003 to 2006 (Fig. 3).

The numbers of perennial and annual species were similar during the first year (1997) with, respectively, 11 and 17 species. The number of perennials significantly increased during the survey period (Spearman $\rho = 0.88$; $P < 0.001$) reaching a maximum value of 18 species in 2003, 2004, and 2006. The number of annuals also increased significantly (Spearman $\rho = 0.89$; $P < 0.001$) reaching a maximum value of 50 in 2003. The increase of the annual species showed two distinct phases: a fast increase until 2001 reaching 39 followed by inter-annual fluctuations around a mean value of 41 species. Over the 10 years of survey the contribution of the annual species to the total flora of the pool rose from 60 to

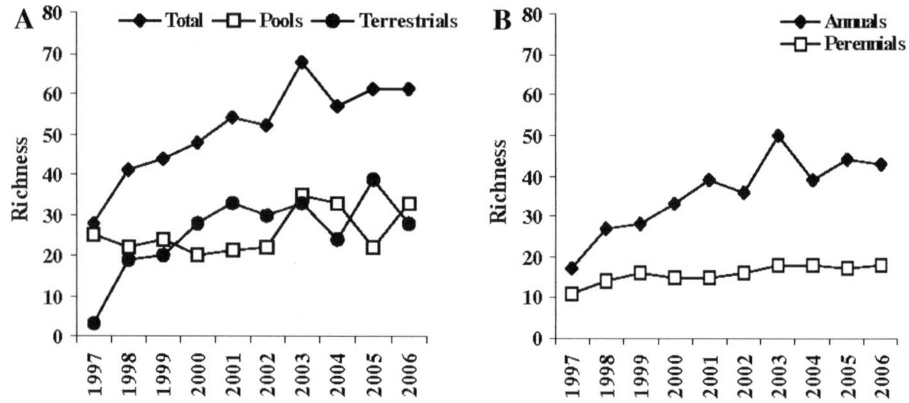

Fig. 3 Inter-annual variations of the number of species: **A** characteristic pool species, terrestrials, and total; **B** perennial and annual species

70%. There was no significant correlations between the numbers of annual and perennial species found every year ($P > 0.05$).

Among the 37 Pool species, 14 were significantly more frequent during the wet years and two during the dry years (*Polypogon monspeliensis* and *Hypericum tomentosum*) (Table 1). The annual number of Pool species did not show a significant correlation with the number of the Opportunistic species ($P > 0.05$).

Discussion

Species composition of the vegetation of the pool

The vegetation of the studied pool is representative of the oligotrophic Mediterranean temporary pools of Morocco (Rhazi et al., 2001a) with notably high richness in annual species (>65% of the total). The abundance of annual species, often with very short life cycles, is characteristic of the Mediterranean temporary pools of the Old-World (e.g., Nègre, 1956; Boutin et al., 1982; Médail et al., 1998) and of the vernal pools of California (Zedler, 1987) where it is usually interpreted as an adaptive strategy to the unpredictability of the environmental conditions (e.g., Keeley & Zedler, 1998; Deil, 2005; Grillas et al., 2004), notably the hydrology.

Two groups of species have been distinguished in this study: Pool species and Opportunistic species. The Pool species group includes the species usually found in the plant communities of the Isoeto-Nanojuncetea class (*Isoetes velata*, *Isoetes histrix*, *Juncus bufonius*, *Juncus pygmaeus*, *Juncus capitatus*, etc.) which is considered as a characteristic of Mediterranean temporary pools (Deil, 2005). More generally it contains species which are most generally found in wetland habitats such as *Ranunculus baudotii*, *Bolboschoenus maritimus* (Supplementary material—Annex 1). In this group, six species were rare or threatened in Morocco (*Elatine brochonii*, *Pilularia minuta*, *Lythrum thymifolia*, *Myriophyllum alterniflorum*, *Isoetes velata*, and *Exaculum pusillum*) (Fennane & Ibn Tattou, 1998). The Opportunistic species group is made of terrestrial species encompassing both forests species (e.g., *Cistus monspeliensis*) and species found in a wide range of habitats (e.g., *Trifolium campestre*, *Cynodon dactylon*). The Pool species contributed only to 39% of the total species richness during the 10 years and the majority of the species encountered in the pool were Opportunistic species. In each given year, the numbers of Opportunistic and Pool species were similar except in the first years when Pool species dominated. This large contribution of non-Pool species to about half of the total species richness is striking suggesting that the pool receives a significant seed or propagule rain from the surrounding habitats. These Opportunistic, terrestrial, species found in the pool a suitable habitat during the dry phases (Keeley & Zedler, 1998) where their establishment is probably further favored by the generally low cover of vegetation. At the edge of the pool, flooding occurs only during wet years and the terrestrial vegetation, including perennial species, such as *Cistus* spp., intolerant to flooding can develop during several years (e.g., five dry years after 1997 flood). The invasibility of the plant communities of the vernal pool vegetation has been highlighted in California by the encroachment of exogenous invasive species (Gerhardt & Collinge, 2003).

Table 1 Results of the tests for significant differences in the occurrence of species between wet (4) and dry years (6); each species is characterized as Pool/Non-Pool (P/NP), with the number of years the species was present during the 10 years (Total), during the four wet years (Wet) and the six dry years (Dry), the preference for dry or wet years, the value of χ^2 (Kruskall–Wallis) and the probability [* $P < 0.05$; ** $P < 0.01$; (*) P is very near to 0.05]

Species	Pool/Non-Pool	Occurrence (years)			Test		
		Total	Wet	Dry	Preference	χ^2	P
Damasonium stellatum	P	4	4	0	Wet	8.37	**
*Elatine brochonii**	P	4	4	0	Wet	8.37	**
*Pilularia minuta**	P	4	4	0	Wet	8.3	**
*Lythrum thymifolium**	P	5	4	1	Wet	6.93	**
Lythrum hyssopifolium	P	8	4	4	Wet	6.75	**
Eleocharis palustris	P	10	4	6	Wet	6.62	**
Heliotropium supinum	P	9	4	5	Wet	6.62	**
Scirpus pseudocetaceus	P	3	3	0	Wet	5.62	*
Juncus capitatus	P	3	3	0	Wet	5.62	*
Lythrum borysthenicum	P	6	4	2	Wet	5.36	*
*Exaculum pusillum**	P	6	4	2	Wet	4.39	*
Cerastium glomeratum	NP	4	3	1	Wet	4.3	*
Mentha pulegium	P	8	4	4	Wet	3.79	*
*Isoetes velata**	P	10	4	6	Wet	4.12	*
Spergula arvense	NP	9	3	6	Dry	5.36	*
Polypogon monspeliensis	P	9	3	6	Dry	4.15	*
Scilla autumnalis	NP	9	3	6	Dry	5.07	*
Hypericum tomentosum	P	10	4	6	Dry	4.6	*
Anthirinum orontium	NP	4	0	4	Dry	3.8	(*)

For the pools of California, Keeley & Zedler (1998) distinguished different patterns of plant dynamics between endemic Pool-obligate species and cosmopolitan wetland species. Our approach here, and thus the way the groups were made, differs in the sense that our focus was on the understanding of the dynamics of communities considering the processes rather than the biogeographical issues. More detailed groups could have been used such as the seven functional groups identified by Brock & Casanova (1997); however, the needed information was not available for most species. Furthermore, the lower number of species that would have resulted from more groups would have limited the statistical power of the analyses although it would have probably also resulted in lower intra-group variance.

The results of the CA are consistent with the grouping of the species, the two groups being well separated on axis 1. This axis separates the terrestrial species (*Cistus* spp., *Sanguisorba minor*, etc.) from those found in the center of the pool (*Myriophyllum alterniflorum*, *Callitriche brutia*, *Ranunculus peltatus*). The most characteristic species of the Mediterranean temporary pool are found in the center of the graphs which results from both their intermediate position in the depth gradient and their frequency throughout the pool.

Inter-annual fluctuations of the vegetation

The survey period (1997–2006) was characterized by an alternation of contrasted dry and humid years. The first year (1997) was exceptionally wet with maximum water levels recorded in the pool (54 cm). Four analogous rainfall cycles (mostly dry years with the exceptional extremely wet years) characterized the climate between 1962 and 1995 in the region of Benslimane (Fig. 1). The rain distribution pattern for the survey period (10 years) is representative of the last 45 years, with similar average rainfall (439 mm against 458 mm), but very dry years (<321 mm) which are more frequent (five in 10 years against 13 in 45). This 45-year rainfall pattern can be divided into two successive phases, with a transition toward drier conditions from the beginning of the 1980s. Eleven of the 13 driest year of the period 1960–2006 occurred after 1980. The rainfall pattern (Fig. 1) is consistent with the predictions for climatic changes (Bates et al., 2008), i.e., an increase in temperature, a reduction of the annual rainfall leading to an increase in the frequency and intensity of droughts, and the occurrence of unusually humid years (Jalil, 2001; Elouali, 2008).

The species composition of the vegetation in the pool differed from 1 year to the next with large difference in species richness. In any given year could be found only 30–70% of the total number of species found during the 10 years of survey (Fig. 3). The variations of the total species richness showed an inter-annual dynamic that followed two distinct patterns: (1) an alternation between dry and humid years and (2) a directional dynamic accompanied by an increase of the specific richness. These variations clearly showed that the specific composition of the pool vegetation is not steady and varies over time according to rainfall conditions (Jeffries, 2008).

The number of Pool species was correlated with the amount of rainfall, but did not show any significant trend over the 10 years. Within this ecological group, there was however, a large variation in the frequency of the different species corresponding to different life strategies more particularly at the establishment phase (Brock & Casanova, 1997):

- Some species, such as *Damasonium stellatum*, *Elatine brochonii**, *Pilularia minuta**, *Scirpus pseudocetaceus*, and *Juncus capitatus*, were solely present in the wet years (Table 1). These species, of which two are rare and of high patrimonial interest for Morocco (*), are restricted to the edges of the pool. The survival of these species implies the adoption of adaptive strategies related to the germination/reproduction and spatial displacements in search of favorable niches. These species are considered as more threatened by increasing drought as it is likely to decrease the frequency of reproductive success and therefore increase stochastic extinction risk.
- Species, such as *Isoetes velata**, *Exaculum pusillum**, *Lythrum thymifolium**, *Lythrum hyssopifolium*, *Lythrum borysthenicum*, *Eleocharis palustris*, and *Heliotropium supinum*, of which three are rare (*), were present in both dry and wet years, but showed a preference for the latter (Table 1). These species are likely to be threatened by climatic change similarly as the previous group.
- The remaining species, which include *Polypogon monspeliensis* and *Hypericum tomentosum*, were present in both wet and dry years, but showed a preference for the latter (Table 1). Unlike the others, these species could be, on one hand, favored by climate changes due to the rise in drought frequency and intensity.

10 years dynamics of communities: hydrological control, resilience, or drift?

Beyond inter-annual fluctuations, a directional process (Holland & Jain, 1981) has been identified over the 10 years. The high water levels in 1997 probably explain the very low frequency of Opportunistic (terrestrial) species (in particular perennials). A fast re-colonization of the edges of the pool by Opportunistic species was observed during the five following years which were dry (Fig. 1). The number of Opportunistic species increased until it reached a plateau at about 30 species (Fig. 3) from 2001. This phase appears as a recovery phase after a major disturbance. It was only in 2004 that the number of Opportunistic species showed a clear decline after two successive wet years. The inter-annual dynamics of the vegetation from 2004 seems to enter into a new inter-annual dynamics with opposite changes in the respective patterns of Pool and Opportunistic species apparently driven by rainfall and flooding conditions. Both perennial and annual species richness increased significantly over time, but the increase of the annual species richness was more accentuated (Fig. 3B).

The number of Pool species showed lower inter-annual variations than the Opportunistic species suggesting the former species constitute rather stable components of the communities, while Opportunistic species are temporary colonizers during the dry years (Keeley & Zedler, 1998). No negative interaction was found between the groups of species (Annual/Perennial and Pool/Opportunistic) suggesting that competition was not a major factor influencing the richness of the plant communities. The stress intensity, resulting from the climate fluctuations, seems to be the main factor controlling the species richness of the vegetation (Jeffries, 2008). However, the lack of correlation could be the result of the large increase of the number of Opportunistic species during the first years of survey. After a recovery stage, when species richness returns to high values, competitive interactions are expected in the community (Rhazi et al., 2001b).

The dynamics of the vegetation in the studied temporarily flooded pool provide clues for the consequences of climatic changes in Morocco. The results show large variations in the species composition of the vegetation with two groups of species contrasting in their dynamics: Pool species and terrestrial Opportunistic species. These two groups may separate "niche" structured communities while the second could be the result of stochastic processes. The former group constitutes the core of the plant communities adapted to the harsh environmental conditions met in temporary pools. The latter group contains species randomly recruited from surrounding habitats. Because of the contrast between the vegetation of the pool and the vegetation of surrounding dry habitats, temporary pool are probably valuable habitats for the identification of the pattern controlling the dynamics of communities (De Meester et al., 2005). These preliminary results obtained on one single pool should be confirmed with more sites. Large changes in the species composition are common in temporary flooded pools. Increasing drought may lead to important changes in the species composition of the plant communities. Long-term monitoring of the vegetation and water levels in the pool as well as a detailed analysis of the different border and center communities, with possible spatial displacements of the species along the topographic gradient, would lead to a better understanding of the climate change phenomena and of the species responses.

Acknowledgments We thank Deirdre Flanagan for correcting the English, Serge D. Muller (University of Montpellier 2) for providing material through the Egide Volubilis program (AI.-MA/07-172), and Florence Daubigney for her logistical and technical support and the three anonymous referees for their suggestions and critiques that helped improving this work. This project has been achieved with the financial support of the EGIDE-CMIFM program (PHC Volubilis AI- No. MA/07/172) and was partly funded by the Fondation Sansouire and Fondation MAVA.

References

Alibou, J., 2002. Impacts des changements climatiques sur les ressources en eau et les zones humides du Maroc. Rapport CERSHE-EHTP, Ministère de l'Aménagement du territoire, de l'Urbanisme, de l'Habitat et de l'Environnement, Département de l'Environnement: 42 pp.

Barbéro, M., J. Giudicelli, R. Loisel, P. Quézel & E. Terzian, 1982. Etude des biocénoses des mares et ruisseaux temporaires à éphémérophytes dominants en région méditerranéenne française. Bulletin d'Ecologie 13: 387–400.

Barbour, M. G. & M. Major, 1977. Terrestrial Vegetation of California. John Wiley and Sons, New York.

Bates, B., Z. W. Kundzewicz, S. Wu & J. Palutikof, 2008. Climate Change and Water IPCC Technical Paper: 214 pp.

Bauder, E. T., 2005. The effects of an unpredictable precipitation regime on vernal pool hydrology. Freshwater Biology 50: 2129–2135.

Been, C. M., J. Heeg & M. Seaman, 1993. Wetlands of Africa: South Africa. In Whigham, D. F., D. Dykyjova & S. Hejny (eds), Wetlands of the World I.: Inventory, Ecology, and Management. Kluwer, Dordrecht: 47–79.

Biggs, J., P. Williams, P. Whitfield, P. Nicolet & A. Weatherby, 2005. 15 years of pond assessment in Britain: results and lessons learned from the work of Pond Conservation. Aquatic Conservation: Marine and Freshwater Ecosystems 15: 693–714.

Bliss, S. A. & P. H. Zedler, 1998. The germination process in vernal pools: sensitivity to environmental conditions and effects on community structure. Oecologia 113: 67–73.

Bonis, A., 1993. Dynamique des communautés et mécanismes de coexistence des populations de macrophytes immergées en marais temporaires. Thèse de Doctorat de l'Université de Montpellier II: 173 pp.

Boutin, C., L. Lesne & A. Thiéry, 1982. Ecologie et typologie de quelques mares temporaires à Isoetes d'une région aride du Maroc occidental. Ecologia Mediterranea 8: 31–56.

Braun-Blanquet, J., 1936. Un Joyau floristique et phytosociologique, l'Isoetion méditerranéen. SIGMA, Communication 42.

Brock, M. A. & M. T. Casanova, 1997. Plant life at the edges of wetlands; ecological responses to wetting and drying patterns. In Klomp, N. & I. Lunt (eds), Frontiers in Ecology: Building the Links. Elsevier Science, Oxford, UK: 181–192.

Brooks, R. T., 2005. A review of basin morphology and pool hydrology of isolated ponded wetlands: implications for seasonal forest pools of the northeastern United States. Wetlands Ecology and Management 13: 335–348.

Brooks, R. T. & M. Hayashi, 2002. Depth-area-volume and hydroperiod relationships of ephemeral (vernal) forest pools in Southern New England. Wetlands 22(2): 247–255.

Chevassut, G. & P. Quézel, 1956. Contribution à l'étude des groupements végétaux des mares transitoires à *Isoetes velata* et des dépressions humides à *Isotes histrix* en Afrique du nord. Bulletin Société d'Histoire Naturelle d'Afrique du Nord 47: 59–73.

De Meester, L., S. Declercks, R. Stoks, G. Louette, F. V. De Meutter, T. De Bie, E. Michels & L. Brendonck, 2005. Ponds and pools as model systems in conservation biology, ecology and evolutionary biology. Aquatic Conservation: Marine and Freshwater Ecosystem 15: 715–725.

Deil, U., 2005. A review on habitats, plant traits and vegetation of ephemeral wetlands – a global perspective. Phytocoenologia 35: 533–705.

Elouali, A., 2008. Seconde communication nationale du Maroc sur les changements climatiques/vulnérabilité et adaptation du Maroc: volet climat/Etat de référence Ministère de l'Energie de Mine et de l'Environnement: 65 pp.

Fennane, M. & M. Ibn Tattou, 1998. Catalogue des plantes endémiques, rares ou menacées du Maroc. Bocconea 8: 1–243.

Fennane, M., M. Ibn Tattou, J. Mathez, A. Ouyahya & J. El Oualidi (eds), 1999. Flore Pratique du Maroc. Manuel de Détermination des Plantes Vasculaire, Vol. 1. Travaux de l'Institut Scientifique, Série Botanique, 36, Rabat.

Fennane, M., M. Ibn Tattou, A. Ouyahya & J. El Oualidi (eds), 2007. Flore pratique du Maroc. Manuel de Détermination des Plantes Vasculaire, Vol. 2. Travaux de l'Institut Scientifique, Série Botanique, 38, Rabat.

Gerhardt, F. & S. K. Collinge, 2003. Exotic plant invasions of vernal pools in the Central Valley of California, USA. Journal of Biogeography 30: 1043–1052.

Grillas, P. & G. Battedou, 1998. Effects of the date of flooding on the biomass, species composition and seed production of submerged macrophyte beds in temporary marshes in the Camargue (S. France). Proceedings of the Intecol Conference, Perth, September 1996. In McComb, A. J. & J. A. Davis (eds), Wetlands for the Future. INTECOL'S V International Wetland Conference: 207–218.

Grillas, P., P. Gauthier, N. Yavercovski & C. Perennou (eds), 2004. Mediterranean Temporary Pools: Volume 1. Issues Relating to Conservation, Functioning and Management. Tour du Valat, Arles.

Holland, R. F. & S. K. Jain, 1981. Spatial and temporal variation in plant species diversity in vernal pools. In Jain, S. & P. Moyle (eds), Vernal Pools and Intermittent Streams. Institute of Ecology, University of California, Davis: 198–209.

Hulsmans, A., B. Vanschoenwinkel, C. Pyke, J. Bruce, B. J. Riddoch & L. Brendonck, 2008. Quantifying the hydroregime of a temporary pool habitat: a modelling approach for ephemeral rock pools in SE Botswana. Ecosystems 11: 89–100.

Jacobs, S. W. L. & M. A. Brock, 1993. Wetlands of Australia: Southern (Termperate) Australia. In Whigham, D. F., D. Dykyjova & S. Hejny (eds), Wetlands of the World I. Inventory, Ecology, and Management. Kluwer, Dordrecht: 244–305.

Jacob, C., G. Poizat, M. Veith, A. Seitz & A. J. Crivelli, 2003. Breeding phenology and larval distribution of amphibians in Mediterranean pond network with unpredictable hydrology. Hydrobiologia 499: 51–61.

Jalil, M., 2001. Première communication nationale du Maroc sur les changements climatiques/vulnérabilité et adaptation du Maroc: volet climat Ministère de l'Eau, de Mine d'Energie et de l'Environnement: 55 pp.

Jeffries, M., 2008. The spatial and temporal heterogeneity of macrophyte communities in thirty small, temporary ponds over a period of ten years. Ecography 31: 765–775.

Keddy, P. A., 2000. Wetland Ecology: Principles and Conservation. Cambridge University Press, Cambridge, UK.

Keeley, J. E. & P. H. Zedler, 1998. Characterization and global distribution of vernal pools. In Witham, C. W., et al. (eds), Ecology, Conservation, and Management of Vernal Pool Ecosystems. Proceedings from a 1996 Conference. California Native Plant Society, Sacramento, CA: 1–14.

Maire, R. (ed.), (1952–1987). Flore de l'Afrique du Nord, 16 Vol. Lechevalier, Paris.

Médail, F., H. Michaud, J. Molina, G. Paradis & R. Loisel, 1998. Conservation de la flore et de la végétation des mares temporaires dulçaquicoles et oligotrophes de France méditerranéenne. Ecologia Mediterranea 24: 119–134.

Metge, G., 1986. Etude des écosystèmes hydromorphes (dayas et merjas) de la meseta occidentale marocaine. Thèse de Doctorat, Université Aix-Marseille III, Marseille: 280 pp.

Nègre, R., 1956. Notes sur la végétation de quelques dayas des Jbilets orientaux et occidentaux. Bulletin de la Société des Sciences Naturelles du Maroc 36: 229–241.

Oertli, B., N. Indermuehle, S. Angélibert, H. Hinden & A. Stoll, 2008. Macroinvertebrate assemblages in 25 high alpine ponds of the Swiss National Park (Cirque of Macun) and relation to environmental variables. Hydrobiologia 597: 29–41.

Pyke, C., 2005. Assessing climate change impacts on vernal pool ecosystems and endemic branchiopods. Ecosystems 8: 95–105.

Quézel, P., 1998. La végétation des mares transitoires à *Isoetes* en région méditerranéenne, intérêt patrimonial et conservation. Ecologia Mediterranea 24: 111–117.

Rhazi, L., P. Grillas, A. Mounirou Touré & L. Tan Ham, 2001a. Impact of land use and activities on water, sediment and vegetation of temporary pools in Morocco. Comptes Rendus de l'Académie des Sciences: Life Sciences 324: 165–177.

Rhazi, L., P. Grillas, L. Tan Ham & D. El Khyari, 2001b. The seed bank and the between years dynamics of the vegetation of a Mediterranean temporary pool (NW Morocco). Ecologia Mediterranea 27: 69–88.

Rhazi, L., M. Rhazi, P. Grillas & D. El Khyari, 2006. Richness and structure of plant communities in temporary pools from western Morocco: influence of human activities. Hydrobiologia 570: 197–203.

Saber, E., 2006. Approche de la dynamique spatio-temporelle des geosystèmes foret et mares temporaires de la province de Benslimane (Maroc Occidental). Thèse d'Université de Provence (Aix Marseille I): 188 pp.

Thiéry, A., 1991. Multispecies coexistence of branchiopods (Anostraca, Notostraca, Spinicaudata) in temporary ponds of chaouia plain (Western Morocco) – sympatry of syntopy between usually allopatric species. Hydrobiologia 212: 117–136.

Thomas, C. D., A. Cameron, R. E. Green, M. Bakkenes, L. J. Beaumont, Y. C. Collingham, B. F. N. Erasmus, M. F. de Siqueira, A. Grainger, L. Hannah, L. Hughes, B. Huntley, A. S. van Jaarsveld, G. F. Midgley, L. Miles, M. A. Ortega-Huerta, A. T. Peterson, O. L. Phillips & S. E. Williams, 2004. Extinction risk from climate change. Nature 427: 145–148.

Waterkeyn, A., P. Grillas, B. Vanschoenwinkel & L. Brendonck, 2008. Invertebrate community patterns in Mediterranean temporary wetlands along hydroperiod and salinity gradients. Freshwater Biology 53: 1808–1822.

Williams, D. D., 2006. The Biology of Temporary Waters. Oxford University Press, Oxford.

Williams, P., M. Whitfield, J. Biggs, S. Bray, G. Fox, P. Nicolet & D. Sear, 2004. Comparative biodiversity of rivers, streams, ditches and ponds in an agricultural landscape in Southern England. Biological Conservation 115: 329–341.

Zedler, P. H., 1987. The ecology of Southern California vernal pools: a community profile. U. S. Fish & Wildlife Service Biological Report, 85: 7–11.

Hydrobiologia (2009) 634:195–208
DOI 10.1007/s10750-009-9898-2

POND CONSERVATION

Modelling hydrological characteristics of Mediterranean Temporary Ponds and potential impacts from climate change

E. Dimitriou · E. Moussoulis · F. Stamati · N. Nikolaidis

Published online: 29 July 2009

Abstract 'Mediterranean Temporary Ponds' (MTP) constitutes a priority, substantially vulnerable and unstable habitat (Natura code: 3170*). In this article, the influences of climate change on the hydroperiod of two MTPs in Crete, have been quantitatively explored by using: (i) a physically based, semi-distributed lake basin model of Lake Kourna, where the hydrology of the lake is directly related to that of the adjacent MTP and (ii) a conceptual/mathematical model of an MTP in Omalos plateau. A water balance model was also set up to estimate net groundwater inflows for Lake Kourna and the basin. The water balance estimates and GIS tools were then used to set up the physically based model which was calibrated for the hydrological year 2005–2006 and validated for two periods: April–September 2005 and the hydrological year 2006–2007 (two split-sample tests). Calibration of the mathematical model was very good, while for the physically based model calibration was satisfactory. The two models were then setup and simulated for two future Intergovernmental Panel for Climate Change (IPCC) scenarios: A2 (pessimistic) and B2 (more optimistic). The responses of Lake Kourna and Omalos MTP water levels and their hydroperiods were then predicted. Results for IPCC B2 and A2 climate scenarios show longer hydroperiod and smaller decreases in the future for Omalos MTP than in Lake Kourna MTP. Results for Lake Kourna MTP demonstrated a hydroperiod decrease of more than 52 days after the application of the IPCC scenarios. Scenario A2 does not present a significantly different higher impact on the MTPs' hydroperiod.

Keywords Lake Kourna · Omalos · Mediterranean Temporary Ponds · Hydroperiod · Climate change

Guest editors: B. Oertli, R. Cereghino, A. Hull & R. Miracle
Pond Conservation: From Science to Practice. 3rd Conference of the European Pond Conservation Network, Valencia, Spain, 14–16 May 2008.

E. Dimitriou (✉) · E. Moussoulis
Institute of Inland Waters, Hellenic Centre for Marine Research, 19013 Anavissos Attikis, Greece
e-mail: elias@ath.hcmr.gr

F. Stamati · N. Nikolaidis
Department of Environmental Engineering, Technical University of Crete, 73100 Chania, Crete, Greece

Introduction

The targets of the National Strategy for Sustainable Development, regarding the management of water resources in Greece, are related to the sustainable use of water resources, the efficient protection of water ecosystems and the attainment of high quality standards for all the surface and ground water bodies by the year 2015. The main sectors of action are climate change abatement; water resources management;

combating desertification; protection of biodiversity and natural ecosystems (MEPPW, 2004).

Mediterranean Temporary Ponds (MTPs) constitute a priority habitat (Natura code: 3170*) in Annex I of the Directive 92/43/EC. This substantially vulnerable and unstable habitat exists in areas that, due to their specific characteristics, are under significant human and natural pressures and have become prone to extinction. The hydrology of MTPs can be characterised as self-adjusting and presents significant variations not only at the length of the ponds' hydroperiod but also at the start of their flooding period. These habitats often occur in small depressions with impermeable substratum and usually belong to a relatively small catchment area. They may also occur in karstic areas where groundwater flow originating from their catchment area results in the ponds' water level rise (Dimitriou et al., 2006). The point at which the pond passes into the dry phase depends on when the last significant rainfall or snowfall occurs in the region. According to Stamati & Nikolaidis (2006), the pond retains very small volumes of water for a day even with rainfall as low as 2 mm/day. The length of the hydroperiod defines the developing flora and fauna. These hydrological alterations are totally natural and as the water volume changes, the developing aquatic vegetation and invertebrates change, as well. During the pond's flooding, the aquatic habitat has available trophic resources, and the predatorial faunal activity is low. During the drought period, the higher faunal density leads to higher competition and appropriate conditions for predatorial activity (Collinson et al., 1995; Warwick & Brock, 2003; Grillas et al., 2004).

The threats that MTPs in Western Crete face are mainly due to human activities and interventions. Omalos MTP is frequented daily by many sheep and goats which use the pond for both watering and grazing, while Lake Kourna MTP is threatened by water abstractions as well as agricultural and grazing sources of pollution. Therefore, climate change constitutes an important factor to investigate to assess its relative potential impact on the MTPs' hydroperiod.

Hydrological models provide a framework to conceptualise and investigate the relationships between climate, human activities (e.g. land use changes in agriculture, urban areas, etc) and water resources and also to assess different management alternatives as well as land use and climate change scenarios.

In this article, the main objective is to quantitatively explore the influences of climate change on the hydroperiod of two MTPs in Crete. For this purpose, two different hydrologic models have been used: (i) a physically based, distributed lake basin model in the case of Lake Kourna, where the hydrology of the lake is directly related to that of the adjacent MTP and (ii) a conceptual/mathematical model of an MTP in Omalos based on previous study by Stamati & Nikolaidis (2006). The effects of two IPCC climate scenarios on Lake Kourna and Omalos MTP water levels will then be investigated.

Study areas

Lake Kourna is located in northern Crete (latitude 35°20′, longitude 24°16′ and altitude of 19 m ASL), at the foot of Lefka Ori mountains, about 2.5 km from the sea. Lake Kourna is the only big natural lake of Crete and the most southern lake of Europe, with a surface area of 0.6 km^2. The deepest point is found at 3.5 m below sea level, while the maximum water elevation is at 22 m. It is notable that two fish species which occur in Lake Kourna (*Blennious fluviatilis* and *Atherina boyeri*) do not exist in other freshwaters of Crete. The MTP habitat is located adjacent to the northern lake coastline, at roughly 50–100-m distance and is hydrologically directly connected to Lake Kourna, in the sense that when there is a water level rise above 19.5 m, the pond gets flooded with lake water.

The hydrologic basin of Lake Kourna has a surface area of approximately 19.7 km^2 and mean altitude 152 m above sea level. Its northeastern part is flat with altitude ranging from sea level to less than 100 m and slopes of less than 4%, while the southwest part of the basin is mountainous with altitudes reaching 1,200 m and slopes above 10% (Dimitriou et al., 2006) (Figs. 1, 2).

The region's climate is typically Mediterranean with dry-hot summers and mild winters. For the Lake Kourna region, historical data show that annually local rainfall has a mean 1,100 mm with values fluctuating from roughly 700 to 1,800 mm. The month with the highest rainfall is January (19% of the total), followed by December (18% of the total) while the lowest rainfall is observed during the period July–August (0.4% of the total) (Dimitriou et al., 2006). Temperature in the region is also relatively high with values

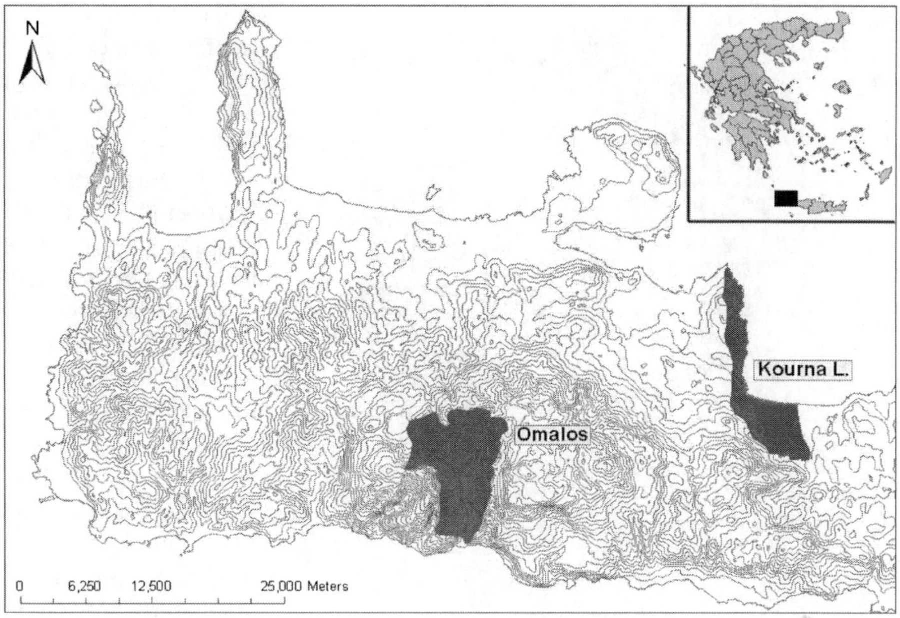

Fig. 1 Hydrologic basins of Omalos and Lake Kourna

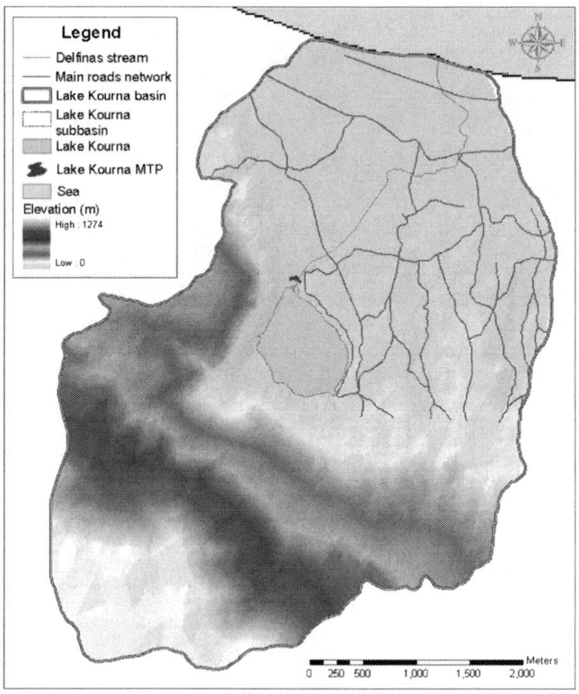

Fig. 2 Topographic map of Lake Kourna basin

fluctuating from 12 to 27°C (January and July, respectively) while mean annual temperature is approximately 19°C. The case of relatively high temperatures is expected in the region, since it concerns a coastal, mostly lowland area, influenced mainly by hot air masses from the Cretan Sea (Dimitriou et al., 2006).

Omalos plateau is located in the middle of Chania prefecture in the mountains of Lefka Ori. The surface extent and average elevation of the basin reach 26.7 km^2 and 1,183 m, respectively, while average slope of the basin is approximately 12°. Omalos MTP is located at the central part of the basin at an altitude of 1,050 m, and covers an area of 5.9 ha (Fig. 3).

Omalos basin is characterised by high rainfall/snowfall, therefore presenting longer hydroperiod compared to other MTPs in western Crete (Stamati & Nikolaidis, 2006). Observations showed that the aquifer's head elevation in the winter is marginally at the same height with the MTP, thus relative interaction occurs between groundwater and the pond's surface water. The time at which the pond passes from the wet to the dry phase depends mainly on the time that the last snowfall occurs, and therefore the point up to which snowmelt occurs (Stamati & Nikolaidis, 2006).

Mean precipitation in the region is 1,600 mm, while higher values (more than 300 mm per month) have been observed in December and November. The dry period occurs between May and September, where monthly rainfall does not exceed 25 mm, while in June, July and August, rainfall is almost absent. The significant amounts of precipitation in this particular region leads to extended MTP

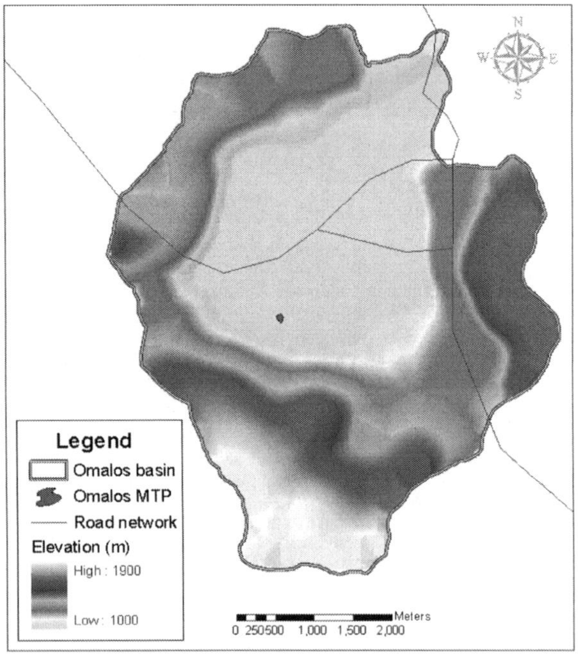

Fig. 3 Topographic map of Omalos basin

hydroperiod, which can go up to 10 months per year. February and March are the colder months with average minimum temperature of −0.6°C. July is the hottest month with an average maximum temperature of 23.2°C. The relatively low temperatures combined with lack of strong winds, because of the geomorphology (plateau), leads to low levels of evaporation which also contributes to the extended MTP hydroperiod (Stamati & Nikolaidis, 2006).

Materials and methods

In the absence of a model that can adequately describe both the MTPs, two different models have been used to illustrate the significant differences in their scale, hydrogeological regime and dynamics.

Model for the MTP in Lake Kourna

The pond in the area of Kourna is located on the shore of the lake and is greatly dependent on the lake, as it is mainly supplied by the flood runoff of the lake when lake's water level increases during the winter. The sediment in the area of the pond presents limited infiltration capacity, and thus, the main hydrologic processes are lake's flood runoff, rainfall, evaporation and infiltration. Therefore, the MTP's hydroperiod in Lake Kourna principally depends on the fluctuation of the lake's water level, since the lake's flood discharge constitutes the MTP's main water supply source. It was, therefore, considered that if lake's water elevation was over the altitude for 100% coverage of the MTP (at 19.5 m), then the MTP was in flooding phase.

For Lake Kourna MTP, MIKE SHE modelling software has been used (Abbott et al., 1986) which is a comprehensive, deterministic, physically based, spatially distributed hydrological model that has been widely used to study a variety of water resource and environmental problems under diverse climatological and hydrological regimes (Refsgaard & Storm, 1995; in Thompson et al., 2004).

Figure 4 presents the main hydrological processes affecting the MTP's hydroperiod.

The simulation period covered one hydrological year (2005–2006) for which hydrological (lake's water levels) and meteorological data were available. Daily time steps were considered for the hydrological model. The guiding principle in the parameterisation was to construct a simple model with as few parameters, subject to calibration, as possible. Topography, land use, soil and geological maps were preprocessed in GIS software and imported into the model. Precipitation, potential evaporation/evapotranspiration and pumping time series were used to define temporal variability.

The basin model domain was uniformly distributed into a finite-difference grid of fine cell size of 50 × 50 m, so that the ratio of grid cell area to basin surface area was from 1 to ~7,800, which allowed for realistic representation of hydrological variables without, at the same time, placing excessive demands on computational time. The Digital Bathymetric Model (DBM) of the lake has been developed by elaborating in GIS the lake's bathymetry contours (2-m resolution obtained by topographic maps 1:5,000, produced by the Greek Military Geographical Service. The Digital Elevation Model (DEM) of the basin has been developed by combining in GIS the DBM and topographical contours (20 m produced by the Greek Military Geographical Service) of the land part of the basin. For each land use in the basin (derived from the CORINE 2000 database) an appropriate vegetation/crop/land use type was

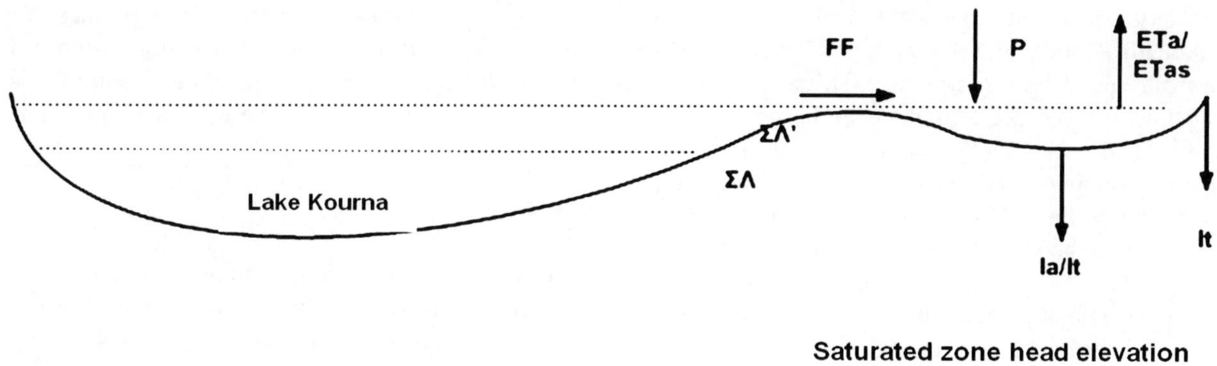

Fig. 4 Schematic diagram of the conceptual model describing the hydrologic cycle of the MTP in Lake Kourna (Adapted from Stamati & Nikolaidis, 2006). *P* precipitation, *ETa* evapotranspiration—aquatic, *ETs* evapotranspiration—terrestrial, *Ia* infiltration—aquatic, *It* infiltration—terrestrial, *FF* flood flow

selected from a vegetation/crop/land use property database along with its associated time series of Leaf Area Index (LAI) and Root Depth (RD).

Geological units were grouped in terms of their hydrogeological characteristics and specifically based on their permeability (high, medium and low) to keep the model as simple as possible. The saturated zone was defined using one impermeable flysch layer (350–1,000 m deep), on top of which three geological layers were placed: one thick limestone layer (350 m deep) of high permeability, covering the whole basin; one thin alluvial layer (80 m deep) of medium to high permeability extending at the central and north part of the basin; one thin layer consisting of silty and marly lake sediments (20 m). The geometry of these geological units was defined based on available geological cross-sections and geological maps.

Lake water levels were used as calibration targets. Model calibration was conducted by altering the hydraulic conductivity and storativity values until the simulated lake water levels matched closely the observed ones. Fifteen reference points evenly distributed ($\sim$200 m distance) within the lake ($\sim$1 point/3.8 ha) were used since it was not possible to account for all temporal variations of water level in every grid cell (230 cells within the lake). Additional check points were used to test the model's performance across the land part of the basin and throughout the calibration procedure. These fifteen calibration points inside the lake area have been used because the modelled lake's water levels present slight differences in the fluctuations at a spatial scale. This is due to the large water volumes entering the lake in winter and late spring, mainly through submerged springs. This particular lake is considered to be an extension of the local groundwater body, and therefore the respective water level fluctuations are partially transferred to the lake. However, the spatial differences in the lake's water level are at the magnitude of few centimetres, and the 15 calibration points have been chosen to ensure elimination of potential inconsistencies.

The model was calibrated for the hydrological period 2005–2006 and temporally validated for two split sample tests: (i) the period from April to September 2005 (6 months prior to the start of the original calibration period); (ii) the next hydrological year (2006–2007). Correlation-based and error-based numerical criteria were subsequently used to assess model's calibration and validation. The correlation-based performance criteria used in this study include the Correlation Coefficient (R) and the Nash–Sutcliffe Correlation Coefficient (R^2) while the error-based measures include the root mean squared error (RMSE), the mean absolute error (MAE), the mean error (ME) and the Standard Deviation of the Residuals (STDres) (Legates & McCabe, 1999).

Model for the MTP in Omalos

In the area of Omalos where the pond is located, the water table depth during the summer is found at approximately 18 m, while during the winter at 4 m below the ground surface. Thus, no direct interaction occurs between groundwater and pond's surface water. The pond's proximate drainage basin is small due to local topography (low slopes), and therefore overland inflow is expected to have minimum contribution to the MTP's water storage. Infiltration capacity experiments conducted in situ by Stamati & Nikolaidis (2006) showed that at the distance of 8 m from the pond's wet perimeter, infiltration velocity was low, in the order of 0.014 cm/min, while on the shore of the pond infiltration velocity was nil. This finding justifies observations that Omalos pond retains water during the summer months (longer hydroperiod). Thus, although infiltration/percolation processes cannot be omitted, the main hydrological processes are rainfall/snowfall, evaporation, evapotranspiration from the terrestrial part of the pond and overland inflow (Fig. 5). Thus, the model assumes three compartments: that of snow, terrestrial sediment (proximate direct drainage basin) and the pond (Fig. 5).

Figure 5 presents the main hydrological processes affecting the MTP's hydroperiod.

The conceptual approach of the MTP hydrologic cycle in Omalos allowed the development of a mathematical model named HPM in Matlab software for the determination of the MTP's hydroperiod. Input parameters of the HPM model such as surface extent and volumes of the terrestrial sediment and the pond's water resulted from the bathymetry/topography 3D model developed in GIS by Stamati & Nikolaidis (2006). At first, two relationships could be established between the pond's surface extent and water level, and between the pond's volume and water level. Then, the relationship between the pond's surface and the pond's volume was explicitly defined and presented in Fig. 11. Thus, the pond's surface and water level may then be calculated from the corresponding volume that results from the HPM model output.

The model solves the water storage balances for three compartments (snow, terrestrial sediment, pond) with the method of Euler on a daily basis. The equations that describe the hydrologic processes for each compartment are given below.

Snow, S

The water balance for the compartment of snow appears in Eq. 1. The change of snow storage (V_s) with time equals the difference of snowmelt from snowfall. When mean daily temperature (T) is lower than the temperature below which precipitation is in the snow phase (T_s), then snowmelt does not occur ($M_s = 0$) during that day and snowfall occurs ($P_s = \mathrm{cs} \times P$), equal to the rainfall recorded that day multiplied by a correction factor (cs). When mean daily temperature (T) is higher than temperature, T_s, then snowfall does not occur ($P_s = 0$). If snowfall occurs then, if no rainfall occurs, snow melts according to Eq. 2 or, if rainfall occurs, snow melts according to Eq. 3. Factor k (1.8 TEMPC)$^{(n+1)}$ characterises the day (degree day factor) while n usually takes the value of 0.25. Finally, if snow that is able to melt is more than storage, V_s, then snowmelt is corrected at this volume, V_s (Stamati & Nikolaidis, 2006)

$$\frac{dV_S}{dt} = \underbrace{P_s \times A_{sf}}_{\text{snowfall}} - \underbrace{\mathrm{cm} \times M_s \times A_s}_{\text{snowmelt}} \quad (1)$$

$$M_s = \left((T \times 1.8)^{(n+1)}\right) \times k/1000, \quad (2)$$

$$M_s = (((P \times 0.007 + 0.074) \times T \times 1.8 + 0.05) \times 0.254)/1000. \quad (3)$$

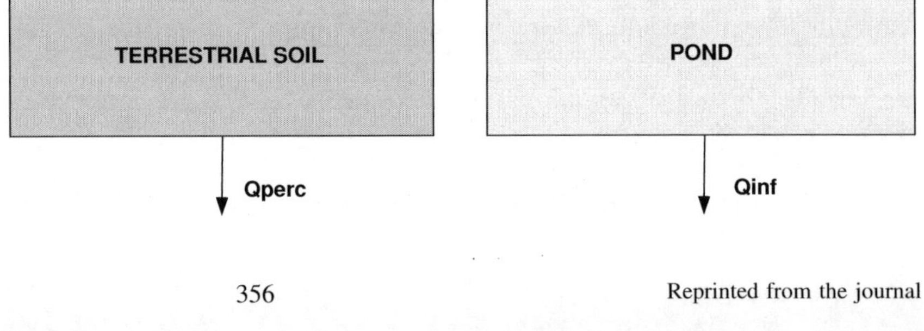

Fig. 5 Schematic diagram of the conceptual model describing the hydrologic cycle of the MTP in Omalos (Stamati & Nikolaidis, 2006). *P*/P_s rainfall/snowfall, *M* snowmelt, *PE* evaporation—aquatic, *ET* evapotranspiration—terrestrial, Q_{inf} infiltration—aquatic (central or peripheral), Q_{perc} infiltration/percolation—terrestrial, Q_{over} overland inflow, *Q* flood discharge

Terrestrial sediment, T

The water balance for the compartment of terrestrial sediment is given by Eq. 4, while the equations for the calculation of storages of percolation, evapotranspiration and overland flow from the terrestrial sediment are given by Eqs. 5, 6 and 7, respectively. The change of water storage in the terrestrial sediment (V_t) equals the difference of evapotranspiration, percolation and overland flow from the sum of rainfall and snowmelt (Stamati & Nikolaidis, 2006). The following conditions apply:

- when snowfall occurs ($P_s > 0$) during that day, rainfall, evapotranspiration and percolation are nil;
- when no inflow (rainfall or snowmelt) occurs in the compartment and the existing volume of water is equal to minimum ($V_{t_{min}}$: porosity volume multiplied by minimum humidity), evapotranspiration and percolation are nil;
- when the volume of water is more than minimum volume ($V_{t_{min}}$) and the active volume ($V - V_{t_{min}}$) is less than evapotranspiration and percolation, then the latter parameters are corrected to this volume with the corresponding percentages;
- similarly, when rainfall and/or snowmelt occur and the existing active volume of water along with the inflows are less than evapotranspiration and percolation, then the latter parameters are corrected to the volume given by the existing active volume of water with the inflows, with the corresponding percentages;
- finally, if the resulting volume is more than the maximum volume of terrestrial sediment ($V_{t_{max}}$: porosity volume), then the surplus volume gives additional overland flow (Ot).

$$\frac{dV_T}{dt} = \underbrace{M_{st}}_{\text{snowmelt}} + \underbrace{P_t}_{\text{rainfall}} - \underbrace{E_t}_{\text{evapotranspiration}} - \underbrace{PR_t}_{\text{percolation}} - \underbrace{O_t}_{\text{overland flow}} \quad (4)$$

$$PR_t = c_p \times f_c \times A_t \quad (5)$$

$$E_t = c_{et} \times E_t \times A_t \quad (6)$$

$$O_t = P \times A_{tf} + (M_s - f_c) \times A_t \quad (7)$$

where f_c: field's infiltration velocity (cm/min), c_p: percolation coefficient, A_t: surface of the terrestrial sediment, c_{et}: evaporation coefficient, A_{tf}: flat surface of the terrestrial sediment, M_s: snowmelt, and E_t: evapotranspiration.

Pond, P

The water balance for the pond compartment is given by Eq. 8, while the equations for calculating the storages of infiltration and evaporation from the pond's surface are given by Eqs. 9 and 10, respectively. The change of the pond's water storage (V_p) with time equals to the difference of evaporation, infiltration and outflow (flood runoff) from the sum of rainfall and snowmelt (Stamati & Nikolaidis, 2006).

The following conditions apply:

- when snowfall occurs ($P_s > 0$), at a particular day, rainfall, evaporation and infiltration are nil;
- on the other hand, when no inflows occur (rainfall or snowmelt) in the compartment and the existing water volume is nil, then evaporation and infiltration are nil;
- in the same case, when there is water storage available, if this is less than evaporation and infiltration, then the latter parameters are corrected to this volume with the corresponding percentages;
- similarly, when rainfall and/or snowmelt occur and the existing water volume along with the inflows are less than the sum of potential evaporation and infiltration, then the latter parameters are corrected down to the sum of the existing water volume and the inflows, with the corresponding percentages.

$$\frac{dV_P}{dt} = \underbrace{M_{st}}_{\text{snowmelt}} + \underbrace{P_P}_{\text{rainfall}} + \underbrace{O_t}_{\text{overlandinflow}} - \underbrace{E_P}_{\text{evapotranspiration}} - \underbrace{I_t}_{\text{infiltration}} - \underbrace{Q_P}_{\text{flood discharge}} \quad (8)$$

$$I_p = c_i \times f_c \times (\text{Apwet} - A_c) \quad (9)$$

$$E_p = c_{pe} \times P_e \times \text{Apwet} \quad (10)$$

where I_p: infiltration from the pond, c_i: infiltration coefficient, A_c: surface of the central part of the pond, Apwet: surface of the pond's wet area, c_{pe}: evaporation coefficient and P_e: evaporation.

Scenarios

In order to have an indication of the impact of climate change on Lake Kourna water level (and consequently on the adjacent MTP) and on the MTP water level in Omalos, two future climate scenarios were applied according to the IPCC (2007) climate predictions:

- One 'pessimistic' A2 IPCC scenario for 3.5°C increase in temperature (with respective increase of evaporation and evapotranspiration) and 0.25 mm/day decline of precipitation and
- one more 'optimistic' B2 IPCC scenario for 2.5°C increase in temperature (with respective increase of evaporation and evapotranspiration) and 0.25 mm/day decline of precipitation.

The only variables changed in the model were precipitation, evaporation and evapotranspiration. The results were then assessed by comparing the simulated water levels with the baseline-current scenario.

Results

Kourna lake MTP

The average hydroperiod value for the hydrological year 2005–2006 is 72 days, while for the next hydrological year 2006–2007 was estimated at 213 days (Fig. 6). Table 1 shows the hydroperiod values for calendar years 1996–1999 and for the hydrological years 2005–2006 and 2006–2007.

Table 1 Estimation of the hydroperiod of Lake Kourna MTP

Simulation period	Hydroperiod (days)
1996[a]	70
1997[a]	77
1998[a]	21
1999[a]	76
2005–2006	72
2006–2007	213

[a] Stamati & Nikolaidis (2006)

Figure 7 shows the simulated Lake Kourna water level (at calibration target point L02) and the observed water level for the calibration period 1 October 2005–30 September 2006. Numerical criteria tested on the calibration receptors (15 points in the lake) revealed satisfactory calibration of the model with an average of 72% for the R correlation coefficient and 31% for the R^2 (Nash–Sutcliffe) coefficient, while the average values for the error-based criteria MAE, RMSE and STDres were 0.68, 0.83 and 0.66, respectively. Apart from calibration receptors L02 and L05, which presented the lowest R correlation coefficient of 69 and 53%, respectively, the model worked well on simulating Lake Kourna water level.

The model was also tested on an annual basis to ensure its bulk performance by comparing the simulated overland (lake) storage change water balance output with the expected lake storage change estimated using the bathymetry/topography model of the lake in GIS. The results were similar and therefore, on an average, satisfactory behaviour of

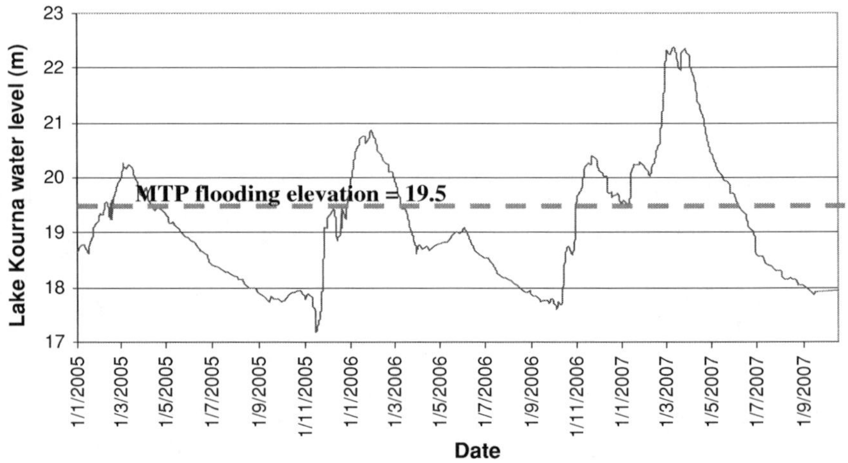

Fig. 6 Lake's Kourna water level fluctuation and the threshold water level above which the MTP is flooded (MTP's wet phase)

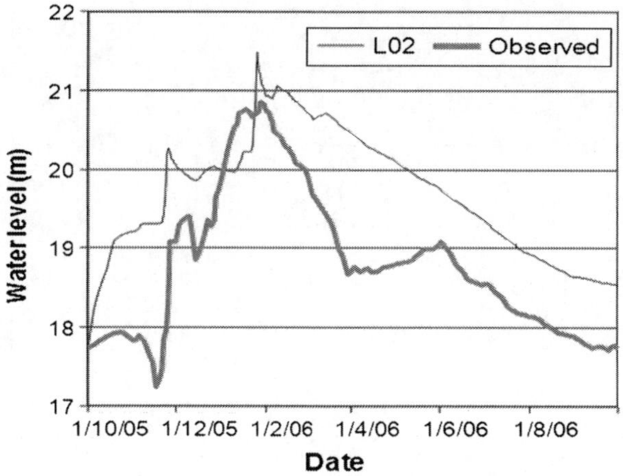

Fig. 7 Comparison between the simulated Lake Kourna water level (at 15 calibration target points within the lake) and the observed water level fluctuation (*double width line*) during the calibration period (hydrological year 2005–2006)

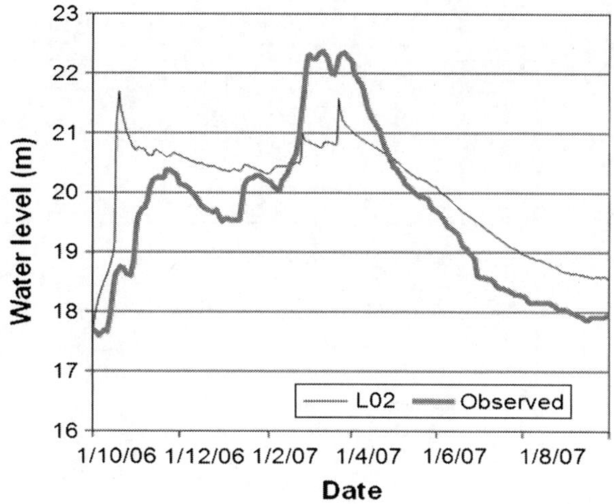

Fig. 8 Comparison between the simulated Lake Kourna water level (at 15 calibration target points within the lake) and the observed water level fluctuation (*double width line*) during the validation period of the 2006–2007 hydrological year (2nd split-sample test)

the hydrological model across the surface of Lake Kourna was achieved.

Numerical criteria tested on the validation receptors revealed good validation of the model with an average of 98% for the R correlation coefficient and 55% for the R^2 (Nash–Sutcliffe) coefficient, while the average values for the error-based criteria MAE, RMSE and STDres were 0.36, 0.42 and 0.23, respectively.

Figure 8 shows the simulated Lake Kourna water level (at L02 validation target point) and the observed water level for the validation period 1 October 2006–30 September 2007. Numerical criteria tested on the validation receptors revealed satisfactory validation of the model with an average of 77% for the R correlation coefficient and 32% for the R^2 (Nash–Sutcliffe) coefficient, while the average values for the error-based criteria MAE, RMSE and STDres were 0.84, 1.10 and 0.89, respectively. However, as it is evident from Fig. 8, the model is unable to simulate adequately the peaks in March and April 2007, as it underestimates the peaks by approximately 1 m. In addition, it overestimates observed values at the start of the simulation, whereas it follows water levels well during the recession months May–September.

The impact of climate change on Lake Kourna water level (and consequently on the adjacent MTP) and on the MTP water level in Omalos, was assessed by applying two future climate scenarios (IPCC B2 and A2) both on the calibrated 2005–2006 model (Baseline scenario 1) and the validated 2006–2007 model (Baseline scenario 2). Figures 9 and 10 present a pictorial view of the results from the application of the scenarios on Lake Kourna water levels for the periods 2005–2006 (receptors L08 and L11) and 2006–2007 (receptors L07 and L10), respectively (Tables 2, 3).

Tables 4 and 5 present a comparison between the predicted water levels and hydroperiod by the B2 and A2 IPCC scenarios and the baseline scenarios for the calibration and validation periods 2005–2006 and 2006–2007, respectively. Results show, for both the baseline scenarios tested, a reduction of 52 days for IPCC scenario B2, while there is close agreement for IPCC scenario A2, since a reduction of 55 and 67 days are predicted for baseline scenarios 1 and 2, respectively. Results also demonstrate close agreement between water level reduction predicted by the B2 IPCC scenario for baseline scenarios 1 (39 cm) and 2 (31 cm), as well as by the A2 IPCC scenario (43 and 41 cm, respectively, for baseline scenarios 1 and 2). The large difference between hydroperiod (or water level) values for the two baseline scenarios 1 and 2 and the agreement of the relative predicted reduction of B2 and A2 hydroperiod (or water level) values provide another indication of the satisfactory calibration and validation of the model (Fig. 11).

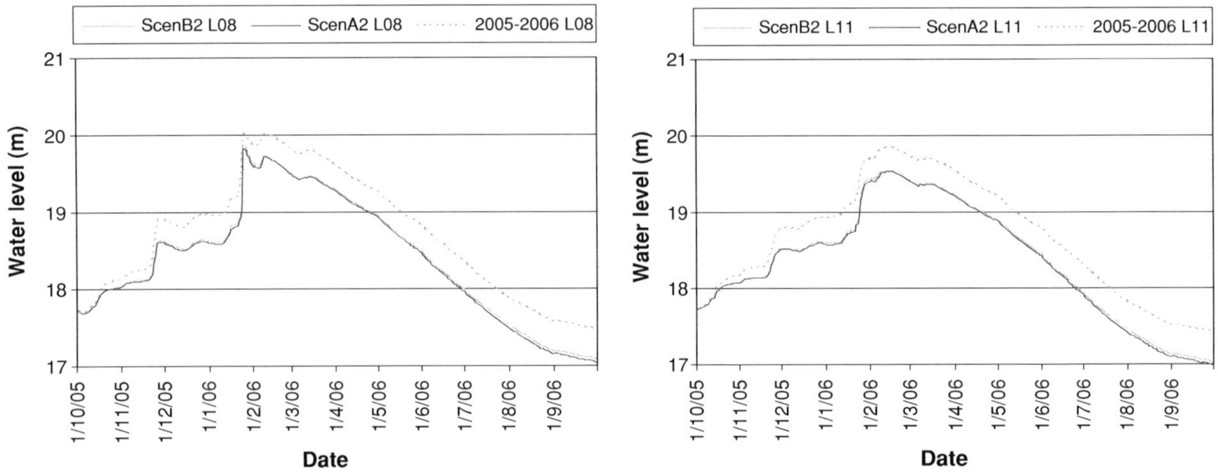

Fig. 9 Comparison of simulated Lake Kourna water level for the calibration period 2005–2006 with IPCC predictions for scenarios B2 and A2 at calibration receptors L08 and L11

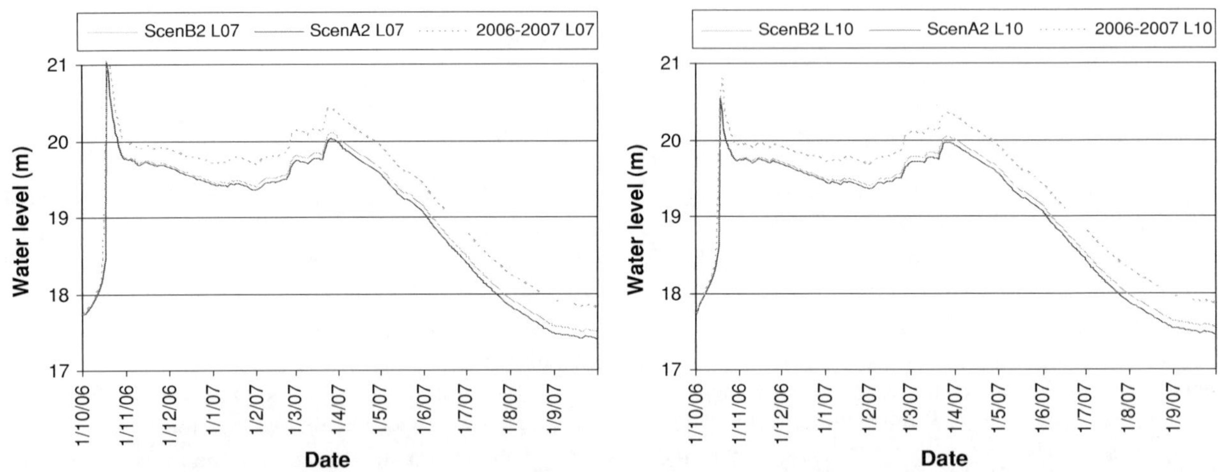

Fig. 10 Comparison of simulated Lake Kourna water level for the validation period 2006–2007 with IPCC predictions for scenarios B2 and A2 at calibration receptors L07 and L10

Table 2 Comparison of simulated and observed Kourna MTP water level and hydroperiod values for the calibration period (hydrological year 2005–2006) at calibration receptors (L08, L10, L11, L13)

Location in lake	Observed 2005–2006			Model 2005–2006		
	WL end of simulation (m)	Hydroperiod (days)	WL difference end-start of simulation (m)	WL end of simulation (m)	WL difference end-start of simulation (m)	Hydroperiod (days)
Mean	17.746	72	−0.013	17.48	−0.17	75

Omalos conceptual model (HPM) calibration and validation

Figure 12 show the volume of water in the pond (calibration target) simulated by the HPM model as opposed to that observed in the field during the calibration (2005–2006) period. Numerical criteria tested for both periods show good correlation of the model with the observed measurements with R correlation and R^2 (Nash–Sutcliffe) coefficients

Table 3 Comparison of simulated and observed Kourna MTP water level and hydroperiod values for the validation period (hydrological year 2006–2007)

Location in lake	Observed 2006–2007			Model 2006–2007		
	WL end of simulation (m)	Hydroperiod (days)	WL difference end-start of simulation (m)	WL end of simulation (m)	WL difference end-start of simulation (m)	Hydroperiod (days)
Mean	17.926	213	0.193	17.92	0.19	224

Table 4 Comparison of simulated 2005–2006 model and IPCC predictions (scenarios B2 and A2) for Kourna MTP water level and hydroperiod

Location in lake	B2			A2		
	WL difference B2 Baseline scenario 1 end of simulation (m)	Hydroperiod (days)	Hydroperiod decrease from Baseline scenario 1 (days)	WL difference A2 Baseline scenario 1 end of simulation (m)	Hydroperiod (days)	Hydroperiod decrease from Baseline scenario 1 (days)
Mean	−0.39	23	−52	−0.43	20	−55

Table 5 Comparison of simulated 2006–2007 model and IPCC predictions (scenarios B2 and A2) for Kourna MTP water level and hydroperiod

Location in lake	B2			A2		
	WL difference B2 Baseline scenario 2 end of simulation (m)	Hydroperiod (days)	Hydroperiod decrease from Baseline scenario 2 (days)	WL difference A2 Baseline scenario 2 end of simulation (m)	Hydroperiod (days)	Hydroperiod decrease from Baseline scenario 2 (days)
Mean	−0.31	172	−52	−0.41	156	−67

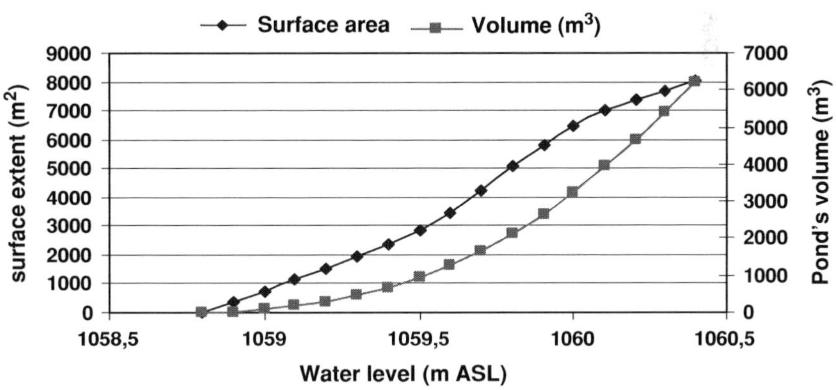

Fig. 11 Relationship between Omalos pond's surface, volume and water level (*ASL* above sea level) (Stamati & Nikolaidis, 2006)

approaching perfect model value of 1, both for the calibration (99.9%) and the validation period (98.2%) (Table 6). Error-based measures present lower values for the calibration period since during this period more measurements were available.

Comparison between the baseline scenarios (2005–2006: calibration period and 2006–2007: validation period) and IPCC scenarios is illustrated in Fig. 13. Results for 2005–2006 show there is a reduction in the pond's water level, especially the peaks from February to April, and the hydroperiod by 10–20 cm and approximately 20 days, respectively. Results for the hydrological year 2006–2007 (although the model was validated only for the first 3 months) also indicate a

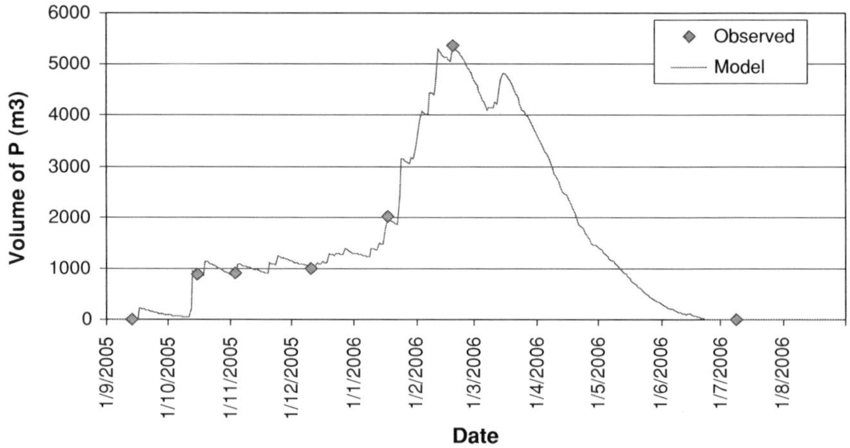

Fig. 12 Omalos MTP's simulated water volume as opposed to the observed volume (estimated using the 3D bathymetry/topography model) for the simulation period 1 September 2005–31 August 2006

Table 6 Numerical criteria for the calibration and validation of the HPM model for the hydrological year 2005–2006 and the period 1 September 2006–31 December 2006, respectively

Name	ME	MAE	RMSE	STDres	*R* (Correlation)	R^2 (Nash–Sutcliffe)
2005–2006	4.92	27.23	13.43	35.19	0.997	0.9996
September 2006–December 2006	−180.60	180.60	128.78	23.50	0.995	0.982

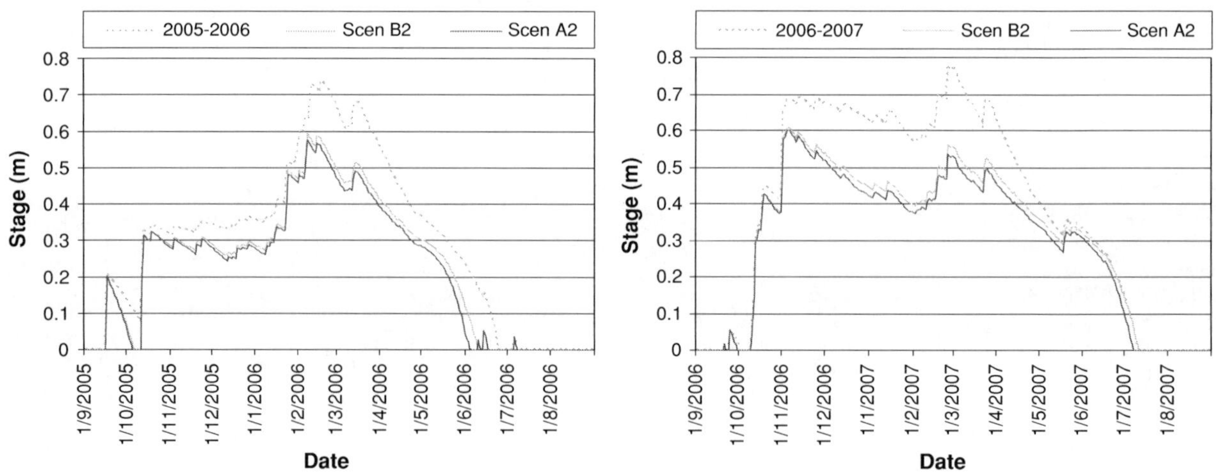

Fig. 13 Comparison of Omalos MTP water level for the calibration period 2005–2006 with IPCC predictions for scenarios B2 and A2

reduction in the pond's water level, although reduction in the pond's hydroperiod is much smaller (Table 7) as a result of higher precipitation.

Table 7 shows model results for the hydroperiod of the MTP in Omalos. The MTP presents stationarity in its hydroperiod (281–282 days) between the calibration and validation period (assuming the validated model is extended to apply during the whole hydrological year), although there is much higher precipitation observed during the validation period. Reduction of the MTP's hydroperiod is comparatively (to Lake Kourna MTP) small, with only 16 and 24 days' decrease in hydroperiod as a result of IPCC scenarios B2 and A2, respectively.

Table 7 Comparison of simulated (2005–2006 and 2006–2007) HPM models and IPCC predictions (scenarios B2 and A2) for Omalos MTP hydroperiod

Omalos MTP	Hydroperiod (days)	Hydroperiod for IPCC B2 scenario (days)	Hydroperiod decrease for B2 (days)	Hydroperiod for IPCC A2 scenario (days)	Hydroperiod decrease for A2 (days)
2005–2006	281	265	−16	257	−24
2006–2007	282	279	−3	274	−8

Discussion and conclusions

Different approaches in methodology for the determination of the hydroperiod of the two MTPs were adopted in this particular study. Incorporation of a conceptual view of the hydrologic cycle of the MTPs in Lake Kourna and Omalos has assisted to determine the respective modelling approaches that were adopted to simulate the MTPs hydroperiod and subsequently predict their changes based on IPCC climate change scenarios. Thus, with regard to the MTP in Lake Kourna, modelling of its hydrologic cycle and estimation of its hydroperiod presupposes modelling of the lake's hydrology, since there is direct communication between the two water bodies, while the lake's flood runoff contributes to the MTP's main inflow.

A physically based distributed model of Lake Kourna was, therefore, applied to determine the MTP's hydroperiod. A critical factor in the setup of the model was, as accurately as possible, the representation of the catchment groundwater inflows boundary condition, which was achieved using the lake's water balance and recharge coefficients for the geological units of the lake's sub basin. Numerical criteria tested on the calibration receptors revealed satisfactory calibration of the model with an average of 72% for the R correlation coefficient and 31% for the R^2 (Nash–Sutcliffe) coefficient, while the majority of the model points falls between the ±5% error bounds (83%) for the perfect model line. Results also show close agreement between the modelled average hydroperiod value of 75 days and the observation of 72 days. Numerical criteria tested on the validation receptors for the first split-sample test (April–September 2005) revealed good validation of the model with an average of 98% for the R correlation coefficient and 55% for the R^2 (Nash–Sutcliffe) coefficient with 91% of model points falling between the ±5% error bounds for the perfect model line. Numerical Satisfactory validation of the model has been recorded for the second split-sample test (hydrological year 2006–2007) with an average of 77% for the R correlation coefficient and 32% for the R^2 (Nash–Sutcliffe) coefficient, although the model was unable to simulate adequately the peaks in March and April 2007.

Nevertheless, 72% of model points fall between the ±5% error bounds for the perfect model line, while results show close agreement between the modelled average hydroperiod value of 224 days and the observation of 213 days. It is noteworthy that the model for both split-sample tests performed slightly better than the calibration.

In the case of Omalos, a more simplistic approach was followed with application of the mathematical representation of the conceptual model of the MTP using the mathematical software Matlab, developed by Stamati & Nikolaidis (2006). Numerical criteria tested for the calibration and the validation periods (hydrological years 2005–2006 and 2006–2007) show good correlation of the model with the observed measurements with R correlation (99.9%) and R^2 (Nash–Sutcliffe) (98.2%) coefficients.

The impact of climate change on Lake Kourna water level (and consequently on the adjacent MTP) and on the MTP water level in Omalos, was assessed by applying two future climate scenarios. Results for IPCC B2 and A2 climate scenarios show longer hydroperiod and smaller decreases in the future for Omalos MTP than in Lake Kourna MTP. Results for Lake Kourna MTP demonstrated a hydroperiod decrease of more than 52 days after the application of the IPCC scenarios. Scenario A2 does not present a significantly differentiated-higher impact on the MTPs' hydroperiod. In particular, a difference of 3–15 days and 5–8 days compared with IPCC scenario B2 predictions was estimated for the MTP in the case of Lake Kourna and in Omalos, respectively.

Thus, the lowland MTP proved to be far more vulnerable to climate change in relation to the mountainous one since the percentage decrease of its hydroperiod reaches 68% which could be detrimental for the pond fauna, and flora if it occurs. This difference between the lowland and the mountainous pond is consistent with results from similar climate change impact studies (Blenckner, 2005) that indicated different responses in various lakes depending on the local geographic and biological conditions. There are no similar applications focussing on modelling hydroperiod of MTPs in the literature since this habitat is not yet well studied. Nevertheless, temporary water bodies and particularly this priority habitat (MTP) are highly vulnerable to hydrologic disturbances including climate change since potential, significant changes in their hydroperiod will lead to the alteration of their typical ecological characteristics. Thus, there is a need for more similar studies regarding climate change impact assessment on temporary water bodies to design and undertake the appropriate restoration activities. The particular modelling approaches used in this effort could be easily adapted for similar applications in areas hosting temporary water bodies.

References

Abbott, M., J. Bathrust, J. Cunge, P. O'Connell & J. Rasmussen, 1986. An introduction to the European Hydrological System-Systeme Hydrologique Europeen, "SHE", 2: modelling system. Journal of Hydrology 87: 61–77.

Blenckner, Th., 2005. A conceptual model of climate-related effects on lake ecosystems. Hydrobiologia 533: 1–14.

Collinson, N. H., J. Biggs, A. Corfield, M. J. Hodson, D. Walker, M. Whitfield & P. J. Williams, 1995. Temporary and permanent ponds: an assessment of the effects of drying out on the conservation value of aquatic macroinvertebrate communities. Biological Conservation 74: 125–133.

Dimitriou, E., E. Moussoulis, E. Kolobari & A. Diapoulis, 2006. Hydrological study of MTP catchments in Crete. In Dimitriou, E. & A. Diapoulis (eds), Actions for the Conservation of the Mediterranean Temporary Ponds in Crete, Final Report, Project Life-Nature 2004. HCMR, Anavissos Attikis.

Grillas, P., P. Gauthier, N. Yavercovski & C. Perennou, 2004. Mediterranean Temporary Pools; Volume 1 – Issues Relating to Conservation, Functioning and Management. Station biologique de la Tour du Valat. Technical Report.

Intergovernmental Panel on Climate Change (IPCC), 2007. Climate Change: Working Group I: The Scientific Basis [http://www.grida.no/climate/ipcc_tar/wg1/008.htm].

Legates, D. R. & G. J. McCabe, 1999. Evaluating the use of ''goodness-of-fit'' measures in hydrologic and hydroclimatic model validation. Water Resources Research 35(1): 233–241.

MEPPW (Ministry for the Environment, Physical Planning and Public Works), 2004. Country Profile: Greece, National Reporting to the Twelfth Session of the Commission on Sustainable Development of the United Nations (UN CSD 12), Athens, March 2004.

Stamati, F. & N. Nikolaidis, 2006. Hydrology and Geochemistry of the Mediterranean Temporary Ponds of W. Crete, Actions for the Conservation of the Mediterranean Temporary Ponds in Crete, Project Life-Nature 2004. Laboratory of Hydrogeochemical Engineering and Remediation of Soils, Technical University of Crete. Technical Report.

Thompson, J. R., H. R. Sørenson, H. Gavina & A. Refsgaard, 2004. Application of the coupled MIKE SHE/MIKE 11 modelling system to a lowland wet grassland in southeast England. Journal of Hydrology 293: 151–179.

Warwick, N. W. M. & M. A. Brock, 2003. Plant reproduction in temporary wetlands: the effects of seasonal timing, depth, and duration of flooding. Aquatic Botany 77: 153–167.

Hydrobiologia (2009) 634:209–217
DOI 10.1007/s10750-009-9901-y

Monitoring the invasion of the aquatic bug *Trichocorixa verticalis verticalis* (Hemiptera: Corixidae) in the wetlands of Doñana National Park (SW Spain)

Héctor Rodríguez-Pérez · Margarita Florencio · Carola Gómez-Rodríguez · Andy J. Green · Carmen Díaz-Paniagua · Laura Serrano

Published online: 29 July 2009

Abstract We have detected the presence of the North American native corixid *Trichocorixa verticalis verticalis* (Fieber, 1851) in Doñana wetlands (SW Spain). We have collected data from different research projects done in the area during the period of 2001–2007. We have sampled 134 different sites in Doñana and we found the exotic corixid in 66 occasions. We have found two reproductive populations that might act as sources for the colonization of other waterbodies in the area. When reproduction occurred *T. v. verticalis* outcompeted native corixids. Its presence out of the waterbodies where we detected reproduction was in small numbers and probably due to vagrant individuals.

Keywords Corixid · *Trichocorixa* · Exotic species · Invasive · Doñana

Guest editors: B. Oertli, R. Cereghino, A. Hull & R. Miracle
Pond Conservation: From Science to Practice. 3rd Conference of the European Pond Conservation Net work, Valencia, Spain, 14–16 May 2008

Héctor Rodríguez-Pérez and Margarita Florencio have contributed equally to this work.

H. Rodríguez-Pérez (✉) · M. Florencio · C. Gómez-Rodríguez · A. J. Green · C. Díaz-Paniagua
Department of Wetland Ecology, Estación Biológica de Doñana-CSIC, C/Americo Vespucio, s/n, 41092 Sevilla, Spain
e-mail: hectorrodriguez@vet.ucm.es

Present Address:
H. Rodríguez-Pérez
Department of Animal Production. Veterinary Faculty, Complutense University of Madrid, Av. Puerta de Hierro, 28040 Madrid, Spain

L. Serrano
Department of Plant Biology and Ecology. Biology Faculty, Sevilla University, 41080 Sevilla, Spain

Introduction

The spread of exotic species occurrence worldwide is one of the major causes of global change (Vitousek et al., 1996; Ricciardi, 2006). The establishment of exotic invasive species within an ecosystem usually has strong consequences, affecting ecological functions (Ricciardi et al., 1997; Maezono & Miyashita, 2003), and causing a loss of indigenous biodiversity (Witte et al., 2000). Several scenarios have been described once an exotic species arrives to an ecosystem before it becomes an invasive species, and it controls ecological processes (Carlton, 2003).

In this article, we consider the status and possible impact of an exotic aquatic insect recently detected in the Doñana wetlands in south-west Spain. *Trichocorixa verticalis verticalis* (Fieber, 1851) is a small predaceous corixid (<5.5 mm) (Heteroptera) naturally distributed along the Atlantic coast of North America and on some Caribbean islands. It now occurs as an exotic species in South Africa, New Caledonia, Portugal, Morocco, and Spain (Kment, 2006; Jansson & Reavell, 1999; L'Mohdi et al., submitted). This species is the only aquatic alien Heteroptera recorded in Europe (Rabitsch, 2008). Adult males are easily distinguished from European native corixid species by their left abdominal asymmetry and a tibial elongation over the pala (Günter, 2004). This species is halobiont and usually inhabits brackish and saline waterbodies, even occurring in the open sea (Hutchinson, 1931). This ability to tolerate a broad salinity range is probably a key feature of its success as an invader. *Trichocorixa verticalis verticalis* is widespread in the Portuguese Algarve which begins 80 km to the west of Doñana (Sala & Boix, 2005). In Doñana itself, it has previously been cited in two locations with a hydrological connection to the Guadalquivir Estuary (Günter, 2004; Millán et al., 2005), although these records are predated by some of our own observations.

Study area

Doñana is located in the south-west of Spain in the mouth of the Guadalquivir River (Fig. 1A) and holds a great variety of waterbodies, including natural temporary ponds, natural permanent ponds, artificial permanent ponds, temporary marshes and ricefields (García-Novo and Marín, 2006; Serrano et al., 2006). These wetlands represent one of the most important areas for waterbirds in Europe (Rendón et al., 2008), and the core area dominated by natural, temporary wetlands is protected as a National Park, Biosphere Reserve and UNESCO World Heritage site. Surrounding fish ponds, salt ponds, and ricefields are also partly protected and included within a Ramsar site and an EU Specially Protected Area.

The climate is Mediterranean with an Atlantic influence. The flooding regime of temporary ponds and marshland is highly variable among years owing to rainfall fluctuations. Mean annual precipitation is 542 mm/year with a range of 170–1,032 mm/year. There are up to 26,000 Ha of temporary marshes mainly fed by freshwater (rainfall and runoff) and currently isolated from the tidal influence of the Guadalquivir estuary. The marshes and temporary ponds usually begin to fill by late autumn when rainfall starts (Fig. 1B) and dry out completely in summer. Salinity varies from oligohaline to mesohaline according to the frequency and the duration of flooding, with a wide spatial and temporal variation depending on distance from freshwater sources,

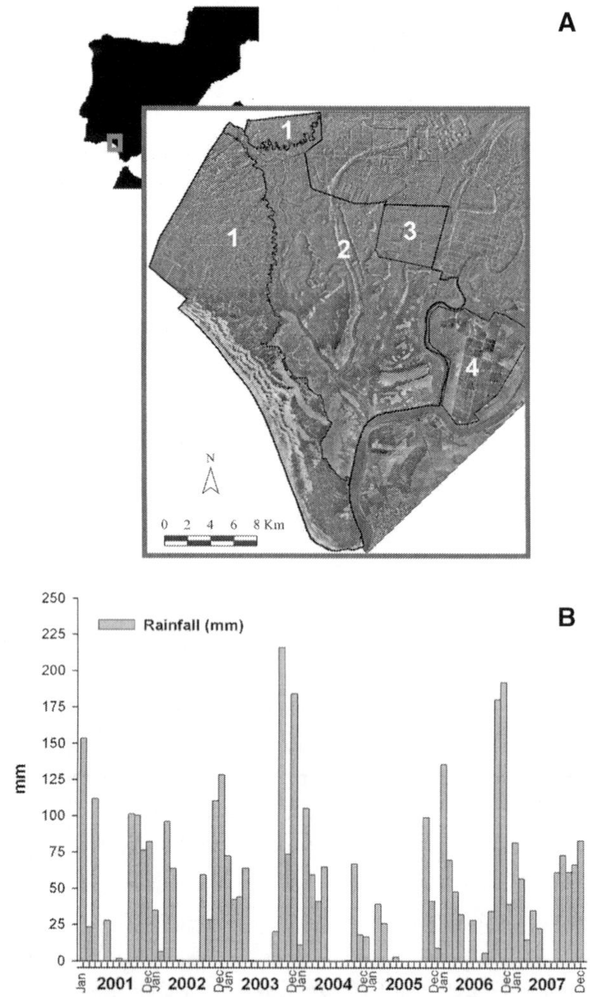

Fig. 1 **A** Map of the study area showing the main different areas where samples were taken: (1) Dunes and stabilized sands, (2) Natural temporary marshes, (3) Caracoles estate (restored marshland) and (4) Veta la Palma estate with fish ponds (transformed marshland). The thinest black lines are the boundaries of each area, the thickest black line is the limit of Doñana National Park. The figure was done with a digital orthophotography, *source*: Junta de Andalucía. 2003. Digital Orthophotografy of Andalusia. Consejería de Obras Públicas y Transportes. Instituto de Cartografía de Andalucía. Junta de Andalucía. **B** Rainfall record for the period of study in the area (2001–2007). Rainfall data were gathered at and provided by Doñana Biological Reserve (EBD-CSIC)

depth, etc. (García-Novo and Marín, 2006; Serrano et al., 2006). Recently, areas of former marshland previously drained for agriculture have been restored by removing dykes and drainage networks. The Caracoles estate (Fig. 1A) is one such area included in our study, in which 96 temporary ponds were created during restoration in 2004 (Frisch & Green, 2007). Conductivity in these newly created ponds range between 7.14 and 51.6 mS/cm.

Elsewhere in Doñana National Park a large network of more than 3,000 temporary ponds occurs in an area of mobile dunes and stabilized sands (Fig. 1A; Fortuna et al., 2006; see a detailed description in Díaz-Paniagua et al., accepted; and also in Gómez-Rodríguez et al., 2009). In this area, there are also some permanent, artificial ponds made as waterholes for livestock. These *zacallones* (local name) were usually made by digging a deep hole near a natural pond or even inside the pond bed itself. Conductivity in these ponds ranged from 0.08 to 9.8 mS/cm.

Large, permanent fish ponds are located to the east of the National Park in the Veta la Palma estate (Fig. 1A), which contains 52 regular ponds. The ponds were constructed in 1992–1993 on top of what was natural marshland in the Guadalquivir estuary. All the ponds are shallow (average 30 cm, maximum depth 50 cm) and flat-bottomed with a total combined surface area of 2997 ha (see Frisch et al., 2006; Rodríguez-Pérez & Green, 2006, for more details). Each pond is dried out under rotation approximately every 2 years to extract fish. Ponds are interconnected via canals and a permanent flow of water taken from the Guadalquivir estuary maintains high-dissolved oxygen levels. Salinity during our study varied from 10.3 mS/cm during winter months of high rainfall to 22.1 mS/cm at the end of September, after the dry summer months. pH ranged from 9.3 to 10.4.

Our study did not include salt pans in Sanlúcar de Barrameda where *Trichocorixa v. verticalis* was initially recorded (Günter, 2004).

Materials and methods

We studied the distribution and abundance of *T. verticalis* in an ad hoc fashion from 2001 to 2007, taking advantage of several research projects designed for different purposes and using different sampling methodologies.

In 2001 and 2002, we sampled 11 ponds in Veta la Palma estate every 3 months (Fig. 1A.). We used a quantitative sampling methodology; a PVC pipe section of 20 cm diameter was inserted vertically down into the sediments to isolate the water within. Using a plastic jar, all the water was then scooped out and sieved through a 250 μm mesh, taking care not to extract sediments. The sieved material was then fixed with formaldehyde. Corixids were later identified and counted.

A Doñana monitoring team (Equipo de Seguimiento de Procesos Naturales de la Reserva Biológica de Doñana (http://www-rbd.ebd.csic.es/Seguimiento/medio biologico.htm)) took samples from marshes, and permanent artificial ponds in 2003, 2004, and 2005. They sampled with eel nets (5 mm mesh size) placed for 24 h, and by dip netting (1 mm mesh size) for ca. 1.5 m while trampling sediments on the bed of the wetland. Samples were preserved in ethanol (70%) and later examined for the presence of *T. verticalis.*

While sampling 14 new ponds for zooplankton from April to May 2006, some corixids were incidentally included in the samples. Twenty liters of water was taken from a transect along the pond and filtered (see Frisch & Green, 2007) before being placed in ethanol (70%). All corixids were later counted and identified. These ponds had flooded for the first time in January 2006 and dried out before July.

Finally, we sampled the natural temporary ponds and zacallones located in the stabilized sands; 64 ponds in 2006 and 90 ponds in 2007. We sampled with a dip net (1 mm mesh size), sampling in the same way as the Doñana monitoring team did. All these natural ponds represented a wide hydroperiod gradient. All ponds were sampled once each year, except 19 temporary ponds, sampled monthly. In 2006, we identified species in situ recording only the presence or absence of each species. On the other hand, in 2007, all captured corixids (or at least 75% of them when there were too many individuals) were retrieved and fixed with ethanol (70%) and later quantified and identified with a microscope at the laboratory. When large corixid adults occurred, we recorded its presence in situ as *Corixa affinis* in order to its largest size and previous sampling in the area. Furthermore, we retrieved few large corixid individuals per pond in each sampling to make a correct identification under a microscope at the laboratory in order to avoid the possible confusion with *Corixa panzeri*. In any case, we have never found

C. panzeri in these ponds in any sampling, so we can assume confidently that all large corixids found were *C. affinis*. Both species are sympatric in the Iberian Peninsula but *C. panzeri* seems to be less frequent (Niesser et al., 1994). We made relative proportion of *T. v. verticalis* out of the total corixids captured in the pond (except *C. affinis*) during the entire sampling season with 2007 data.

Results

Overall, we sampled 134 different sites in the Doñana wetlands situated within a polygon of 54,000 Ha. Some of these points were sampled during several years. Sampled sites included artificial ponds (11 fish ponds, 14 shallow, new temporary ponds and 30 deep waterholes [*zacallones*]) and natural waterbodies (four streams, 12 points in temporary marshes and 63 natural ponds). We detected the presence of *T. v. verticalis* on 66 occasions (Fig. 2), more than half of these (53%) being in artificial ponds. In contrast, artificial waterbodies were only 41% of the total points sampled.

Veta la Palma fish ponds and new temporary ponds in Caracoles estate

Veta la Palma fish ponds and new temporary ponds were the only two areas where reproductive populations of *T. v. verticalis* were recorded. In both areas *T. v.*

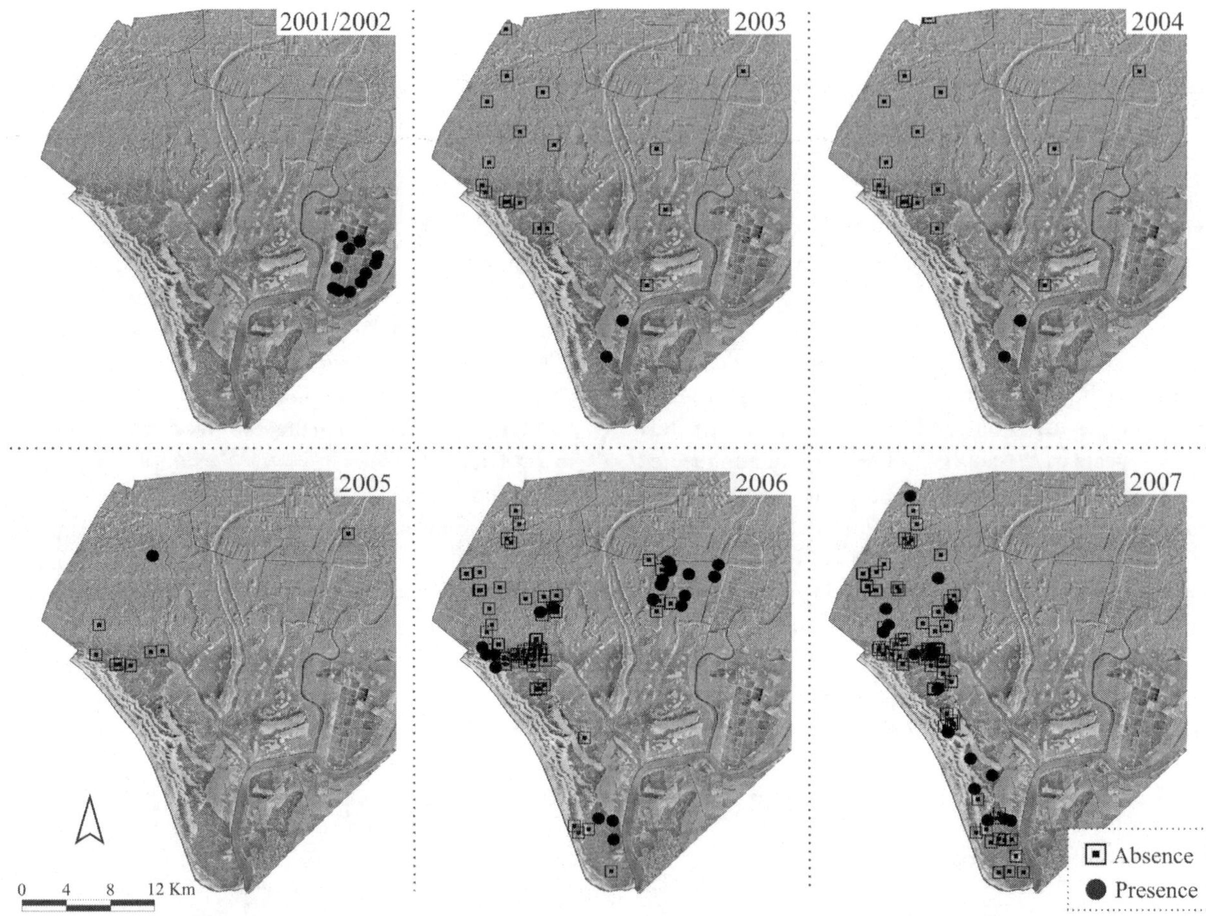

Fig. 2 Maps of the study area for each year of study with the position of the sampling sites. Black dots show the presence of *T. verticalis*, and *open squares* show sampled places where we did not detect *T. verticalis*. We show the results of 2001 and 2002 in the same map because there were not differences in the presences of the exotic corixid. The figure was done with a digital orthophotography, *source*: Junta de Andalucía. 2003. Digital Orthophotografy of Andalusia. Consejería de Obras Públicas y Transportes. Instituto de Cartografía de Andalucía. Junta de Andalucía

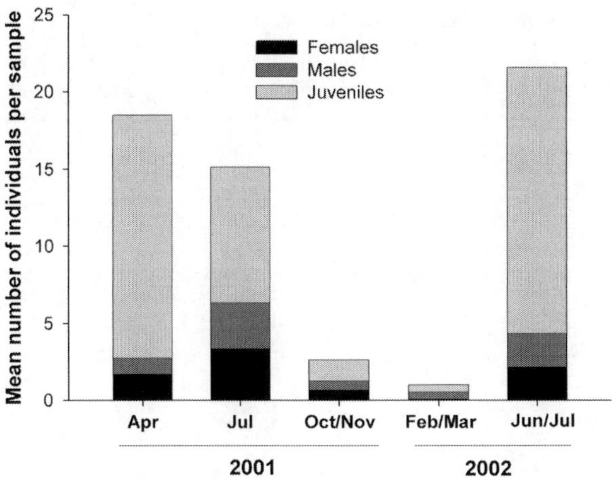

Fig. 3 The figure shows the mean number of adult males and females per sample of *T. verticalis* in each sampling campaing in Veta la Palma fish ponds during 2001 and 2002. Juveniles were mostly *T. verticalis*, but we did not identified all juveniles individuals

verticalis was the dominant species, apparently outcompeting native ones. In Veta la Palma, sampled during 2001 and 2002, 179 samples were gathered with a total of 738 adult corixids, 96% of which were *T. v. verticalis*, the remaining adults being *Sigara stagnalis* and *S. scripta*. Abundance peaked in spring and summer, with the lowest densities in autumn and winter (Fig. 3). There was a highly significant difference in densities between seasons (Rodríguez-Pérez, 2006). The presence of juveniles suggests that reproduction continues throughout the year in this site (Fig. 3).

In 2006, we detected a second reproductive population in the new temporary ponds in Caracoles (see Figs. 1, 2), where 307 adult corixids were retrieved from 14 ponds of which 92% were *T. v. verticalis*. The other three species that occurred in this area were *Sigara lateralis* (4%), *S. stagnalis* (2%), and *S. scripta* (2%). Both here and in Veta la Palma, we also identified freshly moulted adult and juvenile corixids that were surely *T. v. verticalis*. In the three new temporary ponds with the highest *T. verticalis* density (30 in our sample), conductivity was particularly high, ranging from 17.3 to 54.6 mS/cm.

Ponds in stabilized sands and temporary marshes

In the other places where *T. v. verticalis* occurred, no matter the year, only adults were detected, surely these individuals were vagrant adults (we captured juvenile corixids but we identified them as other species). In these occasions *T. v. verticalis* always occurred in small numbers, and making a small proportion of the total corixids sampled. In 2007, we captured and retrived 1881 adult corixids throughout the year but only 37 of those were *T. v. verticalis.* In Table 1 we show the relative proportion of *T. v. verticalis* out of the other corixid species present but only in the ponds where the exotic species occurred. *T. v. verticalis* was detected coexisting with another seven species of corixids in the area (Table 2). Only one other corixid species was present in more waterbodies than *T. v. verticalis* in 2006, and only three other species in 2007. *Paracorixa concinna* was the only species that was never observed coexisting with *T. v. verticalis* in the same pond. *T. v.verticalis* was more likely to be found in ponds than the rarer native corixids (*Sigara scripta*, *S. stagnalis*, and *S. selecta*). It is remarkable that *T. v. verticalis* has not been recorded in the main body of the temporary marsh in the samples studied as yet.

Discussion

The dataset that we have used for this work encompasses 7 years of sampling, and because we did not use the same standardized methodology in every sampling we cannot conclude conclusively that the populations of this invasive species are increasing their occurrence in the area. On the other hand, the strengths of this dataset are the 7 years of data itself, the high number of points that we have visited throughout the 7 years in a restricted territory (54,000 Ha), and that we have sampled every kind of aquatic habitat that occurs in Doñana National Park. Despite the noted weaknesses of the dataset, we show in this work evidence suggesting that an ongoing invasion is happening in the wetlands of this protected area. This fact has strong consequences for the conservation of the ponds and marshes in Doñana National Park, and it adds to other invasion events of aquatic organisms in the aquatic ecosystems of Doñana: i.e., the copepod *Acarthia tonsa* (Frisch et al., 2006), the crayfish *Procambarus clakii* (Geiger et al., 2005), the gastropod *Potamopyrgus antipodarum* (Rodríguez-Pérez 2006), the fishes *Gambusia affinis* and *Lepomis gibbosa* (García-Berthou et al.,

Table 1 Percentages of presence of each corixid species recorded in the ponds where *T. verticalis* occurred in 2007

	Trichocorixa verticalis (%)	*Sigara laterallis* (%)	*Sigara stagnalis* (%)	*Sigara scripta* (%)	*Sigara selecta* (%)	*Micronecta scholzi* (%)	*N* of total corixids
Caño Arenilla	25	50	0	25	0	0	8
Laguna Estratificada	33	67	0	0	0	0	3
Zacallón Mahón	3	29	53	16	0	0	31
Zacallón de la Angostura	67	33	0	0	0	0	3
Punta de Zalabar	13	25	31	31	0	0	16
Laguna Larga o del Carrizal	1	60	39	0	0	0	277
Canal al norte sombrío	100	0	0	0	0	0	1
Zacallón pozo salinas	4	90	2	5	0	0	300
Navazo de la Higuera	3	86	8	3	0	1	76
Camino de Martinazo	20	80	0	0	0	0	5
Orfeon	100	0	0	0	0	0	1
Poli	50	25	25	0	0	0	4
Moral	100	0	0	0	0	0	1
Jiménez	25	75	0	0	0	0	4
Lagunan del Caño Martinazo	50	0	0	0	50	0	2
Adyacete al Navazo del Toro	100	0	0	0	0	0	1
Raya del Pinar	100	0	0	0	0	0	1
Leña	11	79	5	5	0	0	19

Corixa affinis was a frequent and abundant specie and coexisted with *Trichocorixa verticalis verticalis* in five ponds in 2006 and 16 ponds in 2007

2007), or the fern *Azolla filiculoides* (García-Murillo et al., 2007).

Four subspecies of *Trichocorixa verticalis* occur naturally in the brackish and saline waters of North America, covering a broad geographical range from the Caribbean and Atlantic coast to the Pacific coast, and from Mexico to Central Provinces in Canada (Jansson, 2002; Kment, 2006). This trait of one species with highly differentiated populations distributed along its native range has been identified as an indicator of a species with high invasive potential (Lee & Gelembiuk, 2008).

Within the Iberian Peninsula, this species was first detected in samples collected in Algarve (South Portugal) in 1990s (Sala & Boix, 2005). The first evidence of its presence in Doñana is from our samples in Veta la Palma fish ponds in 2001. Given the shortage of detailed studies of corixids in Doñana and other parts of the south-west of Spain, it is impossible to know its date of arrival in Doñana while the limits of its current distribution beyond Doñana remain unclear. Given its abundance in Veta la Palma, it seems likely that this species colonized the fish ponds shortly after their creation in the early 1990s.

Sala & Boix (2005) suggested two different hypotheses to explain the introduction of *T. v. verticalis* in Europe. First, the corixid may have arrived with the introduction of *Fundulus heteroclitus* and *Gambusia hoolbroki* in the area. These two species are sympatric of *T. v. verticalis*. Secondly, there may have been a natural dispersion via the marine current between the Atlantic coasts of North America and Europe, since *T. v. verticalis* has been observed in the open sea (Hutchinson, 1931; Gunter & Christmas, 1959). Alternatively, *T. v. verticalis* has been recently detected in Morocco (L'Mohdi et al., submitted). The populations in the north of Morocco and the ones in the south of Spain might be related, being the North African populations the origin of European ones or vice versa.

Over 7 years, we have sampled most kinds of aquatic habitat occurring in Doñana National Park and its surroundings. It is likely that the species has increased its area of distribution in Doñana over our study period, but we have not been able to

Table 2 Number of sampling sites where each corixid species was detected out of the total number of points sampled in 2006 and 2007

	2006 (n = 76)	2007 (n = 90)
Trichocorixa verticalis	21	18
Paracorixa concinna	0	4
Sigara laterallis	13	50
Sigara stagnalis	2	23
Sigara scripta	4	16
Sigara selecta	0	4
Micronecta scholzi	0	5
Corixa affinis	25	78
Without corixids	24	11

demonstrate that conclusively owing to the ad hoc nature of our sampling regime. The exceptions are the new ponds in Caracoles in which *T. verticalis* immediately established itself as the dominant corixid.

The invasive character of *T. v. verticalis* in the Veta la Palma and Caracoles estates seems clear. At these sites, the species has dominant reproductive populations and has overwhelmed native corixid species. Elsewhere in our study area, we did not confirm reproduction, and further work is required to establish whether the species can be considered invasive or not (see Carlton, 2003). At least in the more brackish parts of Doñana, it seems likely that *T. v. verticalis* will have a significant impact on the abundance of native species and may replace *Sigara lateralis* as the most frequent and abundant corixid in the community. *T. v. verticalis* may also benefit from the increase in salinity projected for the Iberian Peninsula owing to global warming (Rahel & Olden, 2008).

Our results suggest that *T. v. verticalis* in Doñana is currently most abundant in areas with relatively high salinities and artificial areas that are relatively permanent. Permanent sites such as fish ponds or waterholes might act as reservoirs of *T. v. verticalis* populations during the summer, facilitating the colonization of temporary ponds and marshes when they flood in the autumn or winter. There is some evidence to suggest that, with the National Park, the temporary ponds situated closest to the fish ponds or Sanlúcar salt ponds colonized by *T. v. verticalis* are more likely to have been colonized by this species (Fig. 3), supporting the idea that these permanent, artificial sites act as a main source of exotic species colonizing surrounding areas. The Veta la Palma fish ponds similarly seem to act as a reservoir for other exotics such as the copepod *Acartia tonsa* or the gastropod *Potamopyrgus antipodarum* (Frisch et al., 2006; Rodríguez-Pérez, 2006). These ponds are widely recognized as of high value for waterbird conservation (Rendón et al., 2008). However, this benefit for biodiversity conservation in Doñana needs to be balanced against the cost of the role the ponds play in facilitating the expansion of exotic species.

Most heteropterans are extremely good dispersers and have not developed other strategies to resist the drying phase, dispersing to permanent waterbodies as adults (Wiggins et al., 1980; Williams, 2006; Bilton et al., 2001). However, *Trichocorixa verticalis interiores* and *T. v. verticalis* have been reported to develop resistant resting eggs, allowing them to survive ice, hypersalinity or desiccation of pools (Tones, 1977; Kelts, 1979). If this is the case, it raises the possibility that *T. verticalis* will be able to withstand summer droughts in Doñana as extremely durable eggs in sediments, re-emerging during the next hydroperiod. In this scenario, the species will not be so dependent on fish ponds as a source for recolonisation of temporary waterbodies in Doñana.

Doñana contains some of the most important and diverse wetlands in Europe (García-Novo & Marín, 2006; Rendón et al., 2008). Although *T. verticalis* is the first case of an alien aquatic insect, Doñana has already been invaded by a considerable number of other exotic aquatic species (Frisch et al., 2006; García-Berthou et al., 2007). Research is required into the invasion biology of *Trichocorixa verticalis*, particularly its impact on native corixids and prey species and its dispersal biology, as well as more extensive surveys to establish and monitor its distribution in south-west Spain.

Acknowledgments Margarita Florencio was supported by a I3P-CSIC fellowship (European Union Social Fund). We would like to thank Hugue Lefranc (Doñana Monitoring Team. Equipo de Seguimiento de Procesos Naturales-CSIC) for providing us with the samples of 2003, 2004 and 2005. Caracoles samples were prepared by Arantza Arechederra and identified by Frank Van de Meutter. Carlos Marfil Daza, Azahara Gómez Flores and Alexandre Portheault colaborated to sample during 2006-2007. Andrés Millán helped us to identify some specimens.

References

Bilton, D. T., J. R. Freeland & B. Okamura, 2001. Dispersal in freshwater invertebrates. Annual Review in Ecology and Systematics 32: 159–181.

Carlton, J. T., 2003. Bioinvasion ecology: assessing invasion impact and scale. In Leppäkoski, E., S. Gollash & S. Oein (eds), Invasive Aquatic Species of Europe, Distribution Impacts And Management. Kluwer Academic Press, Dordrecht, The Netherlands: 7–19.

Díaz-Paniagua, C., R. Fernández-Zamudio, M. Florencio, P. García Murillo, C. Gómez-Rodríguez, A. Portheault, L. Serrano & P. Siljeström. Temporary ponds from the Doñana National Park: a system of natural habitats for the preservation of aquatic flora and fauna. Limnetica (accepted).

Fortuna, M. A., C. Gómez-Rodríguez & J. Bascompte, 2006. Spatial network structure and amphibian persistence in stochastic enviroments. Proceedings of the Royal Society B-Biological Sciences 273(1592): 1429–1434.

Frisch, D. & A. J. Green, 2007. Copepods come in first: rapid colonization of new temporary ponds. Fundamental and Applied Limnology 168: 289–297.

Frisch, D., H. Rodriguez-Perez & A. J. Green, 2006. Invasion of artificial ponds in Donana Natural Park, southwest Spain, by an exotic estuarine copepod. Aquatic Conservation-Marine and Freshwater Ecosystems 16: 483–492.

García-Berthou, E., D. Boix & M. Clavero, 2007. Non-indigenous animal species naturalized in Iberian inland waters. In Gherardi, F. (ed.), Biological Invaders in Inland Waters: Profiles, Distribution and Threats. Springer, The Netherlands.

García-Murillo, P., R. Fernández-Zamudio, S. Cirujano, A. Sousa & J. M. Espinar, 2007. The invasion of Doñana National Park (SW Spain) by the mosquito fern (*Azolla filiculoides*, Lan). Limnetica 26(2): 243–250.

García-Novo, F. & C. Marín (eds), 2006. Doñana. Water and Biosphere—Confederación Hidrográfica del Guadalquivir. Ministerio de Medio Ambiente, Madrid.

Geiger, W., P. Alcorlo, A. Baltanás & C. Montes, 2005. Impact of an introduced crustacean on the trophic webs of Mediterranean wetlands. Biological Invasions 7(1): 49–73.

Gómez-Rodríguez, C., C. Díaz-Paniagua, L. Serrano, M. Florencio & A. Portheault, 2009. Mediterranean temporary ponds as amphibians breeding habitats: The importance of preserving pond networks. Aquatic Ecology. doi: 10.1007/s10452-009-9235-x.

Günter, H., 2004. *Trichocorixa verticalis verticalis* (Fieber), eine nearktische Ruderwanze in Europa (Heteroptera: Corixidae). Mitteilungen des Internationalen Entomologischen Vereins 29: 45–49.

Gunter, G. & J. Y. Christmas, 1959. Corixids insects as part of the offshore fauna of the sea. Ecology 40: 724–725.

Hutchinson, G., 1931. On the occurrence of *Trichocorixa* Kirkdaly (Corixidae, Hemiptera–Heteroptera) in salt water and its zoo-geogaphical significance. American Naturalist 65: 573–574.

Jansson, A., 2002. New records of Corixidae (Heteroptera) from northeastern USA and Eastern Canada, with one new synonymy. Entomologica Fennica 13: 85–88.

Jansson, A. & P. E. Reavell, 1999. North American species of *Trichocorixa* (Heteroptera: Corixidae) introduced into Africa. African Entomology 7(2): 295–297.

Kelts, L. J., 1979. Ecology of a tidal marsh corixid, *Trichocorixa verticalis* (Insecta, Hemiptera). Hydrobiologia 64: 37–57.

Kment, P., 2006. A contribution to the faunistics of aquatic and semiaquatic Bugs (Heteroptera: Nepomorpha, Gerromorpha) in Portugal, with the review of biology of the Neartic corixid *Trichocorixa verticalis* (Fieber, 1851). Boletín de la Sociedad Entomológica Aragonesa 38: 359–361.

Lee, C. E. & G. W. Gelembiuk, 2008. Evolutionary origins of invasive populations. Evolutionary Applications 1(3): 427–448.

L'Mohdi, O., N. Bennas, K. Hajji, M. El Haissoufi & A. Millán. *Trichocorixa v. verticalis* (Fieber, 1851) (Hémiptère, Corixidae): une nouvelle espèce exotique au Maroc. Boletín de la Sociedad Entomológica Aragonesa (submitted).

Maezono, Y. & T. Miyashita, 2003. Community level impacts induced by introduced largemouth bass and bluegill in farm ponds in Japan. Biological Conservation 109(1): 111–121.

Millán, A., C. Hernándo, P. Aguilera, A. Castro & I. Ribera, 2005. Los coleópteros acuáticos y semiacuáticos de Doñana: Reconocimiento de su biodiversidad y prioridades de conservación. Boletín de la Sociedad Aragonesa de Entomología 36: 157–164.

Niesser, N., M. Baena, J. Martinez-Aviles & A. Millán, 1994. Claves para la identificación de los heterópteros acuáticos (nepomorpha & gerromorpha) de la Península Ibérica-Con notas sobre las especies de las Islas Azores, Baleares, Canarias y Madeira. Asociación Española de Limnología. Claves de identificación de la Flora y Fauna de la Aguas Continentales de la Península Ibérica (n° 5).

Rabitsch, W., 2008. Alien true bugs of Europe (Insecta: Hemiptera: Heteroptera). Zootaxa 1827: 1–44.

Rahel, F. J. & J. D. Olden, 2007. Assessing the effects of climate change on aquatic invasive species. Conservation Biology 22(3): 521–533.

Rendón, M. A., A. J. Green, E. Aguilera & P. Almaraz, 2008. Status, distribution and long-term changes in the waterbird community wintering in Donana, south-west Spain. Biological Conservation 141: 1371–1388.

Ricciardi, A., 2006. Are modern biological invasions an unprecendent form of global change? Conservation Biology 21(2): 329–336.

Ricciardi, A., F. G. Whoriskey & J. B. Rassmussen, 1997. The role of the zebra mussel (*Dreissena polymorpha*) in structuring macroinvertebrate communities on hard substrata. Canadian Journal of Fisheries and Aquatic Sciences 54(11): 2596–2608.

Rodríguez-Pérez, H., 2006. Efectos de las aves acuáticas sobre los macrófitos y los invertebrados en las marismas de Doñana. PhD Thesis. Universidad Autónoma de Madrid.

Rodríguez-Pérez, H. & A. J. Green, 2006. Waterbird impacts on widgeongrass *Ruppia maritima* in a Mediterranean wetland: comparing bird groups and seasonal effects. Oikos 112: 525–534.

Sala, J. & D. Boix, 2005. Presence of the nearctic water boatman *Trichocorixa verticalis verticalis* (Fieber, 1951)

(Heteroptera, Corixidae) in the Algarve region (S Portugal). Graellsia 61(1): 31–36.
Serrano, L., M. Reina, G. Martin, I. Reyes, A. Arechenderra, D. León & J. Toja, 2006. The aquatic systems of Doñana (SW Spain): watersheds and frontiers. Limnetica 25(1–2): 11–32.
Tones, P. I., 1977. The life cycle of *Trichocorixa verticalis interioris* Sailer (Hemiptera, Corixidae) with special reference to diapause. Freshwater Biology 7: 31–36.
Vitousek, P., C. M. D'Antonio, L. L. Loope & R. Westbrooks, 1996. Biological invasions as global environmental change. American Scientist 84(5): 468–478.
Wiggins, G. B., R. J. Mackay & I. M. Smith, 1980. Evolutionary and ecological strategies of animals in annual temporary pools. Archive für Hydrobiologie Supplement 58(1/2): 97–206.
Williams, D. D., 2006. The Biology of Temporary Waters. Oxford University Press, Oxford.
Witte, F., B. S. Msuku, J. H. Wanink, O. Seehausen, E. F. B. Katunzi, P. C. Goudswaard & T. Goldsmidt, 2000. Recovery of cichild species in lake Victoria: an examination of factors leading differential extinction. Reviews in Fish Biology and Fisheries 10: 233–241.

Hydrobiologia (2009) 634:219–230
DOI 10.1007/s10750-009-9889-3

POND CONSERVATION

Copepods and branchiopods of temporary ponds in the Doñana Natural Area (SW Spain): a four-decade record (1964–2007)

K. Fahd · A. Arechederra · M. Florencio · D. León · L. Serrano

Published online: 22 July 2009

Abstract The Doñana Natural Area includes a large array of wetlands with the highest degree of environmental protection in Spain, and so it has long attracted many studies. We present a cumulative list of zooplankton taxa (Copepods and Branchiopods) based on a collection of 18 publications (1964–2007) and 4 unpublished studies. Seventy-eight taxa have been recorded in a set of 55 ponds, and 72 taxa at 38 sites spread over the Doñana marshland. In total, 96 taxa have been recorded, including 50% of all branchiopod species reported for the whole Iberian Peninsula. Taxa composition was significantly segregated between ponds and marshland during floods (ANOSIM test, $R = 0.929$, $P < 0.01$), but this segregation disappeared at a larger spatio-temporal scale when a non-metric MDS ordination produced a gradient from ponds to marshland (ANOSIM test, $R = 0.272$, $P < 0.01$). The lack of segregation between ponds and marshland sites, and among ponds with different hydroperiods, was not due to a large number of cosmopolitan species, but to a random distribution of a large number of low-occurrence species (67% of total taxa occurred with a frequency <15%). Long-hydroperiod ponds occupy a key position among the Doñana wetlands in terms of biodiversity as these ponds accumulated a high crustacean richness over time. They also supported a significantly higher cumulative number of cladoceran and harpacticoid taxa, while short-hydroperiod ponds accumulated the lowest number of diaptomid taxa. Our data indicate the need for recording biodiversity in the long term as richness on a short-temporal scale is not a good indicator of the number of crustacean species that would be encountered with a longer sampling period in Mediterranean temporary wetlands.

Keywords Cumulative richness · Cladocerans · Copepods · Temporary ponds · Fluctuations · Desiccation

Electronic supplementary material The online version of this article (doi:10.1007/s10750-009-9889-3) contains supplementary material, which is available to authorized users.

Guest editors: B. Oertli, R. Cereghino, A. Hull & R. Miracle
Pond Conservation: From Science to Practice. 3rd Conference of the European Pond Conservation Network, Valencia, Spain, 14–16 May 2008

K. Fahd
Fundación Centro de las Nuevas Tecnologías del Agua (CENTA), Av. Américo Vespucio, 5-A, 41092 Sevilla, Spain

D. León · L. Serrano (✉)
Department of Plant Biology and Ecology, University of Sevilla, P.O. Box 1095, 41080 Sevilla, Spain
e-mail: serrano@us.es

A. Arechederra · M. Florencio
Estación Biológica de Doñana, Pabellón de Perú, Av. María Luisa s/n, 41013 Sevilla, Spain

Introduction

Fridley et al. (2006) stated that, until recently, the fact that the spatial and temporal aspects of biodiversity are not independent (Preston, 1960) has rarely been recognized explicitly. In temporary ponds, however, the constraint of the spatio-temporal scale on the assessment of species richness is difficult to avoid because of the fluctuating nature of these aquatic environments. In these aquatic systems, cumulative richness over the scale of a few years can generate a much larger sample size than over a short-time scale and, hence, would indicate biodiversity with more accuracy. Moreover, long-term monitoring of biological communities is essential for the study of temporary aquatic systems because it provides a suitable frame to address the composition of assemblages and cumulative richness across a wide spatio-temporal scale.

The wetlands of the Doñana Natural Area (SW Spain) have long attracted many faunistic studies even before any protection policy came into place. Although these early studies were largely devoted to vertebrates, a few included some aquatic invertebrates owing to the enthusiasm of pioneering conservationists in promoting Doñana through various international scientific expeditions. As a result of the expeditions held in 1959, 1962 and 1965, several collections of invertebrates were gathered, and crustacean samples were sent to various experts (Dussart, 1964, 1967; Bigot & Marazanof, 1965; Marazanof, 1967). These and later collections (Estrada, 1973; Margalef, unpublished; Miracle, unpublished) were first reviewed by Armengol (1976) who focused on the distribution of diaptomids in the Doñana ponds. Seasonal monitoring of zooplankton started with Furest & Toja (1981) and was further extended to the present. Early studies showed that the Doñana ponds support a rich and dynamic community of zooplankton (Mazuelos et al., 1993; Galindo et al., 1994a; Serrano & Toja, 1998) including several species that are endemic to the Iberian-Balear region (Alonso, 1996) and a new rotifer species: *Lecane donyanaensis* (Galindo et al., 1994b). More recently, some exotic crustacean species have been reported across Doñana, such as the cladoceran *Daphnia parvula* in freshwater ponds (Serrano & Fahd, 2005) and an estuarine calanoid in marshland (Frisch et al., 2006b).

This wealth of data on zooplankton in Doñana requires an in-depth revision to produce a reliable inventory of taxa for further ecological assessments. In particular, different zooplankton assemblages are expected in different wetlands, such as freshwater ponds versus marshland and ephemeral ponds versus semi-permanent ponds. Among the variety of aquatic systems in the Doñana region, freshwater ponds and marshland are contrasting landscapes because they are located in two different hydro-geomorphological units. This segregation is mainly due to the lower conductivity, higher water depth and longer hydroperiod of the former during the intense floods of 1996–1998 (Espinar & Serrano, 2009). Will ponds and marshland segregate according to zooplankton composition during floods? Will these differences persist over a longer time scale when environmental conditions change? The spatio-temporal scale of observation has a strong influence on the assessment of environmental variability (Levin, 1992) and to a greater degree in semiarid regions due to the unpredictability of annual rainfall. In the Doñana area, Serrano & Fahd (2005) showed that the zooplankton richness of temporary ponds has been affected differently by hydrologic variability depending of the temporal scale of observation: it was weakly affected in the very short term and strongly affected in the long term because ponds with a longer hydroperiod had more chance to change and so cumulative richness increased positively with pond hydroperiod. The element of chance in pond organisms was introduced by Talling (1951), who considered chance as the indeterminacy of multiple factors involved in dispersal and colonization processes. In the present study, we have assessed whether there is a pattern of cumulative richness in the Doñana wetlands at a large scale by considering a record of taxa based on a long period of studies (1964–2007) of copepods and branchiopods. We have further compared taxa composition to assess whether zooplankton assemblages differ in different wetland types at different spatio-temporal scales: (a) temporary ponds versus marshland sites, both during floods and in the cumulative record; (b) among ponds with different hydroperiods in the cumulative record and (c) in one temporary pond over time (1964–2004).

Materials and methods

Study area

The Doñana Natural Area covers about 1,200 km^2 and is located on the Atlantic coast of south-western Spain (37°N, 6°W). It comprises both a National Park and a Natural Park (Fig. 1). Wetlands in this area exhibit the highest degree of environmental protection in Spain as they are of utmost relevance for the preservation of aquatic wildfowl in Western Europe. The Doñana National Park (ca. 54,000 ha) has been designated a Biosphere Reserve, a World Heritage Site by UNESCO and a Wetland Site of International Importance by the Ramsar Convention, though it entered onto the Montreux Record of Ramsar sites under threat in 1990 (Marín & García-Novo, 2006).

The Doñana region has a Mediterranean climate with Atlantic influence and is generally classified as dry subhumid. This region was formed in the Quaternary age when the estuary of the Guadalquivir River was enlarged and reshaped by the formation of sandy spits after the last postglacial transgression. Alluvial deposition of fine materials brought about the filling of the former estuary and progressively isolated it from the sea (Rodríguez in Marín & García-Novo, 2006). The central plain presents a saline silty-clay layer partially covered by an aeolian sand mantle. Consequently, the permeability of each geomorphological unit is quite different: the aeolian sand mantle corresponds to an unconfined aquifer (with a shallow water table and several flow systems), while groundwater is confined below the silty-clay deposits of the floodplain. Both units comprised an aquifer system of about 3,400 km^2 sealed by impermeable marine marls (Llamas, 1990). Nowadays, the continental marshland covers about 230 km^2 and is practically isolated from tidal influence. The substrate is made of silty-clay, calcareous and saline sediments (Clemente et al., 1998). The aeolian sand mantle is composed of several dune generations originally deposited by marine drift (Vanney & Ménanteau, 1985). Hundreds of small

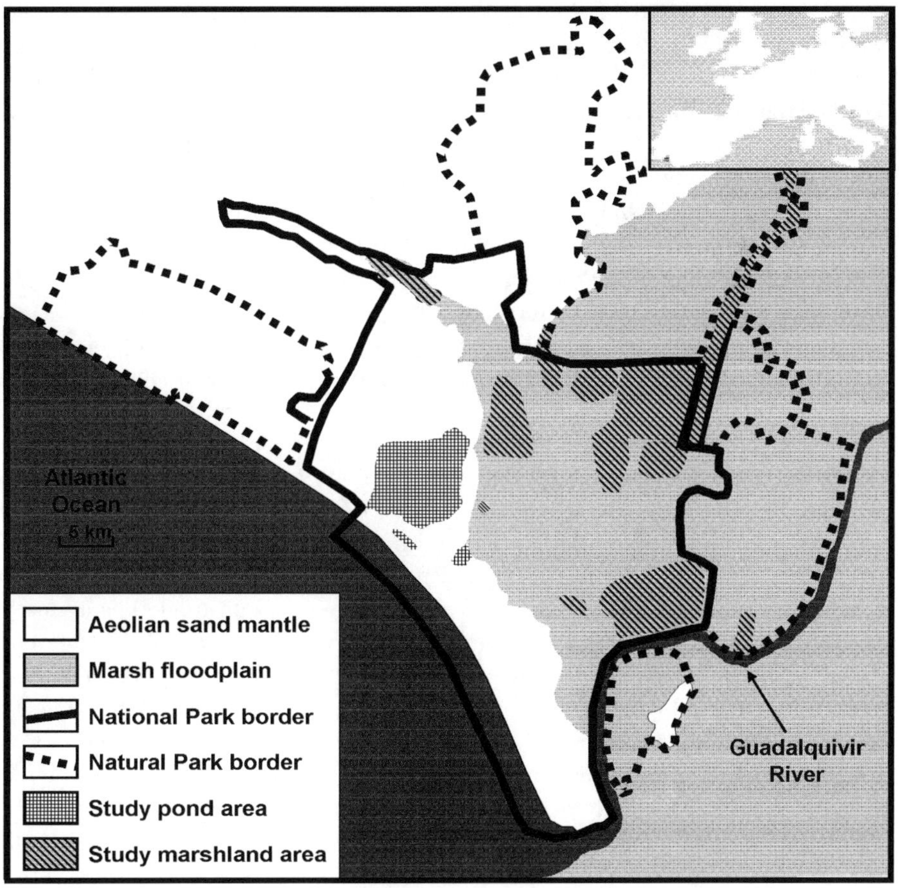

Fig. 1 Map of the Doñana Natural Area (National and Natural Parks) and location of study sites

ponds appear amid depressions of the sand mantle when the water table rises above the topographical surface during heavy rains. Their substrate is mainly sandy with low clay and silt content (Clemente et al., 1998). These ponds are fed by freshwater (rainfall, runoff and groundwater discharge) and have no surface or groundwater connection to the sea. They receive salts of marine origin through airborne deposition. The groundwater feed is relatively complex due to changes in recharge and topographic boundaries that modify their connection to different aquifer flow systems over time (Sacks et al., 1992). They vary widely in size (from rain puddles to shallow lakes of 100 ha) and in flooding duration (from days to decades), but all have been reported to dry out eventually. Hence, all are temporary water bodies exhibiting wide fluctuations of water level. Pond hydroperiod categories were based on flooding occurrence recorded since 1989 (Serrano & Zunzunegui, 2008; Serrano et al., 2008). Our study included 6 long-hydroperiod ponds (which flooded in at least 16 years out of 19), 35 intermediate-hydroperiod ponds (flooding in 10–15 years) and 14 short-hydroperiod ponds (with flooded in fewer than 6 years).

Zooplankton sampling and analyses

Three data sets were used: (i) the total cumulative record of 55 ponds and 38 marshland sites from 1964 to 2007 (the actual number of sites in this collection of published and unpublished studies was higher, but those sites that had records of three or less species were deleted, and yet, the cumulative number of taxa was not reduced); (ii) a set of 19 ponds (Serrano & Fahd, 2005) and 6 marshland sites (Arechederra, unpublished) to test for differences on a smaller spatio-temporal scale and (iii) the cumulative record of Taraje pond to show the taxa accumulation curve in a pond where sampling effort could be estimated. The methodology for the collection and identification of samples can be found in each study (see Appendix—Supplementary Material). In the unpublished work, sample collection was done as follows: 48 ponds were sampled for zooplankton in January and February 1990 (Mazuelos et al., unpublished), but only rotifer distribution was published (Mazuelos et al., 1993). Zooplankton samples were collected with a 63-μm pore conical net of 13 cm diameter along two transects at each pond (one from the littoral to the open water and another along the littoral area), and samples were fixed with 4% formalin for later identification. Six marshland sites along the lower stretch of the Guadiamar River ('Entremuros-Lucio del Cangrejo Grande') were thoroughly monitored from December 2002 to September 2004 (Serrano et al., 2006): zooplankton samples were collected by filtering 8–30 l of water through a nytal mesh of 63 μm pore size in duplicate at each sampling station (Arechederra, unpublished). Each study site included 2–4 sampling stations that were sampled every 2–3 months. Samples were preserved in 4% formalin and later were identified and counted in the laboratory. Most samples were examined in their entirety; however, when populations were dense, samples were sub-sampled. Eighteen zooplankton samples were collected monthly at Taraje pond from December 2002 to November 2004 (León, unpublished): 5–20 l of water was filtered through a nytal mesh of 63 μm pore size for each sample along a linear transect from the littoral to the open water. Samples were preserved in 4% formalin and later were identified and counted in the laboratory. Large branchiopods were sampled monthly at two ponds ('Jimenez' and 'Navazo del Toro') using a dipnet (1 mm mesh size) from November 2006 to May 2007 (Florencio, unpublished). At each pond, samples were collected and pooled together across a linear transect from the littoral to the open water; the number of samples ranged from 1 to 14 according to changes in pond size over the year. Additionally, other distinct habitats outside the linear transect were also sampled and pooled together. Qualitative samples of large branchiopods were preserved in 70% ethanol for identification in the laboratory, except for *Triops cancriformis* which was identified in situ.

Multivariate statistical analyses were performed with the software PRIMER 6 (Plymouth Routines in Multivariate Ecological Research) obtained from PRIMER-E Ltd, Plymouth. Similarity matrices were constructed using the Bray–Curtis similarity index of the cumulative presence/absence using standardized data. ANOSIM tests were carried out using 9999 permutations and non-metric MDS with 999 restarts. Values of stress for MDS analyses were not high enough to discard the ordinations because a high number of data points were used (>50) and ordinations were based on the presence/absence similarity matrices. A multivariate dispersion index (MVDISP) was performed using the similarity matrix of pond zooplankton composition to quantify the relative multivariate variability within each type of short-, intermediate- and long-hydroperiod ponds.

Two pair comparisons for cumulative richness were done with Mann–Whitney *U* test and multiple pair comparisons with Dunn's method after the Kruskal–Wallis test using SYGMASTAT 3.5. A taxa accumulation curve was plotted for the most sampled pond (Taraje pond) using the cumulative number of samplings over the study period to estimate sampling effort (Moreno & Halffter, 2000). Data were fitted to a logarithmic and an exponential regression to model both extreme cases when estimating total species richness in an area (Soberón & Llorente, 1993).

Results

A collection of 18 publications (1964–2007) and 4 unpublished studies (see Appendix—Supplementary Material) were used to obtain the number of zooplankton taxa (copepods and branchiopods) recorded so far in Doñana. Globally, 78 taxa have been recorded in a set of 55 ponds (45 cladocerans, 13 cyclopoids, 8 diaptomids, 5 harpacticoids and 7 large branchiopods, Table 1). Seventy-two taxa were recorded in 38 sites spread over the marshland (Table 1). A total of 96 taxa were recorded in Doñana (48 cladocerans, 20 cyclopoids, 13 diaptomids, 8 harpacticoids and 7 large branchiopods). As a result of rechecking some sample collections and also due to lack of detection in subsequent studies, 22 taxa were eliminated from the present cumulative record though they were originally reported by other authors (Table 2).

Twenty-five species were restricted to the ponds, while 18 taxa were recorded only in marshland sites (Tables 1, 2). Despite this apparent dissimilarity in zooplankton composition between the two types of wetlands, most restricted taxa had a very low occurrence as only four of them occurred with a frequency >20% (*Daphnia hispanica* and *Diaptomus cyaneus* in ponds; *Moina salina* and *Metacyclops planus* in marshland). No clear segregation was observed between the two types of wetlands according to an ANOSIM test performed on the similarity matrix of taxa composition at each site ($R = 0.272$,

Table 1 Cumulative taxa recorded in a collection of 55 ponds and 38 marshland sites

Taxa	Ponds (%)	Marsh (%)
Copidodiaptomus numidicus (Gurney, 1909)	73	26
Chydorus sphaericus (Müller, 1776)	67	26
Megacyclops viridis sensu lato	53	55
Acanthocyclops americanus (Marsh, 1893)	51	45
Alona rectangula (Sars, 1862)	45	34
Ceriodaphnia quadrangula (Müller, 1785)	44	32
Diacyclops bicuspidatus sensu lato	44	13
Simocephalus vetulus (Müller, 1776)	42	24
Daphnia cf. *similis*	38	3
Simocephalus exspinosus (De Geer, 1778)	38	18
Daphnia longispina (Müller, 1776)	35	13
Dussartius baeticus (Dussart, 1967)	33	8
Daphnia hispanica Glagolev & Alonso, 1990	31	–
Canthocamptus staphylinus (Jurine, 1820)	31	11
Alonella excisa (Fischer, 1854)	29	5
Ceriodaphnia reticulata (Jurine, 1820)	29	32
Daphnia magna Strauss, 1820	29	61
Dunhevedia crassa King, 1853	29	45
Diaptomus kenitraensis Kiefer, 1926	27	5
Eucyclops serrulatus (Fischer, 1851)	27	13
Moina brachiata (Jurine, 1820)	27	61
Acanthocyclops robustus (Sars, 1863)	25	29
Hemidiaptomus roubaoui (Richard, 1988)	25	3
Scapholeberis rammneri Dummont & Pensaert, 1983	25	11
Diaptomus cyaneus Gurney, 1909	24	–
Bosmina longirostris (Müller, 1776)	24	8
Ceriodaphnia laticaudata (Müller, 1776)	24	24
Arctodiaptomus wierzejskii (Richard, 1888)	22	47
Daphnia cf. *pulicaria*	22	11
Leydigia acanthocercoides (Fischer, 1854)	20	16
Treptocephala ambigua (Lilljeborg, 1901)	18	3
Alona azorica Frenzel & Alonso, 1988	15	13
Diaphanosoma brachyura (Liévin, 1848)	15	8
Attheyella trispinosa (Brady, 1880)	13	–
Macrothrix rosea (Jurine, 1820)	13	18
Triops cancriformis (Lamarck, 1801)	13	11
Macrothrix laticornis (Jurine, 1820)	11	–
Moina micrura Kurz, 1875	11	3
Estatherosporus gauthieri Alonso, 1990	9	–
Metacyclops minutus (Claus, 1863)	9	39
Oxyurella tenuicaudis (Sars, 1862)	9	11
Cyzicus grubei (Simon, 1886)	7	–

Table 1 continued

Taxa	Ponds (%)	Marsh (%)
Leydigia leydigii (Schödler, 1862)	7	–
Tanymastix stagnalis (Linnaeus, 1758)	7	–
Chirocephalus diaphanus Demarest, 1823	7	3
Macrothrix hirsuticornis Norman & Brady, 1867	7	24
Alonella nana (Baird, 1843)	5	–
Branchipus schaefferi Fischer de Waldheim, 1834	5	–
Ceriodaphnia dubia Richard, 1894	5	–
Cryptocyclops bicolor (Sars, 1863)	5	–
Graptoleberis testudinaria (S. Fischer, 1848)	5	–
Megacyclops gigas(Claus, 1857)	5	–
Ephemeropus margalefi Alonso, 1987	5	3
Eudiaptomus vulgaris (Schmeil, 1896)	5	3
Alona quadrangularis (Müller, 1776)	4	–
Branchipus cortesi Alonso & Jaume, 1991	4	–
Paracyclops fimbriatus sensu lato	4	–
Thermocyclops dybowskii (Lande, 1890)	4	–
Tropocyclops prasinus (Fischer, 1860)	4	–
Alona affinis (Leydig, 1860)	4	5
Alona costata Sars, 1862	4	3
Alona salina Alonso, 1995	4	13
Cletocampus retrogressus (Schmankevitch, 1875)	4	16
Macrocyclops albidus (Jurine, 1820)	4	11
Pleuroxus aduncus (Jurine, 1820)	4	3
Scapholeberis mucronata (Müller, 1776)	4	21
Acroperus harpae (Baird, 1836)	2	–
Bryocamptus pygmaeus (Sars, 1863)	2	–
Daphnia mediterranea Alonso, 1985	2	–
Eurycercus lamellatus (Müller, 1776)	2	–
Maghrebestheria maroccana Thiéry, 1988	2	–
Attheyella crassa (Sars, 1862)	2	5
Daphnia atkinsoni (Baird, 1859)	2	11
Daphnia parvula Fordyce, 1901	2	5
Diacyclops bisetosus (Rehberg, 1880)	2	3
Ilyocryptus silvaeducensis Romijin, 1919	2	3
Megafenestra aurita (Fischer, 1849)	2	5
Mixodiaptomus incrassatus (Sars, 1903)	2	16
Moina salina Daday, 1888	–	24
Metacyclops planus (Gurney, 1909)	–	21
Arctodiaptomus salinus (Daday, 1885)	–	11
Hemidiaptomus maroccanus (Kiefer, 1954)	–	11
Horsiella brevicornis (Van Douwe, 1904)	–	11
Neolovenula alluaudi (Guerne & Richard, 1890)	–	11
Bryocamptus cf. *minutus*	–	8
Halycyclops brevispinosus sensu lato	–	8
Oithona sp.	–	8
Calanipeda aquaedulcis Kritschagin, 1873	–	5
Ilyocryptus sordidus (Liévin, 1848)	–	5
Nitocra lacustris (Schmankevitch, 1875)	–	5
Pleuroxus letourneuxi (Richard, 1888)	–	5
Acartia tonsa Dana, 1848	–	3
Halycyclops magniceps (Lilljeborg, 1853)	–	3
Halycyclops neglectus Kiefer, 1935	–	3
Halycyclops rotundipes Kiefer, 1935	–	3
Mesocyclops leuckarti Claus, 1857	–	3

Occurrence (%) was estimated as the proportion of locations where each taxon was recorded within either ponds or marshland. Absent taxa are indicated by an *en-dash*

$P < 0.01$). The cumulative number of taxa per site was not significantly different between ponds and marshland sites, but ponds accumulated a significantly higher number of diaptomids per site (Mann–Whitney U test, $P < 0.05$).

A non-metric MDS ordination produced a gradient from ponds to marshland sites (Fig. 2) except for one site on the marshland which had been artificially fed by ground water ('Lucio El Bolin'). The long record included a large number of low-occurrence taxa (67% of total taxa occurred with a frequency <15%), and some of them showed a very large disparity as indicated by a non-metric MDS ordination of taxa (Fig. 3). In particular, *Halycyclops magniceps*, *H. rotundipes* and *H. neglectus* were clearly segregated from the rest of the taxa (Fig. 3). These species were found at one site only and can be considered accidental catches due to the proximity of this site to the estuary of the Guadalquivir River. The remaining taxa did not disperse randomly but followed a concentric pattern according to their occurrence (Fig. 3). In contrast, over a smaller spatio-temporal scale, the composition of taxa of ponds and marshland significantly segregated during floods according to an ANOSIM test performed on the similarity matrix of taxa composition ($R = 0.929$, $P < 0.01$) between a set of 19 ponds (Serrano &

Table 2 List of taxa as originally recorded that has been changed or eliminated in the present cumulative record (see Appendix—Supplementary Material for coded citation)

Taxa	Code	Changed to	Reason
Acanthocyclops cf. *kieferi*	14,18,20	*A. americanus*	Alekseev (personal communication)
Acanthocyclops sp.	7,10,11, 13,19	*A. americanus*	Alekseev (personal communication)
Acanthocyclops vernalis	14	*A. americanus*	Collection checked
Alona sp.	2,4,7,11	*Alona salina*	Alonso (1996)
Alona sp.	19	*Alona rectangula*	Collection checked
Attheyella sp.	20	*Attheyella crassa*	Collection checked
Ceriodaphnia setosa	2	Eliminated	Alonso (1996)
Chydorus ovalis	4	Eliminated	Alonso (1996)
Cyclops furcifer	7	Eliminated	Alonso (1998)
Cyclops sp.	19	*Diacyclops bicuspidatus*	Collection checked
Daphnia bolivari	19	*D. atkinsoni*	Petrusek (personal communication)
Diacyclops languidoides	19	Eliminated	Alonso (1998)
Diacyclops languidus	19	*D. bicuspidatus*	Collection checked
Diacyclops sp.	19	*D. bicuspidatus*	Collection checked
Diaptomus castaneti	1	*Dussartius baeticus*	Dussart (1967)
Diaptomus castor	19	*Dussartius baeticus*	Dussart (1967)
Diaptomus sp.	1	*Dussartius baeticus*	Dussart (1967)
Eudiaptomus steueri	2	*C. numidicus*	Dussart (1967)
Graeteriella sp.	19	Eliminated	Alekseev (personal communication)
Metacyclops lusitanus	14	*M. minutus*	Collection checked
Alona tenuicaudis	2	*Oxyurella tenuicaudis*	Alonso (1996)
Paracyclops sp.	20	Eliminated	Collection checked
Pleuroxus sp.	19	Eliminated	Collection checked

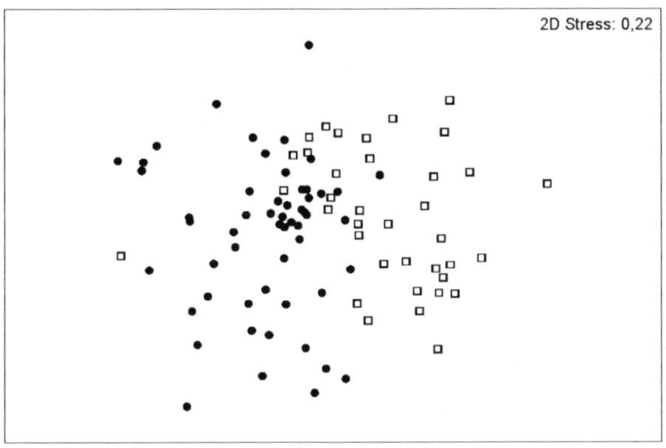

Fig. 2 Non-metric MDS ordination performed on the similarity matrix among sites: ponds (*filled circles*) and marshland (*open squares*)

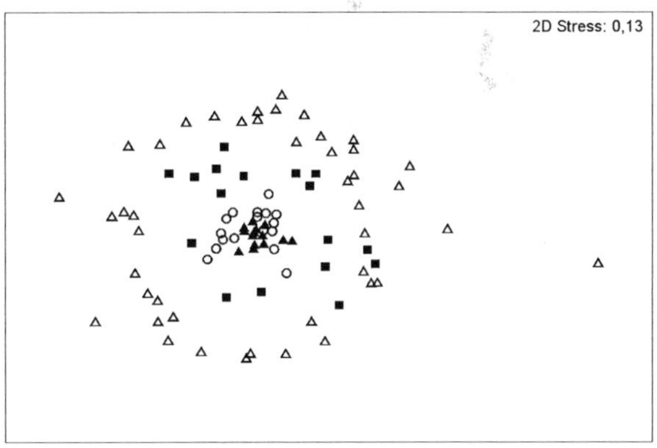

Fig. 3 Non-metric MDS ordination performed on the similarity matrix among taxa according to occurrence: >30% (*filled triangles*), between 29 and 15% (*open circles*), between 14 and 5% (*filled squares*) and <5% (*open triangles*)

Fahd, 2005) and 6 marshland sites (Arechederra, unpublished).

Taking into account only the zooplankton species recorded in the study ponds, this 42-year record also included a large number of low-occurrence taxa: 58% of all taxa occurred with a frequency <15%. In most ponds, at least 30% of the cumulative richness was due to taxa with an occurrence <25%. This wide distribution of uncommon taxa was probably the cause of the lack of segregation among ponds as ponds with different hydroperiods (short, intermediate and long) were not segregated according to their taxa composition (ANOSIM test, $R < 0.150$, $P > 0.05$). The multivariate dispersion value (MVDISP) for taxa composition was much lower in the long-hydroperiod ponds (0.168) than in the intermediate- and short-hydroperiod ponds (MVDISP = 0.987 and 1.221, respectively) even though long-hydroperiod ponds were sampled more than the other ponds. Consequently, long-hydroperiod ponds showed a significantly higher cumulative taxa number per pond (average: 25.3 taxa) than both intermediate- and short-hydroperiod ponds (average: 12.7 and 9.1, respectively), according to a multiple pair-wise comparison procedure (Dunn's method after the Kruskal–Wallis test). In terms of the main taxonomic groups, long-hydroperiod ponds accumulated the largest number of cladocerans and harpacticoids, while short-hydroperiod ponds had the lowest number of diaptomids in the cumulative record (Table 3). Large branchiopods were almost absent from long-hydroperiod ponds; only *Triops cancriformis* was recorded just once in one of them. The diaptomid *Dusartius baeticus* was absent in the long-hydroperiod ponds despite this species showed an occurrence of 31% in the Doñana ponds.

As expected, the most frequently sampled pond (Taraje pond) had the highest cumulative richness (36 taxa after 61 sampling occasions). The exponential regression, which assumes a linear decrease of the probability of finding a new species on increasing the number of samplings, provided a poor fitting to the observed data ($R^2 = 0.778$) and the estimated

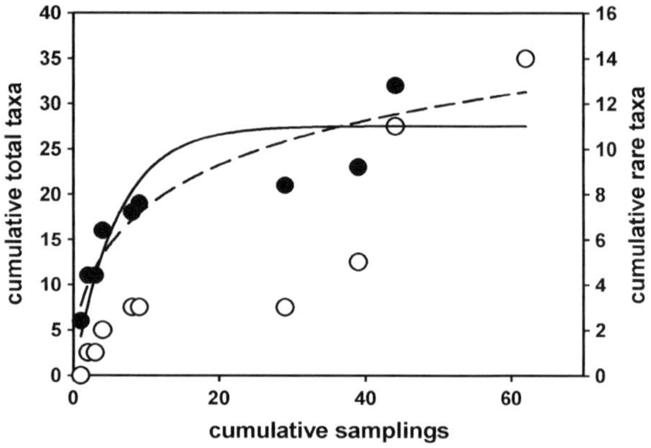

Fig. 4 Cumulative number of total taxa (*filled circles*, left *Y* axis) and rare taxa (*open circles*, right *Y* axis) plotted against the cumulative number of samplings in Taraje pond. Taxa accumulation curves modelled by simple exponential (*solid line*) and logarithmic regressions (*dashed line*)

asymptote was below the last data point (Fig. 4). A better fit was provided by a logarithmic regression ($R^2 = 0.886$). In addition, the number of rare taxa recorded in this pond did not progress linearly (Fig. 4). The first three rare species were recorded in very particular conditions: *Triops cancriformis* was found 4 years after the red swamp crayfish started invading the Doñana marshland (Furest & Toja, 1981), *Scapholeberis rammneri* was only detected after hatching from sediment laboratory incubations (Serrano & Toja, 1998) and *Alona salina* occurred after spring 1992 once water conductivity had reached 22 mS cm^{-1}. These specific conditions suggest that chance also plays an important role when recording cumulative richness in the long-run, particularly in such changing environments as the Mediterranean temporary ponds. The fluctuation of the water table illustrates the magnitude of change endured by this temporary pond during the last two decades (Fig. 5a). Following water table fluctuations, water electrical conductivity also varied widely from 0.1 to 22.0 mS cm^{-1}, and changed considerably from year to year (Fig. 5b).

Table 3 Average number of cumulative taxa within each taxonomic group according to pond hydroperiod

	Long	Intermediate	Short
Cladocerans	17.50*	7.11	5.21
Cyclopoids	3.50	2.29	2.0
Diaptomids	2.83	2.40	1.07*
Harpacticoids	1.50*	0.46	0.21
Large branchiopods	0.17	0.43	0.64

* $P < 0.05$ (Dunn's method after Kruskal–Wallis test)

Discussion

The present cumulative list has recorded 96 taxa, but this figure is likely to be an underestimation of the real cumulative richness of the Doñana Natural Area

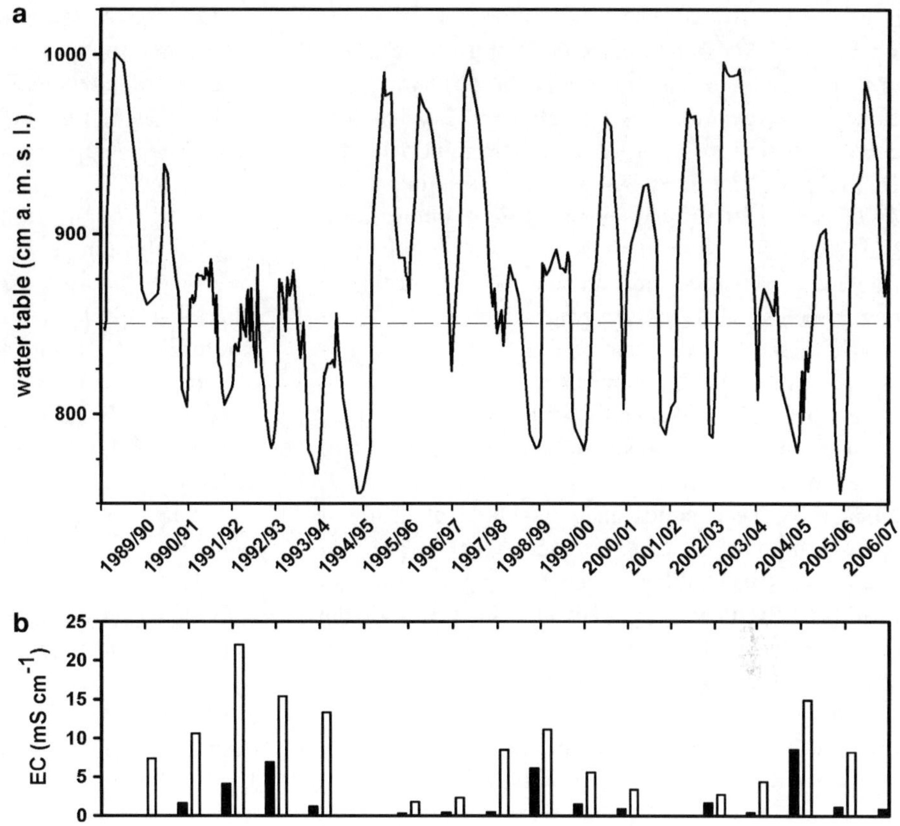

Fig. 5 **a** Water table fluctuations during each hydrological cycle from 1989/1990 to 2006/2007 in Taraje pond; *horizontal dashed line* indicates the altitude of the pond bed. **b** Maximum (*white*) and minimum (*black*) values of electrical conductivity in the water during the same period

for the following reasons: (a) it is an open record; (b) it has been built in a very conservative way, eliminating 22 previously recorded taxa (Table 2) and (c) cryptic diversity is common within some genera. In terms of crustacean biodiversity, the present cumulative record contained two classes (Branchiopoda and Maxillopoda) that included 6 Orders, 22 Families, 57 Genera and 92 identified species (Martin & Davies, 2001). Identification to a lower level than species has not been considered here because only a few authors displayed their species list in that way. Therefore, *Megacyclops viridis* var. *clausi* was pooled with *M. viridis* sensu lato; *Diacyclops bicuspidatus odessanus* with *D. bicuspidatus* sensu lato; *Halycyclops brevispinosus meridionalis* with *H. brevispinosus* sensu lato and *Paracyclops fimbriatus chiltoni* with *P. fimbriatus* sensu lato. When in doubt, we checked the sample in the collections, whenever possible, and, consequently, modified or eliminated some taxa from the list; in other cases, we had to use indirect evidence to make these changes through reviews and expert opinions (Table 2). Therefore, the present list of taxa improves the previous inventory of the micro-crustaceans of the Doñana wetlands (Marín & García-Novo, 2006), and despite now being shorter, the number of species in the present cumulative record of the Doñana wetlands corresponds to 50% of the total branchiopod species reported for the Iberian Peninsula (Alonso, 1996).

Recent unpublished work has produced new citations for Doñana though they were not included in the present study, because our aim was to produce a robust collection for further statistical analyses. For example, *Daphnia curvirostris* was recorded in some imprecise locations within the Doñana marshland in 2004 (Petrusek, personal communication), while *Streptocephalus torvicornis* was detected in 2007 at two previously unrecorded sites in zooplankton studies: one pond near ‘Lucio de Marismillas’ and another near ‘El Puntal’ (Florencio, personal communication). Laboratory hatching incubations of marsh sediment have also yielded larval stages of *Triops*, *Cyzicus*, *Chirocephalus* and *Streptocephalus* (Korn, personal communication). Additionally, cryptic diversity can be found within some species such as *Ceriodaphnia quadrangula* and *Moina micrura* (Petrusek, personal communication). Two lineages of *Daphnia atkinsoni* may exist in Doñana, and genetic studies are in progress to determine whether *Daphnia* cf. *similis* is, in fact, another lineage of *D. hispanica* (Petrusek, personal communication). We are also aware that several morphotypes of *Eucyclops serrulatus* have recently been described as new species, such as *Eucyclops albufera* Alekseev sp. n., including specimens found in Doñana (Miracle, personal communication).

In spite of the limitations of this open list of zooplankton taxa from Doñana, we have been able to give some evidence on how taxa composition changes between ponds and marshland at different spatio-temporal scales. However, what degree of accuracy exists in the significant segregation of taxa during floods in view of the opposite result in the long-term record? The extent of environmental variability in areas dominated by the Mediterranean climate is so vast that it could explain the wide distribution of cosmopolitan species (Alonso, 1998). However, the lack of segregation between ponds and marshland was basically due to a majority of low-occurrence taxa and not to a large number of widely distributed taxa. The concentric pattern of the MDS (Fig. 3) revealed the two-dimensional ordination of a high number of taxa that eventually co-occurred with a few widely distributed species at the same site, and those sites corresponded to the long-hydroperiod ponds which formed the group with the most homogeneous taxa composition. Therefore, long-hydroperiod ponds occupy a key position among the Doñana wetlands in terms of biodiversity as they are able to accumulate a high crustacean richness over time. Nonetheless, 27 and 45 taxa were absent from the six long-hydroperiod ponds studied here, compared with the remaining ponds and marshland, respectively; thus, indicating the need to protect the whole array of Doñana wetlands if crustacean biodiversity is to be preserved.

The present cumulative record could not provide an estimation of total species richness because most publications lacked detailed information on sampling effort. In the most sampled pond, however, sampling effort was estimated using the cumulative number of samplings over the study period (Moreno & Halffter, 2000), but no asymptote could be predicted because the taxa accumulation curve was best described by a logarithmic regression (Fig. 4). The fact that species which were unrecorded during cumulative samplings were hatched from dry sediment (Serrano & Toja, 1998) also supports the idea that low-occurrence species will eventually be sampled only after a sufficiently long-cumulative sampling period, because the longer the sampling period the higher the chance for change and the likelihood of detecting additional species. Hatching of diapausing eggs of Rotifera and Cladocera from dry sediments could provide a rapid assessment of species richness though it is often biased by incubation conditions and differential egg production among species (Vandekerkhove et al., 2005).

The next question that arises is how many years of sampling are needed to assess species richness in aquatic environments that may be dry for several consecutive years. In other words, if the cumulative sampling effort required is too long, new impacts will always make it difficult to obtain a full species record. For example, the invasion of the red swamp crayfish has had a large impact on the Doñana wetlands since 1974 (Alcorlo et al., 2004). Whether the absence of large branchiopods in the long-hydroperiod ponds was due to the establishment of large populations of red swamp crayfish remains to be determined, although the sensitivity of large branchiopods to the presence of predators is well known (Alonso, 1996). Desiccation is a more recent impact on the Doñana ponds, which has already produced significant changes in vegetation (Serrano & Zunzunegui, 2008; Serrano et al., 2008) and threatens to reduce water quality due to an increase in eutrophication (Serrano et al., 2006). Strong floods, on the other hand, produce a huge flushing effect that depletes primary production and drastically reduces water ionic concentrations in these ponds (Serrano & Toja, 1995). As a result, the Doñana ponds have been reported to significantly segregate from marshland sites below a threshold conductivity value of 1.6 mS cm^{-1} after strong floods (Espinar & Serrano, 2009). In contrast, the wide range of conductivity, recorded at the same pond (0.1 to 22.0 mS cm^{-1}), illustrates the rapid alternation between floods and drought over a period of only 3 years. Many other variables fluctuated in this pond, such as total alkalinity and chloride (0.2–23.0 and 3–140 meq l^{-1}, respectively, Serrano & Toja, 1995), submerged macrophyte biomass (0–240 g d.w. m^{-2},

Serrano, 1994) and the presence of fish (Fahd et al., 2007). Consequently, zooplankton samples taken during floods lacked some species that are encountered during drier periods. In fact, this pond supported a wide range of zooplankton assemblages (Alonso, 1998): from typically freshwater (*H. roubai*, *D. cyaneus*) and euryhaline species (*A. wierzejski*, *M. brachiata*) to those indicative of highly mineralized waters (*Alona salina*). Nonetheless, some differences in taxonomic groups among wetlands may be associated with particular features of each wetland type. Long-hydroperiod ponds had a significantly higher cumulative number of cladoceran and harpacticoid taxa than the rest of the ponds, while short-hydroperiod ponds showed the lowest cumulative number of diaptomid taxa; this could be due to the higher availability of resources in the former and a stricter adaptation to desiccation in the latter (Alonso, 1998). Overall, ponds showed a higher cumulative richness of diaptomids compared with the marshland. A larger number of diaptomid species were also found in water bodies from subhumid regions compared to semiarid environments across the Iberian Peninsula (Alonso, 1998). The fact that the Doñana ponds exhibit a significantly longer aquatic phase than the marshland during the same flooding period (Espinar & Serrano, 2009) could account for this difference.

Our results contrast with the diversity–stability relationship found in temperate lakes for zooplankton (Shurin et al., 2007) as zooplankton richness in the Doñana temporary ponds greatly differed across time scales. It has been reported to be weakly affected in the very short term and strongly affected in the long term because ponds with a longer hydroperiod have more chance to change and thus cumulative richness increases positively with pond hydroperiod (Serrano & Fahd, 2005). In contrast to Shurin et al. (2007), the present study indicates that richness over a short-temporal scale is not a good indicator of the number of species that would be encountered with longer sampling in temporary wetlands. This is hardly surprising given that biodiversity in temperate lakes is not comparable to that of Mediterranean temporary wetlands (Alonso, 1998): the cumulative richness of seven temperate lakes over a combined 160-year record (Shurin et al., 2007) was less than that found in the Doñana pond of Taraje, and about half that of the combined six long-hydroperiod ponds over 42 years in Doñana. Our results, however, suited the intermediate-disturbance hypothesis, which predicts that recurrent disturbance can enhance species diversity by contributing to the elimination of dominant species and favouring resource partitioning (Denslow, 1985). Thus, it is essential to maintain a degree of recurrent disturbance by letting the environment fluctuate (alternation of floods and droughts), but preventing impacts that will eventually preclude fluctuations, such as directional changes produced by desiccation. This conclusion is particularly gloomy for the biodiversity of the Doñana wetlands given that several studies have reported that some long-hydroperiod ponds have significantly reduced the length of the aquatic phase in the last decade (Zunzunegui et al., 1998; Serrano & Zunzunegui, 2008; Serrano et al., 2008).

Acknowledgements We are very grateful to Marta Reina, Eli Reyes, Gonzalo Martín, Carmen Díaz-Paniagua, Carola Gómez, Alex Portehault and Azahara Gómez for their help in fieldwork and to Nick Brodie for the map of the study area. Borja Nebot and Ignacio Sanz also collaborated in the taxonomic determination of samples collected in 1990.

References

Alcorlo, P., W. Geiger & M. Otero, 2004. Feeding preferences and food selection of the red swamp crayfish, *Procambarus clarkii*, in habitats differing in food item diversity. Crustaceana 77: 435–453.

Alonso, M., 1996. Crustacea, Branchiopoda. In Ramos, M. A. (ed.), Fauna Ibérica, Vol. 57. Museo Nacional de Ciencias Naturales, CSIC, Madrid, Spain: 1–486.

Alonso, M., 1998. Las lagunas de la España Peninsular. Limnetica 15: 1–176.

Armengol, J., 1976. Crustáceos del Coto de Doñana. Oecologia Acuatica 2: 93–97.

Bigot, L. & F. Marazanof, 1965. Considérations sur l'écologie des invertébrés terrestres et aquatiques des Marismas du Guadalquivir. Vie et Milieu 16: 411–473.

Clemente, L., L. V. García & P. Siljeström, 1998. Suelos del Parque Nacional de Doñana. Ministerio de MedioAmbiente, Madrid.

Denslow, J. S., 1985. Disturbance-mediated coexistence of species. In Pickett, S. J. A. & P. S. White (eds), The Ecology of Natural Disturbance and Patch Dynamics. Academic Press, London: 307–321.

Dussart, B. H., 1964. Copépodes d'Espagne. Bulletin de la Société Zoologique de France 89(2/3): 117–125.

Dussart, B. H., 1967. Contribution à l'étude des Copépodes d'Espagne, Vol. 42. Publicaciones del Instituto de Biologia Aplicada, Barcelona: 87–105.

Espinar, J. L. & L. Serrano, 2009. A quantitative hydrogeomorphic approach to the classification of temporary wetlands in the Doñana National Park (SW Spain). Aquatic Ecology 43: 323–334.

Estrada, M., 1973. Nota sobre diaptòmids del Coto de Doñana. Treballs de la Societá Catalana de Biologia 32: 127–134.

Fahd, K., M. Florencio, C. Keller & L. Serrano, 2007. The effect of the sampling scale on zooplankton community assessment and its implication for the conservation of temporary ponds in south-west Spain. Aquatic Conservation: Marine and Freshwater Ecosystems 17: 175–193.

Fridley, J. D., R. K. Peet, E. van der Maarel & J. H. Willems, 2006. Integration of local and regional species–area relationships from space–time species accumulation. American Naturalist 168: 133–143.

Frisch, D. & A. Green, 2007. Copepods come in first: rapid colonization of new temporary ponds. Fundamental and Applied Limnology 168: 289–297.

Frisch, D., E. Moreno-Ostos & A. Green, 2006a. Species richness and distribution of copepods and cladocerans and their relation to hydroperiod and other environmental variables in Doñana, south-west Spain. Hydrobiologia 556: 327–340.

Frisch, D., H. Rodríguez-Pérez & A. Green, 2006b. Invasion of artificial ponds in Doñana Natural Park, south-west Spain, by an exotic estuarine copepod. Aquatic Conservation: Marine and Freshwater Ecosystems 16: 483–492.

Furest, A. & J. Toja, 1981. Ecosistemas acuáticos del Parque Nacional de Doñana: distribución del zooplancton. Simposio del Agua en Andalucía I: 151–165.

Furest, A. & J. Toja, 1987. Tipificación de lagunas andaluzas según sus comunidades de crustáceos. Oxyura IV: 89–100.

Galindo, M. D., A. J. Mata, N. Mazuelos & L. Serrano, 1994a. Microcrustacean and rotifer diversity and richness relating to water temporality in dune ponds of the Doñana National Park. Verhandlungen der Internationalen Vereinigung für Theoretische und Angewandte Limnologie 25: 1350–1356.

Galindo, M. D., L. Serrano, H. Segers & N. Mazuelos, 1994b. *Lecane donyanaensis* n. sp. (Rotifera: Monogononta, Lecanidae) from the Doñana National Park (Spain). Hydrobiologia 284: 235–239.

Levin, S. A., 1992. The problem of pattern and scale in ecology. Ecology 73: 1943–1967.

Llamas, R., 1990. Geomorphology of the eolian sands of the Doñana National Park (Spain). Catena Supplement 18: 145–154.

López, T., J. Toja & N. A. Gabellone, 1991. Limnological comparison of two peridunar ponds in the Doñana National Park (Spain). Archiv für Hydrobiologie 120: 357–378.

Marazanof, F., 1967. Ostracodes, Cladocères, Hétéroptères et Hydracariens noveaux pour les Marismas du Guadalquivir (Andalousie). Données écologiques Annales de Limnologie 3: 47–64.

Marín, C. & F. García-Novo, 2006. Doñana. Water and Biosphere. Ministerio de Medio Ambiente, Madrid.

Martin, J. W. & G. E. Davies, 2001. An updated classification of the recent crustacea. Natural History Museum of Los Angeles County, Science Series 39: 1–123.

Mazuelos, N., J. Toja & C. Guisande, 1993. Rotifers in ephemeral ponds of Doñana National Park. Hydrobiologia 255/256: 429–243.

Moreno, C. E. & G. Halffter, 2000. Assessing the completeness of bat biodiversity inventories using species accumulation curves. Journal of Applied Ecology 37: 149–158.

Preston, F. W., 1960. Time and space and the variation of species. Ecology 41: 611–627.

Sacks, L. A., J. S. Herman, L. F. Konikow & A. L. Vela, 1992. Seasonal dynamics of groundwater–lake interactions at Doñana National Park, Spain. Journal of Hydrology 136: 123–154.

Serrano, L., 1994. Sources, abundance and disappearance of polyphenolic compounds in temporary ponds of Doñana National Park (South-western Spain). Australian Journal of Marine and Freshwater Research 45: 1555–1564.

Serrano, L. & K. Fahd, 2005. Zooplankton communities across a hydroperiod gradient of temporary ponds in the Doñana National Park (SW Spain). Wetlands 25: 101–111.

Serrano, L. & J. Toja, 1995. Limnological description of four temporary ponds in the Doñana National Park (SW, Spain). Archiv für Hydrobiologie 133: 497–516.

Serrano, L. & J. Toja, 1998. Interannual variability in the zooplankton community of shallow temporary ponds. Verhandlungen der Internationalen Vereinigung fur Theoretische und Angewandte Limnologie 26: 1575–1581.

Serrano, L. & M. Zunzunegui, 2008. The relevance of preserving temporary ponds during drought: hydrological and vegetation changes over a 16-year period in the Doñana National Park (south-west Spain). Aquatic Conservation: Marine and Freshwater Ecosystems 18: 261–279.

Serrano, L., M. Reina, G. Martín, I. Reyes, A. Arechederra, D. León & J. Toja, 2006. The aquatic systems of Doñana (SW Spain): watersheds and frontiers. Limnetica 25: 11–32.

Serrano, L., M. P. Esquivias-Segura & M. Zunzunegui, 2008. Hydrologic and vegetation changes over a 16-year period in the Doñana N. P. (SW Spain). Limnetica 27: 65–78.

Shurin, J. B., S. E. Arnott, H. Hillebrand, A. Longmuir, B. Pinel-Alloul, M. Winder & N. D. Yan, 2007. Diversity–stability relationship varies with latitude in zooplankton. Ecology Letters 10: 127–134.

Soberón, J. & J. Llorente, 1993. The use of species accumulation functions for the prediction of species richness. Conservation Biology 7: 480–488.

Talling, J. F., 1951. The element of chance in pond populations. The Naturalist 1951: 157–170.

Vandekerkhove, J., S. Declerck, L. Brendonck, J. M. Conde-Porcuna, E. Jeppesen, L. S. Johansson & L. De Meester, 2005. Uncovering hidden species: hatching diapausing eggs for the analysis of cladoceran species richness. Limnology & Oceanography: Methods 3: 399–407.

Vanney, J. R. & L. Ménanteau, 1985. Physiographic map of the Atlantic littoral of Andalousia 1/50000. Junta de Andalucia, Sevilla, Spain.

Zunzunegui, M., M. C. Díaz-Barradas & F. García-Novo, 1998. Vegetation fluctuation in Mediterranean dune ponds in relation to rainfall variation and water extraction. Applied Vegetation Science 1: 151–160.